T0183254

Lecture Notes in Computer Science   9725

Commenced Publication in 1973
Founding and Former Series Editors:
Gerhard Goos, Juris Hartmanis, and Jan van Leeuwen

More information about this series at http://www.springer.com/series/7407

Gert-Martin Greuel · Thorsten Koch
Peter Paule · Andrew Sommese (Eds.)

Mathematical Software – ICMS 2016

5th International Conference
Berlin, Germany, July 11–14, 2016
Proceedings

 Springer

Editors

Gert-Martin Greuel
Universität Kaiserslautern
Kaiserslautern
Germany

Thorsten Koch
Zuse Institute Berlin
Berlin
Germany

Peter Paule
Johannes Kepler University Linz
Hagenberg
Austria

Andrew Sommese
University of Notre Dame
Notre Dame, IN
USA

ISSN 0302-9743   ISSN 1611-3349   (electronic)
Lecture Notes in Computer Science
ISBN 978-3-319-42431-6   ISBN 978-3-319-42432-3   (eBook)
DOI 10.1007/978-3-319-42432-3

Library of Congress Control Number: 2016944479

LNCS Sublibrary: SL1 – Theoretical Computer Science and General Issues

Printed on acid-free paper

This Springer imprint is published by Springer Nature
The registered company is Springer International Publishing AG Switzerland

Preface

The 5th International Congress on Mathematical Software (ICMS 2016) was held during July 11–14, 2016, at the Zuse Institute in Berlin, Germany. There were four invited plenary talks and 139 contributed talks. From the submitted extended abstracts, 68 were accepted for the present proceedings.

The scope of mathematics is as broad as human thought. With the ever-increasing power and pervasiveness of technology, computation has emerged as a tool of mathematics rivaling (and many would say surpassing) proof. Many problems and models of wide interest in engineering and science are too complicated for step-by-step human solution in any reasonable time frame, but require solution immediately (although often not completely certain). We agree with the following statement in the preface to the ICMS 2014 volume:

> We in the International Conference of Mathematical Software believe that the appearance of mathematical software is one of the most important modern developments in mathematics, and this phenomenon should be studied as a coherent whole. Our vision for ICMS is to serve as the major forum for mathematicians, scientists, programmers, and developers who are interested in software. Software is not static: Anyone who uses software knows that its typical "halflife" is frustratingly short: But it compiled properly just last year! There is constant renewal, development, and disruptive changes. It is partly caused by new mathematical advances, but often the pressure is from technological changes, e.g., the appearance of graphics processing units (GPUs).

Jack Dongarra in his invited paper discusses the major (and disruptive) changes that moving to "extreme scale" computing will cause. Exoscale computing and the importance of locality are leading to a new generation of algorithms and software.

Computation is a tool not only for applications to other subjects, but also for the modern development of mathematics. As in the wider world of engineering and science, computation that forms a proof as well as computation that constructs objects and reaches conclusions that are not completely certain have important roles. In his plenary talk, Wolfram Decker discusses the challenges to and progress in the integration of a number of powerful open source computer algebra software packages into a next-generation computer algebra system.

The knowledge base of mathematics, including theorems, precise definitions of objects, and a huge constellation of examples, is the bedrock for the development of new algorithms and software. Stephen Watt described in his plenary talk the progress toward the goal of an international mathematical knowledge base useable by individuals and software systems alike.

But computational mathematics and software goes even deeper, right down to the roots of the logical and constructive foundations of mathematics. In 2009 Vladimir Voevodsky added the concept of h-level (homotopy level) and the "univalence axiom" to homotopy type theory, where the basic objects are homotopy types and which replaces set-theoretic formalizations of mathematics. This so-called univalent foundation had far-reaching consequences. It can be regarded as a new foundation for mathematics in general but it also allows mathematical proofs to be formalized, i.e., translated into the formal language that can be processed by a computer proof assistant, much more easily than before. In his plenary talk, Voevodsky explained how "univalent models" of type theories allow for the direct formalization of constructive mathematics and the formalization of classical mathematics by adding the axiom of excluded middle and the axiom of choice for types of certain h-levels. A considerable amount of mathematics has been formalized in the Coq library UniMath using the univalent point of view.

We are thankful to all the individuals whose effort and support made ICMS 2016 possible: the plenary speakers, all the contributors of abstracts and extended abstracts, the special session organizers, and the LNCS team at Springer under the leadership of Alfred Hofmann. We benefited from the experience of the ICMS Advisory Board chaired by Professor Nobuki Takayama and from Professors Chee Yap, Hoon Hong, and Deok-Soo Kim. Last but not least, we acknowledge the work of the local chair, Professor Thorsten Koch, and his committee at the Zuse Institute.

May 2016

Gert-Martin Greuel
Thorsten Koch
Peter Paule
Andrew Sommese

Bylaws of ICMS

The motivation for these bylaws is to guide the future directions and governance of the International Congress of Mathematical Software (ICMS). Ultimately, we hope to build a community of researchers and practitioners centered around the aims of the previous ICMSs, namely "mathematical software" viewed as a scientific activity. Such a community is closely allied with areas such as algorithms and complexity, software engineering, computational sciences, and of course all of constructive mathematics including computer algebra and numerical computation. But mathematical software has unique (evolving) characteristics, which ICMS aims to foster and support. To build such a community, we need continuity and some rules governing the central activity of any research community, namely, the ICMS conference. The following proposal is based on, and consistent with, the historical patterns observed in the first four ICMSs (2002, 2006, 2010, 2014). The proposal is deliberately minimal and under-specified. Therefore, their interpretations should be guided by historical patterns.

Bylaw 1: Composition of Organizers

Each ICMS Conference shall have the following organizational positions:

1. Advisory Board
2. General Chair
3. Program Chair
4. Local Chair
5. Secretary
6. Program Committee
7. Local Committee

Chair could also mean Co-chairs.

Bylaw 2: Appointments

1. The Advisory Board shall consist of the General Chair, Program Chair, and Local Chair of the previous two conferences, and any other members that they shall appoint. The General Chair of the last-but-one ICMS shall serve as the chair for the current Advisory Board. All appointments to the Advisory Board last two ICMS conferences.
2. The Advisory Board appoints the Secretary.
3. The Advisory Board appoints the General Chair for the next conference.
4. The General Chair, in consultation with the Advisory Board, appoints the Program Chair and Local Chair.

5. The Program Chair, in consultation with the General Chair and the Advisory Board, appoints the Program Committee members.
6. The Local Chair, in consultation with the General Chair, appoints the Local Committee members.

Bylaw 3: Duties

1. The Advisory Board Chair will hold an ICMS business meeting during each conference.
2. The Secretary shall maintain a permanent ICMS website for past activities, and also a list of names and e-mails of attendees of past ICMS conferences.

Bylaw 4: Amendments

1. The bylaws can be amended by ballot, either at the ICMS business meeting or by email.
2. Persons who have registered for at least one of the three preceding ICMS Conferences are eligible to vote.

Appendix: Remarks on the Bylaws

1. The bylaw is self-described as minimal and under-specified; both are viewed as positive qualities. This appendix will comment on the bylaws using their historical (nonbinding) interpretations. It will also motivate the exclusion of certain items in the bylaw.
2. Historically, ICMS was organized as a satellite of ICM. Like ICM, ICMS is held every 4 years. But even in our short history, there was a break in this pattern in 2010. Looking forward, there are good arguments to have biennial meetings (e.g., this is better for community building).
3. We do not specify the format of ICMS, believing it to be the prerogative of the General Chair and the Program Chair to shape it to best serve the community. Historically, the program has centered around plenary speakers and special sessions organized by experts in the area of interest.
4. The term "software" is a unique characteristic of ICMS that distinguishes it from the allied areas mentioned in the bylaw. We are not only interested in "paper algorithms" but also in their implementation and in their software environment. We want to foster software development as a scientific activity, to promote the publication of software like paper publications, and to establish standards for such activities. Past ICMSs have an important component of software tutorial and demonstrations and distribution of free software (e.g., Knoppix CD).
5. The ICMS positions listed in the bylaw do not exclude additional positions: the positions of a treasurer and a "documentation chair" for software have been suggested. But we refrain from mandating such positions in the bylaw.

6. Term limit for appointments to ICMS posts is generally a good thing. Again, we do not encode this into the bylaw, as we recognize many good reasons to take exceptions, e.g., a competent "document chair" should probably be given a life appointment.
7. The idea of a permanent repository for ICMS is assumed in the bylaw. Nobuki Takayama has a website that might be considered as the starting point. The General Chair and Program Chair should each deposit a report for the activities of their particular ICMS in this repository.
8. The database swMATH is a freely accessible information service for mathematical software, currently maintained by the research campus MODAL, Zuse Institute Berlin (ZIB), and FIZ Karlsruhe. With its philosophy to provide access to an extensive database of information on mathematical software together with a systematic linking of software packages with relevant mathematical publications, it meets the needs of the mathematical software community.

Organization

Executive Committee

General Chair

Gert-Martin Greuel   University of Kaiserslautern, Germany

Program Chairs

Peter Paule	Johannes Kepler University Linz and RISC, Austria
Andrew Sommese	University of Notre Dame, USA

Local Chair

Thorsten Koch   Zuse Institute Berlin, Germany

Program Committee

Mohamed Barakat	University of Siegen, Germany
Moulay A. Barkatou	University of Limoges, France
Daniel Bates	Colorado State University, USA
Johannes Bluemlein	DESY, Germany
Daniel Brake	University of Notre Dame, USA
Bruno Buchberger	Johannes Kepler University Linz and RISC, Austria
Thomas Cluzeau	University of Limoges, France
Wolfram Decker	University of Kaiserslautern, Germany
Claus Fieker	University of Kaiserslautern, Germany
Ambros M. Gleixner	Zuse Institute Berlin, Germany
Gert-Martin Greuel	University of Kaiserslautern, Germany
Jonathan Hauenstein	University of Notre Dame, USA
Max Horn	University of Giessen, Germany
Tudor Jebelean	Johannes Kepler University Linz and RISC, Austria
Fredrik Johansson	Inria and Institut de Mathématiques de Bordeaux, France
Michael Joswig	TU Berlin, Germany
Mastaka Kaneko	Toho University, Japan
Christian Kirches	IWR Heidelberg and TU Braunschweig, Germany
Michael Kohlhase	Jacobs University Bremen, Germany
Ulrich Kortenkamp	University of Potsdam, Germany
Christoph Koutschan	RICAM, Austria
Temur Kutsia	Johannes Kepler University Linz and RISC, Austria
Suzy S. Maddah	Fields Institute, Canada
John Mitchell	Rensselaer Polytechnic Institute, USA
Chenqi Mou	Beihang University, China

Peter Paule (Chair)	Johannes Kepler University Linz and RISC, Austria
Marc Pfetsch	TU Darmstadt, Germany
Gerhard Pfister	University of Kaiserslautern, Germany
Ted Ralphs	Lehigh University, USA
Carsten Schneider	Johannes Kepler University Linz and RISC, Austria
Andrew Sommese (Chair)	University of Notre Dame, USA
Wolfram Sperber	FIZ Karlsruhe, Germany
Setsuo Takato	Toho University, Japan
Olaf Teschke	FIZ Karlsruhe, Germany
Vladimir Voevodsky	IAS Princeton, USA
Dongming Wang	Beihang University, China
Stephen Watt	University of Waterloo, Canada
Wolfgang Windsteiger	Johannes Kepler University Linz and RISC, Austria

Topical Session Organizers

1. Univalent Foundations and Proof Assistants
   (Vladimir Voevodsky)
2. Software for Mathematical Reasoning and Applications
   (Bruno Buchberger, Tudor Jebelean, Temur Kutsia, and Wolfgang Windsteiger)
3. Computational Number Theory Meets Computational Algebraic Geometry
   (Wolfram Decker and Claus Fieker)
4. Algebraic Geometry in Applications
   (Gerhard Pfister)
5. Computational Aspects of Homological Algebra, Group,
   and Representation Theory
   (Mohamed Barakat and Max Horn)
6. Software of Polynomial Systems
   (Chenqi Mou and Dongming Wang)
7. Software for the Symbolic Study of Functional Equations
   (Moulay A. Barkatou, Thomas Cluzeau, and Suzy S. Maddah)
8. Symbolic Integration
   (Christoph Koutschan)
9. Symbolic Computation and Elementary Particle Physics
   (Carsten Schneider and Johannes Bluemlein)
10. Software for Numerically Solving Polynomial Systems
    (Daniel Bates, Jonathan Hauenstein, and Daniel Brake)
11. High-Precision Arithmetic, Effective Analysis, and Special Functions
    (Fredrik Johansson)
12. Mathematical Optimization
    (Ambros M. Gleixner, Christian Kirches, John Mitchell, and Ted Ralphs)
13. Interactive Operation to Scientific Artwork and Mathematical Reasoning
    (Setsuo Takato, Mastaka Kaneko, and Ulrich Kortenkamp)

14. Information Services for Mathematics: Software, Services, Models, and Data
    (Wolfram Sperber and Michael Kohlhase)
15. SemDML: Towards a Semantic Layer of a World Digital Mathematical Library
    (Michael Kohlhase, Olaf Teschke, and Stephen Watt)
16. Polyhedral Methods in Geometry and Optimization
    (Michael Joswig and Marc Pfetsch)
17. General Session
    (Gert-Martin Greuel, Peter Paule, and Andrew Sommese)

Local Committee

Thorsten Koch (Chair)	Zuse Institute Berlin, Germany
Wolfgang Dalitz	Zuse Institute Berlin, Germany
Ambros M. Gleixner	Zuse Institute Berlin, Germany
Winfried Neun	Zuse Institute Berlin, Germany

Advisory Board

Nobuki Takayama (Chair)	Kobe University, Japan
Chee Yap	NY University, USA
Komei Fukuda	ETH Zurich, Switzerland
Joris Van der Hoeven	École Polytechnique, France
Michael Joswig	TU Berlin, Germany
Hoon Hong	NC State University, USA
Masayuki Noro	Kobe University, Japan
Deok-Soo Kim	Hanyang University, South Korea

Sponsoring Institutions

Research Center Matheon – Mathematics for key technologies
MODAL - Mathematical Optimization and Data Analysis Laboratories
GAMS Development Corporation
Gurobi Optimization
MOSEK ApS
Springer

Abstracts of the Invited Talks

Invited Plenary Speakers and Talks

Invited Plenary Speakers

Wolfram Decker   University of Kaiserslautern, Germany
Jack Dongarra   University of Tennessee, USA
Vladimir Voevodsky   IAS Princeton, USA
Stephen Watt   University of Waterloo, Canada

Abstracts of Invited Plenary Talks

1. *Current Challenges in the Development of Open Source Computer Algebra Software*
   Wolfram Decker (University of Kaiserslautern)

   Computer algebra is facing new challenges as more and more of the abstract concepts of pure mathematics are made constructive, with interdisciplinary methods playing a significant role. On the mathematical side, while we wish to provide cutting-edge techniques for applications in various areas, the implementation of an advanced and more abstract computational machinery often depends on a long chain of more specialized algorithms and efficient data structures at various levels. On the software development side, for cross-border approaches to solving mathematical problems, the efficient interaction of systems specializing in different areas is indispensable. In this talk, I will report on ongoing collaboration between groups of developers of several well-established open source computer algebra system specializing in commutative algebra and algebraic geometry, group and representation theory, convex and polyhedral geometry, and number theory. The ultimate goal of this collaboration is to integrate the systems, together with other packages and libraries, into a next generation computer algebra system surpassing the combined mathematical capabilities of the underlying systems.

2. *With Extreme Scale Computing the Rules Have Changed*
   Jack Dongarra (University of Tennessee)

   In this talk we will look at the current state of high performance computing and look at the next stage of extreme computing. With extreme computing there will be fundamental changes in the character of floating point arithmetic and data movement. In this talk we will look at how extreme scale computing has caused algorithm and software developers to changed their way of thinking on how to implement and program certain applications.

3. *UniMath - a library of mathematics formalized in the univalent style*
   Vladimir Voevodsky (IAS Princeton)

The univalent style of formalization of mathematics in the type theories such as the ones used in Coq, Agda or Lean is based on the discovery in 2009 of a new class of models of such type theories. These "univalent models" led to the new intuition that resulted in the introduction into the type theory of the concept of h-level (homotopy level). This most important concept implies in particular that to obtain good intuitive behavior one should define propositions as types of h-level 1 and sets as types of h-level 2. Instead of syntactic Prop one then *defines* a type hProp(U) - the type of types of h-level 1 in the universe U and the type hSet(U) - the type of types of h-level 2 in U. With types of h-level 1 and 2 one can efficiently formalize all of the set-theoretic mathematics. With types of h-level 3 one can efficiently formalize mathematics at the level of categories etc. Univalent style allows to directly formalize constructive mathematics and to formalize classical mathematics by adding the excluded middle axiom for types of h-level 1 and the axiom of choice for types of h-level 2. UniMath is a growing library of constructive mathematics formalized in the univalent style using a small subset of Coq language.

4. *Toward an International Mathematical Knowledge Base*
   Stephen Watt (University of Waterloo)

The notion of a comprehensive digital mathematics library has been a dream for some decades. More than in many other areas, results in mathematics have lasting value – once proven, always true. It is not uncommon for a research article to have primary references to work decades earlier. Another quality of mathematics is its precision; there is a clarity to mathematical definitions and results. This makes mathematics an ideal subject for mechanized treatment of knowledge. This talk shall outline the challenges and opportunities in transforming the complete mathematical literature into a knowledge base to be used by mathematicians and software systems alike.

Contents

High-Precision Arithmetic, Effective Analysis and Special Functions

Mathematical Optimization

Interactive Operation to Scientific Artwork and Mathematical Reasoning

Information Services for Mathematics: Software, Services, Models, and Data

SemDML: Towards a Semantic Layer of a World Digital Mathematical Library

Miscellanea

Invited Talk

With Extreme Scale Computing the Rules Have Changed

Jack Dongarra[1,2,3(✉)]

[1] University of Tennessee, Knoxville, USA
dongarra@icl.utk.edu
[2] Oak Ridge National Laboratory, Oak Ridge, USA
[3] The University of Manchester, Manchester, UK

1 Challenges

Science priorities lead to scientific models, and models are implemented in the form of algorithms. Algorithm selection is based on various criteria, such as accuracy, verification, convergence, performance, parallelism, and scalability. Models and associated algorithms are not selected in isolation but must be evaluated in the context of the existing computer hardware environment. Algorithms that perform well on one type of computer hardware may become obsolete on newer hardware, so selections must be made carefully and may change over time. Moving forward to exascale will put heavier demands on algorithms in at least two areas: the need for increasing amounts of data locality in order to perform computations efficiently, and the need to obtain much higher factors of fine-grained parallelism as high-end systems support increasing numbers of compute threads. As a consequence, parallel algorithms must adapt to this environment, and new algorithms and implementations must be developed to exploit the computational capabilities of the new hardware. The transition from current sub-petascale and petascale computing to exascale computing will be at least as disruptive as the transition from vector to parallel computing in the 1990's.

We now describe some of the particular challenges ahead in the use of high performance computers.

1.1 New Algorithms for Multicore Architectures

Multicore processors, in which a single chip contains two or more independent processing units called cores, are now ubiquitous on the desktop through to HPC systems. Scalable multicore systems bring a growing cost of communication relative to computation. Within a node (a single multicore processor) data transfer between cores is relatively inexpensive, but across nodes the cost of data transfer is becoming very large. This trend is addressed by new approaches such as communication-avoiding algorithms, algorithms that support simultaneous computation and communication, and algorithms that vectorize well and have a large volume of functional parallelism.

© Springer International Publishing Switzerland 2016
G.-M. Greuel et al. (Eds.): ICMS 2016, LNCS 9725, pp. 3–6, 2016.
DOI: 10.1007/978-3-319-42432-3_1

1.2   Adaptive Response to Load Imbalance

Adaptive multiscale algorithms are an important part of many applications because they apply computational power precisely where it is needed. However, they introduce dynamically changing computation that results in load imbalances from a static distribution of tasks. As we move towards systems with billions of processors, even naturally load-balanced algorithms on homogeneous hardware will present many of the same daunting problems with adaptive load balancing that are observed in today's adaptive codes. For example, software-based recovery mechanisms for fault-tolerance or energy-management will create substantial load-imbalances as tasks are delayed by rollback to a previous state or correction of detected errors. Scheduling based on a directed acyclic graphs (DAGs) also requires new approaches to optimize resource utilization without compromising spatial locality. These challenges require development and deployment of sophisticated software approaches to rebalance computation dynamically in response to changing workloads and conditions of the operating environment.

1.3   Multiple Precision Algorithms and Software

One instance of the increasingly adaptive nature of libraries is the capability to recognize and exploit the presence of mixed precision arithmetic. Motivation comes from the fact that, on modern architectures, 32-bit (single precision) floating-point operations can execute at least twice as fast as 64-bit (double precision) operations. The performance of algorithms for solving linear systems or computing eigenvalues or singular values can be significantly enhanced by applying a given method in single precision then using a few steps of in double precision to elevate the accuracy of the result from single to double precision. This technique can be applied not only to conventional processors but also to other technologies such as graphics processing units (GPUs), and so can more effectively utilize heterogeneous hardware. The use of mixed precision exploits not only the greater speed of single precision arithmetic but also the reduce storage and memory traffic of single versus double precision arrays.

1.4   Communication Avoiding Algorithms

Algorithmic complexity is usually expressed in terms of the number of operations performed rather than the quantity of data movement within memory. However, in modern systems memory movement is increasingly expensive compared with the cost of computation. It is therefore necessary to develop algorithms that reduce communication to a minimum while not unduly increasing the amount of computation. A general approach is to derive bandwidth and latency lower bounds for various dense and sparse linear algebra algorithms on parallel and sequential machines, e.g., by extending the well-known lower bounds for the usual $O(n^3)$ matrix multiplication algorithm, and then to seek new algorithms that (nearly) attain these lower bounds. The study of communication-avoiding algorithms is in it infancy, but it is already leading to new algorithmic ideas and approaches.

1.5   Auto-Tuning

Numerical libraries need to have the ability to adapt to the possibly heterogeneous environment in which they have to operate in order to achieve good
performance, energy efficiency, load balancing, and so on. The objective is to
provide a consistent library interface that remains the same for users independent of scale and processor heterogeneity, but which achieves good performance
and efficiency by binding to different underlying code, depending on the configuration. In addition, the auto-tuning has to be extended to frameworks that
go beyond library limitations, and are able to optimize data layout (such as
blocking strategies for sparse matrix kernels), stencil auto-tuners (since stencil
kernels, which update array elements according to a fixed pattern, are diverse
and not amenable to library calls) and even tuning of the optimization strategy
for multigrid solvers (optimizing the transition between the multigrid coarsening
cycle and course grid solver to minimize run time). Adding heuristic search techniques and combining them with traditional compiler techniques will enhance
the ability to address generic problems extending beyond linear algebra.

1.6   Fault Tolerance and Robustness for Large-Scale Systems

Modern PCs may run for weeks without rebooting and most data servers are
expected to run for years. However, because of their scale and complexity, today's
supercomputers run for only a few days before a reboot is needed. The major
challenge in fault tolerance is that faults in extreme scale systems, with their
millions of processors, will be continuous rather than exceptional events. This
requires a major shift from today's software infrastructure. On today's supercomputers every failure kills the application running on the affected resources.
These applications have to be restarted from the beginning or from their last
checkpoint. The checkpoint/restart technique will not scale to highly parallel
systems because a new fault may occur before the application can be restarted,
causing the application to become stuck in a state of constant restarts. New fault
tolerant paradigms need to be developed and integrated into both the system
software and user applications.

1.7   Building Energy Efficiency into Algorithm Foundations

Energy consumption is becoming a major issue in HPC, with energy costs for
the some of the largest machines already exceeding a million dollars per year.
Power and energy consumption must now be added to the traditional goals of
algorithm design, namely correctness and performance. The emerging metric of
merit is performance per watt. Energy reduction depends on software as well as
hardware., so it is essential to build power and energy awareness, control and
efficiency into the foundations of numerical libraries.

1.8 Sensitivity Analysis

As the high fidelity solution of models becomes possible, the next challenge is to study the sensitivity of the model to parameter variability and uncertainty and to seek an optimal solution over a range of parameter values. The most basic form, the forward method for either local or global sensitivity analysis, simultaneously runs many instances of the model or its linearization, leading to an embarrassingly parallel execution model. Such high-throughput computing tasks are well suited to using spare cycles on pools of PCs, for example running at night or weekends.

2 Outlook

The move to extreme-scale computing will require collaboration between hardware architects, systems software experts, designers of programming models, and implementers of the science applications that provide the rationale for these systems. The various issues discussed in this article will need to be considered from a whole system perspective, and the different tools will need to interoperate. As new ideas and approaches are identified and pursued, some will fail. As with past experience, there may be breakthroughs in hardware technologies that result in different micro and macro architectures becoming feasible and desirable, and these will require rethinking of algorithms and system software.

Univalent Foundations
and Proof Assistants

Some Wellfounded Trees in UniMath
Extended Abstract

Benedikt Ahrens[(✉)] and Anders Mörtberg

Institute for Advanced Study, Princeton, USA
{ahrens,amortberg}@ias.edu

Abstract. UniMath, short for "Univalent Mathematics", refers to both a language (a.k.a. a formal system) for mathematics, as well as to a computer-checked library of mathematics formalized in that system. The UniMath library, under active development, aims to coherently integrate machine-checked proofs of mathematical results from many different branches of mathematics.

The UniMath language is a dependent type theory, augmented by the univalence axiom. One goal is to keep the language as small as possible, to ease verification of the theory. In particular, general inductive types are not part of the language.

In this work, we partially remedy this lack by constructing some inductive (families of) sets. This involves a formalization of a standard category-theoretic result on the construction of initial algebras, as well as a mechanism to conveniently use the inductive sets thus obtained.

The present work constitutes one part of a construction of a framework for specifying, via a signature, programming languages with binders as nested datatypes. To this end, we are going to combine our work with previous work by Ahrens and Matthes (itself based on work by Matthes and Uustalu) on an axiomatisation of substitution for languages with variable binding. The languages specified via the framework will automatically be equipped with the structure of a monad, where the monadic operations and axioms correspond to a well-behaved substitution operation.

Keywords: Proof assistant · Univalent type theory · Inductive datatypes · UniMath · Initial algebras

1 Introduction

The term "UniMath", short for "Univalent Mathematics", refers to both a language (a.k.a. a formal system) for mathematics, as well as to a computer-checked library of mathematics formalized in that system.

The UniMath language is meant to be a "core" univalent type theory, in the sense that its set of basic, primitive, constructions is minimal. In short, it is an intensional Martin-Löf type theory with dependent sums, dependent products, identity types, natural numbers, and a universe $\mathcal{U}$ of types for which

© Springer International Publishing Switzerland 2016
G.-M. Greuel et al. (Eds.): ICMS 2016, LNCS 9725, pp. 9–17, 2016.
DOI: 10.1007/978-3-319-42432-3_2

the univalence axiom is assumed. For a precise description, we refer to [12] and [4, Sect. 2.1]. We only point out one important definition: among all types of $\mathcal{U}$, some can be shown to satisfy a principle called "uniqueness of identity". This principle can be defined (and, for some types, proved) within type theory as

$$\mathsf{uip}(X) := \prod_{(x:X)} \prod_{(p:x=x)} p = \mathsf{refl}(x) \ .$$

The type of all types that satisfy uip forms a "subuniverse" of the universe $\mathcal{U}$. This subuniverse is the type of objects of a category Set where arrows are given by type-theoretic functions; this category will be of interest in the following.

The purpose of minimality of the underlying theory of UniMath is twofold:

- any construction in the UniMath language should embed directly into any other computer proof assistant implementing a univalent type theory, in particular, into the newly developed *cubical type theory* [6,8];
- the verification of the consistency of UniMath should be as easy as possible.

One language feature that has been deliberately omitted from UniMath, but which is usually present in other proof assistants, is *general inductive and coinductive datatypes*.

Inductive datatypes are finite data structures that can be used to store information in a systematic manner. Examples include the data structures of lists (over a given base type) and n-ary trees, but also more complicated, *heterogeneous* families of data types, used to represent programming languages with variable binding such as the lambda calculus—see [7] for such a representation.

Coinductive datatypes are *not necessarily finite* data structures. An example of such a datatype is the type of streams (a.k.a. infinite lists) of elements of a given base type.

In the semantics of type theory, inductive and coinductive types correspond to intial algebras and terminal coalgebras in a suitable sense: the usual 1-categorical notions only apply to types that are *discrete*, that is, to types that satisfy uniqueness of identity, or, equivalently, Axiom K. Indeed, the "subuniverse" of discrete types of a fixed universe of types constitutes a category (as defined in [3]), and inductive and coinductive sets are reasonably defined as initial algebras and terminal coalgebras of endofunctors on that category.

The present work is only concerned with such discrete types, and we do not attempt to construct types that are not sets.

The characterisation of inductive and coinductive types as initial algebras and terminal coalgebras is completely dual; this suggests that their syntactic behaviour is dual as well. However, that impression is somewhat deceptive:

As for *coinductive* datatypes—in the principled formulation of (indexed) M-types—, those are *derivable* in the UniMath language [2]. However, the expected computation rules only hold up to propositional identity, not judgmentally, due to the implementation of function extensionality as an axiom. We expect the computation rules to hold judgmentally for the same construction done in cubical type theory, a type theory where function extensionality is provable instead of an axiom.

As for *inductive* types, the construction of a class of them is the subject of the work we report on in this extended abstract: In the present work, we show that a class of inductive "sets" is derivable in UniMath from just the "prototypical" inductive set of natural numbers. This is done by combining two results:

- a classical category-theoretic result saying that an initial algebra of an ω-cocontinuous functor can be constructed as colimit of a certain chain [1]
- the constructibility of colimits in the category of sets (a.k.a. discrete types) in UniMath as a consequence of the constructibility of set quotients.

The construction of set level quotients was done by Voevodsky [12]. It is a prime example of the new possibilities that the univalence axiom and its consequences provide for the formalization of (set-level) mathematics compared to the type theories implemented by Coq or Agda without the univalence axiom.

Our contribution consists of a formalization of

- some well-known theorems in category theory about existence and preservation of (co)limits
- the construction of colimits from quotients in the category of sets
- proofs that various functors are ω-cocontinuous.

and combining those results in order to obtain some inductive sets in UniMath.

As such, the results presented in this article are *not surprising at all*—it is our hope, however, that their formalization will be *useful*.

1.1 Related work

We do not give, in this extended abstract, a full description of the univalent foundations; instead, we point to Voevodsky's article [12]. A brief introduction to univalent foundations is also given in [4, Sect. 2.1].

Most of the mathematics formalized in this project concerns category theory; category theory in univalent foundations has been studied in [3]. The distinction between precategories and (univalent) categories emphasized there is not of importance in our project.

A characterisation of inductive types in type theory with function extensionality (not necessarily with the univalence axiom) has been studied in [5]. In contrast to the present work, their goal is not the construction of inductive sets, but a proof of equivalence of several different definitions, given within type theory, of inductive types.

In [4], the authors formalize, in UniMath, the notion of "heterogeneous substitution system", a categorical notion axiomatizing substitution for programming languages with variable binding. One of the main results formalized there says that any initial algebra of a given functor gives rise to a monad. The existence of inductive sets (or, more generally, existence of initial algebras for a given functor) is taken as a hypothesis in that result. We can plug our construction of inductive sets into that theorem and thus, e.g., obtain a monad structure on the lambda calculus, see the example below.

2   Overview of the Mathematics Formalized in This Project

In this section we give an overview of the mathematical results that we have formalized in the course of this project. It is outside the scope of this extended abstract to give a complete list; for details follow the pointers given in Sect. 3.

We also explain how those results help in the construction of inductive sets.

We are not concerned here with giving a general definition of inductive types, nor do we give a precise specification of the class of inductive types that we construct. This is due to the fact that the class is easily extensible by extending the library of mathematical theorems formalized in our library, a statement that we illustrate with the examples in this section.

2.1   Inductive Sets as Initial Algebras

In the present work, we do not attempt to construct general inductive types, but only inductive *sets*. We thus define

Definition 1. *The inductive set generated by an endofunctor F : Set $\to$ Set is the initial algebra of F.*

This definition is expressible within type theory, we refer to [3] for details. The use of the definite article ("*the* initial algebra") is justified by the fact that any two initial algebras are isomorphic, and hence their carriers are propositionally equal as a consequence of the univalence axiom.

Here, the category Set is the category the objects of which are discrete types of a fixed universe. It would thus be more precise to call the category Set($\mathcal{U}$), where $\mathcal{U}$ is the fixed universe.

An example of an inductive set is the set of natural numbers, which is an initial algebra of the functor $F(X) := 1 + X$.

Another example is the inductive set of lists of elements of a fixed base set A. The defining endofunctor is given by

$$F_A(X) := 1 + A \times X \ .$$

There are more complicated inductive definitions, examples of which are used to model programming languages with variable binding [7]. We call those inductive families:

Definition 2. *The inductive family of sets generated by an endofunctor F : [Set, Set] $\to$ [Set, Set] is the initial algebra of F.*

An example of an inductive family is given by the datatype of untyped lambda terms with an algebraic variant of De Bruijn variables—see [7] for an explanation. The functor Λ : [Set, Set] $\to$ [Set, Set] specifying this datatype is given by

$$\Lambda(G) := \mathsf{Id} + G \times G + G \circ \mathsf{option} \ .$$

Here, the functor option : Set $\to$ Set is defined on objects by option$(X) := 1 + X$.

2.2   Existence of Initial Algebras

Obviously, not every functor as above gives rise to an initial algebra. The following theorem, due to Adámek [1], is our main tool for the construction of initial algebras.

Theorem 1. (existence of initial algebras of ω-cocontinuous functors).
Let $F : \mathcal{C} \to \mathcal{C}$ be an ω-cocontinuous endofunctor, i.e., an endofunctor that preserves colimits of chains $A_0 \to A_1 \to A_2 \to \dots$. Then, provided its existence, the colimit of

$$0 \xrightarrow{\ !\ } F0 \xrightarrow{F!} F^2 0 \xrightarrow{F^2!} \dots$$

is an initial algebra of F.

The above theorem is applicable to construct inductive sets as in Definition 1 as per the following lemma:

Lemma 1. *The category* Set *has all colimits.*

The colimits in the category Set are given by the usual formula: For a diagram D, its colimit is given by

$$\left(\sum_{g:G} D(g) \right) / \sim$$

with $\sim$ being the smallest equivalence relation containing the relation

$$(g, x) \sim' (g', y) \quad \text{iff} \quad \exists e : g \to g' \text{ with } D(e)(x) = y \ .$$

It is in the definition of this colimit that the univalence axiom helps: as shown by Voevodsky [12], set-level quotients of arbitrary types, but in particular of sets, modulo an equivalence relation, can be constructed in univalent type theory.

The construction of inductive families as in Definition 2 is helped by the lifting of colimits to functor categories:

Lemma 2. *If $\mathcal{A}$ has all colimits, then also $[\mathcal{C}, \mathcal{A}]$ has all colimits.*

However, Theorem 1 only guarantuees the existence of initial algebras of ω-*cocontinuous* functors. This is discussed in the next section.

2.3   Preservation of Colimits

While Theorem 1 is crucial for us, its formulation and proof do not constitute the bulk of work necessary for the construction of inductive sets. Instead, it is the proof that various functors are ω-cocontinuous—hypothesis of Theorem 1—that requires most of our time and efforts.

Fortunately, there is a class of endofunctors on categories with binary products and coproducts that is built from a small set of constructions, where each construction is, or preserves, ω-cocontinuity. This class consists of "finite polynomials" of the general form

$$F(X) = B_0 + B_1 X + B_2 X^2 + \ldots + B_n X^n .$$

Our first example, the datatype of natural numbers, is given by such a finite polynomial, as is the second example of lists of elements of a given set A.

In order to show that such functors are ω-cocontinuous, it is sufficient to show that $F + G$ and $F \times G$ are ω-cocontinuous whenever F and G are, and that constant functors and the identity functor are ω-cocontinuous. We omit the precise statements, and pass to the more interesting situation of *heterogeneous* datatypes.

Heterogeneous datatypes are specified by endofunctors on functor categories. More specifically, what makes them heterogeneous is a "summand" of the form

$$F_K(X) := X \circ K$$

for an endofunctor K. In the case of the example of the lambda calculus, $K =$ option is the addition of a distinguished variable in the "context".

The ω-cocontinuity of the functor F_K is ensured as long as the target category has limits. More precisely:

Theorem 2 [10, Sect. X.3]. *Let $K : \mathcal{M} \to \mathcal{C}$ be a functor, and let $\mathcal{A}$ be a category with (specified) limits. Then the functor*

$$\mathcal{A}^K : [\mathcal{C}, \mathcal{A}] \to [\mathcal{M}, \mathcal{A}] , \qquad F \mapsto F \circ K$$

is a left adjoint (a.k.a., global right Kan extensions along K exist).

Theorem 3 [10, Sect. V.5]. *If $F : \mathcal{C} \to \mathcal{A}$ is a left adjoint, then it preserves any colimit.*

Combining those two results, we obtain that the functor F_K preserves colimits, in particular, it is ω-cocontinuous. This result hence allows the construction of heterogeneous datatypes in `UniMath`.

We note that a more direct proof of the fact that F_K preserves colimits is also possible. However, the construction of right Kan extensions is important also in the context of the related project [4].

Every summand of the functor Λ above corresponds to one language constructor. Note that a generalization to infinitely many summands is straightforward, so that languages with infinitely many language constructors can also be modeled by our inductive families of sets.

2.4   Connection to Work on Heterogeneous Substitution Systems

Matthes and Uustalu [11] introduce the notion of *(heterogeneous) substitution system* in order to give a categorical axiomatisation of *substitution* for languages with variable binding, such as the lambda calculus. Ahrens and Matthes [4] extend the axiomatisation by devising a notion of morphism of substitution systems, and formalize results of [11] as well as some new results in UniMath. One main result formalized in [4] equips a given initial algebra—whose existence is assumed—of an endofunctor on functor categories with a substitution structure [4, Theorem 28], and shows that the resulting substitution system is initial in a suitable category [4, Theorem 29].

We instantiate Theorem 28 of [4] with inductive families of sets, e.g., the lambda calculus specified by the functor Λ, and thus obtain a monad structure on that inductive family for free. Afterwards, we implement a convenient way of specifying languages via a syntactic notion of signature, where, e.g., the lambda calculus is specified by the signature $\{[0,0],[1]\}$. Here, the numbers indicate the number of variables bound in each argument of the corresponding language constructor. We hence obtain a framework which allows to concisely specify a programming language with binders, and which, from such a specification, automatically produces a certified formalization of that language.

While the property of being an initial algebra is for free and automatic for the languages obtained via the machinery we have formalized, we do not know at the moment of an automated way to generate, from initiality, the type-theoretic constructors and suitable recursion schemes. Doing this automatically probably requires some engineering on the meta-level of the proof assistant Coq, e.g., writing a plugin to that program.

Details about the project sketched in this section will be explained in a forthcoming article.

3   Some Details on the Formalization

In practice, the UniMath language is given by a subset of the language implemented in the proof assistant Coq. It is up to the formalizers to make sure that they respect this restriction—there is no mechanic check at the moment. Formalizing in UniMath hence consists in writing Coq files, using only the constructions mentioned in the introduction as being part of the UniMath language. At the time of writing, the UniMath library of formalized mathematics consists of about 46k lines of code. The library of formalized mathematics written for this project consists of approximately 4500 lines of code. The code has been integrated into an already existing library of category theory in UniMath.

In general, there are two ways of implementing a language construct:

Firstly, it can be *internal* to the language, as in the present work. Then, the construct is merely an abbreviation for its expansion into the primitive notions.

Secondly, the implementation can be *external*, i.e., on the meta-level. This option allows to refer to concepts that cannot be referred to within the language, such as *convertibility* (a.k.a. *judgmental equality*). For instance, the native inductive types of the Coq proof assistant are implemented on the meta-level.

Our approach has two disadvantages compared to an implementation on the meta-level of inductive sets:

- the inductive sets we construct will generally not satisfy computation rules *judgmentally*, but only *propositionally*
- the inductive sets we construct will generally not have a normal form.

The lack of judgmental computation rules is reminiscent of the *deliberate* blocking of computation in the SSReflect [9] extension of the Coq proof assistant.

On the other hand, the advantage of our approach to inductive sets is clear: since it does not extend the language, its consistency is immediate.

4   Conclusions

In this work, we report on a construction of some inductive sets in UniMath. The inductive sets we construct suffer from some defects, but have the advantage of not requiring any language extension. It remains to be seen if the defects get in the way when doing proofs involving inductive sets—the fact that SSReflect deliberately imposes a similar restriction indicates that the practical implications of the defects will be minor.

Acknowledgments. We thank Dan Grayson, Ralph Matthes, Paige North and Vladimir Voevodsky for helpful discussions on the subject matter.

This material is based upon work supported by the National Science Foundation under agreement Nos. DMS-1128155 and CMU 1150129-338510. Any opinions, findings and conclusions or recommendations expressed in this material are those of the author(s) and do not necessarily reflect the views of the National Science Foundation.

References

1. Adámek, J.: Free algebras and automata realizations in the language of categories. Comment. Math. Univ. Carol. **15**(4), 589–602 (1974)
2. Ahrens, B., Capriotti, P., Spadotti, R.: Non-wellfounded trees in homotopy type theory. In: Altenkirch, T. (ed.) 13th International Conference on Typed Lambda Calculi and Applications, TLCA, Warsaw, Poland, Schloss Dagstuhl - Leibniz-Zentrum fuer Informatik, 1–3 July 2015. LIPIcs, vol. 38, pp. 17–30 (2015)
3. Ahrens, B., Kapulkin, K., Shulman, M.: Univalent categories and the Rezk completion. Math. Struct. Comput. Sci. **25**, 1010–1039 (2015)
4. Ahrens, B., Matthes, R.: Heterogeneous substitution systems revisited. CORR, abs/1601.04299 (2016)
5. Awodey, S., Gambino, N., Sojakova, K.: Inductive types in homotopy type theory. In: Proceedings of the 27th Annual IEEE Symposium on Logic in Computer Science, LICS 2012, Dubrovnik, Croatia, 25–28 June 2012, pp. 95–104. IEEE Computer Society (2012)
6. Bezem, M., Coquand, T., Huber, S.: A model of type theory in cubical sets. In: Matthes, R., Schubert, A. (eds.) 19th Conference on Types for Proofs and Programs, TYPES 2013, Schloss Dagstuhl - Leibniz-Zentrum fuer Informatik. LIPIcs, vol. 26, pp. 107–128 (2013)
7. Bird, R.S., Paterson, R.: De Bruijn notation as a nested datatype. J. Funct. Prog. **9**(1), 77–91 (1999)

8. Cohen, C., Coquand, T., Huber, S., Mörtberg, A.: Cubical Type Theory: a constructive interpretation of the univalence axiom (2015, Preprint)
9. Gonthier, G., Mahboubi, A.: An introduction to small scale reflection in Coq. J. Formaliz. Reason. **3**(2), 95–152 (2010)
10. Mac Lane, S.: Categories for the Working Mathematician. Graduate Texts in Mathematics, vol. 5, 2nd edn. Springer, New York (1998)
11. Matthes, R., Uustalu, T.: Substitution in non-wellfounded syntax with variable binding. Theor. Comput. Sci. **327**(1–2), 155–174 (2004)
12. Voevodsky, V.: An experimental library of formalized mathematics based on the univalent foundations. Math. Struct. Comput. Sci. **25**, 1278–1294 (2015). http://arxiv.org/pdf/1401.0053.pdf

Exercising Nuprl's Open-Endedness

Vincent Rahli[✉]

SnT, University of Luxembourg, Luxembourg City, Luxembourg
vincent.rahli@gmail.com

Abstract. Nuprl is an interactive theorem prover that implements an extensional constructive type theory, where types are interpreted as partial equivalence relations on closed terms. Nuprl is both computationally and type-theoretically open-ended in the sense that both its computation system and its type theory can be extended as needed by checking a handful of conditions. For example, Doug Howe characterized the computations that can be added to Nuprl in order to preserve the congruence of its computational equivalence relation. We have implemented Nuprl's computation and type systems in Coq, and we have showed among other things that it is consistent. Using our Coq framework we can now easily and rigorously add new computations and types to Nuprl by mechanically verifying that all the necessary conditions still hold. We have recently exercised Nuprl's open-endedness by adding nominal features to Nuprl in order to prove a version of Brouwer's continuity principle, as well as choice sequences in order to prove truncated versions of the axiom of choice and of Brouwer's bar induction principle. This paper illustrate the process of extending Nuprl with versions of the axiom of choice.

Keywords: Nuprl · Coq · Semantics · Open-Endedness · Axiom of choice · Choice sequences · Bar induction · Continuity

1 Introduction

Nuprl. The Nuprl interactive theorem prover [3,13] implements a dependent type theory called *Constructive Type Theory* (CTT), which is based on an untyped functional programming language. It has a rich type theory including identity (or equality) types, a hierarchy of universes, W types, quotient types [14], set types, union and (dependent) intersection types [28], image types [32], partial equivalence relation types [4], approximation and computational equivalence types [35], and partial types [16,38]. CTT "mostly" differs from other similar constructive type theories such as the ones implemented by Agda [1,10], Coq [8,15], or Idris [11,26], in the sense that CTT is an *extensional* theory (i.e., propositional and definitional equality are identified [20]) with types of partial functions [16,38]. For example, the fixpoint $\text{fix}(\lambda x.x)$ diverges. It is nonetheless a

---

V. Rahli—Partially supported by the SnT and by the National Research Fund Luxembourg (FNR), through PEARL grant FNR/P14/8149128.

G.-M. Greuel et al. (Eds.): ICMS 2016, LNCS 9725, pp. 18–27, 2016.
DOI: 10.1007/978-3-319-42432-3_3

member of many types such as $\overline{\mathbb{Z}}$, which is the type of integers and diverging terms (essentially the integer type of ML-like programming languages such as OCaml). In Nuprl, type checking is undecidable but in practice this is mitigated by type inference and type checking heuristics implemented as tactics.

Formalization of Nuprl's Metatheory in Coq. Following Allen's semantics [2], CTT types are interpreted as Partial Equivalence Relations (PERs) on closed terms. We have formalized Nuprl's metatheory in Coq [5,6]. Our implementation includes: (1) an implementation of Nuprl's computation system; (2) an implementation of Howe's computational equivalence relation [22], and a proof that it is a congruence; (3) a definition of Allen's PER semantics of CTT [2,16]; (4) definitions of Nuprl's derivation rules, and proofs that these rules are valid w.r.t. Allen's PER semantics; (5) and a proof of Nuprl's consistency [5,6]. Our implementation is available at https://github.com/vrahli/NuprlInCoq, and additional information can be found at http://www.nuprl.org/html/Nuprl2Coq/.

Exploring Type Theory. Using our implementation of CTT in Coq, we are exploring type theory. For example, (1) we are reformulating CTT using a smaller core of primitive types, by allowing the theory to directly represent PERs as types [4]. We conjecture that dependent product and sum types will be definable in this new theory. (2) We have proved the validity of truncated (or squashed—see Sect. 2.2 for a discussion of truncation/squashing) versions of Brouwer's continuity principle [12,17,27,37,40,41,44]. w.r.t. Nuprl's PER semantics [34]. For that, following Longley's method [31], we used named exceptions as a probing mechanism to compute the modulus of continuity of a function. (3) We have proved the validity of versions of Brouwer's bar induction principle [12,17,21,27,37,41,43], which we are using to build parametrized families of W types from parametrized families of co-W types [9]. For that we added "choice sequences" to Nuprl's term language [36]. A choice sequence of type T is a Coq function from natural numbers to terms of type T. They are similar to the infinite sequences in [7], which are used to prove that the negative translation of the Axiom of Choice (AC) is realizable. They are also similar to Howe's set-theoretical functions in [23–25], which he used to provide a set-theoretical semantics of both Nuprl and HOL, allowing the shallow embedding of HOL in Nuprl. (4) We have proved the validity of truncated versions of AC [34, Sect. 5.3].

Open-Endedness. Because Nuprl was designed to be open-ended, i.e., theorems about computations and types "hold for a broad class of extensions to the system" [19], adding new features to the system often did not require much modifications to the existing properties of its computation system or to the statements and proofs of its inference rules. We illustrate this here with AC. Sect. 4 presents true and false versions of AC, which we have proved by extending CTT's computation and type systems.

2   Background on Nuprl

2.1   Constructive Type Theory

Nuprl's programming language is an untyped (à la Curry), lazy and applied (with pairs, injections, a fixpoint operator,...) λ-calculus. Its term language is open-ended in the sense that it contains all possible terms that follow a given structure described, for example, in [22], and mentioned below. Therefore, most terms do not have any operational semantics according to Nuprl's computation system. A type is a value of the computation system. Among other things, Allen's PER semantics associates PERs to these values, i.e., types are interpreted as partial equivalence relations (PERs) on closed terms [2]. As illustrated in [6, Fig. 6], Allen's PER semantics can be seen as an inductive-recursive definition of: (1) an inductive relation $T_1 \equiv T_2$ that expresses type equality; and (2) a recursive function $a \equiv b \in T$ that expresses equality in a type.

Type equality is mostly intentional in the sense that two types that are interpreted by the same PER are not necessarily equal. Most notably, identity types of the form $a =_T b$, which expresses that a and b are equal members of the type T, can only be inhabited by the constant $\star$, i.e., they do not have any computational content as opposed to HoTT [42]. However, the two types $0 =_\mathbb{N} 0$ and $1 =_\mathbb{N} 1$ are not equal, even though they have the same PER. Note that Uniqueness of Identify Proofs (UIP) is true by definition in Nuprl. As mentioned above, because of this treatment of equality, Nuprl is also extensional in the sense that propositional and definitional equality are identified [20]. Also, function extensionality is true by definition of dependent product (function) types.

It turns out that Nuprl's type system is not only closed under computation but more generally under Howe's computational equivalence $\sim$, which he proved to be a congruence [22]. For example, one can prove that $\lambda x.(x+1)+0 \sim \lambda x.x+1$ without requiring one to infer a type for x. In any context C, when $t \sim t'$ we can rewrite t into t' without having to prove anything about types. We rely on this relation to prove equalities between programs (bisimulations) without concern for typing [35]. See Sect. 2.3 below for more details.

As mentioned above, we have implemented Nuprl's term language, its computation system, Howe's $\sim$ relation, and Allen's PER semantics in Coq [5,6]. We have also showed that Nuprl is consistent by (1) proving that Nuprl's inference rules are valid w.r.t. Allen's PER semantics, and (2) proving that `False` is not inhabited. Using these two facts, we derive that there cannot be a proof derivation of `False`, i.e., Nuprl is consistent (see [5,6,33, Appendix A] for more details). We are using our Coq formalization to prove the validity of all the inference rules of Nuprl, and have already verified a large number of them.

2.2   Squashing

It is sometimes necessary to *truncate* or *squash* types for them to be true or consistent with Martin-Löf-like type theories such as Nuprl. For example, the non-truncated version of Brouwer's continuity principle, i.e., where the existential quantifier is interpreted constructively, is false in type theories such as Agda or Nuprl [18,29,34,36,39], while its truncated version is true in Nuprl [34]. Similar results hold about AC as discussed in Sect. 4, as well as about Brouwer's bar induction principle [36].

In Nuprl, there are various ways of *squashing* or *truncating* a type. The most widely used squashing operator in Nuprl throws away the evidence that a type is inhabited and squashes it down to a single inhabitant using, e.g., set types: $\downarrow T = \{\texttt{Unit} \mid T\}$ (as defined in [13, p. 60]). The only member of this type is the constant $\star$, which is $\texttt{Unit}$'s single inhabitant, and which is similar to () in languages such as OCaml, Haskell or SML. The constant $\star$ inhabits $\downarrow T$ if T is true/inhabited, but we do not keep the proof that it is true. See [33, Appendix F] for more information regarding squashing. Using the HoTT terminology, we also sometimes truncate types at the *propositional level* [42, p. 117]. In Nuprl propositional truncation corresponds to *squashing* a type down to a single equivalence class, i.e. all inhabitants are equal, using, e.g., quotient types [14]: $\downarrow T = T//\texttt{True}$. $\downarrow T$ is a proof-irrelevant type. Its members are the members of T, and they are all equal to each other in $\downarrow T$ because if $x, y \in T$ then $(x =_T y \iff \texttt{True})$. Note that the implication $\downarrow T \rightarrow \downarrow T$ is true because it is inhabited by $\lambda x.\star$, but we cannot prove the converse because to prove $\downarrow T$ we have to exhibit an inhabitant of T, which $\downarrow T$ does not give us because only $\star$ inhabits $\downarrow T$.

2.3 Howe's Computational Equivalence

Howe's computational equivalence is defined on closed terms as follows: $t \sim u$ if $t \preccurlyeq u \land u \preccurlyeq t$. Howe coinductively defines the approximation (or simulation) relation $\preccurlyeq$ as the largest relation R on closed terms such that $R \subset [R]$, where $[\cdot]$ is the following closure operator (also defined on closed terms): $t [R] u$ if whenever t computes to a value $\theta(\overline{b})$, then u also computes to a value $\theta(\overline{b'})$ such that $\overline{b} R \overline{b'}$. We write $\theta(\overline{b})$ for the term with outer operator θ and subterms $\overline{b}$, where each subterm is essentially a pair of a list of binding variables and a term. For example $\lambda x.x$ has one subterm that has one binding variable x. See [5,6,22] for details. By definition, one can derive, e.g., that $\bot \preccurlyeq t$ for all closed term t.

To prove that $\sim$ is a congruence, Howe first proves that $\preccurlyeq$ is a congruence [22]. Unfortunately, this is not easy to prove directly. Howe's "trick" was to define another inductive relation $\preccurlyeq^*$, which is a congruence and contains $\preccurlyeq$ by definition. To prove that $\preccurlyeq^*$ and $\preccurlyeq$ are equivalent and therefore that $\preccurlyeq$ and $\sim$ are congruences, it suffices to prove that $\preccurlyeq^*$ respects computation, i.e., given that $t \preccurlyeq^* u$, if t computes to a value of the form $\theta(\overline{b})$ then u also computes to a value $\theta(\overline{b'})$ such that $\overline{b} \preccurlyeq^* \overline{b'}$. Howe defined a condition called *extensionality* [22, Definition 5] that non-canonical (i.e., non-values) operators of lazy computation systems have to satisfy for $\preccurlyeq^*$ to imply $\preccurlyeq$. Essentially, a non-canonical operator is extensional if it never reduces a term by making a decision based on non-canonical subterms, which is the case about Nuprl's non-canonical operators. For example, the following "bad" non-canonical operator is not extensional: if we allow bad($f(a)$), where $f(a)$ is the application of f to a, to reduce to a, and bad(v), where v is a value, to reduce to v, then bad$((\lambda x.x + 1)\ 1)$ would reduce to 1, and $(\lambda x.x + 1)\ 1$ would reduce to 2, while bad(2) would reduce to 2, which is different from 1.

3   Open-Endedness and Exploration

One can extend CTT by either adding new computations, new types, or new inference rules. Typically, to add a new computation, one simply has to prove that it satisfies various preservation properties such as: if a term t_1 computes to a term t_2 then the free variables of t_2 are included in the free variables of t_1 and $t_1[x\backslash u]$ (the substitution of x for u in t_1) computes to $t_2[x\backslash u]$. Also, in the case of non-canonical operators, one has to prove that they are extensional.

We have recently added several operators to Nuprl: (1) named exceptions [34]; (2) a try/catch operator [34]; (3) a fresh operator to generate fresh names [34]; (4) choice sequences; (5) an eager application operator; as well as (6) various values denoting types such as our PER types [4]. In the case of exceptions, which are some sorts of values in the sense that they do not compute further, we had to modify one inference rule [33, Appendix C]. The proofs that our try/catch and eager application operators are extensional were standard. However, proving that our fresh operator is extensional required modifying the definition of $\preccurlyeq^*$ as discussed in [34, Sect. 4.2]. Adding choice sequences made us loose the decidability of several relations such as α-equality or even syntactic equality, which it turned out we did not need [36, Sect. 4.1]. We also had to modify one inference rule. Because types are values of the computation system, when adding a type we usually do not have to modify theorems and proofs about computations. However, we have to (1) provide an interpretation for the type: essentially a PER. (2) Then, because a type system has to satisfy some properties, as explained in [6, Sect. 6.3], such as: PERs respect computation, we have to prove that the new type constructor satisfies these properties. We can then start stating and proving the validity of type inference rules regarding the new type constructor.

4   The Axiom of Choice

We now illustrate the process of extending Nuprl's computation system and type theory in order to validate axiom of choice type inference rules.

4.1   Squashed or Non-squashed?

The axiom of choice (where A and B are types, and P is of type $A \to B \to \mathbb{P}$)

$$\Pi a{:}A.\Sigma b{:}B.P\ a\ b \;\Rightarrow\; \Sigma f{:}B^A.\Pi a{:}A.P\ a\ f(a)$$

follows from the usual inference rules of the universal (dependent product) and existential (dependent sum) quantifiers [34, Sect. 5.3]. However, this non-squashed version of AC is not always enough because existential quantifiers cannot always be interpreted as Σ but sometimes as truncated Σ's (see for example [18, 34, 36]). Therefore, we sometimes need instances of AC where Σ is either ⫫-squashed or ↓-squashed. In that case it is not obvious anymore which instances of AC are consistent with or provable in Nuprl. This section provides some answers.

We showed in [34, Sect. 5.3] that we can directly prove in Nuprl the following ⫫-squashed versions of $AC_{0,B}$ and $AC_{1,B}$, where $B = \mathbb{N}^{\mathbb{N}}$:

$$\Pi n{:}\mathbb{N}.{\downarrow}\Sigma f{:}B.P\ n\ f\ \Rightarrow\ {\downarrow}\Sigma f{:}B^{\mathbb{N}}.\Pi n{:}\mathbb{N}.P\ n\ f(n)$$

$$\Pi n{:}B.{\downarrow}\Sigma f{:}B.P\ n\ f\ \Rightarrow\ {\downarrow}\Sigma f{:}B^{B}.\Pi n{:}B.P\ n\ f(n)$$

We mentioned in [36, Appendix B] that we have proved the validity of the following ↓-squashed version of $AC_{0,\text{NBase}}$ in our Coq framework using classical logic and choice sequences of terms of type $\text{NBase} = \{t : \text{Base} \mid (t : \text{Base})\#\}$, where $(t : T)\#$ says that the term t is in the type T and does not contain any name, and Base is the type of closed terms with $\sim$ as its equality:

$$\Pi n{:}\mathbb{N}.{\downarrow}\Sigma f{:}\text{NBase}.P\ n\ f\ \Rightarrow\ {\downarrow}\Sigma f{:}\text{NBase}^{\mathbb{N}}.\Pi n{:}\mathbb{N}.P\ n\ f(n)$$

We showed in [36, Appendix B] that the ↓-squashed version of Brouwer's weak continuity principle (WCP) and the negation of its unsquashed version, which are provable in Nuprl, imply the negation of the following ↓-version of $AC_{2,0}$ (where T is a non-empty type):

$$\Pi P{:}\mathbb{N}^{B} \to T \to \mathbb{P}.$$
$$(\Pi f{:}\mathbb{N}^{B}.{\downarrow}\Sigma n{:}\mathbb{N}.\ P\ f\ n)\ \Rightarrow\ {\downarrow}\Sigma N{:}\mathbb{N}^{B} \to T.\ \Pi f{:}\mathbb{N}^{B}.P\ f\ (N\ f)$$

Let us repeat the proof here. It suffices to prove that $\mathbb{N}^{B}$ does not have the following choice principle (while B and $\mathbb{N}$ do):

$$\text{ChoicePrinciple}(T) = \Pi P{:}T \to \mathbb{P}.(\Pi t{:}T.{\downarrow}P(t)) \iff ({\downarrow}\Pi t{:}T.P(t))$$

which follows easily from both the facts that the ↓-version of WCP is true in Nuprl and its unsquashed version is false. Let us prove $\neg\text{ChoicePrinciple}(\mathbb{N}^{B})$, i.e., assuming the hypothesis $\text{ChoicePrinciple}(\mathbb{N}^{B})$ we have to prove $\textbf{False}$. We instantiate this hypothesis with the following function, which we call C (where $\mathbb{N}_k$ is the type of natural numbers strictly less than k):

$$\lambda F.\Sigma M{:}\mathbb{N}^{B}.\Pi f, g{:}B.f =_{(\mathbb{N}_{M(f)} \to \mathbb{N})} g \to F(f) =_{\mathbb{N}} F(g)$$

We now get to assume $(\Pi t{:}\mathbb{N}^{B}.{\downarrow}C(t)) \iff ({\downarrow}\Pi t{:}\mathbb{N}^{B}.C(t))$. Because the ↓-squashed version of WCP is true in Nuprl, we also get to assume $\Pi t{:}\mathbb{N}^{B}.{\downarrow}C(t)$. From the above double implication, we obtain ${\downarrow}\Pi t{:}\mathbb{N}^{B}.C(t)$. Because we are proving $\textbf{False}$, we can unsquash this new hypothesis, i.e., we get to assume $\Pi t{:}\mathbb{N}^{B}.C(t)$, which is the unsquashed version of WCP, which is false.

4.2   Choice Sequences

As mentioned above, in order to prove the validity of $AC_{0,\text{NBase}}$, a $\downarrow$-squashed version of AC, we added choice sequences of terms to Nuprl's term syntax:

```
Inductive Term := vterm (v : Var) | sterm (s : nat → Term) | oterm (op : Opid) (bs : list BTerm)
with BTerm := bterm (vs : list Var) (t : Term).
```

Our choice sequences are Coq functions from natural numbers to Nuprl terms. Additionally, we require that such choice sequences do not contain free variables or names [36, Sect. 4.2]. This addition had interesting consequences such as: most relations on terms became undecidable such as syntactic equality and α-equality. Also, because of this additional limit constructor in the definition of terms, the return types of functions that are recursively applied to choice sequences had to be turned into W-like types with limit constructors. For example, we had to change the statement of our general induction principle on terms. Originally, this lemma went by induction on the size of term, which was simply a natural number. With choice sequences, the size of a term is now an ordinal number with a limit operator for the size of sequences:

```
Fixpoint osize (t : Term) : ord :=
  match t with
  | vterm _ ⇒ OS OZ
  | sterm f ⇒ OS (OL (fun x ⇒ osize (f x)))
  | oterm op bterms ⇒ OS (oaddl (map osize_bterm bterms))
  end
with osize_bterm (bt : BTerm) : ord := match bt with bterm lv nt ⇒ osize nt end.
```

where oaddl is an addition operation on lists of ordinals, which are defined as follows: `Inductive ord := OZ | OS (o : ord) | OL (s : nat → ord)`. We also had proved equalities between terms, where now in order to still prove an equality, we need to use in addition the function extensionality axiom to prove that two choice sequences are equal. We leave for future work the investigation of whether we can do without additional axioms using custom equality relations.

5   Conclusion

Much remains to be done to bridge the gap between our Coq implementation and Nuprl's current implementation. Following the footsteps of [30], we would like to synthesize a version of Nuprl from our implementation. Nevertheless, our Coq implementation of CTT already turned out to be very useful to investigate extensional type theory on a large scale, which often was made relatively easy by the fact that Nuprl is open-ended in many ways. However, we are sometimes making decisions that limit Nuprl's open-endedness. For example, because we have now added exceptions to Nuprl, it is not clear how or even whether we could add a parallel operator to Nuprl as mentioned in [34, Sect. 8]. As another example, we have often used the fact that Nuprl's computation system is deterministic in our implementation, which will prevent us from adding non-deterministic operators. It is not clear yet whether we can do without this property, and most importantly it is not clear how to avoid such accidental limitations.

References

1. The Agda Wiki. http://wiki.portal.chalmers.se/agda/pmwiki.php
2. Allen, S.F.: A non-type-theoretic semantics for type-theoretic language. Ph.D. thesis, Cornell University, (1987)
3. Allen, S.F., Bickford, M., Constable, R.L., Eaton, R., Kreitz, C., Lorigo, L., Moran, E.: Innovations in computational type theory using Nuprl. J. Appl. Logic **4**(4), 428–469 (2006). http://www.nuprl.org/
4. Anand, A., Bickford, M., Constable, R.L., Rahli, V.: A type theory with partial equivalence relations as types. Presented at TYPES 2014 (2014)
5. Anand, A., Rahli, V.: Towards a formally verified proof assistant. Technical report, Cornell University (2014). http://www.nuprl.org/html/Nuprl2Coq/
6. Anand, A., Rahli, V.: Towards a formally verified proof assistant. In: Klein, G., Gamboa, R. (eds.) ITP 2014. LNCS, vol. 8558, pp. 27–44. Springer, Heidelberg (2014)
7. Berardi, S., Bezem, M., Coquand, T.: On the computational content of the axiom of choice. J. Symb. Log. **63**(2), 600–622 (1998)
8. Bertot, Y., Casteran, P.: Interactive Theorem Proving and Program Development. Springer, Heidelberg (2004). http://www.labri.fr/perso/casteran/CoqArt
9. Bickford, M., Constable, R.: Inductive construction in Nuprl type theory using bar induction. Presented at TYPES 2014 (2014). http://nuprl.org/KB/show.php?ID=723
10. Bove, A., Dybjer, P., Norell, U.: A brief overview of Agda – a functional language with dependent types. In: Berghofer, S., Nipkow, T., Urban, C., Wenzel, M. (eds.) TPHOLs 2009. LNCS, vol. 5674, pp. 73–78. Springer, Heidelberg (2009)
11. Brady, E.: IDRIS: systems programming meets full dependent types. In: PLPV 2011, pp. 43–54. ACM (2011)
12. Bridges, D., Richman, F.: Varieties of Constructive Mathematics. London Mathematical Society Lecture Notes Series. Cambridge University Press, Cambridge (1987)
13. Constable, R.L., Allen, S.F., Bromley, H.M., Cleaveland, W.R., Cremer, J.F., Harper, R.W., Howe, D.J., Knoblock, T.B., Mendler, N.P., Panangaden, P., Sasaki, J.T., Smith, S.F.: Implementing Mathematics with the Nuprl Proof Development System. Prentice-Hall Inc, Upper Saddle River (1986)
14. Constable, R.L.: Constructive mathematics as a programming logic I: some principles of theory. In: Karpinski, M. (ed.) Fundamentals of Computation Theory. LNCS, vol. 158, pp. 64–77. Springer, London (1983)
15. The Coq Proof Assistant. http://coq.inria.fr/
16. Crary, K.: Type-theoretic methodology for practical programming languages. Ph.D. thesis, Cornell University, Ithaca, NY, August 1998
17. Dummett, M.A.E.: Elements of Intuitionism, 2nd edn. Clarendon Press, Oxford (2000)
18. Escardó, M.H., Chuangjie, X.: The Inconsistency of a Brouwerian Continuity principle with the curry-howard interpretation. In: TLCA 2015, vol. 38, pp. 153–164. LIPIcs. Schloss Dagstuhl - Leibniz- Zentrum fuer Informatik (2015)
19. Allen, S.F., Constable, R.L., Howe, D.J.: Reflecting the open-ended computation system of constructive type theory. In: Bauer, F.L. (ed.) Logic, Algebra and Computation. NATO ASI Series, vol. 79, pp. 265–280. Springer, Heidelberg (1990)
20. Hofmann, M.: Extensional concepts in intensional type theory. Ph.D. thesis, University of Edinburgh (1995)

21. Howard, W.A., Kreisel, G.: Transfinite induction and bar induction of types zero and one, and the role of continuity in intuitionistic analysis. J. Symb. Log. **31**(3), 325–358 (1966)
22. Howe, D.J.: Equality in lazy computation systems. In: LICS 1989, pp. 198–203. IEEE Computer Society (1989)
23. Howe, D.J.: Importing mathematics from HOL into Nuprl. In: von Wright, J., Harrison, J., Grundy, J. (eds.) TPHOLs 1996. LNCS, vol. 1125, pp. 267–282. Springer, Heidelberg (1996)
24. Howe, D.J.: On computational open-endedness in Martin-Löf's type theory. In: LICS 1991, pp. 162–172. IEEE Computer Society (1991)
25. Howe, D.J.: Semantic foundations for embedding HOL in Nuprl. In: Nivat, M., Wirsing, M. (eds.) AMAST 1996. LNCS, vol. 1101, pp. 85–101. Springer, Heidelberg (1996)
26. Idris. http://www.idris-lang.org/
27. Kleene, S.C., Vesley, R.E.: The Foundations of Intuitionistic Mathematics, Especially in Relation to Recursive Functions. North-Holland Publishing Company, Amsterdam (1965)
28. Kopylov, A.: Type theoretical foundations for data structures, classes, and objects. Ph.D. thesis, Cornell University, Ithaca, NY (2004)
29. Kreisel, G.: On weak completeness of intuitionistic predicate logic. J. Symb. Logic **27**(2), 139–158 (1962)
30. Kumar, R., Arthan, R., Myreen, M.O., Owens, S.: Self-formalisation of higher-order logic - semantics, soundness, and a verified implementation. J. Autom. Reason. **56**(3), 221–259 (2016)
31. Longley, J.: When is a functional program not a functional program? In: ICFP 1999, pp. 1–7. ACM (1999)
32. Nogin, A., Kopylov, A.: Formalizing type operations using the "Image" type constructor. Electr. Notes Theor. Comput. Sci. **165**, 121–132 (2006)
33. Rahli, V., Bickford, M.: A nominal exploration of intuitionism. Extended version of our CPP. 2016 paper (2015). http://www.nuprl.org/html/Nuprl2Coq/continuity-long.pdf
34. Rahli, V., Bickford, M.: A nominal exploration of intuitionism. In: CPP, pp. 130–141. ACM (2016)
35. Rahli, V., Bickford, M., Anand, A.: Formal program optimization in Nuprl using computational equivalence and partial types. In: Blazy, S., Paulin-Mohring, C., Pichardie, D. (eds.) ITP 2013. LNCS, vol. 7998, pp. 261–278. Springer, Heidelberg (2013)
36. Rahli, V., Bickford, M., Constable, R.L.: A story of bar induction in Nuprl (2015). Extended version http://www.nuprl.org/html/Nuprl2Coq/bar-induction-long.pdf
37. Rathjen, M.: Constructive set theory and brouwerian principles. J. UCS **11**(12), 2008–2033 (2005)
38. Smith, S.F.: Partial objects in type theory. Ph.D. thesis, Cornell University, Ithaca, NY (1989)
39. Troelstra, A.S.: A note on non-extensional operations in connection with continuity and recursiveness. Indagationes Mathematicae **39**(5), 455–462 (1977)
40. Troelstra, A.S.: Aspects of constructive mathematics. In: Barwise, J. (ed.) Handbook of Mathematical Logic, pp. 973–1052. North-Holland Publishing Company, Amsterdam (1977)
41. Troelstra, A.S., van Dalen, D.: Constructivism in Mathematics an Introduction. Studies in Logic and the Foundations of Mathematics, vol. 121. Elsevier, North Holland (1988)

42. The Univalent Foundations Program. Homotopy Type Theory: Univalent Foundations of Mathematics. Institute for Advanced Study (2013). http://homotopytypetheory.org/book
43. Veldman, W.: Brouwer's real thesis on bars. Philosophia Scientiæ **CS6**, 21–42 (2006)
44. Veldman, W.: Understanding and using Brouwer's Continuity principle. In: Schuster, P., Berger, U., Osswald, H. (eds.) Reuniting the Antipodes Constructive and Nonstandard Views of the Continuum. Synthese Library, vol. 306, pp. 285–302. Springer, Netherlands (2001)

Formalizing Double Groupoids and Cross Modules in the Lean Theorem Prover

Jakob von Raumer$^{(\boxtimes)}$

University of Nottingham, Nottingham, UK
psxjv4@nottingham.ac.uk
http://www.cs.nott.ac.uk/~psxjv4/

Abstract. Lean is a new open source dependently typed theorem prover which is mainly being developed by Leonardo de Moura at Microsoft Research. It is suited to be used for proof irrelevant reasoning as well as for proof relevant formalizations of mathematics. In my talk, I will present my experiences doing a formalization project in Lean. One of the interesting aspects of homotopy type theory is the ability to perform synthetic homotopy theory on higher types. While for the first homotopy group the choice of a suitable algebraic structure to capture the homotopic information is obvious – it's a group –, implementing a structure to capture the information about both the first and the second homotopy group (or groupoid) of a type and their interactions is more involved. Following Ronald Brown's book on Nonabelian Algebraic Topology, I formalized two structures: Double groupoids with thin structures and crossed modules on groupoids. I furthermore attempted to prove their equivalence. The project can be seen as a usability and performance test for the new theorem prover.

Keywords: Formalization of mathematics · Algebraic topology

1 Introduction

Making mathematical definitions and theorem proofs readable and verifiable by computers has become increasingly important in the last years, not only since there are proofs that are hard or impossible to be checked by a single person due to their size (one example being Tom Hales' proof of the Kepler conjecture). With the rise of formally verified software, one also wants the same level of trust for the mathematical theories whose soundness guarantee the correct functionality of the program. Fields where formal verification has been successfully used to certify computer programs include cryptography and aerospace industry. These rely heavily on results from algebra and calculus and differential equations.

Homotopy type theory (HoTT) can serve as a foundation of mathematics that is better suited to fit the needs of formalizing certain branches of mathematics, especially the ones of *topology*. In traditional, set-based approaches to formalizing the world of mathematical knowledge, topological spaces and their properties

© Springer International Publishing Switzerland 2016
G.-M. Greuel et al. (Eds.): ICMS 2016, LNCS 9725, pp. 28–33, 2016.
DOI: 10.1007/978-3-319-42432-3_4

have to be modeled with much effort by referring to the type of real numbers. In contrast to this, homotopy type theory, in a certain sense, contains topologically motivated objects like fibrations and homotopy types as primitives. This makes it much easier and more natural to reason about topological properties of these objects. Homotopy type theory is a relatively new field but it already has produced several useful implementations and libraries in interactive theorem provers like Agda and Coq. One important feature of homotopy type theory is that it is *constructive* and thus allows to extract programs from definitions and proofs.

Homotopy type theory is *proof relevant* which means that there can be distinct (and internally distinguishable) proofs for one statement. This leads to the fact that types in HoTT bear the structure of a higher groupoid in their identities. The essential problem in the field of *homotopy* is to analyze this structure of paths and iterated paths between paths in topological spaces or, in the world of HoTT, in higher types. This happens by considering the algebraic properties of the homotopy groups or *homotopy groupoids* of the spaces resp. types.

In his book "Nonabelian Algebraic Topology" [1], Ronald Brown introduces the notion of *double groupoids with thin structures* and *crossed modules over groupoids* to describe the interaction between the first and the second homotopy groupoid of a space algebraically. Brown's approach, preceding the discovery of homotopy type theory by a few decades, is formulated entirely classically and set-based.

I will describe how I translated some of the central definitions and lemmas from his book to dependently typed algebraic structures in homotopy type theory, made them applicable to the analysis of 2-truncated types by creating the notion of a *fundamental double groupoid of a presented 2-type*, and then formalized them in the newly built interactive theorem proving system Lean [2].

2   Double Categories and Double Groupoids

Seeing a (small) category as a tuple of object set, morphism set, domain and codomain functions, identity function and composition, Brown defines double categories similar to the following:

A **double category** D is given by the following data: Three sets D_0, D_1, and D_2, the elements of which are respectively called **0-, 1- and 2-cells**, together with maps ∂^-, ∂^+, ϵ, $\circ_D$, ∂_1^-, ∂_1^+, ϵ_1, $\circ_1$, ∂_2^-, ∂_2^+, ϵ_2, and $\circ_2$ that make these sets form three categories:

- a category $(D_0, D_1, \partial^-, \partial^+, \epsilon, \circ_D)$ on D_0, often called the **(1-)skeleton** of the double category,
- a **vertical category** $(D_1, D_2, \partial_1^-, \partial_1^+, \epsilon_1, \circ_1)$, and
- a **horizontal category** $(D_1, D_2, \partial_2^-, \partial_2^+, \epsilon_2, \circ_2)$.

The mentioned maps are required to satisfy the following **cubical identities:**

$$\partial^- \circ \partial_1^+ = \partial^- \circ \partial_2^-,$$
$$\partial^- \circ \partial_1^+ = \partial^+ \circ \partial_2^-,$$
$$\partial^+ \circ \partial_1^- = \partial^- \circ \partial_2^+,$$
$$\partial^+ \circ \partial_1^+ = \partial^+ \circ \partial_2^+,$$

$$\partial_1^- \circ \epsilon_2 = \epsilon \circ \partial^-,$$
$$\partial_1^+ \circ \epsilon_2 = \epsilon \circ \partial^+,$$
$$\partial_2^- \circ \epsilon_1 = \epsilon \circ \partial^-,$$
$$\partial_2^+ \circ \epsilon_1 = \epsilon \circ \partial^+, \text{ and}$$
$$\epsilon_1 \circ \epsilon = \epsilon_2 \circ \epsilon =: 0.$$

The boundary and degeneracy maps of the vertical category are furthermore assumed to be a homomorphism with respect to the composition of the horizontal category, and vice versa:

$$\partial_2^- (v \circ_1 u) = \partial_2^- (v) \circ_D \partial_2^- (u),$$
$$\partial_2^+ (v \circ_1 u) = \partial_2^+ (v) \circ_D \partial_2^+ (u),$$
$$\partial_1^- (v \circ_2 u) = \partial_1^- (v) \circ_D \partial_1^- (u),$$
$$\partial_1^+ (v \circ_2 u) = \partial_1^+ (v) \circ_D \partial_1^+ (u),$$
$$\epsilon_2 (g \circ_D f) = \epsilon_2(g) \circ_1 \epsilon_2(f), \text{ and}$$
$$\epsilon_1 (g \circ_D f) = \epsilon_1(g) \circ_2 \epsilon_1(g),$$

for each $f, g \in D_1$ and $u, v \in D_2$ where the compositions are defined.

A last condition, called the **interchange law**, has to be fulfilled: For each $u, v, w, x \in D_2$,

$$(x \circ_2 w) \circ_1 (v \circ_2 u) = (x \circ_1 v) \circ_2 (w \circ_1 u)$$

has to hold wherever it is well-defined.

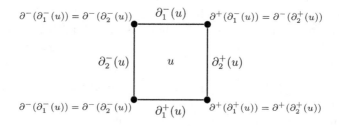

$$\partial^- (\partial_1^- (u)) = \partial^- (\partial_2^- (u)) \qquad \partial_1^- (u) \qquad \partial^+ (\partial_1^- (u)) = \partial^- (\partial_2^+ (u))$$

$$\partial_2^- (u) \qquad u \qquad \partial_2^+ (u)$$

$$\partial^- (\partial_1^- (u)) = \partial^- (\partial_2^- (u)) \qquad \partial_1^+ (u) \qquad \partial^+ (\partial_1^+ (u)) = \partial^+ (\partial_2^+ (u))$$

Fig. 1. A square $u \in D_2$, its faces, and its corners.

In more pictorial words, double categories do not only contain objects and morphisms (lines), but also square-shaped two cells (see Fig. 1). These can be

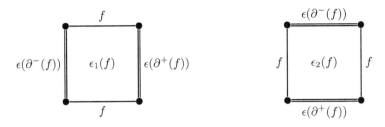

Fig. 2. Degenerate squares of the vertical and horizontal category for a given line $f \in D_1$. Degenerate lines are drawn as double lines.

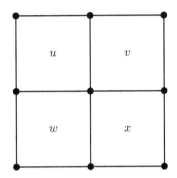

Fig. 3. The grid we use to illustrate the composition $(x \circ_2 w) \circ_1 (v \circ_2 u)$ as well as $(x \circ_1 v) \circ_2 (w \circ_1 u)$, which are identical by the interchange law.

composed vertically or horizontally, given that their edges match. There are squares which act as the identity with respect one of the composition (see Fig. 2), and when composing in a 2-by-2 grid, it doesn't matter whether we give precedence to vertical or to horizontal composition (see Fig. 3) To prevent the necessity of the composition of a partial function, we make the the type of two-cells depend on its boundary when we translate the definition to one in type theory:

We define a **double category** to be a record containing the following:

- The object set $D_0 :$ Set,
- A *precategory* (here, "pre" means that isomorphic in that category are not necessarily equal) on D_0, consisting of:
  - A type family of morphisms $D_1 : \prod_{(a,b:D_0)}$ Set.
  - The composition of morphisms

$$\circ : \prod_{a,b,c:D_0} D_1(b,c) \to D_1(a,b) \to D_1(a,c).$$

  - An identity operator id $: \prod_{(a:D_0)} D_1(a,a)$.
  - A witness ensuring associativity for all morphisms:

$$\prod_{a,b,c,d:D_0} \prod_{h:D_1(c,d)} \prod_{g:D_1(b,c)} \prod_{f:D_1(a,b)} h \circ (g \circ f) = (h \circ g) \circ f$$

- Witnesses that the identity morphisms are neutral with respect to composition from the left and from the right:

$$\prod_{a,b:D_0} \prod_{f:D_1(a,b)} (\mathrm{id}(b) \circ f = f) \times (f \circ \mathrm{id}(a) = f)$$

– A set family of two-cells:

$$D_2 : \prod_{a,b,c,d:D_0} \prod_{f:D_1(a,b)} \prod_{g:D_1(c,d)} \prod_{h:D_1(a,c)} \prod_{i:D_1(b,d)} \mathrm{Set}$$

We will always leave the first four parameters implicit and write $D_2(f, g, h, i)$ for the type of two-cells with f as their upper face, g as their bottom face, h as their left face, and i as their right face.

– The vertical composition operation: For all $a, b, c_1, d_1, c_2, d_2 : D_0$ and $f_1 : D_1(a, b)$, $g_1 : D_1(c_1, d_1)$, $h_1 : D_1(a, c_1)$, $i_1 : D_1(b, d_1)$, $g_2 : D_1(c_2, d_2)$, $h_2 : D_1(c_1, c_2)$, and $i_2 : D_1(d_1, d_2)$ the composition of two cells

$$v \circ_1 u : D_2(g_1, g_2, h_2, i_2) \to D_2(f, g_1, h_1, i_1) \to D_2(f_1, g_2, h_2 \circ h_1, i_2 \circ i_1).$$

– The vertical identity $\mathrm{id}_1 : \prod_{(a,b:D_0)} \prod_{(f:D_1(a,b))} D_2(f, f, \mathrm{id}(a), \mathrm{id}(b))$.

– For all $w : D_2(g_2, g_3, h_3, i_3)$, $v : D_2(g_1, g_2, h_2, i_2)$, and $u : D_2(f, g_1, h_1, i_1)$ a witness for the associativity of the vertical composition $\mathsf{assoc}_1(w, v, u)$ in

$$\mathsf{assoc}(i_3, i_2, i_1)_*(\mathsf{assoc}(h_3, h_2, h_1)_*(w \circ_1 (v \circ_1 u))) = (w \circ_1 v) \circ_1 u,$$

where assoc is the associativity proof in the 1-skeleton. The transport is required since the cells at the left and right side of the equation do not definitionally have the same set of faces.

– Horizontal composition $\circ_2$ and horizontal identity id_2.

– Finally, we need witnesses that the axioms of a double category, as stated in the definition above, hold. Note that there are only four of these rules which are not yet expressed by the types of composition and identity.

3   The Fundamental Double Groupoid

How can we now use this structure to characterize types in homotopy type theory? The role of the basepoint in the consideration of the fundamental group of a type or a set of basepoints for the fundamental double groupoid, we need a *presentation* relative to which we express the fundamental double groupoid:

We define a **presented 2-type** to be a triple (X, A, C) of types $X, A, C : \mathcal{U}$ together with functions $\iota : C \to A$ and $\iota' : A \to X$ where X is a 2-type, A is a 1-type, and C is a set.

From each presented 2-type (X, A, C) we receive its **fundamental double category** G by defining

$$G_0 :\equiv C$$
$$G_1(a, b) :\equiv \iota(a) =_A \iota(b)$$
$$G_2(f, g, h, i) :\equiv \mathsf{ap}_{\iota'}(h) \cdot \mathsf{ap}_{\iota'}(g) =_X \mathsf{ap}_{\iota'}(f) \cdot \mathsf{ap}_{\iota'}(i)$$

for all $a, b : C$, $f : \iota(a) = \iota(b)$, $g : \iota(c) = \iota(d)$, $h : \iota(a) = \iota(c)$, and $i : \iota(b) = \iota(d)$.

Omitted from this abstract, we can then account for the symmetry of the identity relation by extending the definition to the one of a **(weak) double groupoid** and for the fact that each commuting square boundary gives rise to a homotopically degenerate square filler by equipping the double groupoid with a **thin structure** or, equivalently, **connections**. From the category of double groupoids we can then switch to a more "flat" representation by transforming these into **crossed modules**. Future work will include a formulation of the statement and proof of a Seifert-van Kampen theorem which yields the fundamental double groupoid of certain pushouts of types.

References

1. Brown, R., Higgins, P.J., Sivera, R.: Nonabelian algebraic topology: filtered spaces, crossed complexes, cubical homotopy groupoids. European Mathematical Society (2011)
2. de Moura, L., Kong, S., van Doorn, F., von Raumer, J.: The lean theorem prover (system description). In: Felty, A.P., Middeldorp, A. (eds.) CADE-25. LNCS, vol. 9195, pp. 378–388. Springer International Publishing, Switzerland (2015)

Software for Mathematical Reasoning
and Applications

Towards the Automatic Discovery of Theorems in GeoGebra

Miguel Abánades[1], Francisco Botana[2(✉)], Zoltán Kovács[3], Tomás Recio[4],
and Csilla Sólyom-Gecse[5]

[1] Universidad Rey Juan Carlos, Móstoles, Spain
`miguelangel.abanades@urjc.es`
[2] Universidad de Vigo, Vigo, Spain
`fbotana@uvigo.es`
[3] The Private University College of Education of the Diocese of Linz, Linz, Austria
`zoltan@geogebra.org`
[4] Universidad de Cantabria, Santander, Spain
`tomas.recio@unican.es`
[5] Babeş-Bolyai University, Cluj-Napoca, Romania
`solyom_csilla@yahoo.com`
`http://webs.uvigo.es/fbotana/`, `http://sites.google.com/site/kovzol`,
`http://www.recio.tk`, `http://www.geogebra.org/solyom-gecse+csilla`

Abstract. Considerable attention and efforts have been given to the implementation of automatic reasoning tools in interactive geometric environments. Nevertheless, the main goal in such works focused on theorem proving, cf. Java Geometry Expert or GeoGebra. A related issue, automatic discovery, remains almost unexplored in the field of dynamic geometry software.

This extended abstract sketches our initial results towards the incorporation into GeoGebra, a worldwide spread software with tenths of millions of users, of automatic discovery abilities. As a first result, currently available in the official version, we report on a new command allowing the automatic discovery of loci of points in diagrams. Besides the standard mover-tracer locus finding, the approach also deals with loci constrained by implicit conditions. Hence, our proposal successfully automates a kind of bound dragging in dynamic geometry, the 'dummy locus dragging'. In this way, the cycle of conjecturing-checking-proving will be accessible for general learners in elementary geometry.

Keywords: Automatic discovery · Dynamic geometry · GeoGebra · Computational algebraic geometry

1 Introduction

We present and discuss the implementation of a new command, included in the most recent version of the dynamic geometry program GeoGebra, for the automatic discovery of elementary geometry statements. While the command is

© Springer International Publishing Switzerland 2016
G.-M. Greuel et al. (Eds.): ICMS 2016, LNCS 9725, pp. 37–42, 2016.
DOI: 10.1007/978-3-319-42432-3_5

currently limited to discovery situations where there is, at most, an unknown point constrained by single conditions, our findings show that it can deal with a huge number of elementary geometric constructions. Furthermore, the underlying protocol used for discovery is transparent with respect to the number of points or conditions, thus being theoretically simple its extension to more general constructions. Our `LocusEquation` command is ready for use under real conditions (say, school context), and, although featured just on the last GeoGebra version, there are already examples of its use by teachers[1].

Section 2 shortly recalls the main concepts and issues related to automatic deduction in geometry, with particular emphasis in dynamic geometry. The protocol for automatic discovery is sketched, addressing the reader to appropriate references. The specific GeoGebra command for discovery is described in Sect. 3, and a canonical example is discussed for illustration. Finally, Sect. 4 lists current shortcomings related to discovery in GeoGebra, and points out some lines of future work.

2   Automatic Deduction in Geometry

Along this note, when we refer to automatic *proving* of elementary geometry theorems, we restrict ourselves to the theorem proving approach through computational algebraic geometry methods, as initiated, forty years ago, in the pioneer work of Wu [9], carefully described and disseminated by the popular book of Chou [3].

The goal of this particular approach to automatic proving is to provide algorithms, using computer algebra methods, for confirming (or refuting) the truth of some given geometric statement. More precisely, the goal is to decide if a given statement is generally true or not, i.e. true except for some degenerate cases, to be described by the algorithm. Hundreds of highly non trivial theorems in elementary geometry have been successfully—and almost instantaneously—verified by a variety of symbolic computation methods, see [3] for an early collection of examples.

Briefly—and very roughly—this approach proceeds by translating geometric facts, say hypotheses H and theses T, into systems of polynomial equations, say S_H, S_T, and, then, considering geometric statements $(H \Rightarrow T)$ as inclusion tests $S_H \subseteq S_T$ between the solutions of the system of equations S_H, translating the geometry described by H, and those of the corresponding system S_T, expressing the theses T. Such inclusion tests are, then, elucidated by some computer algebra tools deciding if a polynomial f is or not an algebraic combination of some given collection of polynomials S, which is—approximately—a way to show if the roots of f form or not a superset of the solutions of the system $S = 0$.

Let us add—because it will be important in what follows—that in this approach to automatic proving it is often the case that the inclusion test $S_H \subseteq S_T$ does not fully happen because of the extended, often unexpected, meaning of

---

[1] E.g. http://tube.geogebra.org/m/DCSFzaph.

the algebraic translation of geometric facts. Rather, it might be the case that there is a 'negligible' portion of the solution set for S_H that is not included in the solution set for S_T: elements in that portion represent the, so called, degenerate cases. 'Negligible' means—and it has been crucial to arrive to a precise expression of this concept—that the solutions for S_H not included in S_T have to verify some non-trivial system of equations in the free variables of the geometric hypotheses H.

A closely related, albeit different, issue is that of the automatic discovery of theorems (see [6] for a large collection of references on this topic). Yet, let us remark that the term 'discovery' is already used in [7] or [8], where it is written (on page 292), under the specific section with title 'Discovering Theorems', that 'a typical example is the automatic discovery of Quin-Heron's formula that represents the area of a triangle in terms of its three sides'. We should remark that automatic discovery (in this particular sense of automatic derivation of statements) is being currently of particular interest, associated to different techniques for 'recognizing' (we would say, 'deriving') geometric facts contained in a diagram (from textbooks, pdf's, human sketches . . .) after translating the graphic information into geometric data by means of the Hough transform, see [2].

Roughly speaking, automatic *proving* deals with establishing (or denying) that some statement holds in most instances, while automatic *discovery* –in its most general conception– addresses the case of statements $H \Rightarrow T$ that are false in most relevant cases. In fact, it aims to automatically produce additional, necessary, hypotheses H' for the statement $(H\&H') \Rightarrow T$ to be correct; and, then testing, by automatic proving methods, if adding these extra hypotheses is also sufficient.

One must remark that the search for complementary hypotheses should be done in terms of the free variables for the construction. For example, suppose that for a given triangle and an arbitrary point P we state that the projections of P on the sides of the triangle are always aligned. It is, obviously, false; what we will like is to discover restrictions on the coordinates of the vertices of the triangle and on the coordinates of P, which are free variables in this formulation, so that the thesis will hold true.

Describing the *implicit geometric locus* of a point subject to some geometric constraints, say, finding the locus of a point P when its projection on the three sides of a given, determined triangle form a triangle of given constant area ([3], Chapter IV, Example 5.8) can be considered as a variant of this 'automatic discovery' approach. In fact, the steps in the construction of the projections of P can be considered as the hypotheses H, while the given constraints over the point P (e.g. requiring that the area of the triangle described by the three projections of P over the sides of the given triangle must be constant) can be considered as the proposed thesis T, one that is false for arbitrary positions of P; finally, the description H' of the locus (for point P to verify that its three projections form a triangle of fixed area) can be understood as the extra, necessary hypotheses required for the given statement to hold true, so that $(H\&H') \Rightarrow T$.

While automatic proving using computer algebra methods has been used in dynamic geometric software[2], similar automatic discovery abilities are not present in software ready for universal use. Following our goal towards the popularization of tools for automatic reasoning in geometry, we have cooperated with other authors providing automatic proving resources to GeoGebra [1]. Continuing this trend, our recent work focuses on GeoGebra discovery capabilities based on the computational approach described in [4,5].

3   The GeoGebra Command `LocusEquation`

Automatic discovery in GeoGebra requires that the user first constructs a geometric diagram with GeoGebra's *drawing tools* or *drawing commands*. Although theoretically all *algebraic* constructions (i.e. those composed of elements that can be expressed by polynomial equations) can serve as initial data for GeoGebra's discovery tool, technical reasons, mainly related to computational time limitations, restrict the applicability of the tool for some involved constructions. Moreover, it is important to note that non-algebraic elements, such as the graph of a sine function, fall out of the scope of the method, algebraic in nature.

After constructing a geometric diagram the user needs to type the command `LocusEquation`[3] with two parameters: the sought thesis T (which must be an atomic Boolean expression) and a free point P 'supporting' the discovery. The Boolean expression defining the discovery plays the part of the *extra condition* that we require our diagram to satisfy. The free point P, second parameter of the command `LocusEquation`, is the point over which the sought extra hypothesis will verse. In algebraic terms, the symbolic coordinates of P will be the variables of the polynomials conforming the necessary conditions obtained as a result of the discovery process. As a result, `LocusEquation[T,P]` will produce a set V (providing its implicit equation) such that "if T is true then $P \in V$". It should be noted that the basic points of the construction—other than P—are fixed, that is, their numerical coordinates will be used in the discovery computation. Recall that P has always symbolic coordinates. Thus, we are discovering on a specific instance of the general construction, and, then, it could happen that we discover some incidental property only related to this particular model. For instance, if we intend to construct a general triangle but we actually draw—without noticing it—an isosceles one, we can find out statements which are true just for this particular kind of triangles. Of course, this confusion will be clarified in the *proving* phase, when intending to check the validity, in general, of the obtained result.

As a simple illustrative example we pose the discovery of Wallace-Simson theorem, a particular case of the implicit locus recalled in the previous Section:

---

[2] See Java Geometry Expert, JGEX, http://www.cs.wichita.edu/~ye/gex.html, for a paradigmatic example.

[3] The command `LocusEquation` was introduced in GeoGebra version 5.0.213.0, distributed since March 12, 2016. Note that the software has at least one new published version every week.

Given a triangle ABC a point P on its plane, find the locus of points P such that their perpendicular projections on the triangle sides are collinear.

As said above, the user constructs an instance of the configuration (Fig. 1) and imposes the collinearity of projections D, E, F in order to find conditions on P for it through the command `LocusEquation[AreCollinear[D,E,F],P]`. The result shows the triangle circumcircle, graphically and providing its equation (see implicit equation in Fig. 1). Let us insist on the fact that points A, B, C are not generally considered, but their numerical coordinates are used in computations. This implementation decision tries to imitate the traditional inductive process, reasoning at a first step on concrete situations. Furthermore, the discovery is not disturbed by the need of studying degenerate conditions (as they will appear for the case where P coincides with a vertex). Once the system has returned the conjecture, the user can test it by dragging any basic element, the triangle vertices in this construction. A final step will allow redefining point P to lie on the circle, and then check the truth of the general statement.

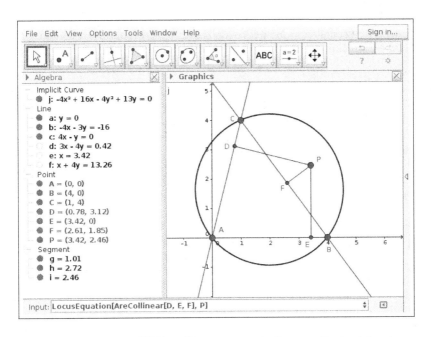

Fig. 1. Discovering Wallace-Simson property for a particular triangle

4   Further Work

Currently, the last statement concerning generalization of the discovered property is not fully implemented in GeoGebra. Ideally, once the conjectural necessary condition is found (i.e. that P lies in the discovered locus) a user should be able to bind the point P to the locus curve and then check for the sufficiency of this condition for the statement correctness, either numerically or formally [1]. Nevertheless, the fully integration of formal proving and discovery subsystems is yet ongoing work. In order to check that correctness with the actual version, the user must repeat the whole construction from the beginning, including—say—the construction of the circumcircle and then verify the Wallace-Simson theorem for a point on this circle. The main obstacle is technical: the locus output is not an acceptable input for the current GeoGebra `Prove` command. Future GeoGebra versions will eliminate this redundant approach.

Another severe limitations deal with the atomic character of Boolean conditions (i.e. we can deal just with one thesis at a time) and the number of discovery points. While none of them introduce theoretical difficulties for discovery, the size and types of uses of GeoGebra require careful considerations before modifying its data structure. For instance, accepting non atomic Boolean conditions implies extending the output of `LocusEquation` to more general data types.

Acknowledgment. First, second and fourth authors partially supported by the Spanish Ministerio de Economía y Competitividad and by the European Regional Development Fund (ERDF), under the Project MTM2014-54141-P.

References

1. Botana, F., Hohenwarter, M., Janičić, P., Kovács, Z., Petrović, I., Recio, T., Weitzhofer, S.: Automated theorem proving in GeoGebra: current achievements. J. Autom. Reasoning **55**, 39–59 (2015)
2. Chen, X., Song, D., Wang, D.: Automated generation of geometric theorems from images of diagrams. Ann. Math. Artif. Intell. **74**, 333–358 (2015)
3. Chou, S.-C.: Mechanical Geometry Theorem Proving, in Mathematics and its Applications. D. Reidel Publ. Comp, Dordrecht (1988)
4. Dalzotto, G., Recio, T.: On protocols for the automated discovery of theorems in elementary geometry. J. Autom. Reasoning **43**, 203–236 (2009)
5. Recio, T., Vélez, M.P.: Automatic discovery of theorems in elementary geometry. J. Autom. Reasoning **23**, 63–82 (1999)
6. Recio, T., Vélez, M.P.: An introduction to automated discovery in geometry through symbolic computation. In: Langer, U., Paule, P. (eds.) Numerical and Symbolic Scientific Computing: Progress and Prospects. Texts and Monographs in Symbolic Computation, pp. 25–271. Springer, Vienna (2011)
7. Wang, D.: A new theorem discovered by computer prover. J. Geom. **36**, 173–182 (1989)
8. Wang, D.: Gröbner bases applied to geometric theorem proving and discovering. In: Buchberger, B., Winkler, F. (eds.) Gröbner Bases and Applications. London Mathematical Society Lecture Notes Series, vol. 251, pp. 281–301. Cambridge University Press (1998)
9. Wen-Tsün, W.: On the decision problem and the mechanization of theorem-proving in elementary geometry. In: Bledsoe, W.W., Loveland, D.W. (eds.) Automated Theorem Proving: After 25 years, pp. 213–234. AMS, Providence (1984)

Automating Free Logic in Isabelle/HOL

Christoph Benzmüller[1,2(✉)] and Dana Scott[3]

[1] Freie Universität Berlin, Berlin, Germany
`c.benzmueller@fu-berlin.de`
[2] Visiting Scholar at Stanford University, Stanford, USA
[3] Visiting Scholar at University of California, Berkeley, USA
`dana.scott@cs.cmu.edu`
`http://www.christoph-benzmueller.de`, `http://www.cs.cmu.edu/~scott/`

Abstract. We present an interactive and automated theorem prover for free higher-order logic. Our implementation on top of the Isabelle/HOL framework utilizes a semantic embedding of free logic in classical higher-order logic. The capabilities of our tool are demonstrated with first experiments in category theory.

Keywords: Free logic · Interactive and automated theorem proving · Model finding · Application to category theory

1 Introduction

Partiality and undefinedness are core concepts in various areas of mathematics. Modern mathematical proof assistants and theorem proving systems are often based on traditional classical or intuitionistic logics and provide rather inadequate support for these challenge concepts. Free logic [5,6], in contrast, offers a theoretically and practically appealing solution. Unfortunately, however, we are not aware of any implemented and available theorem proving system for free logic.

In this extended abstract we show how free logic can be "implemented" in any theorem proving system for classical higher-order logic (HOL) [1]. The proposed solution employs a semantic embedding of free (or inclusive logic) in HOL. We present an exemplary implementation of this idea in the mathematical proof assistant Isabelle/HOL [4]. Various state-of-the-art first-order and higher-order automated theorem provers and model finders are integrated (modulo suitable logic translations) with Isabelle via the Sledgehammer tool [2], so that our solution can be utilized, via Isabelle as foreground system, with a whole range of other background reasoners. As a result we obtain an elegant and powerful implementation of an interactive and automated theorem proving (and model finding) system for free logic.

To demonstrate the practical relevance of our new system, we report on first experiments in category theory. In these experiments, theorem provers were able to detect a (presumably unknown) redundancy in the foundational axiom system of the category theory textbook by Freyd and Scedrov [3].

© Springer International Publishing Switzerland 2016
G.-M. Greuel et al. (Eds.): ICMS 2016, LNCS 9725, pp. 43–50, 2016.
DOI: 10.1007/978-3-319-42432-3_6

2   Free Logic

Terms in classical logic denote, without exceptions, entities in a non-empty domain of (existing) objects **D**, and it are these objects of **D** the universal and existential quantifiers do range over. Unfortunately, however, these conditions may render classical logic unsuited for handling mathematically relevant issues such as undefinedness and partiality. For example in category theory composition of maps is not always defined.

Free logic (and inclusive logic) has been proposed as an alternative to remedy these shortcomings. It distinguishes between a raw domain of possibly non-existing objects **D** and a particular subdomain **E** of **D**, containing only the "existing" entities. Free variables range over **D** and quantified variables only over **E**. Each term denotes in **D** but not necessarily in **E**. The particular notion of free logic as exploited below has been introduced by Scott [6]. A graphical illustration of this notion of free logic is presented in Fig. 1.

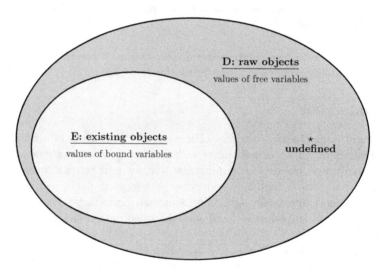

Fig. 1. Illustration of the semantical domains of free logic

3   Implementing Free Logic in Isabell/HOL

We start out with introducing a type i of individuals. The domain of objects associated with this this type will serve as the domain of raw objects **D**, cf. Fig. 1. Moreover, we introduce an existence predicate **E** on type i. As mentioned, **E** characterises the subset of existing objects in **D**. Next, we declare a special constant symbol star ⋆, which is intended to denote a distinguished "non-existing" element of **D**.

typedecl i — the type for indiviuals
consts $fExistence::$ $i{\Rightarrow}bool$ (**E**)   — Existence predicate
consts $fStar::$ i (⋆)   — Distinguished symbol for undefinedness

We postulate that $\star$ is a "non-existing" object in D.

axiomatization where *fStarAxiom*: $\neg\mathbf{E}(\star)$

The two primitive logical connective we introduce for free logic are negation ($\neg$) and implication ($\rightarrow$). They are identified with negation ($\neg$) and implication ($\longrightarrow$) in the underlying Isabelle/HOL logic. The internal names in Isabelle/HOL of the new logical connectives are *fNot* and *fImplies* (the prefix *f* stands for "free"); $\neg$ and the infix operator $\rightarrow$ are introduced as syntactical sugar.[1]

abbreviation *fNot*:: *bool*$\Rightarrow$*bool* ($\neg$)
  where $\neg\varphi \equiv \neg\varphi$
abbreviation *fImplies*:: *bool*$\Rightarrow$*bool*$\Rightarrow$*bool* (**infixr** $\rightarrow$ *49*)
  where $\varphi\rightarrow\psi \equiv \varphi\longrightarrow\psi$

The main challenge is to appropriately define free logic universal quantification ($\forall$) and free logic definite description ($\mathbf{I}$). Again, we are interested to relate these logical operators to the respective operators $\forall$ and *THE* in the Isabelle/HOL logic. Different to the trivial maps for $\neg$ and $\rightarrow$ from above, their mappings are relativized in the sense that the existence predicate $\mathbf{E}$ is utilized as guard in their definitions.

The definition of the free logic universal quantifier $\forall$ thus becomes:

abbreviation *fForall*:: ($i\Rightarrow$*bool*)$\Rightarrow$*bool* ($\forall$)
  where $\forall\Phi \equiv \forall x.\ \mathbf{E}(x)\longrightarrow\Phi(x)$

Apparently, this definitions restricts the set of objects the $\forall$-operator is ranging over to the set of existing objects $\mathbf{E}$. Note that this set can be empty (if desired, we may of course simply postulate that the domain $\mathbf{E}$ is non-empty: $\exists x.\ \mathbf{E}(x)$). The Isabelle framework supports the introduction of syntactic sugar for binding notations. Here we make use of this option to introduce binding notation for $\forall$. With the definition below we can now use the more familiar notation $\forall x.\ \varphi(x)$ instead of writing $\forall(\lambda x.\ \varphi(x))$ or $\forall\varphi$.

abbreviation *fForallBinder*:: ($i\Rightarrow$*bool*)$\Rightarrow$*bool* (**binder** $\forall$ [8] *9*)
  where $\forall x.\ \varphi(x) \equiv \forall\varphi$

Definite description $\mathbf{I}$ in free logic works as follows: Given an unary set $\Phi = \{a\}$, with a being an "existing" element in $\mathbf{E}$, $\mathbf{I}$ returns the single element a of Φ. In all other cases, that is, if Φ is not unary or a is not an element of $\mathbf{E}$, $\mathbf{I}\Phi$ returns the distinguished "non-existing" object denoted by $\star$. With the help of Isabelle/HOL's definite description operator *THE*, $\mathbf{I}$ can thus be defined as follows:

abbreviation *fThat*:: ($i\Rightarrow$*bool*)$\Rightarrow i$ ($\mathbf{I}$)
  where $\mathbf{I}\Phi \equiv$ *if* $\exists x.\ \mathbf{E}(x) \wedge \Phi(x) \wedge (\forall y.\ (\mathbf{E}(y) \wedge \Phi(y)) \longrightarrow (y = x))$
        then THE $x.\ \mathbf{E}(x) \wedge \Phi(x)$
        else $\star$

Analogous to above we introduce binder notation for $\mathbf{I}$, so that we can write $\mathbf{I}x.\ \varphi(x)$ instead of $\mathbf{I}(\lambda x.\ \varphi(x))$ or $\mathbf{I}\varphi$.

---

[1] The numbers in (*infixr* $\rightarrow$ *49*) and (*binder* $\forall$ [8] *9*) (see below) specify structural priorities and thus help to avoid brackets in formula representations.

abbreviation *fThatBinder*:: $(i{\Rightarrow}bool){\Rightarrow}i$ (**binder I** [8] 9)
  where I$x.\ \varphi(x) \equiv \mathbf{I}(\varphi)$

Further logical connectives of free can now be defined in the usual way (and for $\exists$ we again introduce binder notation).

abbreviation *fOr* (**infixr** $\vee$ 51) **where** $\varphi{\vee}\psi \equiv (\neg\varphi){\to}\psi$
abbreviation *fAnd* (**infixr** $\wedge$ 52) **where** $\varphi{\wedge}\psi \equiv \neg(\neg\varphi{\vee}\neg\psi)$
abbreviation *fEquiv* (**infixr** $\leftrightarrow$ 50) **where** $\varphi{\leftrightarrow}\psi \equiv (\varphi{\to}\psi){\wedge}(\psi{\to}\varphi)$
abbreviation *fEquals* (**infixr** = 56) **where** $x{=}y \equiv x{=}y$
abbreviation *fExists* ($\exists$) **where** $\exists\,\Phi \equiv \neg(\forall\,(\lambda y.\neg(\Phi\ y)))$
abbreviation *fExistsBinder* (**binder** $\exists$ [8]9) **where** $\exists\,x.\ \varphi(x) \equiv \exists\,\varphi$

4   Functionality Tests

We exemplarily investigate some example proof problems from Scott's paper [6], pp. 183–184, where a free logic with a single relation symbol **r** is discussed.

consts r:: $i{\Rightarrow}i{\Rightarrow}bool$ (**infixr r** 70)

The implication $x\ \mathbf{r}\ x \to x\ \mathbf{r}\ x$, where x is a free variable, is valid independently whether x is defined (i.e. "exists") or not. In Isabelle/HOL this quickly confirmed by the simplification procedure simp.

lemma $x\ \mathbf{r}\ x \to x\ \mathbf{r}\ x$ **by** *simp*

However, as intended, the formula $\exists\,y.\ y\ \mathbf{r}\ y \to y\ \mathbf{r}\ y$ is not valid, since set of existing objects **E** could be empty. Nitpick quickly presents a respective countermodel.

lemma $\exists\,y.\ y\ \mathbf{r}\ y \to y\ \mathbf{r}\ y$ **nitpick** [*user-axioms*] **oops**

Consequently, also the implication $(x\ \mathbf{r}\ x \to x\ \mathbf{r}\ x) \to (\exists\,y.\ y\ \mathbf{r}\ y \to y\ \mathbf{r}\ y)$ has a countermodel, where **E** is empty.

lemma $(x\ \mathbf{r}\ x \to x\ \mathbf{r}\ x) \to (\exists\,y.\ y\ \mathbf{r}\ y \to y\ \mathbf{r}\ y)$ **nitpick** [*user-axioms*]   **oops**

If we rule out that **E** is empty, e.g. with additional condition $(\exists\,y.\ y = y)$ in the antecedent of the above formula, then we obtain a valid implication. Isabelle trivially proves this with procedure simp.

lemma $((x\ \mathbf{r}\ x \to x\ \mathbf{r}\ x) \wedge (\exists\,y.\ y = y)) \to (\exists\,y.\ y\ \mathbf{r}\ y \to y\ \mathbf{r}\ y)$ **by** *simp*

We analyse some further statements (respectively statement instances) from Scott's paper [6], p. 185. Because of space restrictions we do not further comment these statements here. Altogether they provide further evidence that our implementation of free logic in fact obeys the intended properties.

lemma *S1*: $(\forall\,x.\ \Phi(x) \to \Psi(x)) \to ((\forall\,x.\ \Phi(x)) \to (\forall\,x.\ \Psi(x)))$ **by** *auto*
lemma *S2*: $\forall\,y.\ \exists\,x.\ x = y$ **by** *auto*
lemma *S3*: $\alpha = \alpha$ **by** *auto*
lemma *S4*: $(\Phi(\alpha) \wedge (\alpha = \beta)) \to \Phi(\beta)$ **by** *auto*
lemma *UI-1*: $((\forall\,x.\ \Phi(x)) \wedge (\exists\,x.\ x = \alpha)) \to \Phi(\alpha)$ **by** *auto*

lemma *UI-2*: $(\forall x.\ \varPhi(x)) \rightarrow \varPhi(\alpha)$ **nitpick** [*user-axioms*] **oops** — Countermodel by Nitpick
lemma *UI-cor1*: $\forall y.((\forall x.\ \varPhi(x)) \rightarrow \varPhi(y))$ **by** *auto*
lemma *UI-cor2*: $\forall y.((\forall x.\ \neg(x = y)) \rightarrow \neg(y = y))$ **by** *auto*
lemma *UI-cor3*: $\forall y.((y = y) \rightarrow (\exists x.\ x = y))$ **by** *auto*
lemma *UI-cor4*: $(\forall y.\ y = y) \rightarrow (\forall y.\exists x.\ x = y)$ **by** *simp*
lemma *Existence*: $(\exists x.\ x = \alpha) \longrightarrow \mathbf{E}(\alpha)$ **by** *simp*
lemma *I1*: $\forall y.\ ((y = (\mathbf{I}x.\ \varPhi(x))) \leftrightarrow (\forall x.\ ((x = y) \leftrightarrow \varPhi(x))))$ **by** (*smt fStarAxiom the-equality*)
abbreviation *Star* ($\bigotimes$) **where** $\bigotimes \equiv \mathbf{I}y.\ \neg\ (y = y)$
lemma *StarTest*: $\bigotimes = \star$ **by** *simp*
lemma *I2*: $\neg(\exists y.\ y = (\mathbf{I}x.\ \varPhi(x))) \rightarrow (\bigotimes = (\mathbf{I}x.\ \varPhi(x)))$ **by** (*metis (no-types, lifting) the-equality*)
lemma *ExtI*: $(\forall x.\ \varPhi(x) \leftrightarrow \varPsi(x)) \rightarrow ((\mathbf{I}x.\ \varPhi(x)) = (\mathbf{I}x.\ \varPsi(x)))$ **by** (*smt the1-equality*)
lemma *I3*: $(\bigotimes = \alpha \vee \bigotimes = \beta) \rightarrow \neg(\alpha \text{ r } \beta)$ **nitpick** [*user-axioms*] **oops**— Countermodel by Nitpick

5   Application in Category Theory

We exemplarily employ our free logic reasoning framework from above for an application in category theory. More precisely, we study some properties of the foundational axiom system of Freyd and Scedrov; see their textbook "Categories, Allegories" [3], p. 3. As expected, the composition $x \cdot y$, for morphisms x and y, is introduced by Freyd and Scedrov as a partial operation, cf. axiom *A1* below: the composition $x \cdot y$ exists if and only if the target of x coincides with the source of y. This is why free logic, as opposed to e.g. classical logic, is better suited as a starting point in this mathematical application area.[2]

In the remainder we identify the base type i of free logic with the raw type of morphisms. Moreover, we introduce constant symbols for the following operations: *source* of a morphism x, *target* of a morphism x and *composition* of morphisms x and y. These operations are denoted by Freyd and Scedrov as $\square x$, $x\square$ and $x \cdot y$, respectively. We adopt their notation as syntactic sugar below, even though we are not particularly fond of the use of $\square$ in this context.

consts *source*:: $i \Rightarrow i$ ($\square$- [*108*] *109*)
      target:: $i \Rightarrow i$ (-$\square$ [*110*] *111*)
      composition:: $i \Rightarrow i \Rightarrow i$ (**infix** $\cdot$ *110*)

Ordinary equality on morphisms is defined as follows:

abbreviation *OrdinaryEquality*:: $i \Rightarrow i \Rightarrow bool$ (**infix** $\approx$ *60*)
  where $x \approx y \equiv ((\mathbf{E}\ x) \leftrightarrow (\mathbf{E}\ y)) \wedge x = y$

---

[2] The precise logic setting is unfortunately not discussed in the very beginning of Freyd's and Scedrov's textbook. Appendix B, however, contains a concise formal definition of the assumed logic. Note the special notion of equality used below (which is different from Kleene equality) and also remember that we postulated a 'non-existing' entity.

We are now in the position to model the category theory axiom system of Freyd and Scedrov.

axiomatization *FreydsAxiomSystem* **where**
  A1: $\mathbf{E}(x\cdot y) \leftrightarrow ((x\Box) \approx (\Box y))$ **and**
  A2a: $((\Box x)\Box) \approx \Box x$ **and**
  A2b: $\Box(x\Box) \approx \Box x$ **and**
  A3a: $(\Box x)\cdot x \approx x$ **and**
  A3b: $x\cdot(x\Box) \approx x$ **and**
  A4a: $\Box(x\cdot y) \approx \Box(x\cdot(\Box y))$ **and**
  A4b: $(x\cdot y)\Box \approx ((x\Box)\cdot y)\Box$ **and**
  A5: $x\cdot(y\cdot z) \approx (x\cdot y)\cdot z$

Experiments with our new reasoning framework for free logic quickly showed that axiom *A2a* is redundant. For example, as Isabelle's internal prover metis[3] confirms, *A2a* is implied by *A2b*, *A3a*, *A3b* and *A4a*.

lemma *A2aIsRedundant-1*: $(\Box x)\Box \approx \Box x$ **by** (*metis A2b A3a A3b A4a*)

A human readable and comprehensible reconstruction of this redundancy is presented below. Our handmade proof employs axioms *A2b*, *A3a*, *A3b*, *A4a* and *A5*, that is, this proof could be further optimized by eliminating the dependency on *A5*.

lemma *A2aIsRedundant-2*: $(\Box x)\Box \approx \Box x$
  proof $-$
  have *L1*: $\forall x.\ (\Box\Box x)\cdot((\Box x)\cdot x) \approx ((\Box\Box x)\cdot(\Box x))\cdot x$ **using** *A5* **by** *metis*
  hence *L2*: $\forall x.\ (\Box\Box x)\cdot x \approx ((\Box\Box x)\cdot(\Box x))\cdot x$ **using** *A3a* **by** *metis*
  hence *L3*: $\forall x.\ (\Box\Box x)\cdot x \approx (\Box x)\cdot x$ **using** *A3a* **by** *metis*
  hence *L4*: $\forall x.\ (\Box\Box x)\cdot x \approx x$ **using** *A3a* **by** *metis*
  have *L5*: $\forall x.\ \Box((\Box\Box x)\cdot x) \approx \Box((\Box\Box x)\cdot(\Box x))$ **using** *A4a* **by** *auto*
  hence *L6*: $\forall x\ .\Box((\Box\Box x)\cdot x) \approx \Box\Box x$ **using** *A3a* **by** *metis*
  hence *L7*: $\forall x.\ \Box\Box(x\Box) \approx \Box(\Box\Box(x\Box))\cdot(x\Box)$ **by** *auto*
  hence *L8*: $\forall x.\ \Box\Box(x\Box) \approx \Box(x\Box)$ **using** *L4* **by** *metis*
  hence *L9*: $\forall x.\ \Box\Box(x\Box) \approx \Box x$ **using** *A2b* **by** *metis*
  hence *L10*: $\forall x.\ \Box\Box x \approx \Box x$ **using** *A2b* **by** *metis*
  hence *L11*: $\forall x.\ \Box\Box((\Box x)\Box) \approx \Box\Box(x\Box)$ **using** *A2b* **by** *metis*
  hence *L12*: $\forall x.\ \Box\Box((\Box x)\Box) \approx \Box x$ **using** *L9* **by** *metis*
  have *L13*: $\forall x.\ (\Box\Box((\Box x)\Box))\cdot((\Box x)\Box) \approx ((\Box x)\Box)$ **using** *L4* **by** *auto*
  hence *L14*: $\forall x.\ (\Box x)\cdot((\Box x)\Box) \approx (\Box x)\Box$ **using** *L12* **by** *metis*
  hence *L15*: $\forall x.\ (\Box x)\Box \approx (\Box x)\cdot((\Box x)\Box)$ **using** *L14* **by** *auto*
  then show *?thesis* **using** *A3b* **by** *metis*
  qed

Thus, axiom *A2a* can be removed from the theory. Alternatively, we could also eliminate *A2b* which is implied by *A1*, *A2a* and *A3a*:

lemma *A2bIsRedundant*: $\Box(x\Box) \approx \Box x$ **by** (*metis A1 A2a A3a*)

---

[3] Metis is a trusted prover of Isabelle, since it returns proofs in Isabelle's trusted proof kernel. Initially, however, we have worked with Isabelle's Sledgehammer tool in our experiments, which in turn performs calls to several integrated first-order theorem provers. These calls then return valuable information on the particular proof dependencies, which in turn suggest the successful calls with metis as presented here.

In fact, by a systematic experimentation within our free logic theorem proving framework, we can show that Freyd's and Scedroc's axiomatic theory can be reduced to just the following five axioms:

axiomatization *FreydsAxiomSystemReduced* **where**
B1: $\mathbf{E}(x{\cdot}y) \leftrightarrow ((x\square) \approx (\square y))$ **and**
B2a: $((\square x)\square) \approx \square x$ **and**
B3a: $(\square x){\cdot}x \approx x$ **and**
B3b: $x{\cdot}(x\square) \approx x$ **and**
B5: $x{\cdot}(y{\cdot}z) \approx (x{\cdot}y){\cdot}z$

The dropped axioms can then be introduced as lemmas.

lemma *B2b*: $\square(x\square) \approx \square x$ **by** (*metis B1 B2a B3a*)
lemma *B4a*: $\square(x{\cdot}y) \approx \square(x{\cdot}(\square y))$ **by** (*metis B1 B2a B3a*)
lemma *B4b*: $(x{\cdot}y)\square \approx ((x\square){\cdot}y)\square$ **by** (*metis B1 B2a B3a*)

6   Summary of Technical Contribution and Further Work

We have presented a new reasoning framework for free logic, and we have exemplary applied it for some first experiments in category theory. We have shown that, in our free logic setting, the category theory axiom system of Freyd and Scedrov is redundant and that three axioms can be dropped.

Our free logic reasoning framework is publicly available for reuse: Simply download Isabelle from https://isabelle.in.tum.de and initialize it (respectively import) the file `FreeFOL.thy` from our sources available at www.christoph-benzmueller.de/papers/2016-ICMS.zip. Our category theory experiments are contained in the file `FreydScedrov.thy`.

Comparisons with other theorem provers for free logic are not possible at this stage, since we are not aware of any other existing systems.

We also want to emphasize that this paper has been written entirely within the Isabelle framework by utilizing the Isabelle "build" tool; cf. [8], Sect. 2. It is thus an example of a formally verified mathematical document, where the pdf document as presented here has been generated directly from the verified source files mentioned above.[4]

Further work includes the continuation of our formalization studies in category theory. It seems plausible that substantial parts of the textbook of Freyd and Scedrov can now be formalised in our framework. An interesting question clearly is how far automation scales and whether some further (previously unknown) insights can eventually be contributed by the theorem provers. Moreover, we have already started to compare the axiom system by Freyd and Scedrov with a more elegant set of self-dual axioms developed by Scott. Furthermore, we plan to extend our studies to projective geometry, which is another area where free logic may serve as a suitable starting point for formalisation.

---

[4] By suitably adapting the Isabelle call as contained in file runIsabelle.sh in our zip-package, the verification and generation process can be easily reproduced by the reader.

In addition to our implementation of free logic as a theory in Isabelle/HOL, we plan to support an analogous logic embedding in the new LEO-III theorem prover [9]. The idea is that LEO-III can then be envoked with a specific flag telling it to automatically switch its underlying logic setting from higher-order classical logic to first-order and higher-order free logic, while retaining TPTP TH0 [7] as the common input syntax.

References

1. Benzmüller, C., Miller, D.: Automation of higher-order logic. In: Siekmann, J., Gabbay, D., Woods, J. (eds.) Handbook of the History of Logic, vol. 9. Logic and Computation, Elsevier (2014)
2. Blanchette, J., Böhme, S., Paulson, L.: Extending sledgehammer with SMT solvers. J. Autom. Reason. **51**(1), 109–128 (2013)
3. Freyd, P.J., Scedrov, A.: Categories, Allegories. North Holland, Amsterdam (1990)
4. Nipkow, T., Paulson, L., Wenzel, M.: Isabelle/HOL: A Proof Assistant for Higher-Order Logic. LNCS, vol. 2283. Springer, Heidelberg (2002)
5. Nolt, J.: Free logic. In: Zalta, E.N. (ed.) The Stanford Encyclopedia of Philosophy. Winter 2014 edn. (2014)
6. Scott, D.: Existence and description in formal logic. In: Schoenman, R., Russell, B. (eds.) Philosopher of the Century, pp. 181–200. George Allen & Unwin, London (1967). Reprinted with additions. In: Lambert, K. (ed.) Philosophical Application of Free Logic, pp. 28–48. Oxford Universitry Press, 1991
7. Sutcliffe, G., Benzmüller, C.: Automated reasoning in higher-order logic using the TPTP THF infrastructure. J. Formaliz. Reason. **3**(1), 1–27 (2010)
8. Wenzel, M.: The isabelle system manual, February 2016. https://www.cl.cam.ac. uk/research/hvg/Isabelle/dist/Isabelle2016/doc/system.pdf
9. Wisniewski, M., Steen, A., Benzmüller, C.: LeoPARD — a generic platform for the implementation of higher-order reasoners. In: Kerber, M., Carette, J., Kaliszyk, C., Rabe, F., Sorge, V. (eds.) CICM 2015. LNCS, vol. 9150, pp. 325–330. Springer, Heidelberg (2015)

Efficient Knot Discrimination via Quandle Coloring with SAT and #-SAT

Andrew Fish[1], Alexei Lisitsa[2(✉)], David Stanovský[3], and Sarah Swartwood[1]

[1] School of Computing, Engineering and Mathematics,
University of Brighton, Brighton, UK
[2] Department of Computer Science, The University of Liverpool, Liverpool, UK
a.lisitsa@liverpool.ac.uk
[3] Department of Algebra, Faculty of Mathematics and Physics,
Charles University, Prague, Czech Republic

Abstract. We apply SAT and #-SAT to problems of computational topology: knot detection and recognition. Quandle coloring can be viewed as associations of elements of algebraic structures, called quandles, to arcs of knot diagrams such that certain algebraic relations hold at each crossing. The existence of a coloring (called colorability) and the number of colorings of a knot by a quandle are knot invariants that can be used to distinguish knots. We realise coloring instances as SAT and #-SAT instances, and produce experimental data demonstrating that a SAT-based approach to colorability is a practically efficient method for knot detection and #-SAT can be utilised for knot recognition.

Keywords: Computational topology · Knot detection and equivalence · SAT and #-SAT solving · Quandle coloring

1 Introduction

Advances in methods for detecting knotting or entangling, or distinguishing various forms of knotting, have potential for impact across scientific disciplines, in relation to molecules, interacting particles, DNA strands, or any other objects that can be knotted [Sum90,FN97,BF09]. The longstanding computational topology problem of *unknot detection* asks: given a *knot K* (i.e. a closed loop without self-intersection embedded in 3-dimensional Euclidean space $\mathbb{R}^3$), is it ambient isotopic to the unknot? More generally, the *knot recognition problem* asks: given two knots, are they ambient isotopic?

In computational terms, it is unknown whether unknot detection is in PTIME, but it does lie in NP [HLP99] and in co-NP if the generalized Riemann hypothesis holds [Kup14]. Most practically fast algorithms are incomplete, in the sense that they do not recognize all non-trivial knots. A typical example is use of the Alexander polynomial, which can be calculated in PTIME (and very quickly in practice). Many other classical invariants do not have

---

D. Stanovský—Partially supported by the GAČR grant 13-01832S.

G.-M. Greuel et al. (Eds.): ICMS 2016, LNCS 9725, pp. 51–58, 2016.
DOI: 10.1007/978-3-319-42432-3_7

fast algorithms: for instance, calculating the Jones polynomial is known to be #*P*-hard [JVW90]. The algorithms based on monotone simplifications [Dyn03] provide practically fast recognition of unknots but do not necessarily yield a decision procedure. Several implementations for unknot detection, with varying functionalities, exist. Most notably, algorithms based on *normal surface theory* were implemented in Regina system [BO12]. Another software capable of knot recognition is SnapPy [CDW], which also has a random link generator that we utilise.

In [FL14], a new theorem proving/disproving approach to knot recognition was proposed, based on algebraic objects called involutory quandles. In [FLS15], the unknot certification technique was re-interpreted in terms of *quandle color-ings* of knots [CESY14], leading to a decision procedure for unknot detection, fairly efficient in practice. A reduction of the *colorability* of a knot by a given quandle to the Boolean Satisfiability Problem, SAT, was proposed to address the unknot detection problem.

In the present paper, we build significantly on the initial ideas in [FLS15], providing substantial experimental data and exploring more sophisticated SAT and #-SAT solving techniques, in order to gain insight into and potentially improve performance of our unknot detection and knot recognition methods. In order to certify the in-equivalence of knots, as opposed to just detect knot-tedness, one can utilise a smarter invariant, the *number of colorings* by a given quandle, which can be encoded via #-SAT. This confirms the indication provided by experiments in [CESY14] that computing quandle colorings is a promising avenue.

2 Knots, Quandles and Colorings

A self-contained introduction into quandle coloring and its application to knot recognition can be found in [FLS15]. We refer there for all details and further information, only summarizing the critical concepts here.

Definition 1. A set Q equipped with a binary operation $\triangleright$ is called a *quandle* if the following conditions hold:

Q1 $x \triangleright x = x$ for all $x \in Q$.
Q2 For all $x, y \in Q$, there is a unique $z \in Q$ such that $x = z \triangleright y$.
Q3 For all $x, y, z \in Q$, we have $(x \triangleright y) \triangleright z = (x \triangleright z) \triangleright (y \triangleright z)$.

Fig. 1. A labeled crossing. Any coloring f must satisfy $f(\gamma) = f(\beta) \triangleright f(\alpha)$. (Color figure online)

Let D be a knot diagram and let Q be a quandle. A *coloring* of D by Q is a mapping f assigning to every arc (a continuous segment in the diagram) a color from Q such that for every crossing with arcs labeled α, β, γ, as in Fig. 1, $f(\gamma) = f(\beta) \triangleright f(\alpha)$ holds. A coloring is called *trivial* if it only uses one color. Let $\mathrm{col}_Q(D)$ denote the number of non-trivial colorings of the diagram D by the quandle Q. Then $\mathrm{col}_Q(D)$ is invariant with respect to the Reidemeister moves, and therefore is a knot property, invariant with respect to ambient isotopy. A knot K is non-trivial if and only if $\mathrm{col}_Q(K) > 0$ for some quandle Q.

Given a quandle Q, and a knot diagram D, one formulates the following problems:

Q-colorability. Is $\mathrm{col}_Q(D) > 0$, i.e., is there a non-trivial Q-coloring of D?
Q-coloring number. Compute $\mathrm{col}_Q(D)$.

In computational terms, Q-colorability is in the complexity class NP: given an assignment of colors to arcs, it is easy to check whether it is a non-trivial Q-coloring. Correspondingly, calculating the coloring numbers is in the complexity class #P. To find a non-trivial coloring (all colorings, respectively), one has to solve a system of equations over the quandle Q: for every crossing, as in Fig. 1, we have the equation $x_\beta \triangleright x_\alpha = x_\gamma$, where $x_\alpha, x_\beta, x_\gamma$ are variables that determine the colors of the arcs α, β, γ.

In [FLS15] a natural reduction of coloring to SAT was proposed. Fix a connected quandle $Q = (\{1, \ldots, q\}, \triangleright)$ and a knot diagram D with $|D| = n$, with arcs numbered $\alpha_1, \ldots, \alpha_n$. We consider nq boolean variables $v_{i,c}$ that determine whether the arc α_i has the color c. A SAT instance specifies all required constraints: every arc has a unique color, not all arcs have the same color and colorings of all arcs crossing as in Fig. 1 satisfy quandle condition.

If the quandle Q is homogeneous (i.e., its automorphism group is transitive on Q), we can assume that the arc α_1 has color 1, i.e., add the clause $v_{1,1}$. Since the color of α_1 can be chosen arbitrarily from Q, we have $\mathrm{col}_Q(D) = |Q| \cdot m$, where m is the number of solutions to the SAT problem.

Proposition 1. *Let D be a knot diagram and Q a homogeneous quandle. Then $\mathrm{col}_Q(D) = |Q| \cdot m$, where m is the number of solutions to the SAT problem described above.*

We say that D has m *essential* colorings by Q, and we will record essential colorings as output from the #-SAT solvers.

3   Experimental Set-Up

For our experiments, the following families of quandles and knots were used:

SQ. All 354 simple quandles of size ≤ 47, indexed in accordance to size.
CQ. 26 quandles (each of size ≤ 182), indexed as per Clark et al. [CESY14].
SQ1, SQ2, QQ. Small sets of quandles used for knot recognition (with #-SAT), described in Sect. 5.

K10–K12. All 249, 801 and 2977 prime knots (up to reverse and mirror image) with crossing numbers not exceeding 10, 11 and 12 respectively.

A13. All 34659 alternating minimal projections of prime knots with crossing numbers not exceeding 13.

R. 52 randomly generated diagrams with 25, 50, 75, 100 and 125 crossings.

These choices were influenced by practical concerns of the availability and accessibility of input sources to use in the experiments.

Knot diagrams can be represented by various codes, and the KnotInfo library [CL] provides more than a dozen discrete representations of tabulated knots. The Planar Diagram (PD) notation was used, since it is readily available for all prime knots up to 12 crossings **K12**, and it is straightforward to produce the crossing conditions of Fig. 1 from it (i.e. the quandle presentation). For **A13** [FR], a conversion from Gauss codes was also necessary. For **R** we repeatedly used the random link generator of SnapPy [CDW], only recording results when knots were produced.

MiniSat 2.2.0 and #-SAT 12.08, driven by Perl/Prolog scripts running on Debian Linux VM, hosted on Windows 7 system, were used for the experiments.

4   Experimental Results: Running Time

4.1   Knottedness Certification with MiniSat

In the first series of experiments the knot detection algorithm based on quandle colorability was performed. Given a (PD code of a) knot, a procedure iterates over all quandles from **SQ**, converts the quandle colorability task into a SAT instance and checks its satisfiability by MiniSAT. The iteration proceeds until the first satisfiable case is found. This is a solution to the Q-colorability problem, giving witness to the non-triviality of the knot. For all knots from the **K12** family the detection time was in the interval 0.013–3.31 s with the vast majority of knots being detected under 1 s. Figure 2 presents the cumulative frequency of running times for this case.

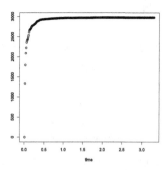

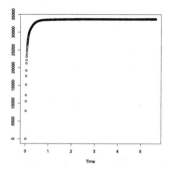

Fig. 2. Cumulative frequency of running times (s) for knot detection for the **K12** family.

Fig. 3. Cumulative frequency of running times (s) for knot detection for the **A13** family.

This compares extremely favourably with the claim of the efficiency of the same set of cases by Regina's algorithm (see [BO12]), which computed each individual case in under 5 min (as compared to 3.31 s). In fact, we also observe that the total time for the completion of the set of all cases, following our approach, was only 3 min and 54 s. For the case of **A13** family the range of times was 0.015–5.654 s, with a mean time of 0.087 s and a median time of 0.039 s. The cumulative frequency graph of running times is presented in Fig. 3. Recognition of knottedness in **R** was virtually instant (much less than 1 s) for 49 of the cases, whilst for the other 3 cases (all of size 125) it took just under 290 s. The size of the smallest quandle by which the knot is colorable has a critical effect on the running time.

For a regular family of *torus* knots $(3, n)$ in **T3** running time for Q-colorability averaged over all quandles in **SQ** is roughly linear in n, and for the case $n = 602$ it is around 13 s.

4.2   Knot Recognition with #-SAT

We give details of the use of #-SAT for quandle colorings for one particular setup: coloring all knots from the **K12** family by each quandle from **SQ**. The histogram shown in Fig. 4 demonstrates the distribution of running times for the **K12** family of knots, where for each knot the time is averaged over all of the 354 quandles. The range of times is 0.006–11.5 s, with a mean time of 0.57 s and a median time of 0.51 s.

The histogram shown in Fig. 5 demonstrates the experimental data from another perspective, focussing on the time taken, on average, to count the number of colorings by a given quandle over the whole family of knots. The range of these average times was found to be 0.095–2.65 s, whilst the mean time was 0.57 s and the median time was 0.27 s. The experimental data for the #-SAT-based quandle coloring of the same knot family **K12** but a different quandle

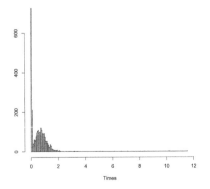

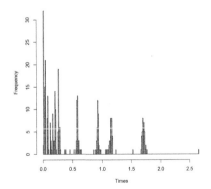

Fig. 4. Distribution of running times (s) for counting Q-colorings for the **K12** family, averaged over all quandles in **SQ**.

Fig. 5. Distribution of running times (s) for counting Q-colorings for the **SQ** family, averaged over all knots in **K12**.

family **CQ**, demonstrates similar features. Further details can be found online at [col15].

5   Experimental Results: Small Set of Distinguishing Quandles

In [CESY14], they calculated a family **CQ** of 26 quandles whose coloring numbers distinguish all knots in **K12**. Most pairs of knots can be distinguished using a fairly small quandle, and therefore relatively quickly, but some pairs require fairly large quandles: the largest one used in **CQ** has 182 elements.

Table 1. Proportion of knots distinguished by quandle colourings, for various sets of quandles.

	K10	K11	K12
Total number	249	801	2977
Proportion recognised by colouring numbers in **SQ** with $\|\mathbf{SQ}\| = 354$	100 %	97.8 %	96.2 %
Proportion recognised by colouring numbers in **SQ1** with $\|\mathbf{SQ1}\| = 30$	100 %	94.4 %	89.9 %
Proportion recognised by colouring numbers in **SQ2** with $\|\mathbf{SQ2}\| = 15$	100 %	92.3 %	85.2 %
Proportion recognised by colouring numbers in **QQ** with $\|\mathbf{QQ}\| = 17$	100 %	100 %	100 %
Proportion recognised by colouring numbers in **CQ** with $\|\mathbf{CQ}\| = 26$	100 %	100 %	100 %

In [FLS15, Sect. 3.3], we argued that for certification of knottedness, one can restrict to colorability by simple quandles. A natural question is, to what extent this is true for the (more general) problem of knot recognition, using the actual number of colorings. The first row in Table 1 shows the proportion of pairs of knots with up to 10, 11 and 12 crossings distinguished by the family **SQ**: we see that knots with up to 10 crossings are distinguished completely, but not so for larger knots. Presumably, this is because quandles in **SQ** are too small.

We will consider two small subsets of **SQ** which distinguish all knots in **K10**: let **SQ1** and **SQ2** be subsets of **SQ** consisting of quandles with indices:

SQ1. 100, 119, 120, 136, 148, 16, 184, 185, 1, 222, 223, 263, 264, 265, 26, 307, 308, 309, 38, 39, 75, 76, 77, 78, 186, 101, 121, 130, 23, 69.
SQ2. 100, 136, 148, 223, 263, 264, 265, 26, 38, 76, 101, 121, 130, 23, 69.

We used the number of essential colorings as a basic heuristic of recognition power of a quandle. The set **SQ1** contains all 25 quandles from **SQ** with 4 or more essential colorings over **K10**; this is insufficient for distinguishing all pairs in **K10**, hence we added 5 additional quandles from **SQ**, distinguishing the remaining pairs. Then we picked **SQ2** to be a minimal subset of **SQ1** (minimal with respect to inclusion) that solves the recognition problem for **K10**. Table 1 shows their performance on larger knots.

To investigate the potential of simple quandles for knot recognition, we extended **SQ2** by adding a small subset of quandles from **CQ** so that the union solves the knot recognition problem for **K12**, and then took a minimal subset that does the job. The result, to be refered as **QQ**, consists of the following 17 quandles:

- from **SQ**, take quandles with indices 23, 26, 38, 69, 100, 148, 223, 263, 264;
- from **CQ**, take quandles with indices 7, 19, 20, 21, 22, 23, 25, 26.

The running times of the knot recognition problem for **K12** using #-SAT are 6511 min using the family **QQ**, compared to 6855 min using **CQ**. In [CESY14], they report that the running time of their implementation is "months" of serial time, but only about 100 min for a parallelized version of their algorithm. This suggests that parallelization could significantly improve coloring by #-SAT.

Our #-SAT experiments also produced coloring data for **K12** with the family **CQ**, which allows for cross-validation with the data computed by [CESY14]. There are some differences in experimental set-up: our #-SAT encoding takes a PD code list from KnotInfo, whilst [CESY14] uses a braid-based input code list. Upon translation of the input format, the coloring numbers match exactly, except for the cases of quandles with indices 16 and 17, and 23 and 24. Here the values on a given knot either coincide, or they are switched on both pairs. The latter cases indicate instances when KnotInfo gives the PD and braid codes for the knot representatives with reversed orientations (see [CESY14, Lemma 3.2(3)]).

6   Conclusion

We have provided evidence that SAT and #-SAT can be used for efficient computation of topological invariants of knots, by the means of quandle coloring. The SAT procedure can be used for fast detection of non-trivial knots, and we demonstrated that: (i) it outperforms existing algorithms (Regina on **K12** with **SQ**); (ii) reasonably efficient performance for alternating knots with up to 13 crossings; (iii) scalability – efficient detection for random large knots, and also for torus knots [FLS15]. Furthermore we have shown that #-SAT can be used for reasonably efficient computation of quandle colourings, and we have provided experimental data for the #-SAT tasks for **K12** with **SQ**. Experimental data will be made available online at [col15].

We found a small set of simple quandles that distinguish all prime knots (up to reverse and mirror image) of up to 10 crossings, and a new set of 17 quandles that distinguish all prime knots (up to reverse and mirror image) of up to 12 crossings, compared to the previous record of 26 quandles [CESY14]. We demonstrated that the new set is more efficient with respect to running time using #-SAT.

References

[BF09]   Buck, D., Flapan, E. (eds.): Applications of Knot Theory. American Mathematical Society Short Course, San Diego, CA, USA, 4–5 January 2008. American Mathematical Society (AMS), Providence, RI (2009)

[BO12]   Burton, B.A., Özlen, M.: A fast branching algorithm for unknot recognizion with experimental polynomial-time behaviour (2012). http://arxiv.org/abs/1211.1079v3

[CDW]   Culler, M., Dunfield, N.M., Weeks, J.R.: SnapPy, a computer program for studying the topology of 3-manifolds. http://snappy.computop.org

[CESY14]   Clark, W.E., Elhamdadi, M., Saito, M., Yeatman, T.: Quandle colorings of knots and applications. J. Knot Theory Ramif. **23**, 1450035 (2014)

[CL]   Cha, J.C., Livingston, C.: Knotinfo: table of knot invariants. http://www.indiana.edu/~knotinfo. Accessed Jan 2015

[col15]   Quandle colouring data (2015). http://cgi.csc.liv.ac.uk/~alexei/quandle_colourings

[Dyn03]   Dynnikov, I.A.: Recognition algorithms in knot theory. Uspekhi Mat. Nauk **58**(6(354)), 45–92 (2003)

[FL14]   Fish, A., Lisitsa, A.: Detecting unknots via equational reasoning, I: exploration. In: Watt, S.M., Davenport, J.H., Sexton, A.P., Sojka, P., Urban, J. (eds.) CICM 2014. LNCS, vol. 8543, pp. 76–91. Springer, Heidelberg (2014)

[FLS15]   Fish, A., Lisitsa, A., Stanovský, D.: A combinatorial approach to knot recognition. In: Horne, R. (ed.) EGC 2015. CCIS, vol. 514, pp. 64–78. Springer, Heidelberg (2015). doi:10.1007/978-3-319-25043-4_7

[FN97]   Faddeev, L., Niemi, A.J.: Stable knot-like structures in classical field theory. Nature **387**, 58–61 (1997)

[FR]   Flint, O., Rankin, S.: Gauss codes for the distinct minimal diagrams for the primealternating knots of 13 crossings. http://www-home.math.uwo.ca/~srankin/knots/knotprint.html. Accessed Jan 2016

[HLP99]   Hass, J., Lagarias, J.C., Pippenger, N.: The computational complexity of knot and link problems. J. Assoc. Comput. Mach. **46**, 185–211 (1999)

[JVW90]   Jaeger, F., Vertigan, D.L., Welsh, D.J.A.: On the computational complexity of the Jones and Tutte polynomials. Math. Proc. Camb. Philos. Soc. **108**, 35–53 (1990)

[Kup14]   Kuperberg, G.: Knottedness is in NP, modulo GRH. Adv. Math. **256**, 493–506 (2014)

[Sum90]   Sumners, D.: Untangling DNA. Math. Intelligencer **12**, 71–80 (1990)

Interactive Proving, Higher-Order Rewriting, and Theory Analysis in Theorema 2.0

Alexander Maletzky[✉]

Doctoral Program "Computational Mathematics" and RISC,
Johannes Kepler University, Linz, Austria
alexander.maletzky@dk-compmath.jku.at
https://www.dk-compmath.jku.at/people/alexander-maletzky/

Abstract. In this talk we will report on three useful tools recently implemented in the frame of the Theorema project: a graphical user interface for interactive proof development, a higher-order rewriting mechanism, and a tool for automatically analyzing the logical structure of Theorema-theories. Each of these three tools already proved extremely useful in the extensive formal exploration of a non-trivial mathematical theory, namely the theory of Gröbner bases and reduction rings, in Theorema 2.0.

Keywords: Computer-assisted mathematical theory exploration · Interactive theorem proving · Theorema

1 Introduction

Theorema[1] is a so-called *mathematical assistant system* supporting its users in all aspects of mathematical theory exploration: inventing new notions and problems, implementing and experimenting with algorithms, making conjectures, and finally proving or disproving them. Theorema 2.0 [1] is the latest version of the system, released roughly two years ago in 2014; as its predecessor, it is still based on *Mathematica*.

The present paper reports on three tools we recently developed for making working with the system more attractive and efficient: a versatile *interactive proof strategy* giving the user full control over proving and complementing the existing automatic strategies, a powerful *rewriting mechanism* for translating first- and higher-order formulas into *Mathematica* transformation rules for rewriting other formulas in proofs, and a simple but nonetheless extremely helpful tool for analyzing the logical structure of Theorema-theories. The development of each of these three tools was motivated by our extensive formal treatment of the theory of Gröbner bases and reduction rings in Theorema, see [3] for details.

This research was funded by the Austrian Science Fund (FWF): grant no. W1214-N15, project DK1.

[1] http://www.risc.jku.at/research/theorema/software/.

© Springer International Publishing Switzerland 2016
G.-M. Greuel et al. (Eds.): ICMS 2016, LNCS 9725, pp. 59–66, 2016.
DOI: 10.1007/978-3-319-42432-3_8

Please note that the tools have not been integrated into the official version of Theorema 2.0 yet, but they are expected to be in the near future. Still, they can easily be installed manually, relying on *Mathematica*'s comfortable package-system.

Small parts of this paper are also contained in [3].

2 Interactive Proving

The first of the three tools we present in this paper is an *interactive proof strategy* (IPS) that, as its name suggests, can be used for developing proofs in Theorema 2.0 fully interactively, in the sense that the human user has full control over what happens at each stage of a proof. This is in contrast to the automatic, or, at least, semi-automatic proof strategies typically available in the system.

As can be seen in a concrete example below, the IPS in Theorema 2.0 is not text-based, as in most other proof assistants, but *dialog-oriented*. This means that whenever a user interaction is required, a dialog window displaying the current proof situation pops up, asking the user to perform an action (by clicking on a button, typing in some text, etc.; see Fig. 1). This, in fact, follows the tradition of interactive proving in Theorema 1, the predecessor version of Theorema 2.0, where the environment for interactive proving developed in [5] is dialog-oriented as well. Note that we did not just migrate said environment from Theorema 1 to Theorema 2.0, but really implemented the new IPS completely from scratch; this seemed to be the more reasonable approach, as the internal architecture of Theorema 2.0 differs considerably from the one of Theorema 1.

Before explaining how the IPS can be used in practical applications, some words on its implementation are in place: the IPS is implemented simply as a Theorema proof strategy, meaning that it essentially is a function taking a *proof situation* (characterized by the current proof goal and a list of assumptions) as input and returning a list of new, ideally simpler proof situations as output; the logical relation between in- and output obviously is that the validity of the input-situation follows from the validity of all of the output-situations. The output is constructed by applying *inference rules* that are themselves independent of the IPS and could well be used together with any other (automatic) proof strategy installed in the system. The main task of the IPS is only to *guide* the application of the inference rules, by specifying which rules shall be applied and how they shall be applied.

2.1 How the Interactive Proof Strategy Works in Practice

Once the IPS is properly installed, it can be selected as the proof strategy of choice just as any other, pre-defined proof strategy when initiating a proof attempt; no further setup by the user is required. Then, whenever a new proof situation p arises during the proof search, the IPS proceeds as follows:

– First, it automatically tries to apply an available *high-priority* inference rule to p. If this is possible, the respective rule is applied and the proof search continues.

– Otherwise, if no high-priority rule is applicable to p, it asks the user how to proceed by displaying a graphical dialog window.

Every inference rule in Theorema has a *priority* attached to it. Automatic proof strategies usually fall back on these priorities for determining the order they try to apply inference rules in. The IPS takes rule priorities into account solely for filtering out the high-priority rules, i. e. those rules whose priorities are above a certain, user-adjustable threshold.[2]

Assume now that no high-priority rule could be applied to p. The user now has a range of possibilities how to proceed, including

– choosing another inference rule to apply to p (or, more precisely, to *try*, since non-applicable rules are not automatically filtered out),
– choosing a different pending proof situation where to continue,
– adjusting various settings, like the current set of inference rules and even the proof strategy (making it possible to switch to an automatic strategy at some point during the proof development),
– inspecting the so-far constructed proof in a nicely-formatted proof document,
– inspecting the internal representation of p as a plain *Mathematica* expression for debugging purposes,
– saving the current proof status to an external file, for creating a "secure point" the proof may be resumed from later, and
– aborting the proof attempt.

Before choosing an inference rule the user may also activate and deactivate formulas appearing in p by marking check-boxes in the dialog window (see Fig. 1). This might affect *how* the chosen rule is applied, e. g. if several cases based on a disjunction in the knowledge base shall be distinguished, but more than one disjunctions appear among the assumptions, the user can specify exactly which one to consider simply by deactivating all others. It must be noted, though, that the information about whether a formula is activated or not might well be ignored by the chosen inference rule; this cannot be influenced by the IPS.

2.2 An Example

As an example, let us consider the interactive proof of the well-known *drinker paradox*: "In every non-empty pub there is someone such that, if he is drinking, everyone else is drinking as well." This is actually no paradox but a theorem in classical logic and may hence be proved in Theorema.

Figure 1 depicts two dialog windows of the IPS arising in the interactive proof of the drinker paradox. The first one corresponds to the case where someone who does *not* drink is assumed to be in the pub (Formula (A#1)), and where the next action to be taken, as specified by the user, is to eliminate the existential quantifier in Formula (A#1) by introducing a new constant that witnesses this person. The resulting proof situation is displayed in the second window.

---

[2] A typical example of a high-priority rule is the inference rule that proves implications by assuming their premises and proving their conclusions.

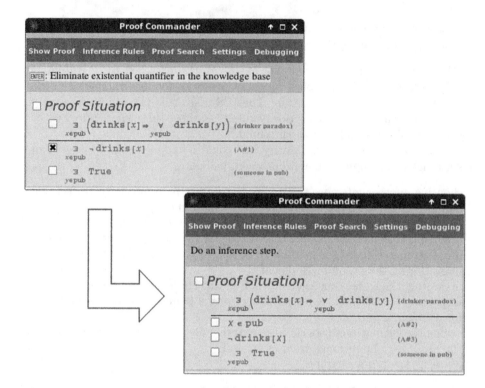

Fig. 1. Two dialog windows arising in the interactive proof of the drinker paradox.

The interactive dialogs display the current proof situation in the bottom part of the respective windows, on light-brown background. In each case, the top-most formula (above the black line) is the current goal, whereas the formulas below the black line are the assumptions. The check-boxes next to the formulas indicate whether the respective formulas have been activated or deactivated by the user. Above the proof situation, the name of inference rule to be applied next, as chosen by the user, is displayed; it is applied by simply hitting the Enter-key.

3 Higher-Order Rewriting

Rewriting constitutes one of the core components of theorem proving in Theorema as well as many other proof assistants: assumptions (equalities, equivalences, implications) are transformed into rewrite-rules which may then be used to rewrite other formulas in the current proof situation. By default, Theorema can only deal with *first-order* formulas and rewrite-rules, respectively, in the sense that the left-hand-side of a rewrite-rule has to match an expression syntactically in order to be applicable; no $\alpha\beta\eta$-equivalence is taken into account. However, a lot of formulas one frequently encounters in mathematical theories are not first- but higher-order (typical examples are induction rules), and should be treated as such for efficiently working in the respective theories in Theorema.

The higher-order rewriting mechanism we describe in this section serves exactly said purpose: it is able to translate (potentially higher-order) rewrite-rules $\rho : l \mapsto r$ originating from Theorema formulas into *Mathematica* transformation rules p :> b that can later be applied by simply calling the standard rule-application-functions from *Mathematica*'s algorithm library (ReplaceAll, ReplaceList, etc.), such that the correctness condition

$$e' \in \mathtt{ReplaceList}[\, e, \, \mathtt{p} :> \mathtt{b}] \;\Rightarrow\; e \to_\rho e' \tag{1}$$

is met (where $e \to_\rho e'$ means that expression e can be rewritten into e' by rule ρ modulo $\alpha\beta\eta$-equivalence). The other direction of (1), though desirable in principle, is out of reach in general if ρ is a higher-order rule: whether higher-order matching (which is one of the key ingredients of rewriting) is decidable or not is still an open problem.[3] Hence, if ρ is higher-order, the *Mathematica* transformation rule p :> b cannot be expected to fully reflect the higher-order nature of ρ *in any case*.

3.1   Main Idea

Sometimes, the strategy to tackle problems related to (possibly) undecidable, infinitary matching in concrete implementations is to restrict the class of left-hand-sides of rewrite-rules to so-called *higher-order patterns*; for instance, the simplifier in the Isabelle proof assistant by default can only handle rules falling into this category (see [7], pp. 205–206). The main idea behind our mechanism is similar but less restrictive: if the compiler (i. e. the function that turns rewrite-rules into *Mathematica* transformation rules) can infer that the matching problem associated to the left-hand-side of a given rule ρ is *unitary*, because bound variables appearing among the arguments of free higher-order variables can be used to uniquely determine the instances of these variables when matching an expression, then ρ is "accepted" and turned into a transformation rule that *exactly* corresponds to ρ. Otherwise, if the compiler cannot infer that the matching problem is unitary (maybe because it simply is not), some free higher-order variables have to be treated just like first-order variables that need to match syntactically, meaning that the resulting transformation rule does not correspond to ρ exactly.

Example 1. The left-hand-side of the rewrite-rule (with P, T and S being free variables)

$$\forall_{i=1,\ldots,|T|+|S|} P(\mathtt{join}(T,S)_i) \mapsto \forall_{i=1,\ldots,|T|} P(T_i) \wedge \forall_{i=1,\ldots,|S|} P(S_i)$$

is no higher-order pattern, but can still be handled without much ado by our mechanism because the occurrence of the bound variable i in the argument of

---

[3] Higher-order matching is known to be decidable under certain restrictions on the types involved [6], as well as for general problems below order five [4].

P on the left-hand-side uniquely determines the instance of P when matched against a concrete expression. In contrast, the free variable P in the (nonsense) rule

$$P(0) \mapsto \exists_x P(x)$$

must be treated like a first-order variable by the compiler, for otherwise the instance of P would not be unique in general: matching $0 < 1$ could be accomplished by instantiating P either by $\lambda_x\ 0 < 1$ or by $\lambda_x\ x < 1$, leading to fundamentally different instances of the right-hand-side.

3.2 Implementation Details

The compiler translates rewrite-rules into ordinary *Mathematica* transformation rules. Hence, since *Mathematica* only supports syntactic matching, all possible higher-order aspects ($\alpha\beta\eta$-equivalence, automatic instantiation by λ-terms, etc.) have to be encoded explicitly in the pattern p and the body b of the resulting transformation rules, e. g. by means of *Mathematica*'s `Condition` function.

Example 2. Consider the higher-order rewrite-rule

$$\sum_{i=1,\ldots,n+1} F(i) - \sum_{i=1,\ldots,n} F(i) \mapsto F(n+1)$$

Ignoring syntactical details of the internal representation of Theorema expressions and technicalities related to capture-avoiding substitutions, the *Mathematica* transformation rule automatically generated by the compiler reads as something like

```
Sum[{i1_, 1, n_+1}, F1_] - Sum[{i2_, 1, n_}, F2_] :>
    substFree[F1, {i1 -> n + 1}] /;
        alphaEquiv[substFree[F1, {i1 -> i2}], F2]
```

As can be seen, higher-order variables (like F in the previous example) are actually never instantiated by concrete λ-terms, but rather the instances of the right-hand-sides of rules are constructed by directly replacing certain sub-expressions (like the bound variable i above). This saves expensive (capture-avoiding) substitutions and explicit β-reductions and, for that reason, is a general principle the compiler adheres to. Moreover, apparently there is no hard-coded, general-purpose higher-order matching algorithm that is attached to every transformation rule, but rather every single transformation rule is equipped with its very own, tailor-made, dynamically generated, optimized algorithm that does not perform any redundant operations. In the example above, only the α-equivalence of two expressions has to be checked in addition to the default syntactic matching carried out by *Mathematica*—a fact the compiler detects and exploits fully automatically when generating the transformation rule.

3.3   More Features

Due to the lack of space, the preceding sections could only provide a glimpse of the higher-order rewriting mechanism, and in particular of the transformation rule compiler; more detailed information can be found in our forthcoming PhD thesis [2]. Still, we want to briefly mention two further features also here:

- Conditional rules, n-ary higher-order variables, and sequence variables are supported as well (to a certain extent). Conditional rules do not cause any difficulties at all, but the presence of free higher-order variables with arity > 1 or free sequence variables complicates matters considerably.
- The compiler by default applies a range of optimizations to the rules it generates for increasing efficiency.
- The condition on matching problems associated to the left-hand-sides of rules being unitary can be relaxed in some situations.

4   Theory Analysis

The third an last tool presented in this paper, called TheoryAnalyzer, enables the automatic analysis of the logical structure of one or several Theorema-theories (i. e. content notebooks together with external proof files). The main idea behind the TheoryAnalyzer is simple enough: read the proof files, and from each proof file store the proof goal and the list of assumptions as the nodes of a graph G that eventually reflects the dependencies between all the formulas thus collected. Namely, a formula φ depends on another formula ψ iff ψ is used as an assumption in a proof of φ; in such a case, G contains a directed edge from ψ to φ.

Once G has been constructed, it can easily be analyzed by means of well-known graph-theoretic functions (like exhaustive search); in particular, it is possible to

- inspect all direct/indirect assumptions/consequences of a given node (corresponding to a formula in the theory),
- detect cycles in the graph, corresponding to circular arguments in the theory,
- find the logical relation between two nodes/formulas, and
- visualize theory dependency graphs and formula statistics diagrams (the latter display the numbers of formulas in each theory); see Fig. 2.

The development of the TheoryAnalyzer was mainly triggered by the practical experience we gained from formalizing Gröbner bases theory in Theorema: it turned out that quite frequently it becomes necessary to re-structure existing parts of formalizations, e. g. by slightly modifying formulas that have already been used as assumptions in proofs. In such situations, the responsibility for maintaining the coherence the formalization exclusively is with the user of Theorema; the system itself does not automatically initiate the re-proving of existing theorems affected by changes in the background theory. Therefore, knowing which theorems *are* affected is of utmost importance—and this is exactly where the TheoryAnalyzer comes into play.

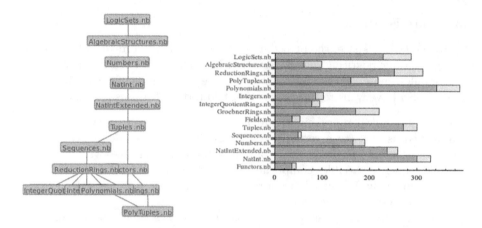

Fig. 2. The theory dependency graph and formula statistics diagram of the Gröbner bases formalization, as automatically generated by the TheoryAnalyzer.

5   Conclusion

In the preceding sections we gave account on three new tools for Theorema 2.0 that already proved extremely useful in practice and are expected to be integrated into the official version of the system in the near future.

There are several directions for further improving the tools: for instance, the dialog-oriented interactive proof strategy could be enhanced by a text-based interface, making it more attractive for people used to such interfaces.

References

1. Buchberger, B., Jebelean, T., Kutsia, T., Maletzky, A., Windsteiger, W.: Theorema 2.0: computer-assisted natural-style mathematics. J. Formaliz. Reason. **9**(1), 149–185 (2016)
2. Maletzky, A.: Computer-Assisted Exploration of Gröbner Bases Theory in Theorema. Ph.D. thesis, Research Institute for Symbolic Computation, Johannes Kepler University Linz, Austria (2016, to appear)
3. Maletzky, A.: Mathematical Theory Exploration in Theorema: Reduction Rings. In: CICM 2016 (2016). Preprint on http://arxiv.org/abs/1602.04339
4. Padovani, V.: Filtrage d'ordre supérieure. Ph.D. thesis, Université Paris 7, Paris, France (1996)
5. Piroi, F., Kutsia, T.: The theorema environment for interactive proof development. In: Sutcliffe, G., Voronkov, A. (eds.) LPAR 2005. LNCS (LNAI), vol. 3835, pp. 261–275. Springer, Heidelberg (2005)
6. Stirling, C.: Decidability of higher-order matching. Log. Methods Comput. Sci. **5**(3), 1–52 (2009)
7. Wenzel, M.: The Isabelle/Isar Reference Manual (2016), part of the Isabelle documentation. https://isabelle.in.tum.de/documentation.html

Automated Deduction in Ring Theory

Ranganathan Padmanabhan and Yang Zhang[(✉)]

Department of Mathematics, University of Manitoba, Winnipeg R3T 2N2, Canada
{Ranganathan.Padmanabhan,Yang.Zhang}@umanitoba.ca
http://www.math.umanitoba.ca/people/faculty.php?id=22
http://www.math.umanitoba.ca/people/faculty.php?id=Yang_Zhang

Abstract. Pover9/Mace4 or its predecessor Otter is one of the powerful automated theorem provers for first-order and equational logic. In this paper we explore various possibilities of using Prover9 in ring theory and semiring theory, in particular, associative rings, rings with involutions, semirings with cancellation laws and near-rings. We code the corresponding axioms in Prover9, check some well-known theorems, for example, Jacobson's commutativity theorem, give some new proofs, and also present some new results.

Keywords: Prover9 · Otter · Commutativity · Associative rings

1 Introduction

It is well-known that automated reasoning tools have been widely used in many areas such as lattice theory, loop algebra, group theory, which provide powerful methods to check or simplify the proofs, prove or construct counter-examples for conjectures, and discover the new theorems, see, for example, McCune and Padmanabhan [11] and Phillips and Stanovsky [13].

Pover9/Mace4 [10] or its predecessor Otter [9] is one of the powerful automated theorem provers for first-order and equational logic. To our best knowledge, only few papers considered how to use Prover9 in ring theory. In non-commutative rings, one of significant difficult points is to construct some conter-examples, which is almost impossible to do that by hand. Now using Mace4, many examples can be constructed quickly.

The purpose of this paper is to explore various possibilities of using Prover9 in ring theory, in particular, associative rings, rings with involutions, rings with derivations and near-rings. We code the corresponding axioms in Prover9, check some well-known theorems, give some new proofs, and also discuss some conjectures.

This paper is self-contained in that we do not assume that the reader is familiar with ring theory or Prover9. We provide some Prover9 commands in ring theory, and expect that other researchers can use Prover9 in their research works easily.

© Springer International Publishing Switzerland 2016
G.-M. Greuel et al. (Eds.): ICMS 2016, LNCS 9725, pp. 67–74, 2016.
DOI: 10.1007/978-3-319-42432-3_9

In Sect. 2, we first code the axioms of associative rings in Prover9, and then prove Jacobson's commutativity theorem, as well as other commutativity theorems. The similar theorems in rings with involutions and derivations as well as near-rings are discussed in Sect. 3. Finally some new results in semirings are presented in Sect. 4.

All results in this paper are tested in iMac 3.06 GHz. Prover9's default term-ordering is LPO but KBO (Knuth-Bendix) is sometimes preferable. For associative rings, KBO usually finds direct proofs (instead of indirect proofs via contradictions).

2   Commutativity in Associative Rings

When we study polynomial identities in non-commutative rings, one interesting question is that a ring will be commutative if its elements satisfy a polynomial identity. The summary of early work can be found in Herstein's books [2,3]. Since then, many people have worked in this area and various identities have been found.

The first famous result was given by Jacobson [6]: Let R be a ring. If for any $x \in R$, there exists an integer $n(x) > 1$ such that $x^{n(x)} = x$, then R is commutative. The proof of this theorem is difficult, and one has to use some deep structure theory in non-commutative rings, not a constructive proof. In past two decades, how to prove this kind of commutativity theorems by a computational point of view has become a challenge question in computer science. Various methods have been explored, for example, by computing GCD (Zhang [16]), Gröbner bases method (Wavrik [15]), and the first-order logic method (Prover9 [10] and Otter [9]).

In the early versions of Otter, it was easy to find the proof of Jacobson's theorem for $n = 2$ but out of the memory for $n = 3$ or bigger. Prover9 has largely improved the performance. In this section, we start from Jacobson theorem, and give more discussions in this direction. In Prover9, "$1 = 0$" could be true and it may induce some unexpected cancellations. Therefore, usually one should add "$1 \mathrel{!}= 0$" as an assumption.

For $n = 2$ and $n = 3$, Prover9 produces proofs within 3 s. When $n = 4$, Prover9 gives out a proof of length 174 and given clauses 707 with the input file in the proof of Theorem 1. Under the weaker condition, Prover9 derived a smart proof which only uses the multiplicative semigroup aspect of the ring as following:

Theorem 1. *If for any $x \in R$, $x^4 = x$ and x^2 belongs to the center of R, then R is commutative.*

Proof.

```
\% -------- Comments from original proof --------
\% Proof 1 at 0.00 (+ 0.00) seconds: commutativity.
\% Length of proof is 9.
```

```
\% Level of proof is 4.
\% Maximum clause weight is 13.
\% Given clauses 10.
```

Input Clauses:

```
10 (x * y) * z = x * (y * z).  [assumption].
14 (x * x) * y = y * (x * x).  [assumption].
15A x * (x * y) = y * (x * x).   [para(10(a,1),14(a,1))].
15 x * (y * y) = y * (y * x).   [copy(15A),flip(a)].
16 x * (x * (x * x)) = x # label("hypothesis x^4=x").  [assumption].
17 c2 * c1 != c1 * c2 # label(commutativity) # answer(commutativity).
```

Goal:

```
x * y = y * x # label(commutativity)  # label(goal).
```

Output Clauses:

```
44A x * (y * (y * (y * y))) = (y * y) * ((y * y) * x).
    [para(15(a,1),15(a,1,2))].
44B x * y = (y * y) * ((y * y) * x).  [para(16(a,1),44A(a,1,2))].
44C x * y = (y * y) * (y * (y * x)).  [para(10(a,1),44B(a,2,2))].
44D x * y = y * (y * (y * (y * x))).  [para(10(a,1),44C(a,2))].
44 x * (x * (x * (x * y))) = y * x.   [copy(44D),flip(a)].
48A x * y = x * ((x * (x * x)) * y).  [para(16(a,1),10(a,1,1))].
48B x * y = x * (x * ((x * x) * y)).  [para(10(a,1),48A(a,2,2))].
48C x * y = x * (x * (x * (x * y))).  [para(10(a,1),48B(a,2,2,2))].
48 x * y = y * x.  [para(44(a,1),48C(a,2))].
49 $F # answer(commutativity).  [resolve(48,a,17,a)].
```

Also Prover9 gives a proof for the following result with 0.70 s and 103 clauses. It is possible for people to pick up necessary clauses and write down a proof by hand.

Theorem 2. *If for any* $x \in R, x^4 = x,$ *then* x^3 *is in the center of R.*

When $n = 5$, no result came out within half hour. Therefore people need to add some hints to allow Prover9 to prove it within a reasonable time. How to choose hints is an interesting open question.

One directly generalized version of Jacobson's theorem is to prove that R is commutative if $x^n - x$ belongs to the center of R for some positive integer $n > 1$. When $n = 2$, we use

$$(x * x + -x) * y = y * (x * x + -x).$$

for the condition in Prover9. The result can be proved for $n = 2, 3$ within 2 min. No result came out for $n = 4$ within 30 min.

Another version is that for any $x \in R$ there exists an n such that $x^n = -x$. Clearly if n is even number, it is equivalent to Jacobson theorem. Using Prover9, we have

Theorem 3. *R is commutative if $x^n = -x$ for $n = 2, 3, 4, 5, 6, 7$.*

The interesting point is that Prover9 can find the proofs for $n = 2, 3$ within one second, for $n = 4, 5, 6$ about 38 s and for $n = 7$ about 288 s. From above results, we guess the following new result is true, and in fact Porver9 produces a very short proof.

Theorem 4. *R is commutative if for any $x \in R$, $x^3 + x^2 + x = 0$. Moreover $xy = 0$ for any $x, y \in R$.*

Also we can prove the following other versions:

Theorem 5. *If R satisfies $x^3 - x^2 + x = 0$ or $x^3 - x^2 - x = 0$, then $xy = 0$ and R is commutative.*

For higher degree, we have already tested some identities, for example, $x^8 + x^2 - x = 0$, $x^6 + x^5 - x = 0$. They all produce $xy = 0$. One may guess that "$xy = 0$" is true for all these kinds of identities. But for $x^2 + x = 0$, Mace4 gives the following example to show that $xy \neq 0$.

-:	0 1
	0 1

+:	0 1
0	0 1
1	1 0

*:	0 1
0	0 0
1	0 1

This naturally leads to the following open problem: For what kind of identities, for example, if R satisfies $x^{n_1} + (-1)^{m_2} x^{n_2} + \cdots + (-1)^{m_l} x^{n_l} = 0$, then R is commutative or $xy = 0$.

Herstein [4] extended Jacobson's theorem to a more general case:

Theorem 6. *R is commutative if $(xy - yx)^n = xy - yx$ for any $x, y \in R$. More general, R is commutative if $(xy - yx)^n - (xy - yx)$ belongs to the center of R.*

We proved it for $n = 2, 3$ cases within 5 min. On the other hand, MacHale [7] proved so-called anti-commutativity theorem as following:

Theorem 7. *If for any $x, y \in R$, there exists $n > 0$ such that $(xy + yx)^n = xy + yx$, then R is anti-commutative, that is, $xy = -yx$.*

Prover9 can prove $n = 2, 3$ cases within 20 min. One interesting question is that if for any $x, y \in R$, there exists an $n > 0$ such that $(xy + yx)^n = xy - yz$ or $(xy - yx)^n = xy + yz$, then what happen for R, commutative or anti-commutative? To our best knowledge, no answer was published for this question.

Due to noncommutative property, it is very difficult to construct some examples by hands. Now we are working on this questions, either prove it or give some counter-examples by Mace4. We also use Mace4 to disprove some conjectures. The following theorem is well-known:

Theorem 8. *R is commutative if for any $x, y \in R$ there exists an integer n such that $(xy)^n = xy^n x$.*

One intuition is that if one swaps the positions of x and y on the right side, that is, $(xy)^n = yx^ny$, then R is anti-commutative. But Mace4 gives us a counter-examples for $n = 2$ as follows:

$-:$	0	1	2
	0	2	1

$+:$	0	1	2
0	0	1	2
1	1	2	0
2	2	0	1

$*:$	0	1	2
0	0	0	0
1	0	1	2
2	0	2	1

This example also provides a counter-example for "$(xy)^2 = y^2x^2$ implies R is anti-commutative".

3   Rings with Some Operations

In this section, we discuss the commutativity properties in rings with some additional conditions. Recall the a ring R is called a **prime ring** if for $x, y \in R$, $xRy = 0$ implies $x = 0$ or $y = 0$, and R is called **semiprime** if $xRx = 0$ implies $x = 0$. The axioms of a prime ring can be coded in Prover9 as

```
all z ((x*z)*y = 0) -> x = 0 | y = 0.
```

In prime or semiprime rings, more commutativity theorems can be proved, for example,

Theorem 9. *Let R be a semiprime ring. R is commutative if for any $x, y \in R$ there exists an integer n such that $(xy)^n = (yx)^n$ (or $(xy)^n = x^ny^n$).*

Rings with involutions is an old but still active research area in associative ring theory and there are many applications in other areas. Let R be a ring. A mapping $' : R \to R$ is called an **involution** of R if for any $x, y \in R$,

$$(x + y)' = x' + y', \qquad (xy)' = y'x', \qquad (x')' = x.$$

Some elementary properties can be found in Herstein [5]. Usually people use "*" for the involution. In Prover9, "*" has already declared as the symbol of multiplication. Therefore we use "'" for the involution. The symmetric element set in R is defined as $\mathbb{S} = \{s \in R \mid s' = s\}$, and the skew-symmetric set is $\mathbb{K} = \{k \in R \mid k' = -k\}$. Both of them play important roles in rings with involutions. Studying commutativity in rings with involutions start from division rings. Hence we have to code the axioms of division rings. In Prover9, we can declare "@" for the inverse of one element. Open "Language Options" and add:

```
op(450, postfix,"@").   redeclare(implication, IMPLIES).
```

and the division ring assumptions are:

```
x != 0   ->   x@ * x = 1.      x != 0   ->   x * x@ = 1.
```

or one may use

```
x != 0 -> exists y (y*x = 1).      x != 0 -> exists y (x*y = 1).
```

Some commutativity theorems in rings with involutions can be proved:

Theorem 10. *Let D be a division ring with an involution.*

(i) If $s^n = s, n > 1$, for every $s \in \mathbb{S}$. Then D is commutative.
(ii) If $k^n = k, n > 1$, for every $k \in \mathbb{K}$. Then D is commutative.

For rings with derivations, the first paper was written by Posner [14] including the important theorem so-called Posner's theorem now. We proved some results using Prover9. For example,

Theorem 11. *Let R be a ring with derivation δ and $a \in R$. If $a\delta(x) = 0$ for all $x \in R$, then $a = 0$ or $\delta = 0$.*

Finally, the axioms defining near-rings can be easily coded in Prover9 by deleting some conditions from the axioms of associative rings, and some well-known theorems can be proved, for example, the famous Zassenhaus-Neumann theorem (for examples, see, [12,17]).

Theorem 12. *The additive group of a division near-ring is abelian.*

4   Commutativity Theorems in Semirings

In this section, we present some complete new results by using Prover9. We remove the identities 1 and 0 from the definition of semirings, and define a class of weak semirings called **cancellation-laws semiring** as following:

Definition 1. *A* **cancellation-laws semiring** *(CL-semiring hereafter) is a set R equipped with two binary operations "+" and "* ", such that: for any $a, b, c \in R$,*

(1) commutative law: $a + b = b + a$.
(2) associative laws: $(a + b) + c = a + (b + c)$ and $(a * b) * c = a * (b * c)$.
(3) distributive laws: (i) $a*(b+c) = (a*b)+(a*c)$ (ii) $(a+b)*c = (a*c)+(b*c)$.
(4) cancellation laws:

$$(i) \quad a + b \mathrel{!=} a + c \mid b = c;$$

$$(ii) \quad a * b \mathrel{!=} a * c \mid a = 0 \mid b = c; \qquad (iii) \quad a * b \mathrel{!=} c * b \mid b = 0 \mid a = c.$$

Next we give a series of theorems to show when a CL semiring is commutative, that is, $a*b = b*a$ for any $a, b \in R$. The commutative theorems in CL-semirings are much more complicate than ones in usual associative rings. We list some new theorems as following:

An one-sided additive commutator $c(x, y)$ in a CL-semiring R is defined as $xy + c(x, y) = yx$, where $x, y \in R$.

Theorem 13. *Let R be a CL-semiring. For any $x, y \in R$,*

(i) $x * c(x, y) = c(x, y) * x$ *implies* $x * y = y * x$.
(ii) $c(x, y) * c(x, y) = c(x, y)$ *implies* $x * y = y * x$.
(iii) $c(x, y) * c(x, z) = c(x, z) * c(x, y)$ *implies* $x * y = y * x$.

For the commutativity of additive group, we can prove more general results:

Theorem 14. *Let R be a set with operations "$+$" and "$*$" and satisfy*

(i) $(R, +)$ *is a cancellative semigroup.*
(ii) $(R, *)$ *is distributive over addition.*

Then $(x * y) + (z * u) = (z * u) + (x * y)$, *for any* $x, y, z, u \in R$. *Furthermore, if there exists a left identity, $1 * x = x$, then the additive group $(R, +)$ is abelian.*

Theorem 15. *If additive commutators are central in a semiring with cancellation then its multiplication is commutative.*

Now we define the additive commutator $\mathcal{C}(x, y) \in R$ satisfying:

$$xy + \mathcal{C}(x, y) = yx, \qquad x\, \mathcal{C}(y, z) = \mathcal{C}(y, z)\, x,$$

for all $x, y, z \in R$. After proving several results by Prover9, we can prove that $\mathcal{C}(x, y) = 0$.

Theorem 16. *Suppose that R is a semiring with cancellation laws. For any $x, y, z \in R$, the following statements hold:*

(i) $x + \mathcal{C}(y, y) = x$, $\mathcal{C}(y, y) + \mathcal{C}(y, y) = \mathcal{C}(y, y)$.
(ii) $x + y = y$ *implies* $\mathcal{C}(z, z) = x$, $\mathcal{C}(x, x) = 0$.
(iii) $\mathcal{C}(x, \mathcal{C}(y, z)) = 0$.
(iv) $\mathcal{C}(x,\ y\, \mathcal{C}(z, u)) = \mathcal{C}(z, u)\, \mathcal{C}(x, y)$.
(v) $\mathcal{C}(xy,\ x) = x\, \mathcal{C}(y,\ x)$.

Theorem 17. *In a semiring with cancellation laws, the additive commutator $\mathcal{C}(x,\ y) = 0$.*

5   Conclusion

Prover9 is very useful in discovering new results in the equational logic of fundamental algebraic systems like semigroups, groups, rings and lattices. It can also handle any number of fundamental operations, relational systems, Skolem functions, implications and, of course, identities. In this paper, we use all the aspects of the software. Prover9 successfully proved the Robbin's conjecture, an open problem in the area of Boolean algebras for some 40+ years (see, [1,8]). Also, Prover9 was successfully employed in finding new single axioms for groups and lattices even though there is no known decision procedure to test whether a specific equation is a single axiom for a given theory. Here we formulate and prove several commutativity theorems in semirings and near-rings generalizing some of the well-known theorems.

Prover9 works in tandem with MACE (models and counter-examples). The so-called "ring-models" were incorporated into the MACE to decide whether a given single axiom defines group theory. We use MACE to disprove certain conjectures in semirings and near-rings.

We use the Knuth-Bendix algorithm in Prover9 to prove new commutativity theorems for rings with involutions and also rings admitting derivatives. Additive commutators are defined without the presence of the unary negative operation and prove several results for additive commutators in this new set-up.

Acknowledgements. This research was partially supported by the grants from the Natural Sciences and Engineering Research Council of Canada (NSERC).

References

1. Fitelson, B.: Using Wolfram's mathematica to understand the computer proof of Robbin's conjecture. Math. Educ. Res. **7**(1), 17–26 (1998)
2. Herstein, I.K.: Noncommutative rings, No. 15. In: Carus Mathematical Monographs. American Mathematical Society (1968)
3. Herstein, I.K.: Topics in Algebra, 2nd edn. Wiley, Toronto (1975). Copyright: Xerox Corporation (1975)
4. Herstein, I.K.: A condition for the commutativity of rings. Canad. J. Math. **9**, 583–586 (1957)
5. Herstein, I.K.: Rings with Involution. University of Chicago, Chicago (1976)
6. Jacobson, N.: Structure theory for algebraic algebras of bounded degree. Ann. Math. **46**(2), 695–707 (1945)
7. MacHale, D.: An anticommutativity consequence of a ring commutativity theorem of Herstein. Amer. Math. Monthly **94**(2), 162–165 (1987)
8. McCune, W.: Solution of Robbin's problem. J. Automat. Reason **19**, 263–276 (1997)
9. McCune, W.: Otter 3.3 Reference Manual and Guide, Argonne National Laboratory Technical Memorandum ANL/MCS-TM-263 (2003)
10. McCune, W.: Prover9, automated reasoning software, and Mace4, finite model builder, Argonne National Laboratory (2005). https://www.cs.unm.edu/~mccune/mace4/
11. McCune, W., Padmanabhan, R. (eds.): Automated Deduction in Equational Logic and Cubic Curves. LNCS (LNAI), vol. 1095. Springer, Heidelberg (1996)
12. Neumann, B.H.: On the commutativity of addition. J. London Math. Soc. **15**, 203–208 (1940)
13. Phillips, J.D., Stanovsky, D.: Automated theorem proving in loop theory. In: Proceedings of ESARM 2008, pp. 42–54 (2008)
14. Posner, E.C.: Derivations in prime rings. Proc. Amer. Soc. **8**(6), 1093–1100 (1957)
15. Wavrik, J.J.: Commutativity theorems: examples in search of algorithms. In: Proceedings of 1999 International Symposium on Symbolic and Algebraic Computations, pp. 31–36. ACM (1999)
16. Zhang, H.: Automated proof of ring commutativity problems by algebraic methods. J. Symbolic Comput. **9**, 423–427 (1990)
17. Zemmer, J.L.: The addition group of an infinite near-field is abelian. J. London Math. Soc. **44**, 65–67 (1969)

Agent-Based HOL Reasoning

Alexander Steen[1(✉)], Max Wisniewski[1], and Christoph Benzmüller[1,2]

[1] Institute of Computer Science, Freie Universität Berlin, Berlin, Germany
{a.steen,m.wisniewski,c.benzmueller}@fu-berlin.de
[2] CSLI, Stanford University, Stanford, USA

Abstract. In the Leo-III project, a new agent-based deduction system for classical higher-order logic is developed. Leo-III combines its predecessor's concept of cooperating external specialist systems with a novel agent-based proof procedure. Key goals of the system's development involve parallelism on various levels of the proof search, adaptability for different external specialists, and native support for reasoning in expressive non-classical logics.

Keywords: Higher-order logic · Automated theorem proving · Reasoning · Non-classical logics

1 Introduction

We present the automated theorem prover Leo-III and its associated system platform. In the DFG funded project a novel agent-based deduction system for classical higher-order logic (HOL) is developed which aims at exploiting massive parallelism at various levels in the reasoning process. The system allows ad-hoc inclusion of independent specialist agents that add advanced functionality to the proof search such as consistency checks of the input axiomatization using model finders or augmented deduction processes for non-classical logics. The latter, very powerful, capability is enabled by semantical embedding of the desired goal logic in HOL. Several of such embeddings will be included in Leo-III, yielding an out-of-the-box automation tool for a great number of (quantified) non-classical logics relevant in mathematics (e.g. inclusive/free logic as used in projective geometry), philosophy (e.g. modal logics) and computer science (e.g. many-valued logics, paraconsistent logics).

In its current state, Leo-III is based on an ordered paramodulation calculus for typed lambda-terms, augmented with special means of extensionality treatment. The employment of agents allows parallelism on the search level by introducing and-/or-splits of the search space. The scheduling of the agents' actions is realized as optimization procedure using combinatorical auction games.

This work has been supported by the DFG under grant BE 2501/11-1 (Leo-III).

G.-M. Greuel et al. (Eds.): ICMS 2016, LNCS 9725, pp. 75–81, 2016.
DOI: 10.1007/978-3-319-42432-3_10

2 Classical Higher-Order Logic

Simple type theory, also referred to as classical higher-order logic (HOL), is an expressive logic formalism that allows for higher-order quantification [Fre79], that is quantification over arbitrary set and function variables. It is based on the simply typed λ-calculus and was, in its current formulation, developed by Church [Chu40]. In the following, we briefly introduce the syntax and semantics of HOL. For thorough discussions we refer to the literature[1].

HOL is a typed logic. The set of *simple types* $\mathcal{T}$ is thereby freely generated using the binary function type constructor $\rightarrow$ and the set of base types T. We assume that T consists of at least two elements $\{o, \iota\} \subseteq T$, where o and ι denote the type of Booleans and some non-empty domain of individuals, respectively.

The *terms* of HOL are then given by the following grammar ($\tau, \nu \in \mathcal{T}$):

$$s, t ::= c_\tau \mid X_\tau \mid (\lambda X_\tau. s_\nu)_{\tau \rightarrow \nu} \mid (s_{\tau \rightarrow \nu}\ t_\tau)_\nu \tag{1}$$

where c_τ denotes a typed constant from the signature Σ and X_τ is a variable. The remaining two cases are called *abstraction* and *application*. The type of a term is explicitly stated as subscript but may be dropped for legibility reasons if obvious from the context. Terms s_o of type o are *formulas*.

We require Σ to contain a complete logical signature. To that end, we choose Σ to consist at least of the primitive logical connectives for disjunction, negation, and, for each type, equality and universal quantification. Hence, we have $\{\vee_{o \rightarrow o \rightarrow o}, \neg_{o \rightarrow o}, =^\tau_{\tau \rightarrow \tau \rightarrow o} \Pi^\tau_{(\tau \rightarrow o) \rightarrow o}\} \subseteq \Sigma$ for all $\tau \in \mathcal{T}$. The remaining logical connectives can be defined as usual, e.g. $s \wedge t := \neg(\neg s \vee \neg t)$.

The semantics of HOL is now briefly addressed. A frame $\{\mathcal{D}_\tau\}_{\tau \in \mathcal{T}}$ is a collection of non-empty sets $\mathcal{D}_\tau$ such that $\mathcal{D}_o = \{T, F\}$ (for truth and falsehood, respectively) and $\mathcal{D}_{\tau \rightarrow \nu} \subseteq \mathcal{D}_\nu{}^{\mathcal{D}_\tau}$ is a collection of functions from $\mathcal{D}_\tau$ to $\mathcal{D}_\nu$. An *interpretation* is a pair $\mathcal{M} = (\{\mathcal{D}_\tau\}_{\tau \in \mathcal{T}}, \mathcal{I})$ where $\{\mathcal{D}_\tau\}_{\tau \in \mathcal{T}}$ is a frame and $\mathcal{I}$ is a function mapping each constant c_τ to some denotation in $\mathcal{D}_\tau$. We assume that the primitive logical connectives are assigned their usual denotation. Given a variable assignment σ we can define a valuation $\|.\|^{\mathcal{M}, \sigma}$ by

$$\begin{aligned}
\|c_\tau\|^{\mathcal{M}, \sigma} &= \mathcal{I}(c_\tau) \\
\|X_\tau\|^{\mathcal{M}, \sigma} &= \sigma(X_\tau) \\
\|s_{\tau \rightarrow \nu}\ t_\tau\|^{\mathcal{M}, \sigma} &= \|s_{\tau \rightarrow \nu}\|^{\mathcal{M}, \sigma} \|t_\tau\|^{\mathcal{M}, \sigma} \\
\|\lambda X_\tau. s_\nu\|^{\mathcal{M}, \sigma} &= \left(f : z \longmapsto \|s\|^{\mathcal{M}, \sigma[z/X_\tau]}\right) \in \mathcal{D}_{\tau \rightarrow \nu}
\end{aligned} \tag{2}$$

A formula s_o is called valid, iff $\|s_o\|^{\mathcal{M}, \sigma} = T$ for every variable assignment σ and every interpretation $\mathcal{M}$. We call $\mathcal{M}$ a *standard model* iff $\mathcal{D}_{\tau \rightarrow \nu}$ is the complete set of total functions, i.e. $\mathcal{D}_{\tau \rightarrow \nu} = \mathcal{D}_\nu^{\mathcal{D}_\tau}$. As a consequence of Gödel's Incompleteness Theorem [God31], HOL with standard semantics is necessarily incomplete. However, if we allow $\mathcal{D}_{\tau \rightarrow \nu}$ to be a proper subset of $\mathcal{D}_\nu^{\mathcal{D}_\tau}$ with the constraint that $\|.\|$ remains total, a meaningful notion of completeness can be achieved [Hen50]. We assume this so-called *Henkin semantics* in the following.

---

[1] Detailed information about typed λ-calculi and formal aspects of HOL can e.g. be found in [BDS13, BM14, Ben15a, BBK04] and references therein.

3   Extensional Paramodulation for HOL

The proof search of Leo-III is guided by a refutation-based calculus which uses the fact that $A_1, \ldots, A_n \vdash C$ if and only if $\{A_1, \ldots, A_n, \neg\, C\}$ is inconsistent. To that end, the initial set of formulas is transformed into equational clausal normal form and saturated until the empty clause is found. A popular method for saturating a given set of clauses is resolution, i.e. as employed by LEO-II [BPST15]. In first-order theorem proving, many successful systems use calculi based on ordered *paramodulation* [BG94] (or its even more restricted form, *superposition*), which improves naive resolution not only by an appropriate handling of equality, but also by using ordering constraints to restrict the number of possible inferences. In HOL, however, finding appropriate term orderings is more involved and only few such orderings exist.

We now sketch a (unordered) paramodulation rule for HOL and then briefly discuss, how ordering restrictions can be employed for the paramodulation-based calculus of Leo-III.

An equation is a pair $s \simeq t$ of terms. A literal is a signed equation, written $[s \simeq t]^\alpha$ where $\alpha \in \{\mathtt{tt}, \mathtt{ff}\}$ is the polarity of the literal. A clause $\mathcal{C}$ is a multiset of literals, denoting its disjunction. For brevity, if $\mathcal{C}$ and $\mathcal{D}$ are clauses and l is a literal, we write $\mathcal{C} \vee l$ and $\mathcal{C} \vee \mathcal{D}$ for the multi-union of $\mathcal{C} \cup \{l\}$ and $\mathcal{C} \cup \mathcal{D}$, respectively. The paramodulation inference can then be stated as

$$\frac{\mathcal{C} \vee [l \simeq r]^{\mathtt{tt}} \qquad \mathcal{D} \vee [s \simeq t]^\alpha}{\mathcal{C} \vee \mathcal{D} \vee [s[r]_\pi \simeq t]^\alpha \vee [s|_\pi \simeq l]^{\mathtt{ff}}} \text{ (Para)}$$

where negative equality literals encode postponed unification tasks, $s|_\pi$ is the subterm of s at position π, and $s[r]_\pi$ denotes the term that is created by replacing the subterm of s at position π by r. Intuitively, paramodulation is a conditional rewriting step that is justified if the unification tasks can be solved. Further calculus rules include equality factoring, unification handling and clausification.

The above rule (Para) is unordered and will, especially in a higher-order setting, produce a lot of irrelevant (redundant) clauses in the search space. In order to restrict the inference rules such as (Para), we are employing a higher-order term ordering primarily investigated for automated termination proofs, called *computability path ordering* (CPO) [BJR15].

In its current state, we successfully use CPO to orient equations and pre-select maximal literals eligible for paramodulation and factorization inferences. However, the employment of full ordered paramodulation constraints, that additionally discard generated clauses that do not match the ordering constraints after unification, is not yet implemented. This is partly due to the complicated nature of higher-order unification where, in general, there exists no most general unifier between two terms. Nevertheless, already at this point, the use of CPO seems promising as a candidate ordering towards a fully ordered paramodulation calculus for HOL.

In addition to the usual paramodulation inference rules above, extensionality aspects need to be considered explicitly as well. This is because equalities in HOL

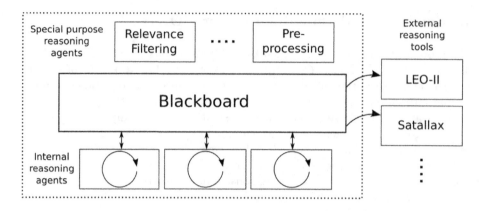

Fig. 1. Leo-III's agent-based proof search cooperation

can occur between terms of any type, in particular between terms of Boolean type or function type. The mere addition of extensionality axioms for each relevant symbol does not suffice, as it leads to a massive explosion of the search space. Hence, we include special means of calculus-level treatment similar to the rules used by LEO-II [BPST15], and combine them with extensionality handling in adequate preprocessing steps [WSKB16].

The overall saturation procedure consisting of the above sketched ingredients is, at the moment, organized as a sequential loop using a variant of the *given-clause* algorithm.

4   Agent-Based Refutation

An *agent* is a software component that can be executed independently of others. Moreover, an agent is given the ability to decide on its own when to execute its functionality. This high amount of autonomy is a key feature of agents [Wei13]. In the Leo-III system, agents are employed as *specialists* for some aspects of the proof search. The underlying architecture of Leo-III is designed as a blackboard which the agents use to collaboratively find a proof. The work of the agents is thereby divided in transactional tasks and organized in auctions, in which it is decided which tasks are performed next in case of interference.[2]

In its current state, Leo-III employs agents in three different scenarios: During the preprocessing phase of the overall proof procedure each agent is responsible (i.e. a specialist) for one sort of normalization to be applied to a formula. Here, the overall goal is to exhaustively apply all normalization procedures [WSKB16] to all clauses. Since normalization is a local problem, an agent can judge solely by observing a particular formula whether it wants to act on it. Due to only little existing interference between the normalization methods, the execution of the different normalization routines can easily be distributed among the agents.

---

[2] Further information can e.g. be found in [WB16, Wis14].

As a second employment scenario, a relevance filter (cf. [MP09]) is implemented that prunes the search space prior to preprocessing. Relevance filtering can be performed similarly to the preprocessing, except that the problem is not local in the above sense since information about other formulas have to be considered as well. As most of the agents of Leo-III will have this kind of non-local dependencies, a reasonable coordination of those agents is one of the main goals of further development. The last employment of the agents is to parallelize heavy weighted proof procedures. Here, the agent-based approach can be applied on the calculus level, e.g. for single clauses, but also for parallelizing one or more (sequential) proof procedures (so-called multi search). In Leo-III, the calculus sketched in Sect. 3 is distributed among multiple agents. Additionally, HOL theorem proving systems such as LEO-II and Satallax [Bro12] are included as external specialists.

In Fig. 1 the connections among the components of Leo-III are visualized. The focus in the current state relies mostly on the last of the three above described cases. We employ sequential proof procedures and external provers to solve the input problem in parallel and wait for the first positive result. These tasks either differ in some parameters of the proof search or in previously applied normalization techniques. For further work, we will experiment with different granularities for the generated tasks and different means of agent coordination. The task sizes can hereby vary from the execution of whole proof procedures to very fine-grained responsibilities (e.g. the application of single inference steps). The coordination is at the lowest level bound to the auction system. On top of that however, additional mechanisms (such as fixed execution priorities, or coalitions and coalition games [CEW11]) can be added.

5   The Leo-III System

In its core, Leo-III is a new higher-order automated theorem prover based on the associated system platform LeoPARD [WSB15]. LeoPARD is a framework for deduction systems (implemented in Scala) providing sophisticated term, search, and indexing data structures for typed λ-terms, as well as an generic agent-based blackboard architecture. Leo-III makes use of these supported data structures and implements the concrete agents as described in Sect. 4 on top of the provided blackboard architecture. The internal reasoning agents implement a proof procedure realizing the calculus depicted in Sect. 3.

During the development of Leo-III, special care was given to providing maximal compatibility with existing systems and conventions of the application area. As input language, for instance, Leo-III supports every standard dialect of the TPTP syntax [Sut09] (including THF, TFF and FOF). For best possible external utilization, Leo-III can output a proof object used for proof reconstruction pointers (e.g. in Isabelle [NPW02]) or proof verification tools (e.g. IDV [Sut09]).

One major goal of Leo-III is to provide native means of reasoning within (and about) non-classical logics including free logic, (quantified) conditional logic, and (quantified) modal logic[3]. Such logics are of strong interest in many different fields of research, for example in mathematics, artificial intelligence, and philosophy. In its current state, our system is already capable of reasoning in that embedded logics and even – with a few modifications – of parsing the syntax representation of these formalisms. The automated transformation of an input problem stated in such a specialized syntax representation into an equivalent HOL formulation is still in development, but can easily be added to Leo-III as a new preprocessing procedure.

6   Conclusion

In this paper, we have presented a new automated theorem prover for HOL, called Leo-III. A rough sketch of its underlying paramodulation calculus and its extensionality handling has been given. A proof procedure based on that calculus is included in the presented agent-based blackboard architecture. Additionally, external deduction systems can be included as agents. Leo-III is, on the long perspective, intended as a platform for *universal reasoning*, offering not only support for reasoning in classical higher-order logics, but also for reasoning within further expressive, non-classical logics such as modal logics, conditional logics or even many-valued logics. This enables our system to serve as a reasoning tool for a wide spectrum of formal scientific disciplines.

A native out-of-the-box automation of input problems stated in specialized syntax of corresponding non-classical logics is further work. Additionally, we will develop and include further specialized agents to allow a more fine-grained parallelization of the proof search. To that end, experiments with various parameters and heuristics for the guidance of the proof search and the organization of the agents will be conducted.

References

[BBK04] Benzmüller, C., Brown, C., Kohlhase, M.: Higher-order semantics and extensionality. J. Symbolic Logic **69**(4), 1027–1088 (2004)

[BDS13] Barendregt, H.P., Dekkers, W., Statman, R.: Lambda Calculus with Types. Perspectives in Logic. Cambridge University Press, Cambridge (2013)

[Ben15a] Benzmüller, C.: Higher-order automated theorem provers. In: Delahaye, D., Woltzenlogel Paleo, B. (eds.) All About Proofs, Proof for All. Mathematical Logic and Foundations, pp. 171–214. College Publications, London (2015)

[Ben15b] Benzmüller, C.: Invited talk: on a (quite) universal theorem proving approach and its application in metaphysics. In: De Nivelle, H. (ed.) TABLEAUX 2015. LNCS, vol. 9323, pp. 213–220. Springer, Heidelberg (2015)

[BG94] Bachmair, L., Ganzinger, H.: Rewrite-based equational theorem proving with selection and simplification. J. Logic Comput. **4**(3), 217–247 (1994)

---

[3] The reasoning in such non-classical logics is enabled by a semantical embedding of the target logic into HOL. Detailed information about this approach can be found, e.g. in [Ben15b] and the references therein.

[BJR15] Blanqui, F., Jouannaud, J.-P., Rubio, A.: The computability path ordering (2015). CoRR, abs/1506.03943

[BM14] Benzmüller, C., Miller, D.: Automation of higher-order logic. In: Gabbay, D.M., Siekmann, J.H., Woods, J. (eds.) Handbook of the History of Logic. Computational Logic, vol. 9, pp. 215–254. Elsevier, North Holland (2014)

[BPST15] Benzmüller, C., Paulson, L.C., Sultana, N., Theiß, F.: The higher-order prover LEO-II. J. Autom. Reasoning 55(4), 389–404 (2015)

[Bro12] Brown, C.E.: Satallax: an automatic higher-order prover. In: Gramlich, B., Miller, D., Sattler, U. (eds.) IJCAR 2012. LNCS, vol. 7364, pp. 111–117. Springer, Heidelberg (2012)

[CEW11] Chalkiadakis, G., Elkind, E., Wooldridge, M.: Computational Aspects of Cooperative Game Theory. Synthesis Lectures on Artificial Intelligence and Machine Learning. Morgan & Claypool Publishers, San Rafael (2011)

[Chu40] Church, A.: A formulation of the simple theory of types. J. Symbolic Logic 5(2), 56–68 (1940)

[Fre79] Frege, G.: Begriffsschrift, eine der arithmetischen nachgebildete Formelsprache des reinen Denkens. Verlag von Louis Nebert, Halle (1879)

[God31] Gödel, K.: Über formal unentscheidbare Sätze der Principia Mathematica und verwandter Systeme. Monatshefte für Mathematik und Physik 38(1), 173–198 (1931)

[Hen50] Henkin, L.: Completeness in the theory of types. J. Symbolic Logic 15(2), 81–91 (1950)

[MP09] Meng, J., Paulson, L.C.: Lightweight relevance filtering for machine-generated resolution problems. J. Appl. Logic 7(1), 41–57 (2009)

[NPW02] Nipkow, T., Paulson, L.C., Wenzel, M.: Isabelle/HOL - A Proof Assistant for Higher-Order Logic. LNCS, vol. 2283. Springer, Heidelberg (2002)

[Sut09] Sutcliffe, G.: The TPTP problem library and associated infrastructure. J. Autom. Reasoning 43(4), 337–362 (2009)

[WB16] Wisniewski, M., Benzmüller, C.: Is it reasonable to employ agents in theorem proving? In: van den Heerik, J., Filipe, J. (eds.) Proceedings of the 8th International Conference on Agents and Artificial Intelligence (ICAART), Rome, Italy, 2016, vol. 1, pp. 281–286. SCITEPRESS - Science and Technology Publications, Lda (2016)

[Wei13] Weiss, G. (ed.): Multiagent Systems. MIT Press, Cambridge (2013)

[Wis14] Wisniewski, M.: Agent-based Blackboard Architecture for a Higher-Order Theorem Prover. Master's thesis, Freie Universität Berlin (2014)

[WSB15] Wisniewski, M., Steen, A., Benzmüller, C.: LeoPARD — a generic platform for the implementation of higher-order reasoners. In: Kerber, M., Carette, J., Kaliszyk, C., Rabe, F., Sorge, V. (eds.) CICM 2015. LNCS, vol. 9150, pp. 325–330. Springer, Heidelberg (2015)

[WSKB16] Wisniewski, M., Steen, A., Kern, K., Benzmüller, C.: Effective normalization techniques for HOL. In: Olivetti, N., Tiwari, A. (eds.) IJCAR 2016. LNCS, vol. 9706, pp. 362–370. Springer, Heidelberg (2016). doi:10.1007/978-3-319-40229-1_25

An Automated Deduction and Its Implementation for Solving Problem of Sequence at University Entrance Examination

Yumi Wada[1], Takuya Matsuzaki[2], Akira Terui[1(✉)], and Noriko H. Arai[3]

[1] University of Tsukuba, Tsukuba, Japan
`wada.yumi.ww@alumni.tsukuba.ac.jp`, `terui@math.tsukuba.ac.jp`
[2] Nagoya University, Nagoya, Japan
`matuzaki@nuee.nagoya-u.ac.jp`
[3] National Institute of Informatics, Tokyo, Japan
`arai@nii.ac.jp`
`http://researchmap.jp/aterui/`,
`http://researchmap.jp/mtzk/`,
`http://researchmap.jp/arai_noriko/`

Abstract. "Todai Robot Project" is a project of artificial intelligence launched by National Institute of Informatics for re-unifying the artificial intelligence field subdivided in 1980s and afterwards. We focus towards attaining a high score in National Center Test for University Admissions, and use Quantifier Elimination (QE) over the real closed fields as a main tool for solving problems in mathematics. However, it is not applicable for several kinds of problems such as one with sequence. In this article, we propose an algorithm for solving problems of sequence at the National Center Test for University Admissions.

Keywords: Automated deduction · Recurrence relations · Sequences

1 Introduction

"Todai Robot Project" is a project of artificial intelligence launched by National Institute of Informatics for re-unifying the artificial intelligence field subdivided in 1980s and afterwards [2]. Our research team has been working since 2011 on the mechanical solving of university entrance examination problems, which includes towards attaining a high score in National Center Test for University Admissions, abbreviated as the "National Center Test". We are working on each subject (Japanese, Mathematics, English, Physics, Japanese history, World history) of examinations by developing solvers independently.

In solving problems in mathematics, first we translate digitalized and annotated texts of a question written in natural language (Japanese) into logical

---

Y. Wada—Current affiliation: Atlas Co., Ltd., Tokyo, Japan. http://www.atlas.jp/en.

© Springer International Publishing Switzerland 2016
G.-M. Greuel et al. (Eds.): ICMS 2016, LNCS 9725, pp. 82–89, 2016.
DOI: 10.1007/978-3-319-42432-3_11

formulas over Zermelo-Fraenkel (ZF) set theory by natural language processing (NLP). Then, by symbolic computation, we re-translate the problem for use
by various solvers for calculating the final answer by computer algebra, where
we use Quantifier Elimination (QE) over real closed fields as a main tool [3,5].
However, it is not applicable for some problems such as ones with sequence.

As related research on solving problems of sequences, there are previous
works on (1) solving recurrence relations and/or difference equations [1] and
(2) automated theorem proving [4]. For solving the most problems of sequence
in National Center Test, we need to solve recurrence relations and algebraic
equations which are given as constraints for the sequence. Although functions
for solving recurrence relations and algebraic equations are available in popular
computer algebra systems, it is not easy to find an answer because we need to
use these functions for solving the given recurrence relations and equations in
appropriate order. In this paper, we propose an algorithm for solving problems
of sequence at the National Center Test.

2   Preliminaries

Let $\mathbb{N}$ be the set of positive integers. Given a set S, denote its cardinality and
power set to $\#S$ and $\mathfrak{P}(S)$, respectively. Define a sequence as a univariate map
from the set of positive integers to the set of real numbers, denoted to $a(n)$. For
m unknown sequences $a_1, \ldots, a_m$, a non-negative number k and a real-valued
function of $m(k + 1) + 1$ variables f, define a recurrence relation of order k
of $a_1, \ldots, a_m$ with an equation $f(n, a_1(n), \ldots, a_1(n + k), \ldots, a_m(n), \ldots, a_m(n + k)) = 0$.

The National Center Test consists of multiple choice questions and grid-in
questions. Mathematics section of the test consists of grid-in questions with
answering alphabetical symbols or numerical answers. Since the most of questions require numerical answers in mathematics section, we assume that the test
has only questions with numerical answers in mathematics section.

Answer columns in mathematics questions have the following format.

1. An answer column is expressed with one or more letters of Japanese *Katakana*
   alphabet,[1] each letter of which corresponds to a single digit non-negative
   integer. If an answer column consists of more than one letter, the leftmost
   letter corresponds to either the negative sign or a single digit non-negative
   integer.
2. Answer columns with the same letter are filled with the same number.

For example, an answer column $\boxed{A}$ corresponds to a single digit non-negative
integer (such that $x \in \{0, \ldots, 9\}$), while an answer column $\boxed{BC}$ corresponds
to either a double digits positive integer (such that $x \in \{10, 11, \ldots, 99\}$) or a
negative integer $x \in \{-9, -8, \ldots, -1\}$.

---

[1] In this paper, for ease of understanding, we express an answer column with English
alphabet.

3 Calculating Inputs for Solving Problems of Sequence

3.1 Characteristics of Questions of Sequence in the National Center Test

Remark 1. By analyzing questions in the real and mock exams in the past, we see that the most of questions of sequence in the National Center Test can be transferred to logical formula(s) with the following properties.

1. Functions appearing in the logical formula consist of sequences, arithmetic operations, exponential functions, finite sum and/or product of sequences (denoted to Σ and Π, respectively), and the maximum and/or the minimum values of a sequence (denoted to "max" and "min", respectively), with the following restrictions.
   (a) No more than two variables (arguments) appear in a sequence.
   (b) Σ and Π are defined as bivariate functions. For example, $\Sigma(a, n)$ is defined as $\sum_{k=1}^{n} a(k)$, where $a(k)$ is a sequence.
   (c) "max" and "min" are defined as trivariate functions with a sequence, a variable and the conjunction of atomic formulas as the variables (the arguments). For example, $\max(a, n, n > 1)$ returns the maximum value of a set $\{a(n) \mid n > 1\}$ for a given sequence $a(n)$.
2. A logical formula may have the following predicates.
   (a) Relational operators: $=$, $<$, $\leq$ and $\neq$.
   (b) "$\mathcal{N}$" defined as $\mathcal{N}(n)$ means that "n is a positive integer (a natural number)".
   (c) "$\mathcal{R}$" defined as $\mathcal{R}(n)$ means that "n is a real number".
   (d) "$\mathcal{S}$" defined as $\mathcal{S}(a)$ means that "a is a sequence".
3. Logical operators appearing in the logical formula consist of $\wedge$ (conjunction), $\rightarrow$ (imply) and $\forall$ (for all).

Example 1. (A question in the National Center Test in 2007). An applicant must fill in the grid-in answer sections in the following paragraph.

> Let (a_n) be a sequence defined with the recurrence relation $a_{n+1} = 3a_n + 60$ for $n = 1, 2, 3, \ldots$, and with the initial term -27. Then, we have $a_n = \boxed{A}^n - \boxed{BC}$. Furthermore, let $S_n = \Sigma_{k=1}^{n} a_k$. Then, we have $S_n = \dfrac{\boxed{D}}{\boxed{E}}\left(\boxed{F}^n - \boxed{G}\right) - \boxed{BC}\,n$, and the least positive integer n satisfying $S_n > 0$ is equal to $\boxed{H}$.

The above paragraph can be translated into a logical formula as

$$\forall a(\mathcal{S}(a) \rightarrow (a(1) = -27 \wedge \forall n(\mathrm{N}(n) \rightarrow a(n+1) = 3a(n) + 60)) \rightarrow$$
$$(\forall n(\mathcal{N}(n) \rightarrow a(n) = \boxed{A}^n - \boxed{BC}) \wedge (\forall n(\mathcal{N}(n) \rightarrow S(n) = \Sigma(a, n)) \rightarrow$$
$$(\forall n(\mathcal{N}(n) \rightarrow$$
$$S(n) = \frac{\boxed{D}}{\boxed{E}}(\boxed{F}^n - \boxed{G}) - \boxed{BC}\,n) \wedge \boxed{H} = \min(n, S(n) > 0))))))). \quad (1)$$

Note that logical formula (1) satisfies all the properties in Remark 1.

Remark 2. From logical formula (1), we have the following conventions.

1. Quantifiers appeared in the formula is only $\forall$, thus it can be omitted.
2. We can omit predicate symbols $\mathcal{R}$, $\mathcal{N}$ and $\mathcal{S}$ because they are used only for specifying scope of variables bounded by quantifiers and we can distinguish whether each variable represents a number or a sequence from the form of formula (for example, positive integers appear only in the arguments in sequences). Let $V_{\mathcal{R}}$, $V_{\mathcal{N}}$ and $V_{\mathcal{S}}$ be the set of variables appeared in a logical formula for real numbers, natural numbers and sequences, respectively. Then, we see that the variables appeared in the logical formula belong to $V_{\mathcal{R}} \cup V_{\mathcal{N}} \cup V_{\mathcal{S}}$ with $V_{\mathcal{R}} \cap V_{\mathcal{N}} \cap V_{\mathcal{S}} = \emptyset$.
3. Let P, Q and R be logical formulas. Then, with equivalences $P \to (Q \wedge R) \equiv (P \to Q) \wedge (P \to R)$ and $P \to (Q \to R) \equiv (P \wedge Q) \to R$, we can transfer logical formula of the form as in Eq. (1) into a logical formula of the form

$$\bigwedge_i (\bigwedge_j A_{i,j} \to \bigwedge_k C_{i,k}), \tag{2}$$

where $A_{i,j}$ and $C_{i,k}$ are propositions without containing logical operators.

Example 2. By Remark 2, logical formula (1) can be transferred into a logical formula as follows.

$$((a(1) = -27 \wedge a(n+1) = 3a(n) + 60) \to a(n) = \boxed{A}^n - \boxed{BC})$$
$$\wedge\, ((a(1) = -27 \wedge a(n+1) = 3a(n) + 60 \wedge S(n) = \Sigma(a, n)) \to$$
$$(S(n) = \frac{\boxed{D}}{\boxed{E}} (\boxed{F}^n - \boxed{G}) - \boxed{BC}\, n \wedge \boxed{H} = \min(n, S(n) > 0))). \tag{3}$$

Remark 3. We see that many questions in the National Center Test and its mock exams that have been transformed into logical formula of the form as in Eq. (2) have the following characteristics (see Eq. (3) for its example).

1. The first argument in functions max and min is usually monotone sequence.
2. $A_{i,j}$ usually contains only equations from which we can derive a closed form of the sequence and/or values of variables.
3. $C_{i,j}$ contains only an equation.
4. If $A_{i,j}$ contains an answer column, then there exist i' and k satisfying $i' < i$ and $C_{i',k}$ contains the same answer column.
5. For every i, integers corresponding to answer columns in $C_{i,k}$ can be derived by putting variables and/or closed form of the sequence calculated from $\bigwedge_j A_{i,j}$ into corresponding variables in $C_{i,k}$.

3.2 Constructing Input Logical Formula

From a logical formula φ of ZF set theory obtained by the Natural Language Processing (NLP), we construct an input logical formula for a sequence solver, as follows. Since, in the NLP process, some clause in the questions are translated directly, the NLP output may contain a logical formula with predicates and/or functions different from those we have introduced in Remark 1. Thus, before constructing input logical formula, we translate such sentences into functions and/or predicates in Remark 1 with assigning appropriate variables (we omit its detail due to lack of space here). Let the resulting logical formula be φ'.

After that, we construct an input for the solver from φ' using the followings:

1. Set $\bar{V}$ and $\bar{S}$ consisting of the elements in $V_{\mathcal{R}}$ and $V_{\mathcal{S}}$, respectively, which appear in φ',
2. Set Z consisting of the answer columns which appear in φ',
3. $A_{i,j}$s and $C_{i,k}$s after transferring φ'.

Example 3. By preprocessing in the above, logical formula (3) is translated into the following input for a solver, where $A_{i,j}$ and $C_{i,k}$ are defined as in Eq. (2).

$$\bar{V} = \{\}, \quad \bar{S} = \{a, S\}, \quad Z = \{A, BC, D, E, F, G, H\},$$
$$A_{1,1} \equiv a(1) = -27, \; A_{1,2} \equiv a(n+1) = 3a(n) + 60, \; C_{1,1} \equiv a(n) = A^n - BC,$$
$$A_{2,1} \equiv a(1) = -27, \; A_{2,2} \equiv a(n+1) = 3a(n) + 60, \; A_{2,3} \equiv S(n) = \Sigma(a, n)),$$
$$C_{2,1} \equiv S(n) = \frac{D}{E}(F^n - H) - BCn, \; C_{2,2} \equiv H = \min(n, S(n) > 0).$$

4   Algorithms for the Sequence Solver

We first discuss for an algorithm for finding values and a sequence satisfying $A_{i,j}$ with the following strategy.

1. Calculate a closed form of the sequence by solving recurrence relation(s).
2. Then, by solving equation(s) with regarding recurrence relation(s) as identity, find values for the variables.

Let *rsolve* be an algorithm for solving recurrence relations. Solution of recurrence relation(s) may have free variables depending on choice of initial value(s). Thus, let V_{st} be the set of free variables found every time solving recurrence relations(s). Similarly, let *solve* be an algorithm for solving equations. Let F_r be a set of free variables in equations if they have. Note that the sets $V_{\mathcal{R}}$, $V_{\mathcal{N}}$, $V_{\mathcal{S}}$, V_{st}, Fr, $\mathbb{Z}$ are pairwisely relatively prime.

For a proposition A, let $I(A)$ be the set of variables appearing in A. For a set of variables B, let $I_B(A) = I(A) \cap B$. Let R be a map from variables in equations and recurrence relations to their solutions and/or answers in the answer columns. Then, we can regard R as a function from a certain subset of $S \cup V \cup V_{st} \cup F_r \cup Z$ to $\mathbb{R}^{\mathbb{N}} \cup \mathbb{R}$. Furthermore, for proposition A, let $R(A)$ be a logical formula obtained by substituting $x \in I_{\text{dom}(R)}(A)$ to $R(x)$, where $\text{dom}(R)$ represents the domain of R. For example, for $A \equiv a(n+1) + b(n) = n$ and a solution of a recurrence relation given as $a(n) = 2n$, then we have $R(A) \equiv 2(n+1) + b(n) = n$ by setting $R(a) = (a(n) \mapsto 2n)$.

We try to solve recurrence relations which has just one sequence. If we can find the solution, then we add it to R and try to solve another recurrence relation, and so on. We show this computation as in the following algorithm.

Algorithm 1 (GetGeneralTerms (GGT)).

Inputs: $\mathcal{A}$: a set of recurrence relations, S: a set of sequences, V: a set of variables, R: a map;

Outputs: $\mathcal{A}' \subset \mathcal{A}$: a set of recurrence relations that have not been used to solve for sequences, $S' \subset S$: a set of sequences whose closed form has not been solved, $V \cup F'$ with $F' \subset F$ is a set of variables representing the initial term in a sequence, $R \cup R'$ with R' is a map from a sequence in S, of which we have solved closed form, to its closed form;

1. $\mathcal{P} \leftarrow \mathfrak{P}(S)\backslash\{\emptyset\}$;
2. Sort the sets in $\mathcal{P}$ with smaller cardinality first;
3. for $P \in \mathcal{P}$ do
   (a) for $\mathcal{A}'' \in \{\mathcal{A}'' \subset \mathcal{A} \mid \bigcup_{A \in \mathcal{A}''} I_S(R(A)) = P \wedge \#\mathcal{A}'' = \#P\}$ do
       i. if rsolve($\{R(A) \mid A \in \mathcal{A}''\}, P$) calculates a closed form R', then
          A. $V' \leftarrow \bigcup_{A \in \mathrm{ran}(R')} I_{V_{\mathrm{st}}}(A)$, where $\mathrm{ran}(R')$ is the range of R';
          B. return GGT($\mathcal{A}\backslash\mathcal{A}'', S\backslash P, V \cup V', R \cup R'$);
4. return $\mathcal{A}, S, V, R$;

We show an algorithm for solving equations for values of variables in V.

Algorithm 2 (GetValues (GV)).

Inputs: $\mathcal{A}_{\mathrm{IN}}$: a set of equations, S: a set of sequences, V: a set of variables, R: a set of substitutions;

Outputs: $\mathcal{A}' \subset \mathcal{A}$: a set of equations never used for evaluation of variables, $V' \subset V$: a set of variables whose value has not been obtained, $R \cup R'$: where R' are substitutions found for (some) elements in V;

1. $\mathcal{P} \leftarrow \mathfrak{P}(V)\backslash\{\emptyset\}$;
2. Sort the sets in $\mathcal{P}$ with smaller cardinality first;
3. for $P \in \mathcal{P}$ do
   (a) for $\mathcal{A}'' \in \{\mathcal{A}'' \subset \mathcal{A} \mid \bigcup_{A \in \mathcal{A}''} I_V(R(A)) = P \wedge \bigcup_{A \in \mathcal{A}''} I_S(R(A)) = \emptyset\}$ do
       i. if solve($\{R(A) \mid A \in \mathcal{A}''\}, P$) found k solutions $R'_1, \ldots, R'_k$ ($k \in \mathbb{N}$) then
          A. for $R' \in \{R'_1, \ldots, R'_k\}$ do
          A-1. $R'' \leftarrow R \cup R'$;
          A-2. if $\{R''(A) \mid A \in \mathcal{A}_{\mathrm{IN}}\}$ does not have *false* then
          A-2-1. $V' \leftarrow \bigcup_{A \in \mathrm{ran}(R')} I_{Fr}(A)$;
          A-2-2. return GV($\mathcal{A}\backslash\mathcal{A}'', \mathcal{A}_{\mathrm{IN}}, S, (V \cup V')\backslash P, R''$);
4. return $\mathcal{A}, V, R$;

With combining Algorithms 1 and 2, we give an algorithm for calculating a closed form of a sequence and solving for variables from $\{A_{i,j}\}_j$, as follows.

Algorithm 3 (GetGeneralTermsAndValues (GGTV)).

Inputs: $\mathcal{A} = \{A_{i,j}\}_j$: a set of conditions, S: a set of sequences, V: a set of variables, R: substitutions from a subset of $\mathbb{Z}$ to $\mathbb{Z}$;

Output: $R \cup R'$: substitutions found for (some) elements in $S \cup V$ from $\{A_{i,j}\}_j$;

1. $\mathcal{A}_{IN} \leftarrow \{A \in \mathcal{A} \mid A : \text{inequalities}\}$;
2. $\mathcal{A}' \leftarrow \mathcal{A} \setminus \mathcal{A}_{IN}$; $S' \leftarrow S$; $V' \leftarrow V$; $R'' \leftarrow R$;
3. repeat
   (a) $\mathcal{A}_{RE} \leftarrow \{A \in \mathcal{A}' \mid A : \text{recurrence relations}\}$;
   (b) $\mathcal{A}' \leftarrow \mathcal{A}' \setminus \mathcal{A}_{RE}$; $S_{tmp} \leftarrow S'$;
   (c) $\mathcal{A}_{RE}, S', V', R'' \leftarrow \text{GGT}(\mathcal{A}_{RE}, S', V', R'')$;
   (d) $changed \leftarrow S_{tmp} \neq S'$;
   (e) $\mathcal{A}' \leftarrow \mathcal{A}_{RE} \cup \mathcal{A}'$; $V_{tmp} \leftarrow V'$;
   (f) $\mathcal{A}', V', R'' \leftarrow \text{GV}(\mathcal{A}', \mathcal{A}_{IN}, S', V', R'')$;
   (g) $changed \leftarrow changed \vee V_{tmp} \neq V'$;
4. until $changed = \text{false}$;
5. return R'';

Next, by using R derived in the above, we calculate integers satisfying the answer columns in $\{C_{i,k}\}_k$. Though it may be solved by a brute-force attack, it may be inefficient for a case with the number of combination of integers for seeking solution in $I_Z(C_{i,k})$ becomes up to more than 10^9, thus we must develop more sophisticated method for finding an answer.

For $z \in Z$, let $n(z) \in \mathbb{Z}$ be a set consisting of integers that are fit in z. For example, for $z = \boxed{A}$, $n(z) = \{1, \ldots, 9\}$, while, for $z' = \boxed{BC}$, $n(z') = \{-9, \ldots, -1, 10, \ldots, 99\}$. Furthermore, for an equation A, let $I^e(A) \subset Z$ be a set of answer columns (which are elements in Z) appearing as exponents.

Before showing an algorithm for finding integers that satisfy $\{C_{i,k}\}_k$, we show a subsidiary algorithm as follows.

Algorithm 4 (IntSolve).

Inputs: R: a set of substitutions, C: an equation satisfying $I^e(R(C)) = \emptyset$;
Output: R': a set of substitutions for the variables in C if C is solved; otherwise, an empty set;

1. $Z' \leftarrow I_Z(R(C))$;
2. $\mathcal{R} \leftarrow \{R' : Z' \rightarrow \mathbb{R} \mid R' : \text{solutions of } R(C) \text{ regarding as identity w.r.t variables which do not belong to } Z'\}$;
3. $\mathcal{R} \leftarrow \{R' \in \mathcal{R} \mid \forall z \in Z', R'(z) \in n(z)\}$;
4. if $\mathcal{R} \neq \emptyset$ then take a $R' \in \mathcal{R}$; else $R' \leftarrow \emptyset$;
5. return R';

Now we show an algorithm for finding integers that satisfy $\{C_{i,k}\}_k$.

Algorithm 5 (FindAnswers).

Inputs: $\mathcal{C} = \{C_{i,k}\}_k$: a set of equations; R: a set of substitutions;
Output: $\mathcal{C}'$: a set of equations that were not solved, $R \cup R'$ with R' is a set of substitutions for variables for which we have solved equations in $\mathcal{C}$;

1. $\mathcal{C}' \leftarrow \emptyset$; $R'' \leftarrow R$;
2. for $C \in \mathcal{C}$ do
   (a) $Z' \leftarrow I_Z(R(C))$;
   (b) if $Z' \neq \emptyset$ then
       i. $Z^e \leftarrow I^e(R(C))$;
       ii. if $Z^e = \emptyset$ then $P \leftarrow \emptyset$; $Q \leftarrow \texttt{IntSolve}(C, R'')$;
       iii. else
            A. for $P \in \prod_{z \in Z^e} n(z)$ do
            A-1. $Q \leftarrow \texttt{IntSolve}(C, R'' \cup P)$;
            A-2. if $Z' = Z^e \vee Q \neq \emptyset$ then break; else $P \leftarrow \emptyset$;
       iv. if $Q \cup P = \emptyset$ then $\mathcal{C}' \leftarrow \mathcal{C}' \cup \{C\}$; else $R'' \leftarrow R'' \cup P \cup Q$;
3. return $\mathcal{C}', R''$;

Summarizing the above, we give an algorithm for the solver, as follows.

Algorithm 6 (SequenceSolver).

Inputs: $\{(\mathcal{A}_i, \mathcal{C}_i)\}_i$: a set of pairs of propositions obtained as in Eq. (2), S: a set of variables for sequences, V: a set of variables for those are not sequences;
Output: R: a set of substitutions from a subset of Z to $\mathbb{Z}$;

1. $R \leftarrow \emptyset$; $unsolved \leftarrow \emptyset$;
2. for $(\mathcal{A}, \mathcal{C}) \in \{(\mathcal{A}_i, \mathcal{C}_i)\}_i$ do
   (a) $R \leftarrow \texttt{GetGeneralTermsAndValues}(\mathcal{A}, S, V, R)$;
   (b) $\mathcal{C}', R \leftarrow \texttt{FindAnswers}(\mathcal{C}, R)$;
   (c) if $\mathcal{C}' \neq \emptyset$ then $unsolved \leftarrow unsolved \cup \{(R, \mathcal{C}')\}$;
   (d) $R \leftarrow R \cap (Z \times \mathbb{Z})$;
3. for $(R', \mathcal{C}') \in unsolved$ do
   (a) for $C \in \mathcal{C}'$ do
       i. $R'' \leftarrow R \cup R'$; $C \leftarrow R''(C)$;
       ii. if $I_Z(C) \neq \emptyset$ then
           A. $solves \leftarrow \{P \in \prod_{z \in I_Z(C)} n(z) \mid P(C) = \text{true}\}$;
           B. if $solves \neq \emptyset$ then take an element $P \in solves$; $R \leftarrow R \cup P$;
4. return R;

References

1. Agarwal, R.P.: Difference Equations and Inequalities: Theory, Methods, and Applications, 2nd edn. Marcel Dekker, New York (2000)
2. Arai, N.H.: The impact of AI: can a robot get into the University of Tokyo? Natl. Sci. Rev. **2**(2), 135–136 (2015)
3. Arai, N.H., Matsuzaki, T., Iwane, H., Anai, H.: Mathematics by machine. In: Proceedings of the 39th International Symposium on Symbolic and Algebraic Computation – ISSAC 2014, pp. 1–8. ACM, New York (2014)
4. Bundy, A.: The automation of proof by mathematical induction. In: Robinson, A., Voronkov, A. (eds.) Handbook of Automated Reasoning, 2nd edn, pp. 845–911. North-Holland, Amsterdam (2001)
5. Matsuzaki, T., Iwane, H., Anai, H., Arai, N.H.: The most uncreative examinee: a first step toward wide coverage natural language math problem solving. In: Proceedings of the Twenty-Eighth AAAI Conference on Artificial Intelligence, pp. 1098–1104 (2014)

Algebraic and Toric Geometry

Bad Primes in Computational Algebraic Geometry

Janko Böhm[1]([⊠]), Wolfram Decker[1], Claus Fieker[1], Santiago Laplagne[2], and Gerhard Pfister[1]

[1] University of Kaiserslautern, 67663 Kaiserslautern, Germany
{boehm,decker,fieker,pfister}@mathematik.uni-kl.de
[2] Universidad de Buenos Aires, Buenos Aires, Argentina
slaplagn@dm.uba.ar

Abstract. Computations over the rational numbers often suffer from intermediate coefficient swell. One solution to this problem is to apply the given algorithm modulo a number of primes and then lift the modular results to the rationals. This method is guaranteed to work if we use a sufficiently large set of good primes. In many applications, however, there is no efficient way of excluding bad primes. In this note, we describe a technique for rational reconstruction which will nevertheless return the correct result, provided the number of good primes in the selected set of primes is large enough. We give a number of illustrating examples which are implemented using the computer algebra system SINGULAR and the programming language JULIA. We discuss applications of our technique in computational algebraic geometry.

Keywords: Modular computations · Algebraic curves · Adjoint ideal

1 Introduction

Many exact computations in computer algebra are carried out over the rationals and extensions thereof. Modular techniques are an important tool to improve the performance of such algorithms since intermediate coefficient growth is avoided and the resulting modular computations can be done in parallel. For this, we require that the algorithm under consideration is also applicable over finite fields and returns a deterministic result. The fundamental approach is then as follows: Compute the result modulo a number of primes. Then reconstruct the result over $\mathbb{Q}$ from the modular results.

Example 1. To compute

$$\frac{1}{2} + \frac{1}{3} = \frac{5}{6}$$

using modular methods, the first step is to apply Chinese remainder isomorphism:

© Springer International Publishing Switzerland 2016
G.-M. Greuel et al. (Eds.): ICMS 2016, LNCS 9725, pp. 93–101, 2016.
DOI: 10.1007/978-3-319-42432-3_12

$$\mathbb{Z}/5 \times \mathbb{Z}/7 \times \mathbb{Z}/101 \;\cong\; \mathbb{Z}/3535$$

$$\tfrac{1}{2} \longmapsto (\; \overline{3} \;,\quad \overline{4} \;,\quad \overline{51}\;)$$

$$+$$

$$\tfrac{1}{3} \longmapsto (\; \overline{2} \;,\quad \overline{5} \;,\quad \overline{34}\;)$$

$$\|$$

$$(\; \overline{0} \;,\quad \overline{2} \;,\quad \overline{85}\;) \longmapsto \overline{590}$$

The second step is to reconstruct a rational number from $\overline{590}$.

2 Rational Reconstruction

Theorem 1 *[8]. For every integer N, the N-**Farey map***

$$\left\{ \tfrac{a}{b} \in \mathbb{Q} \;\middle|\; \begin{array}{l} \gcd(a,b) = 1 \\ \gcd(b,N) = 1 \end{array} \;\; |a|\,,|b| \le \sqrt{(N-1)/2} \right\} \longrightarrow \mathbb{Z}/N$$

$$\tfrac{a}{b} \longmapsto \overline{a} \cdot \overline{b}^{\,-1}$$

is injective.

There are efficient algorithms for computing preimages of the Farey map, see, for example, [8, Sect. 5].

Example 2. We use the computer algebra system SINGULAR [6] to compute the preimage of the Farey map the setting of Example 1:

```
> ring r = 0, x, dp;
> farey(590,3535);
  5/6
```

The basic concept for modular computations is then as follows:

1. Compute the result over $\mathbb{Z}/p_i$ for distinct primes $p_1,\ldots,p_r$.
2. Use the Chinese remainder isomorphism

$$\mathbb{Z}/N \;\cong\; \mathbb{Z}/p_1 \times \ldots \times \mathbb{Z}/p_r$$

   to lift the modular results to $\mathbb{Z}/N$ where $N = p_1 \cdots p_r$.
3. Compute the preimage of the lift with respect to the N-Farey map.
4. Verify the correctness of the lift.

This will yield the correct result, provided N is large enough (that is, the $\mathbb{Q}$-result is contained in the domain of the N-Farey map), and provided none of the p_i is bad.

Definition 1. *A prime p is called **bad** (with respect to a fixed algorithm and input) if the result over $\mathbb{Q}$ does not reduce modulo p to the result over $\mathbb{Z}/p$.*

By convention, this includes the case where, modulo p, the input is not defined or the algorithm in consideration is not applicable.

3   Bad Primes

3.1   Bad Primes in Gröbner Basis Computations

Consider a set of variables $X = \{x_1, \ldots, x_n\}$ and a monomial ordering $>$ on the monomials in X. For a set of polynomials G, write $\mathrm{LM}(G)$ for its set of lead monomials. For $G \subset \mathbb{Z}[X]$ and p prime, write G_p for the image of G in $\mathbb{Z}/p[X]$.

Theorem 2 *[2]. Suppose $F = \{f_1, \ldots, f_r\} \subset \mathbb{Z}[X]$ with all f_i primitive and homogeneous. Let G be the reduced Gröbner basis of $\langle F \rangle \subset \mathbb{Q}[X]$, $G(p)$ the reduced Gröbner basis of $\langle F_p \rangle$, and $G_{\mathbb{Z}}$ a minimal strong Gröbner basis of $\langle F \rangle \subset \mathbb{Z}[X]$. Then p does not divide any lead coefficient in $G_{\mathbb{Z}} \Leftrightarrow \mathrm{LM}(G) = \mathrm{LM}(G(p)) \Leftrightarrow G_p = G(p)$.*

Example 3. Using SINGULAR, we determine the bad primes for a Gröbner basis computation of the Jacobian ideal of a projective plane curve. We compute a minimial strong Gröbner basis over $\mathbb{Z}$:

```
> option("redSB");
> ring R = integer,(x, y, z),lp;
> poly f = x7y5 + x2yz9 + xz11 + y3z9;
> ideal I = groebner(ideal(diff(f, x), diff(f, y), diff(f,z)));
> apply(list(I[1..size(I)]),leadcoef);
137811155278687303447773104646132606 8352191229011351724107460887644460 60
12 4 12 12 45349632 12 1473863040 12 22674816 12 3888 12 12 12 13608 12
108 54 6 2 27 3 1 4 2 2 1 216 1 2 3 1 540 12 108 27 3 1 9 3 1 1 1 1 7 1
5 1 1
```

The bad primes, that is, the primes p with $G_p \neq G(p)$, are then the prime factors

$$p = 2, 3, 5, 7, 11, 13, 257, 247072949, 328838088993550682027$$

of the lead coefficients. In contrast, the lead coefficients of the Gröbner basis over $\mathbb{Q}$ involve only the prime factors $2, 3, 5, 7, 13$, and hence not all bad primes. As shown by the following computation, 257 is indeed a bad prime:

```
> ring R0 = 0,(x, y, z),lp;
> size(lead(groebner(fetch(R,I))));
15
> ring R1 = 257,(x, y, z),lp;
> size(lead(groebner(fetch(R,I))));
14.
```

3.2   Classification of Bad Primes

Bad primes can be classified as follows, see [3, Sect. 3] for details:

- Type 1: The input modulo p is not valid (this poses no problem).
- Type 2: There is a failure in the course of the algorithm (for example, a matrix may not be invertible modulo p; this wastes computation time if it happens).

– Type 3: A computable invariant with known expected value (for example, a Hilbert polynomial) has a wrong value in a modular computation (to detect this we have to do expensive tests for each prime, although the set of bad primes usually is finite, and hence bad primes rarely occur).
– Type 4: A computable invariant with unknown expected value (for example, the lead ideal in a Gröbner basis computation) is wrong (this can be handled by a majority vote, however we have to compute the invariant for each modular result and store the modular results).
– Type 5: otherwise.

The Type 5 case in fact occurs, as is shown by the following example. For an ideal $I \subset \mathbb{Q}[X]$ and a prime p define $I_p = (I \cap \mathbb{Z}[X])_p$.

Example 4. Consider the algorithm $I \mapsto \sqrt{I + \mathrm{Jac}(I)}$ computing the radical of the Jacobian ideal for the curve

$$I = \left\langle x^6 + y^6 + 7x^5z + x^3y^2z - 31x^4z^2 - 224x^3z^3 + 244x^2z^4 + 1632xz^5 + 576z^6 \right\rangle.$$

Note that, with respect to the degree reverse lexicographic order, $\mathrm{LM}(I) = \left\langle x^6 \right\rangle = \mathrm{LM}(I_5)$, that is, 5 is not bad with respect to the input. The following computation in SINGULAR first determines the minimal associated primes of $U(0) = \sqrt{I + \mathrm{Jac}(I)}$ and $U(5) = \sqrt{I_5 + \mathrm{Jac}(I_5)}$.

```
> LIB "primdec.lib";
> ring R0 = 0, (x, y, z), dp;
> poly f = x6+y6+7x5z+x3y2z-31x4z2-224x3z3+244x2z4+1632xz5+576z6;
> ideal U0 = radical(ideal(f, diff(f, x), diff(f, y), diff(f, z)));
> minAssGTZ(U0);
[1]: _[1]=y        [2]: _[1]=y
     _[2]=x+6z          _[2]=x-4z
> ring R5 = 5, (x, y, z), dp;
> poly f =imap(R0,f);
> ideal U5 = radical(ideal(f, diff(f, x), diff(f, y), diff(f, z)));
> minAssGTZ(U5);
[1]: _[1]=y        [2]: _[1]=y
     _[2]=x-z           _[2]=x+z
> minassGTZ(imap(R0,U0));
[1]: _[1]=y
     _[2]=x+z
```

This shows that $U(0)_5 \neq U(5)$, but $\mathrm{LM}(U(0)) = \left\langle y, x^2 \right\rangle = \mathrm{LM}(U(5))$.

4   Error-Tolerant Reconstruction

Our goal is to reconstruct the $\mathbb{Q}$-result $\frac{a}{b}$ from the modular result $\bar{r} \in \mathbb{Z}/N$ in the presence of bad primes. Our basic strategy will be to find an element (x, y) with $\frac{x}{y} = \frac{a}{b}$ in the lattice

$$\Lambda = \langle (N,0), (r,1) \rangle \subset \mathbb{Z}^2.$$

Lemma 1 *[3, Lemma 4.2].* *All $(x,y) \in \Lambda$ with $x^2 + y^2 < N$ are collinear.*

Now suppose $N = N' \cdot M$ with $\gcd(N', M) = 1$. We assume that N' is the product of the good primes with correct result $\bar{s}$, and M is the product of the bad primes with wrong result $\bar{t}$.

Theorem 3 *[3, Lemma 4.3].* *If*

$$\bar{r} \mapsto (\bar{s}, \bar{t}) \quad \text{with respect to} \quad \mathbb{Z}/N \cong \mathbb{Z}/N' \times \mathbb{Z}/M$$

and

$$\frac{a}{b} \bmod N' = s$$

then $(aM, bM) \in \Lambda$. So if $(a^2 + b^2)M < N'$, then (by Lemma 1)

$$\frac{x}{y} = \frac{a}{b} \quad \text{for all } (x,y) \in \Lambda \text{ with } (x^2 + y^2) < N$$

and such vectors exist. Moreover, if $\gcd(a,b) = 1$ and (x,y) is a shortest vector $\neq 0$ in Λ, we also have $\gcd(x,y)|M$.

Hence, if $N' \gg M$, the Gauss-Lagrange-Algorithm for finding a shortest vector $(x,y) \in \Lambda$ gives $\frac{a}{b}$ independently of t, provided $x^2 + y^2 < N$. We use the programming language JULIA[1], to illustrate the resulting algorithm.

```
function ErrorTolerantReconstruction(r::Integer, N::Integer)
  a1 = [N, 0]
  a2 = [r, 1]
  while dot(a1, a1) > dot(a2, a2)
    q = dot(a1, a2)//dot(a2, a2)
    a1, a2 = a2, a1 - Integer(round(q))*a2
  end
  if dot(a1, a1) < N
    return a1[1]//a1[2]
  else
    return false
  end
end
```

The following table shows timings (in seconds), for r and N of bit-length 500, comparing the JULIA-function with implementations in the SINGULAR-kernel (optimized C/C++ code) and the current SINGULAR-interpreter:

SINGULAR-kernel	JULIA	SINGULAR-interpreter
0.001	0.005	0.055

Building on JULIA as a fast mid-level language, a backwards-compatible just-in-time compiled SINGULAR-interpreter is under development.

---

[1] See http://julialang.org/.

Example 5. In the setting of Example 1, we obtain $\frac{5}{6}$ from $\overline{590} \in \mathbb{Z}/3535$ by

```julia
julia> ErrorTolerantReconstruction(590, 3535)
5//6
```

which computes the sequence

$$(3535, 0) = 6 \cdot (590, 1) + (-5, -6),$$
$$(590, 1) = -48 \cdot (-5, -6) + (350, -287).$$

Example 6. Now we introduce an error in the modular results:

$$\mathbb{Z}/5 \times \mathbb{Z}/7 \times \mathbb{Z}/101 \cong \mathbb{Z}/3535$$
$$(\,\overline{1}\,,\quad \overline{2}\qquad \overline{85}\,) \;\mapsto\; \overline{2711}$$

Error tolerant reconstruction computes

$$(3535, 0) = 1 \cdot (2711, 1) + (824, -1),$$
$$(2711, 1) = 3 \cdot (824, -1) + (239, 4)$$
$$(824, -1) = 3 \cdot (239, 4) + (107, -13)$$
$$(239, 4) = 2 \cdot (107, -13) + (25, 30)$$
$$(107, -13) = 1 \cdot (25, 30) + (82, -43)$$

hence yields

$$\frac{25}{30} = \frac{5 \cdot 5}{5 \cdot 6} = \frac{5}{6}.$$

Note that

$$(5^2 + 6^2) \cdot 5 = 305 < 707 = 7 \cdot 101.$$

5 General Reconstruction Scheme for Commutative Algebra

For a given ideal $I \subset \mathbb{Q}[X]$, we want to compute some ideal (or module) $U(0)$ associated to I by a deterministic algorithm. We proceed along the following lines:

1. Over $\mathbb{Z}/p$ compute $U(p)$ from I_p for p in a suitable finite set $\mathcal{P}$ of primes.
2. Replace $\mathcal{P}$ by a subset according to a majority vote on $\mathrm{LM}(U(p))$ (see also [3, Remark 5.7]).
3. For $N = \prod_{p \in \mathcal{P}} p$ compute the coefficient-wise CRT–lift $U(N)$ to $\mathbb{Z}/N$, identifying generators by their lead monomials.
4. Lift $U(N)$ by error tolerant rational reconstruction to U.
5. Test $U_p = U(p)$ for some random extra prime p.
6. Verify $U = U(0)$.
7. If the lift, test or verification fails, then enlarge $\mathcal{P}$ and repeat.

Theorem 4 [3, Lemma 5.6]. *If the bad primes form a Zariski closed proper subset of* $\mathrm{Spec}\,\mathbb{Z}$, *then this strategy terminates with the correct result.*

6   Computing Adjoint Ideals

We discuss an application from algebraic geometry. The goal is to compute adjoint curves, that is, curves which pass with sufficiently high multiplicity through the singularities of a given curve, see Fig. 1. We consider an integral, non-degenerate projective curve $\Gamma \subset \mathbb{P}^r$ with normalization map $\pi : \overline{\Gamma} \to \Gamma$, and a saturated homogeneous ideal I with $I(\Gamma) \subsetneq I \subset k[x_0, ..., x_r]$. We write $\mathrm{Sing}(\Gamma)$ for the singular locus of Γ. Let H be the pullback of a hyperplane, and $\Delta(I)$ the pullback of $\mathrm{Proj}(S/I)$. Then the exact sequence

$$0 \to \widetilde{I}\mathcal{O}_\Gamma \to \pi_*(\widetilde{I}\mathcal{O}_{\overline{\Gamma}}) \to \mathcal{F} \to 0$$

induces, for $m \gg 0$, an exact sequence

$$0 \to I_m/I(\Gamma)_m \xrightarrow{\overline{\varrho_m}} H^0\left(\overline{\Gamma}, \mathcal{O}_{\overline{\Gamma}}(mH - \Delta(I))\right) \to H^0\left(\Gamma, \mathcal{F}\right) \to 0.$$

Definition 2. *The ideal I is an **adjoint ideal** of Γ if $\overline{\varrho_m}$ is surjective for $m \gg 0$.*

Since $h^0\left(\Gamma, \mathcal{F}\right) = \sum_{P \in \mathrm{Sing}(\Gamma)} \mathrm{length}(I_P \overline{\mathcal{O}}_{\Gamma,P}/I_P)$, we obtain:

Theorem 5 *[1]. With notation as above:*

$$I \text{ is an adjoint ideal of } \Gamma \iff I_P \overline{\mathcal{O}}_{\Gamma,P} = I_P \text{ for all } P \in \mathrm{Sing}(\Gamma).$$

The conductor $\mathcal{C}_{\mathcal{O}_{\Gamma,P}}$ of $\mathcal{O}_{\Gamma,P} \subset \overline{\mathcal{O}}_{\Gamma,P}$ is the largest ideal of $\mathcal{O}_{\Gamma,P}$ which is also an ideal in $\overline{\mathcal{O}}_{\Gamma,P}$.

Definition 3. *The **Gorenstein adjoint ideal** of Γ is the largest homogeneous ideal $\mathfrak{G} \subset K[x_0, \dots, x_r]$ with*

$$\mathfrak{G}_P = \mathcal{C}_{\mathcal{O}_{\Gamma,P}} \text{ for all } P \in \mathrm{Sing}(\Gamma).$$

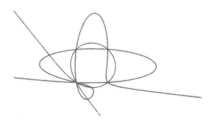

Fig. 1. Degree 3 adjoint curve of a rational curve of degree 5

The Gorenstein adjoint ideal has many applications in the geometry of curves.

Example 7. If Γ be an irreducible plane algebraic curve of degree n, then $\mathfrak{G}_{n-3}$ cuts out the canonical linear series.

Example 8. If Γ is a rational plane curve of degree n, then $\mathfrak{G}_{n-2}$ maps Γ to a rational normal curve of degree $n - 2$ in $\mathbb{P}^{n-2}$.

Example 9. The Gorenstein adjoint ideal can be used in the Brill-Noether-Algorithm to compute Riemann-Roch spaces for singular curves.

The Gorenstein adjoint ideal can be computed via a local-to-global strategy.

Definition 4. *The **local adjoint ideal** of Γ at $P \in Sing\,\Gamma$ is the largest homogeneous ideal $\mathfrak{G}(P) \subset k[x_0, \ldots, x_r]$ with $\mathfrak{G}(P)_P = \mathcal{C}_{\mathcal{O}_{\Gamma,P}}$.*

Lemma 2 *[5, Proposition 5.4].* *With notation as above,*

$$\mathfrak{G} = \bigcap_{P \in \text{Sing}\,\Gamma} \mathfrak{G}(P).$$

Definition 5. *Let A be the coordinate ring of an affine model $C = \text{Spec}\,A$ of Γ and let $P \in \text{Sing}(A)$. A ring $A \subset B \subset \overline{A} \subset \text{Quot}(A)$ is called a **minimal local contribution** to $\overline{A}$ at P if $B_P = \overline{A}_P$ and $B_Q = A_Q$ for all $P \neq Q \in C$.*

The minimal local contribution to $\overline{A}$ at P is unique and can be computed using Grauert-Remmert-type normalization algorithms, see [4,7]. It can be written as $B = \frac{U}{d}$ with an ideal $U \subset A$ and a common denominator $d \in A$.

Algorithm 6 *[5, Algorithm 4].* *With notation as above, $\mathfrak{G}(P) \subset k[x_0, \ldots, x_r]$ is the homogenization of the preimage of $(d : U)$ under $k[x_1, \ldots, x_r] \to k[x_1, \ldots, x_r]/I = A$.*

7   Modular Version of the Algorithm

Applying the general modular strategy gives an algorithm which is two-fold parallel (taking Lemma 2 into account). We use primes p such that the algorithm is applicable to the variety Γ_p defined by $I(\Gamma)_p$. Efficient verification can be realized through a semi-continuity argument, see [5, Theorem 8.14]. Table 1 gives timings (in seconds on a 2.2 GHz processor) for plane curves f_n of degree n with $\binom{n-1}{2}$ singularities of type A_1. Rows LA and IQ refer to global computations of the Gorenstein adjoint ideal via linear algebra [9] and ideal quotients, respectively. The row Maple-IB shows timings for the normalization of the curve via a computation of an integral basis in MAPLE [10]. The row locIQ gives timings for the local-to-global (Lemma 2), and modLocIQ for the modular local-to-global strategy. In square brackets, the number of primes in the modular strategy is shown, in round brackets the number of cores used simultaneously in a parallel computation. We also give timings for the modular probabilistic algorithm obtained by omitting the verification. Observe that, in the example, a local-to-global strategy does not give any benefit when computing over the rationals, since the singular locus does not decompose. However, by Chebotarev's density theorem, the singular locus is likely to decompose when passing to a finite field, as illustrated by the last two rows of the table.

Table 1. Timings

	Parallel	Probabilistic	f_5		f_6		f_7	
`Maple-IB`			5.1		47		318	
`LA`			98		4400		-	
`IQ`			1.3		54		3800	
`locIQ`	■		1.3	(1)	54	(1)	3800	(1)
`modLocIQ`			6.4	[33]	19	[53]	150	[75]
		■	6.2	[33]	18	[53]	104	[75]
	■		.36	(74)	1.6	(153)	51	(230)
	■	■	.21	(74)	0.48	(153)	5.2	(230)

References

1. Arbarello, E., Ciliberto, C.: Adjoint hypersurfaces to curves in $\mathbb{P}^r$ following Petri. In: Commutative Algebra. Lecture Notes in Pure and Applied Mathematics, vol. 84, pp. 1–21, Dekker, New York (1983)
2. Arnold, E.A.: Modular algorithms for computing Gröbner bases. J. Symb. Comput. **35**, 403–419 (2003)
3. Böhm, J., Decker, W., Fieker, C., Pfister, G.: The use of bad primes in rational reconstruction. Math. Comp. **84**, 3013–3027 (2015)
4. Böhm, J., Decker, W., Laplagne, S., Pfister, G., Steenpaß, A., Steidel, S.: Parallel algorithms for normalization. J. Symb. Comp. **51**, 99–114 (2013)
5. Böhm, J., Decker, W., Pfister, G., Laplagne, S.: Local to global algorithms for the Gorenstein adjoint ideal of a curve. Preprint (2015). arXiv:1505.05040
6. Decker, W., Greuel, G.-M., Pfister, G., Schönemann, H.: Singular 4-0-2 - A computer algebra system for polynomial computations (2015). http://www.singular.uni-kl.de
7. Greuel, G.-M., Laplagne, S., Seelisch, S.: Normalization of rings. J. Symb. Comp. **45**(9), 887–901 (2010)
8. Kornerup, P., Gregory, R.T.: Mapping integers and Hensel codes onto Farey fractions. BIT **23**, 9–20 (1983)
9. Mnuk, M.: An algebraic approach to computing adjoint curves. J. Symb. Comput. **23**(2–3), 229–240 (1997)
10. van Hoeij, M.: An algorithm for computing an integral basis in an algebraic function field. J. Symb. Comput. **18**(4), 353–363 (1994)

The Subdivision of Large Simplicial Cones in Normaliz

Winfried Bruns, Richard Sieg$^{(\boxtimes)}$, and Christof Söger

University of Osnabrück, Osnabrück, Germany
{wbruns,risieg,csoeger}@uos.de
http://www.home.uni-osnabrueck.de/wbruns/,
http://www.math.uni-osnabrueck.de/normaliz/

Abstract. Normaliz is an open-source software for the computation of lattice points in rational polyhedra, or, in a different language, the solutions of linear diophantine systems. The two main computational goals are (i) finding a system of generators of the set of lattice points and (ii) counting elements degree-wise in a generating function, the Hilbert Series. In the homogeneous case, in which the polyhedron is a cone, the set of generators is the Hilbert basis of the intersection of the cone and the lattice, an affine monoid.

We will present some improvements to the Normaliz algorithm by subdividing simplicial cones with huge volumes. In the first approach the subdivision points are found by integer programming techniques. For this purpose we interface to the integer programming solver SCIP to our software. In the second approach we try to find good subdivision points in an approximating overcone that is faster to compute.

Keywords: Hilbert basis · Hilbert series · Rational cone · Polyhedron

1 Introduction

Normaliz [3] is a software for the computation of lattice points in rational polyhedra. These are exactly the solutions of linear diophantine systems of inequalities, equations and congruences. It pursues two main computational goals: (i) finding a minimal generating system of the set of lattice points in a polyhedron; (ii) counting elements degree-wise in a generating function, the Hilbert series. In the homogeneous case, in which the polyhedron is a cone, the set of generators is the Hilbert basis of the intersection of the cone and the lattice, which is an affine monoid by Gordan's lemma. For the mathematical background we refer the reader to [2]. The Normaliz algorithms are described in [4,5]. The second paper contains extensive performance data.

Normaliz (present public version 3.1.1) is written in C++ (using Boost and GMP/MPIR), parallelized with OpenMP, and runs under Linux, MacOs and MS Windows. It is based on its C++ library libnormaliz which offers the full functionality of Normaliz. There are file based interfaces for Singular, Macaulay

© Springer International Publishing Switzerland 2016
G.-M. Greuel et al. (Eds.): ICMS 2016, LNCS 9725, pp. 102–109, 2016.
DOI: 10.1007/978-3-319-42432-3_13

2 and Sage, and C++ level interfaces for CoCoA, polymake, Regina and GAP. A C++ level interface to Sage should be available in the near future. There is also the GUI interface jNormaliz.

Normaliz has found applications in commutative algebra, toric geometry, combinatorics, integer programming, invariant theory, elimination theory, group theory, mathematical logic, algebraic topology and even theoretical physics.

2   Hilbert Basis and Hilbert Series

We will first describe the main functionality of Normaliz. For simplicity we restrict ourselves to homogeneous linear systems in the following, or, geometrically speaking, to the intersections of lattices $L \subset \mathbb{Z}^d$ and rational cones $C \subset \mathbb{R}^d$.

Definition 1. *A (rational) polyhedron P is the intersection of finitely many (rational) halfspaces. If it is bounded, then it is called a polytope. If all the halfspaces are linear, then P is a cone.*

The dimension of P is the dimension of the smallest affine subspace $\mathrm{aff}(P)$ containing P.

An affine monoid is a finitely generated submonoid of $\mathbb{Z}^d$ for some d.

By the theorem of Minkowski-Weyl, $C \subset \mathbb{R}^d$ is a (rational) cone if and only if there exist finitely many (rational) vectors $x_1, \ldots, x_n$ such that

$$C = \mathrm{cone}(x_1, \ldots, x_n) = \{a_1 x_1 + \cdots + a_n x_n : a_1, \ldots, a_n \in \mathbb{R}_+\}.$$

If $x_1, \ldots, x_n$ are linearly independent, we call C *simplicial*. For Normaliz, cones C and lattices L can either be specified by generators $x_1, \ldots, x_n \in \mathbb{Z}^d$ or by constraints, i.e., homogeneous systems of diophantine linear inequalities, equations and congruences. Normaliz also offers to define an affine monoid as the quotient of $\mathbb{Z}_+^n$ modulo the intersection with a sublattice of $\mathbb{Z}^n$.

Normaliz puts no restriction on the rational cone C. In the following we will however assume that C is pointed, i.e. $x, -x \in C \Rightarrow x = 0$. This is justified since computations in non-pointed cones are done via the projection to the quotient modulo the maximal linear subspace, which is pointed.

By Gordan's lemma the monoid $M = C \cap L$ is finitely generated. This affine monoid has a (unique) minimal generating system called the *Hilbert basis* $\mathrm{Hilb}(M)$, see Fig. 1 for an example. The computation of the Hilbert basis is the first main task of Normaliz.

One application is the computation of the *normalization* of an affine monoid M; this explains the name Normaliz. The normalization is the intersection of the cone generated by M with the sublattice $\mathrm{gp}(M)$ generated by M. One calls M *normal*, if it coincides with its normalization.

The second main task is to compute the Hilbert (or Ehrhart) series of a graded monoid. A *grading* of a monoid M is simply a homomorphism $\mathrm{deg} : M \to \mathbb{Z}^g$ where $\mathbb{Z}^g$ contains the degrees. The *Hilbert series* of M with respect to the grading is the formal Laurent series

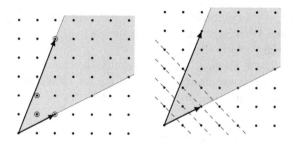

Fig. 1. A cone with the Hilbert basis (circled points) and grading.

$$H(t) = \sum_{u \in \mathbb{Z}^g} \#\{x \in M : \deg x = u\} t_1^{u_1} \cdots t_g^{u_g} = \sum_{x \in M} t^{\deg x},$$

provided all sets $\{x \in M : \deg x = u\}$ are finite. At the moment, Normaliz can only handle the case $g = 1$, and therefore we restrict ourselves to this case. We assume in the following that $\deg x > 0$ for all nonzero $x \in M$ and that there exists an $x \in \mathrm{gp}(M)$ such that $\deg x = 1$. (Normaliz always rescales the grading accordingly.)

Assume that M is a normal and affine monoid. By a theorem of Hilbert and Serre [2, Theorem 6.37], $H(t)$ in the $\mathbb{Z}$-graded case is the Laurent expansion of a rational function at the origin:

$$H(t) = \frac{R(t)}{(1 - t^e)^r}, \qquad R(t) \in \mathbb{Z}[t],$$

where r is the rank of M and e is the least common multiple of the degrees of the extreme integral generators of $\mathrm{cone}(M)$. As a rational function, $H(t)$ has negative degree.

A rational cone C and a grading together define the rational polytope $Q = C \cap A_1$ where $A_1 = \{x : \deg x = 1\}$. In this sense the Hilbert series is nothing but the Ehrhart series of Q.

3   The Primal Algorithm

The *primal* Normaliz algorithm is triangulation based. Normaliz contains a second, *dual* algorithm for the computation of Hilbert bases that implements ideas of Pottier [6]. The dual algorithm is treated in [4], and has not changed much in the last years and we do not discuss it in this article.

The primal algorithm starts from a pointed rational cone $C \subset \mathbb{R}^d$ given by a system of generators $x_1, \ldots, x_n$ and a sublattice $L \subset \mathbb{Z}^d$ that contains $x_1, \ldots, x_n$. Other types of input data are first transformed into this format. The algorithm is composed as follows:

1. Initial coordinate transformation to $E = L \cap (\mathbb{R}x_1 + \cdots + \mathbb{R}x_n)$;
2. Fourier-Motzkin elimination computing the support hyperplanes of C;

3. computation of a triangulation, i.e. a face-to-face decomposition into simplicial cones;
4. evaluation of the simplicial cones in the triangulation;
5. collection of the local data;
6. reverse coordinate transformation to $\mathbb{Z}^d$.

The algorithm does not strictly follow this chronological order, but interleaves steps 2–5 in an intricate way to ensure low memory usage and efficient parallelization.

3.1   Simplicial Cones

We will now focus on step 4 of the primal algorithm, the evaluation of simplicial cones. Let $x_1, \ldots, x_d \in \mathbb{Z}^d$ be linearly independent and $S = \text{cone}(x_1, \ldots, x_d)$. Then the integer points in the *fundamental domain* of S

$$E = \{q_1 x_1 + \cdots + q_d x_d : 0 \le q_i < 1\} \cap \mathbb{Z}^d$$

together with $x_1, \ldots, x_d$ generate the monoid $S \cap \mathbb{Z}^d$ (Fig. 2).

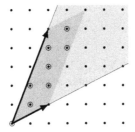

Fig. 2. A cone with a fundamental domain

Every residue class in the quotient $\mathbb{Z}^d / U$, where $U = \mathbb{Z}x_1 + \cdots + \mathbb{Z}x_d$, has exactly one representative in E. Representatives of residue classes can be quickly computed via the elementary divisor algorithm and from an arbitrary representative we obtain the one in E by division with remainder. The integer points of the fundamental domain are candidates for the Hilbert basis of the cone. After their computation they are shrunk to the Hilbert basis by successively discarding elements x which are *reducible*, i.e. there exists an $y \in E, y \ne x$ such that $x - y \in C$. Also the computation of the Hilbert series uses the set E and a *Stanley decomposition* based on it; see [5].

The number of elements in E is given by the (lattice normalized) volume of the simplex:

$$|E| = \text{vol}(S) = \det(x_1, \ldots, x_d).$$

Therefore the determinant of the generators of the simplicial cone has an enormous impact on the runtime of the Normaliz algorithm. The algorithms presented in this paper try to decompose a simplex with big volume into simplices such that the sum of their volumes is considerably smaller. For this purpose we compute integer points from the cone and use them for a new triangulation.

Theoretically the best choice for these points are the vertices of the *bottom* $B(S)$ of the simplex which is defined as the union of the bounded faces of the polyhedron $\mathrm{conv}((S\cap\mathbb{Z}^d)\setminus\{0\})$. In practice, the computation of the whole bottom would equalize the benefit from the small volume or even make it worse.

Therefore, we determine only some points from the bottom. Normaliz employs two methods for this purpose:

(1) computation of subdivision points by integer programming methods,
(2) computation of candidate subdivision points by approximation of the given simplicial cone by an overcone that is generated by vectors of "low denominator".

4   Methods from Integer Programming

For each simplex $S = \mathrm{cone}(x_1, \ldots, x_d)$ in the triangulation with large enough volume we try to compute a point x that minimizes the *sum of determinants*:

$$\sum_{i=1}^{d} \det(x_1, \ldots, x_{i-1}, x, x_{i+1}, \ldots, x_d),$$

which can also be expressed as $N^T x$, where N is a normal vector on the affine hyperplane spanned by $x_1, \ldots, x_d$. Such a point can be found by solving the following integer program:

$$\min\{N^T x : x \in S \cap \mathbb{Z}^d, x \neq 0, N^T x < N^T x_1\}. \tag{$\star$}$$

If the problem has a solution $\hat{x}$, we form a *stellar subdivision* of the simplex with respect to $\hat{x}$: For every support hyperplane H_i (not containing x_i) which does not contain $\hat{x}$ we form the simplex

$$T_i = \mathrm{cone}(x_1, \ldots, x_{i-1}, \hat{x}, x_{i+1}, \ldots, x_d).$$

If the volume of T_i is larger than a particular bound, we repeat this process and continue until all simplices have a smaller volume than this bound or the corresponding integer problems have no solutions. Figure 3 illustrates the algorithm.

After computing a set of integer points $\mathcal{B}$, we triangulate the bottom of $\mathrm{conv}(\mathcal{B} \cup \{x_1, \ldots, x_d\})$ and continue by evaluating this triangulation with the usual Normaliz algorithm.

4.1   Implementation and Results

We use the mixed integer programming solver SCIP [1] via its C++ interface. The algorithm runs in parallel with one SCIP environment for every thread using

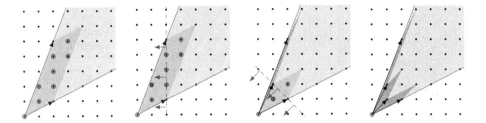

Fig. 3. The integer programming algorithm for a cone

OpenMP. Moreover each SCIP instance has its own time limit $(\log(\mathrm{vol}(S))^2 \sec)$ and feasibility bounds.

The condition that $x \neq 0$ could be implemented by the inequality $N^T x \geq 1$. However this approach is prone to large numbers in N. Therefore we first check, whether all generators are positive in one entry i and thus require $x_i \geq 1$. If this is not the case we make a bound disjunction of the form $(x_i \leq -1 \vee x_i \geq 1)$.

Table 1 presents example data computed on a SUN xFire 4450 with four Intel Xeon X7460 processors, using 20 threads and solving integer programs only for simplices with a volume larger than 10^6.

Table 1. Runtime improvements using integer programming methods

	Hickerson-16	Hickerson-18	Knapsack_11_60
Simplex volume	9.83×10^7	4.17×10^{14}	2.8×10^{14}
Volume under bottom	8.10×10^5	3.86×10^7	2.02×10^7
Volume used	3.93×10^6	5.47×10^7	2.39×10^7
Integer programs solved	4	582016	11621
Improvement factor	25	7.62×10^6	1.17×10^7
Runtime without subdivision	2 s	>12d	>8d
Runtime with subdivision	0.5 s	46 s	5.1 s

The bound on the volume to stop the calculation of a single simplex has a significant effect on the runtime of the algorithm. A smaller bound means that more integer programs have to be solved by SCIP, whereas a large bound prevents a major improvement of the respective volume. Running several experiments, it turns out that 10^6 is a good value in between these two extreme cases. Figure 4 shows a runtime graph illustrating the effect of different choices for this bound. The measured time is a single thread computation of hickerson-18.

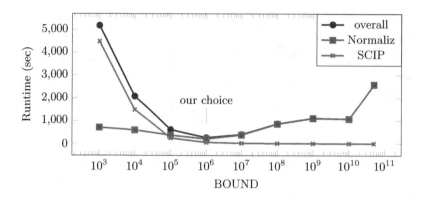

Fig. 4. Runtime graph showing different choices for the bound

5 Approximation

SCIP cannot be employed in all environments. Especially if Normaliz is bundled with another software package it may be undesirable or even impossible to force the link to SCIP.

Our second approach is completely implemented within Normaliz. It first approximates the simplicial cone S by a (not necessarily simplicial) overcone C for which the sets E in a triangulation of C are significantly faster to compute. Then these points are used to decompose the original simplex as before. It is clear that the efficiency depends crucially on the intersection of the sets E with S.

For this purpose we look at the polytope given by the cross section of the simplex at height one, where the height function comes from the normal vector N on the affine hyperplane spanned by the generators. For every vertex of this polytope we triangulate the lattice cube around it using the braid hyperplane arrangement $\{x_i = x_j\}$. We continue by detecting the minimal face containing the vertex and collect its vertices, which are at most d. The approximating cone C is then generated by all vertices found in that way. Figure 5 illustrates the choice of the approximation for a 3-dimensional cone (with a 2-dimensional cross section).

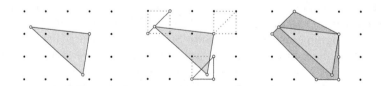

Fig. 5. Approximating cone

As in the usual Normaliz algorithm we create a candidate list for the exterior cone, but keep only those points which lie inside the original simplex S.

The remaining candidates are then reduced as before, which results in a list $\mathcal{B}$ which is used for a recursive decomposition of the simplex as in Sect. 4. Figure 6 illustrates this process for the previous example.

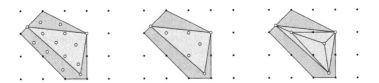

Fig. 6. Decomposition of a simplex after approximation

It might happen for both algorithms that no decomposition point can be found, although the volume of the simplex is still quite large ($>10^9$) and subdivision points exist. In this case, the approximation method is applied again with a higher level of approximation.

Table 2 contains performance data for the examples in Sect. 4.

Table 2. Runtime improvements using the approximation method

	Hickerson-16	Hickerson-18	Knapsack_11_60
Volume used	3.93×10^6	8.42×10^7	9.36×10^7
Improvement factor	25	4.95×10^6	2.99×10^4
Runtime with subdivision	0.4 s	50 s	2 m 30 s

At present we are working on improvements of the approximation method.

References

1. Achterberg, T.: SCIP: solving constraint integer programs. Math. Program. Comput. **1**, 1–41 (2009). http://mpc.zib.de/index.php/MPC/article/view/4
2. Bruns, W., Gubeladze, J.: Polytopes, Rings and K-Theory. Springer, New York (2009)
3. Bruns, W., Ichim, B., Römer, T., Sieg, R., Söger, C.: Normaliz. Algorithms for rational cones and affine monoids. http://www.math.uos.de/normaliz
4. Bruns, W., Ichim, B.: Normaliz: algorithms for affine monoids and rational cones. J. Algebra **324**, 1098–1113 (2010)
5. Bruns, W., Ichim, B., Sger, C.: The power of pyramid decompositions in Normaliz. J. Symb. Comp. **74**, 513–536 (2016)
6. Pottier, L.: The Euclide algorithm in dimension n. Research report, ISSAC 1996. ACM Press (1996)

Extending Singular with New Types and Algorithms

Hans Schönemann[(⊠)]

Department of Mathematics, University of Kaiserslautern, Kaiserslautern, Germany
hannes@mathematik.uni-kl.de
http://www.singular.uni-kl.de/

Abstract. Singular is a comprehensive and steadily growing computer algebra system, with particular emphasis on applications in algebraic geometry, commutative algebra, and singularity theory. Singular can easily be extended by other tools provided as C or C++ library: gfan and polymake will be discussed as an example.

1 Introduction

Singular [DGPS] is a comprehensive and steadily growing computer algebra system, with particular emphasis on applications in algebraic geometry, commutative algebra, and singularity theory.

As for most other computer algebra systems, SINGULAR is organized in two levels. At the lower level, the SINGULAR kernel provides the core algorithms and data types such as the Gröbner basis engine and multivariate polynomial factorization respectively polynomials, ideals, modules, free resolutions. It is written in C/C++, which makes it fast but difficult to extend. At the higher level, more advanced mathematical concepts are realized as libraries, written in the SINGULAR user language. A general concept for more abstract data type which could accompany the abstraction in functions is also needed.

Infrastructure for user-defined data-types (at the level of the SINGULAR language) and easy integration of data-types of external systems (at the kernel level) in SINGULAR has been created: `blackbox` an the C/C++-level, `newstruct` (which is based on `blackbox`) at the level of the SINGULAR language.

2 Extending the Functionality of SINGULAR

For a project dealing with computations with P-divisors (polyhedral formal sums of divisors (P-divisors) where the coefficients are assumed to be polyhedra with fixed tail cone) the need to use the functionality of SINGULAR and POLYMAKE at the same time arose. This is only one example for the need to combine these functionality of special purpose computer algebra systems. In this talk I describe the general mechanism to include C/C++ libraries into SINGULAR in order to provide new data types and new functionality.

© Springer International Publishing Switzerland 2016
G.-M. Greuel et al. (Eds.): ICMS 2016, LNCS 9725, pp. 110–113, 2016.
DOI: 10.1007/978-3-319-42432-3_14

As an example (solving the problem given above) the integration of GFANLIB of Anders Jensen is described, and, based on that the integration of (parts of) POLYMAKE for more operations resp. different algorithms with cones, fans and polytopes.

Another project which required the functionality of SINGULAR for Gröbner base computations and the type `fan` (provided by GFANLIB) is the computation of GIT-fans, torus orbits and GKZ-fans.

3 GFANLIB

This binary library is an interface to the C++ library gfanlib [J] by Anders Jensen, which is distributed together with Singular.

It contains a basic implementation of convex geometry, featuring polyhedral cones, polyhedral fans and polytopes as well as elemental functions on them. For a full list of functions available on the objects, please check the respective documentation on the type.

Moreover, it contains features for computing the Groebner fan, the Groebner complex and the tropical varieties.

4 Integrating GFANLIB

Adding new data types and/or functionality to SINGULAR required many changes of interpreter and at many different places. The main idea behind the introduction of the type `blackbox` and its associated operations was the automating of this process.

Data organized by `blackbox` are represented by a pair: a `void*` pointer and a type identifier. Registering a new type works by filling `struct` with function pointers for the relevant function for the new type and passing that and the name of the new type to `setBlackboxStuff`. For some functions useful defaults are provided:
Required functions:

- create a default object
- destroy an object
- convert an object to a string
- copy an object.

Optional functions:

- print an object (default: print the string representation)
- assign objects of other types to an object (default: raise an error)
- operations where this object is the first operand (default: raise an error)
- serialization (for distributed computations/storing data) (default: raise an error).

GFANLIB contains a basic implementation of convex geometry, featuring polyhedral cones, polyhedral fans and polytopes as well as elemental functions on them. It provides a part of the functionality of GFAN and is based on the same ideas but share only parts of its code: it is designed as C++-library while GFAN is a (set of) stand alone programs.

The interface to GFANLIB is organized as a dynamic module, i.e. an optional loadable part of SINGULAR. Its initialization routine creates via the interface of `blackbox` the types `cone`, `fan` and `polytope` as well as a large list on functions on them. The data are simply pointer to the corresponding classes of GFANLIB. The integration with the data types of SINGULAR is provided by conversion routines to resp. from matrices of integers.

5   Integrating POLYMAKE

POLYMAKE is a computer algebra system for the combinatorics and the geometry of convex polytopes and polyhedra. It is also deals with simplicial complexes, matroids, polyhedral fans, graphs, tropical objects, and other objects.

The interface from SINGULAR currently uses only parts of this functionality: the functions dealing cones,fans and polytopes. Since GFAN 0.3 the main data types (`cone`, `fan`, `polytope`) are compatible with POLYMAKE: the components of the corresponding C++ classes correlate. Therefore, for the definitions of these types in SINGULAR and the conversion routines from the interface to GFANLIB can be used to provide simple conversions.

6   Technical Problems of the Integration

6.1   Data Conversion

One problem of the integration was the use of different data types. As integers (respectively matrices of integers) are the only data types which GFANLIB and SINGULAR share, this was relatively easy. GFANLIB and POLYMAKE share the same general representation for `cone`, `fan` and `polytope`: conversion routines were straight forward.

6.2   Memory Management

SINGULAR uses OMALLOC as its memory manager which is optimized for many small memory blocks of only a few sizes. This memory manager can coexist with others: allocation and deallocation of memory blocks have to use the same manager. This can be a problem at places where the call to them is not explicit, for example with a garbage collector or the automatic call to the destructor of a C++ object.

6.3   Common Libraries

GFANLIB and POLYMAKE both use CDDLIB which contains an initialization part. GFANLIB, POLYMAKE and SINGULAR uses GMP for operations on large integers which allows an initialization: setting the memory management routines. These setting must not be changed and must be done before any use of these libraries: global C++ objects are initialized in an undefined order, but before the start of `main`.

6.4   Alternative Solution

An alternative solution to combine the functionality of SINGULAR and POLYMAKE is the integration of SINGULAR into POLYMAKE as C++ library `libSingular` which is also possible.

References

[DGPS] Decker, W., Greuel, G.-M., Pfister, G., Schönemann, H.: SINGULAR 4-0-3 – a computer algebra system for polynomial computations (2016). http://www.singular.uni-kl.de

[GJ] Gawrilow, E., Joswig, M.: POLYMAKE: a framework for analyzing convex polytopes. In: Polytopes Combinatorics and Computation, Oberwolfach, 1997. DMV Seminar, vol. 29, pp. 43–73. Birkhäuser, Basel. MR1785292 (2001f:52033) (2000). https://polymake.org/

[J] Jensen, A.: GFAN (2011). http://home.math.au.dk/jensen/software/gfan/gfan.html

Algebraic Geometry in Applications

3D Printing Dimensional Calibration Shape: Clebsch Cubic

Janko Böhm[1]([✉]), Magdaleen S. Marais[2], and André F. van der Merwe[3]

[1] Department of Mathematics, University of Kaiserslautern, Kaiserslautern, Germany
boehm@mathematik.uni-kl.de
[2] Department of Mathematics and Applied Mathematics,
University of Pretoria, Pretoria, South Africa
magdaleen.marais@up.ac.za
[3] Department of Industrial Engineering,
Stellenbosch University, Stellenbosch, South Africa
andrevdm@sun.ac.za

Abstract. 3D printing and other layer manufacturing processes are challenged by dimensional accuracy. Various techniques are used to validate and calibrate dimensional accuracy through the complete building envelope. The validation process involves the growing and measuring of a shape with known parameters. The measured result is compared with the intended digital model. Processes with the risk of deformation after time or post-processing may find this technique beneficial. We propose to use objects from algebraic geometry as test shapes. A cubic surface is given as the zero set of a degree 3 polynomial in 3 variables. A class of cubics in real 3D space contains exactly 27 real lines. These lines can be used for dimensional calibration. Due to the thin shape geometry the material required to produce an algebraic surface is minimal. We provide a library for the computer algebra system Singular which, from 6 given points in the plane, constructs a cubic and the lines on it.

Keywords: Clebsch cubic · 3D printing · Applied algebraic geometry

1 Introduction

This paper is the first in a series to investigate whether cubic surface shapes, and specifically the Clebsch cubic, can be used in 3D printing build volume accuracy. A surface shape derived from a cubic offers simplicity to the dimensional comparison process, in that it contains finitely many lines and many other features that can be analytically determined and easily measured using non-digital equipment. For example, the surface contains so-called Eckardt points, in each of which three of the lines intersect, and also other intersection points of pairs of lines. In this initial paper the phases of development are proposed and the mathematical background for calculating with cubic surfaces is described. Various build volumes, growing techniques and materials may require slight adjustments due to its unique characteristics. However, the basic shape and mathematical approach

© Springer International Publishing Switzerland 2016
G.-M. Greuel et al. (Eds.): ICMS 2016, LNCS 9725, pp. 117–126, 2016.
DOI: 10.1007/978-3-319-42432-3_15

remains the same for all variants. The ultimate aim is to have a standard cubic shape which can be grown on any platform in any material, and in any build volume size. The research phases proposed are an initial analytical mathematical model, then an engine which converts from the analytical model to a point cloud, then a digital domain simulated growth, followed by an actual hardware printing phase, and lastly a reverse engineering phase. The initial mathematical model is developed from ground rules to provide the fundamental information for parallel development. The input to the mathematical model, based on the mathematical formulations found by Clebsch and others, is the extent of the build volume. The open source computer algebra system SINGULAR is used for this conversion. The output from the mathematical model is a three-dimensional shape in analytical mathematical formulation, the formulas for 27 lines, the coordinates of the points where the lines cross, and the angles between the lines. After printing, the line straightness is one indicator of the dimensional accuracy. Another indicator is given by the angles between lines and the distances of the cross points. This model is developed in the second part of this paper. The engine which converts the analytical mathematical formulation to a printable point cloud would typically be programmed in Matlab and later in C++. The inputs are the three-dimensional mathematical shape of the Clebsch cubic, and the formulas of the 27 lines that we want to use as part of the dimensional accuracy measurement. Note that the lines will have to be highlighted in some way for the reverse engineering process to pick it up. Several ways could be used to highlight the lines: generating cylinders around the lines with diameter larger than the thickness of the cubic's surface shape, thinning along the lines, or perforating along the lines, are examples. The output of this engine would be a point cloud in an STL format or similar. In phase three of this project we will compare the point cloud with the initial analytical line formulas. This comparison can be done on a CAD platform, but would typically be a manual process. Several alternatives of the previous phases will be evaluated for accuracy. All work up to this point is in the digital domain. This phase is to ensure accuracy and robustness of self-developed tools by comparison with trusted commercially available CAD platforms. The output of this phase is a report which defines constraints and extents within which these techniques are deemed accurate in the digital domain. Phase four will involve the growing of hard copies in various materials on various platforms. This phase will report on any manufacturing issues on any of the platforms using the extent of materials chosen. Phase five will reverse engineer hardware shapes to compare with the initial intended analytical shape. In this phase it will be determined to what extent the use of the 27 lines and their angles are an indication of dimensional accuracy of the process. This phase will seek to propose an economical method of measuring dimensional accuracy of the complete building envelope. This paper starts with the mathematical setup on which the description of the cubic and the lines are based. Then a reference to the mathematical origins of the cubic surfaces is made, followed by the derivation of the surface equation and the lines. Finally an example output from the computer algebra library is given.

2   Algebraic Varieties

We first set the mathematical framework used to describe the cubic and the lines on it. Let K be either the real numbers $\mathbb{R}$ or the complex numbers $\mathbb{C}$. The set of lines through the origin in K^{n+1} is called **projective space** and is denoted $\mathbb{P}_K^n$. We will write $(x_0 : \ldots : x_n)$ for the line with direction $(x_0, \ldots, x_n) \neq 0$. There is an inclusion of usual n-space to projective space

$$K^n \longrightarrow \mathbb{P}_K^n, \quad (x_1, \ldots, x_n) \longmapsto (1 : x_1 : \ldots : x_n).$$

This map is referred to as an **affine chart**. The complement of the image is called the plane at infinity (the horizon in a perspective drawing). An **algebraic variety** $V(f_1, \ldots, f_r) \subset \mathbb{P}_K^n$ is the common zero set of homogeneous polynomials $f_i \in K[x_0, \ldots, x_n]$.

Algebraic varieties are studied in algebraic geometry, which forms a central branch of classical mathematics. It has important applications, e.g., in cryptography, robotics, and computational biology. Algebraic varieties have the advantage over zero sets of non-polynomial equations that they can easily be handled by the means of computer algebra. For computing with polynomials we make use of the open-source computer algebra system SINGULAR [8]. Using projective space in the development of the theory, avoids the problem that some features of an algebraic variety (e.g. a line on it) may be contained in the plane at infinity. For an introduction to algebraic geometry, computer algebra, and its applications see, e.g. [6].

3   Historic Overview and Derivation of the Fundamental Properties of Cubic Hypersurfaces

Starting in the second half of the 19th century, Clebsch, Klein, Salmon, Coble and many other mathematicians investigated cubic surfaces in $\mathbb{P}_\mathbb{C}^3$. These surfaces are given by a single degree three polynomial. In 1849, Arthur Cayley [2] and George Salmon [11] found:

Theorem 1. *Every smooth cubic surface in $\mathbb{P}_\mathbb{C}^3$ contains exactly 27 lines.*

Here smooth means, that C has in every point a well-defined tangent plane. In algebraic geometry there is a process, called blowup, which replaces in a variety a given point by a line and is a $1 : 1$ map everywhere else. In 1871 Alfred Clebsch [4] proved (see also [3]):

Theorem 2. *Every smooth cubic surface in $\mathbb{P}_\mathbb{C}^3$ is the blowup of $\mathbb{P}_\mathbb{C}^2$ in 6 points.*

In the following, let $P_1, \ldots, P_6 \in \mathbb{P}_K^2$ be points in general position, that is, no three are on a line and not all of them on a conic.

Remark 1. The homogeneous linear polynomial

$$l_{i,j}(t) := \det(P_i, P_j, t) := \det \begin{pmatrix} P_{i,0} & P_{j,0} & t_0 \\ P_{i,1} & P_{j,1} & t_1 \\ P_{i,2} & P_{j,2} & t_2 \end{pmatrix} \in K[t_0, t_1, t_2]$$

defines in $\mathbb{P}^2_K$ the line through P_i and P_j.

Proposition 1 [5]. *The blowup* $C = C_{(P_1,\dots,P_6)}$ *of* $\mathbb{P}^2_K$ *in the points* P_i *is the smallest algebraic variety (with respect to inclusion) containing the image of*

$$\varphi_{(P_1,\dots,P_6)} : \mathbb{P}^2_K \setminus \{P_1, \dots, P_6\} \longrightarrow \mathbb{P}^5_K$$
$$(t_0 : t_1 : t_2) \longmapsto (\varphi_0(t) : \dots : \varphi_5(t))$$

(defined on $\mathbb{P}^2_K$ *except at the points* $P_1, \dots, P_6$*), where*

$$\varphi_0 = l_{2,5}l_{1,3}l_{4,6} + l_{5,1}l_{4,2}l_{3,6} + l_{1,4}l_{3,5}l_{2,6} + l_{4,3}l_{2,1}l_{5,6} + l_{3,2}l_{5,4}l_{1,6}$$
$$\varphi_1 = l_{5,3}l_{1,2}l_{4,6} + l_{1,4}l_{2,3}l_{5,6} + l_{2,5}l_{3,4}l_{1,6} + l_{3,1}l_{4,5}l_{2,6} + l_{4,2}l_{5,1}l_{3,6}$$
$$\varphi_2 = l_{5,3}l_{4,1}l_{2,6} + l_{3,4}l_{2,5}l_{1,6} + l_{4,2}l_{1,3}l_{5,6} + l_{2,1}l_{5,4}l_{3,6} + l_{1,5}l_{3,2}l_{4,6}$$
$$\varphi_3 = l_{4,5}l_{3,1}l_{2,6} + l_{5,3}l_{2,4}l_{1,6} + l_{4,1}l_{2,5}l_{3,6} + l_{3,2}l_{1,5}l_{4,6} + l_{2,1}l_{4,3}l_{5,6}$$
$$\varphi_4 = l_{3,1}l_{2,4}l_{5,6} + l_{1,2}l_{5,3}l_{4,6} + l_{2,5}l_{4,1}l_{3,6} + l_{5,4}l_{3,2}l_{1,6} + l_{4,3}l_{1,5}l_{2,6}$$
$$\varphi_5 = l_{4,2}l_{3,5}l_{1,6} + l_{2,3}l_{1,4}l_{5,6} + l_{3,1}l_{5,2}l_{4,6} + l_{1,5}l_{4,3}l_{2,6} + l_{5,4}l_{2,1}l_{3,6}.$$

Remark 2. The **Clebsch cubic**, given in [4, Ch. 16], is obtained by applying this construction to the points in general position

$$P_1 = (0 : 1 : -g) \qquad P_3 = (1 : g : 0) \qquad P_5 = (0 : 1 : g)$$
$$P_2 = (g : 0 : 1) \qquad P_4 = (1 : -g : 0) \qquad P_6 = (-g : 0 : 1),$$

where $g = \frac{1+\sqrt{5}}{2}$ is the golden ratio. These points correspond to the diagonals in an icosahedron. The Clebsch cubic with $K = \mathbb{R}$ contains 27 real lines.

Remark 3. The number

$$|i, j; k, l; m, n| = \det \begin{pmatrix} \det(P_i, P_j, P_m) & \det(P_i, P_j, P_n) \\ \det(P_k, P_l, P_m) & \det(P_k, P_l, P_n) \end{pmatrix},$$

vanishes if the lines defined by $l_{i,j}(t)$, $l_{k,l}(t)$ and $l_{m,n}(t)$ in $\mathbb{P}^2_K$ meet in one point.

Theorem 3 [5,7]. *Consider the skew-symmetric matrix* $A \in K^{6\times 6}$ *with*

$$(A_{i,j}) = \begin{pmatrix} 0 & |1,5;2,4;3,6| & |1,4;3,5;2,6| & |1,2;4,3;5,6| & |2,3;4,5;1,6| & |1,3;5,2;4,6| \\ & 0 & |2,5;3,4;1,6| & |1,3;5,4;2,6| & |1,2;3,5;4,6| & |1,4;2,3;5,6| \\ & & 0 & |1,5;3,2;4,6| & |1,3;2,4;5,6| & |1,2;4,5;3,6| \\ & - & & 0 & |1,4;5,2;3,6| & |2,4;3,5;1,6| \\ & & & & 0 & |1,5;3,4;2,6| \\ & & & & & 0 \end{pmatrix}$$

where the entries are defined as in Remark 3, and write for the sum of the entries of the i-*th row*

$$a_i = \sum_{j=1}^{6} A_{i,j}.$$

Then C is given by the equations

$$x_0^3 + \ldots + x_5^3 = 0$$
$$x_0 + \ldots + x_5 = 0$$
$$a_0 \cdot x_0 + \ldots + a_5 \cdot x_5 = 0.$$

Remark 4. Using the ordering of the P_i from Remark 2, we obtain for the Clebsch cubic surface $a_0 = a_1 = a_2 = a_3 = a_4 = 1$ and $a_5 = -5$.

Remark 5. For a subset $S \subset \mathbb{P}_{\mathbb{C}}^n$ we define $I(S)$ as the ideal of all $f \in \mathbb{C}[x_0, \ldots, x_n]$ with $f(x) = 0$ for all $x \in S$. So $V(I(S))$ is the smallest algebraic variety (with respect to inclusion) containing S. The ideal generated by the φ_i is

$$\langle \varphi_0, \ldots, \varphi_5 \rangle = I(P_1) \cap \ldots \cap I(P_6).$$

With the ring homomorphism

$$\psi_{(P_1, \ldots, P_6)} : \mathbb{C}[x_0, \ldots, x_5] \longrightarrow \mathbb{C}[t_0, t_1, t_2]$$
$$x_i \longmapsto \varphi_i$$

we have

$$I(C) = \ker \psi_{(P_1, \ldots, P_6)} = \left\langle x_0^3 + \ldots + x_5^3,\ x_0 + \ldots + x_5,\ a_0 x_0 + \ldots + a_5 x_5 \right\rangle.$$

Remark 6. Eliminating two variables by the two linear equations, C can be considered as a subset of $\mathbb{P}_K^3$.

Note that a plane intersects C in an irreducible plane cubic, a union of a conic and a line, or in three lines.

Definition 1. *A **tritangent plane** H to C is a plane, such that $H \cap C$ consists out of three lines.*

Remark 7. A tritangent plane H to C is called **generic** if the three lines pairwise intersect in three distinct points. Then H is tangent to C in each of the three points.

If H is not generic, then the three lines on C intersect in a single point. This point is called an **Eckardt point** of C.

Since in an Eckardt point the three lines are tangent to C, they are coplanar, hence, lie on a tritangent plane. So, the Eckardt points are in one-to-one correspondence to the non-generic tritangent planes.

Theorem 4 [2,5]. *There are 45 tritangent planes to C:*

1. Of these, 15 are given by the equations

$$x_i + x_j = 0$$

for $0 \leq i < j \leq 5$.

2. Write M for the set of 2-element subsets of $\{1, \ldots, 6\}$, and $S(M)$ for the set of permutations of M. The remaining 30 tritangent planes are then

$$(m_{i,j} - d_2) \cdot (x_i + x_j) - (m_{k,l} + d_2) \cdot (x_k + x_l) = 0$$

where

$$\in S(M)$$

is a 3-cycle of pairwise disjoint elements of M,

$$d_2 = \det \begin{pmatrix} \det(P_3, P_4, P_1) \cdot \det(P_5, P_6, P_1) & \det(P_5, P_3, P_1) \cdot \det(P_4, P_6, P_1) \\ \det(P_3, P_4, P_2) \cdot \det(P_5, P_6, P_2) & \det(P_5, P_3, P_2) \cdot \det(P_4, P_6, P_2) \end{pmatrix}$$

and

$$m_{i,j} = \sum_{s<t} a_s a_t + 2(a_i^2 + a_j^2 + a_i a_j),$$

where a_i is as defined in Theorem 3.

Remark 8. Possible numbers for Eckardt points are $1, 2, 3, 4, 6, 9, 10, 18$. The Clebsch cubic is the unique cubic with 10 Eckardt points. The **Fermat cubic** $V(x_0^3 + \ldots + x_3^3)$ is the unique cubic with the maximum possible number of 18 Eckardt points, however, only 3 of the lines on the Fermat cubic are defined over $\mathbb{R}$.

Remark 9. Every line on C lies on 5 tritangent planes. Hence, any line on C is the intersection of the planes $x_0 + \ldots + x_5 = 0$, $a_0 x_0 + \ldots + a_5 x_5 = 0$ and two tritangent planes (see Remark 6).

Remark 10. After permuting the coordinates we may assume that $a_5 \neq 0$. Then by eliminating x_4 and x_5 via the two linear equations of C, we obtain $C' = V(F) \subset \mathbb{P}_K^3$ with a homogeneous cubic polynomial $F \in K[x_0, x_1, x_2, x_3]$.

Example 1. The Clebsch Cubic is then given by

$$F = x_0^3 + x_1^3 + x_2^3 + x_3^3 - (x_0 + x_1 + x_2 + x_3)^3.$$

Remark 11. For the Clebsch cubic, as well as cubics "close" to it in the sense of the position of $P_1, \ldots, P_6$, the transformation

$$\begin{aligned} x_0 &= y_0 - y_3 - \sqrt{2}y_1 & x_2 &= y_0 + y_3 + \sqrt{2}y_2 \\ x_1 &= y_0 - y_3 + \sqrt{2}y_1 & x_3 &= -y_0 - y_3 + \sqrt{2}y_2 \end{aligned}$$

of the coordinate system with inverse

$$\begin{aligned} y_0 &= x_0 + x_1 + x_2 - x_3 & y_2 &= \sqrt{2}(x_2 + x_3) \\ y_1 &= \sqrt{2}(-x_0 + x_1) & y_3 &= -x_0 - x_1 + x_2 - x_3 \end{aligned}$$

achieves that all 27 lines, for $K = \mathbb{R}$, are visible in the affine chart

$$K^3 \longrightarrow \mathbb{P}^3_K, \quad (y_1, y_2, y_3) \longmapsto (1 : y_1 : y_2 : y_3).$$

Moreover, they all pass through a ball with radius 6 around 0. In the affine chart we obtain a so-called affine cubic hypersurface $C'' \subset K^3$ given by a single, non-homogeneous degree three polynomial $f \in K[y_1, y_2, y_3]$.

4   Implementation in Singular

We have implemented the constructions above in the library cubic.lib [1] for the open-source computer algebra system SINGULAR [8]. For an introduction to the language of SINGULAR see [10]. Specifically, from 6 points in general position (with coordinates in $\mathbb{Q}$ or an algebraic extension thereof), we give a function to obtain the cubic $C \subset \mathbb{P}^5_K$, its projection $C' \subset \mathbb{P}^3_K$ and the affine cubic hypersurface $C'' \subset K^3$. Moreover, we compute the parametrizations

$$\mathbb{P}^2_K \setminus \{P_1, \ldots, P_6\} \longrightarrow C \longrightarrow C'$$

and an affine parametrization

$$\mathbb{P}^2_K \setminus V(\varphi_0 + \varphi_1 + \varphi_2 - \varphi_3) \longrightarrow C''.$$

Finally, we compute the lines on C, C' and C'' in implicit and parametric form, as well as the Eckardt points. We demonstrate key parts of our library, considering the Clebsch cubic as an example:

Example 2. Our library can be loaded in SINGULAR by:
```
> LIB"cubic.lib";
```
We first create a polynomial ring in 4 variables over the field $\mathbb{Q}[\sqrt{5}]$:
```
> ring R = (0,a),(x0,x1,x2,x3),dp;
> minpoly = a^2-5;
```
We specify a list P with the points $P_1, \ldots, P_6$:
```
> number g = (1 + a)/2;
> list P = vector(0,1,-g), vector(g,0,1), vector(1,g,0),
           vector(1,-g,0), vector(0,1,g), vector(-g,0,1);
```
We compute the equation of C':
```
> poly f = cubic(P);
> f;
-3*x0^2*x1-3*x0*x1^2-3*x0^2*x2-6*x0*x1*x2-3*x1^2*x2-3*x0*x2^2
-3*x1*x2^2-3*x0^2*x3-6*x0*x1*x3-3*x1^2*x3-6*x0*x2*x3-6*x1*x2*x3
-3*x2^2*x3-3*x0*x3^2-3*x1*x3^2-3*x2*x3^2
```
The following command returns a list of all lines on C', each specified by 2 linear equations:
```
> list L = lines(P);
> L[1];
_[1] = x0 + x1
_[2] = x2 + x3
```

We compute a list of Eckardt points, each specified by 3 linear equations:

```
> list E = EckardtPoints(P);
> E[1];
_[1] = x0
_[2] = x1
_[3] = x2 + x3
```

By the commands `affineCubic`, `affineLines` and `affineEckardtPoints`, one can also obtain the affine cubic C'' and the corresponding lines and Eckardt points, respectively. Moreover, the functions `paramLines` and `affineParamLines` compute parametrizations of the lines on C' and C'', respectively.

If, in addition to SINGULAR, the program SURF [9] is installed, C'' can be visualized by:

```
> LIB "surf.lib";
> plot(affineCubic(P));
```

SURF can also plot hyperplane sections of a surface. Hence, we can visualize the lines on the cubic by intersecting with tritangent planes, see Fig. 1.

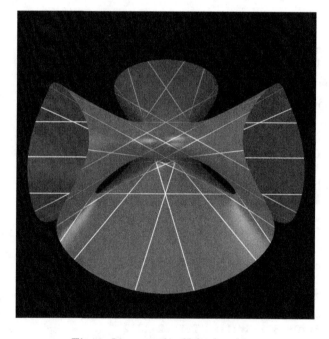

Fig. 1. Lines on the Clebsch cubic,

5 Explicit Data for the Clebsch Cubic

In this section we give the explicit data required for the dimensional comparison process for the Clebsch cubic. In the following let $a = \sqrt{5}$ and $c = \sqrt{2}$. The cubic C'' is the zero set in K^3 of the equation

$$2cy_2^3 + 2y_1^2 y_3 - 8y_2^2 y_3 + 3cy_2 y_3^2 - y_3^3 - 2y_1^2 + 8y_2^2 - 10cy_2 y_3 + 3y_3^2 + 3cy_2 - 3y_3 + 1 = 0.$$

The lines on C'' given in implicit form (by two linear equations each) as well as their parametrizations (specified as maps $K \to K^3, s \mapsto (\psi_1(s), \psi_2(s), \psi_3(s)))$ are listed in Table 1.

Table 1. Lines on the Clebsch cubic in implicit and parametric form

Implicit	Parametric
$\langle y_3 - 1, y_2 \rangle$	$(s, 0, 1)$
$\langle y_3 - 1, y_2 - \frac{2}{c} \rangle$	$(s, \frac{2}{c}, 1)$
$\langle y_3 - 1, y_2 + \frac{2}{c} \rangle$	$(s, -\frac{2}{c}, 1)$
$\langle y_2 - \frac{1}{c} \cdot y_3 + \frac{1}{c}, y_1 - \frac{1}{c} \cdot y_3 - \frac{1}{c} \rangle$	$(\frac{1}{c} \cdot s + \frac{1}{c}, \frac{1}{c} \cdot s - \frac{1}{c}, s)$
$\langle y_2 - \frac{1}{c} \cdot y_3 - \frac{1}{c}, y_1 - \frac{1}{c} \cdot y_3 - \frac{3}{c} \rangle$	$(\frac{1}{c} \cdot s + \frac{3}{c}, \frac{1}{c} \cdot s + \frac{1}{c}, s)$
$\langle y_2 - \frac{1}{c} \cdot y_3 + \frac{3}{c}, y_1 - \frac{1}{c} \cdot y_3 + \frac{1}{c} \rangle$	$(\frac{1}{c} \cdot s - 1, \frac{1}{c} \cdot s - \frac{3}{c}, s)$
$\langle y_2 - \frac{1}{c} \cdot y_3 + \frac{1}{c}, y_1 + \frac{1}{c} \cdot y_3 + \frac{1}{c} \rangle$	$(-\frac{1}{c} \cdot s - \frac{1}{c}, \frac{1}{c} \cdot s - \frac{2}{c}, s)$
$\langle y_2 - \frac{1}{c} \cdot y_3 - \frac{1}{c}, y_1 + \frac{1}{c} \cdot y_3 + \frac{3}{c} \rangle$	$(-\frac{1}{c} \cdot s - \frac{3}{c}, \frac{1}{c} \cdot s + \frac{1}{c}, s)$
$\langle y_2 - \frac{1}{c} \cdot y_3 + \frac{3}{c}, y_1 + \frac{1}{c} \cdot y_3 - \frac{1}{c} \rangle$	$(-\frac{1}{c} \cdot s + \frac{1}{c}, \frac{1}{c} \cdot s - \frac{3}{c}, s)$
$\langle y_2 + \frac{1}{c} \cdot y_3 + \frac{1}{c}, y_1 + \frac{3}{c} \cdot y_3 + \frac{1}{c} \rangle$	$(-\frac{3}{c} \cdot s - 1, -\frac{1}{c} \cdot s - \frac{1}{c}, s)$
$\langle y_2 - \frac{3}{c} \cdot y_3 + \frac{1}{c}, y_1 - \frac{1}{c} \cdot y_3 + \frac{1}{c} \rangle$	$(\frac{1}{c} \cdot s - \frac{1}{c}, \frac{3}{c} \cdot s - \frac{1}{c}, s)$
$\langle y_2 + \frac{1}{c} \cdot y_3 + \frac{1}{c}, y_1 - \frac{3}{c} \cdot y_3 - \frac{1}{c} \rangle$	$(\frac{2}{c} \cdot s + 1, -\frac{1}{c} \cdot s - \frac{1}{c}, s)$
$\langle y_2 - \frac{3}{c} \cdot y_3 + \frac{1}{c}, y_1 + \frac{1}{c} \cdot y_3 - \frac{1}{c} \rangle$	$(-\frac{1}{c} \cdot s + \frac{1}{c}, \frac{3}{c} \cdot s - \frac{1}{c}, s)$
$\langle y_2, y_1 - \frac{1}{c} \cdot y_3 + \frac{1}{c} \rangle$	$(\frac{1}{c} \cdot s - \frac{1}{c}, 0, s)$
$\langle y_2, y_1 + \frac{1}{c} \cdot y_3 - \frac{1}{c} \rangle$	$(-\frac{1}{c} \cdot s + \frac{1}{c}, 0, s)$
$\langle y_2 - \frac{1}{ac} \cdot y_3, y_1 + (\frac{1}{2}ac - 3c) \cdot y_3 - (\frac{1}{2}ac - 2c) \rangle$	$(-(\frac{3}{2}ac - 3c) \cdot s + (\frac{1}{2}ac - 2c), \frac{1}{ac+2c} \cdot s - \frac{1}{2}ac, s)$
$\langle y_2 - \frac{5}{ac} \cdot y_3 + \frac{1}{ac+2c}, y_1 - \frac{11}{ac+4c} \cdot y_3 - \frac{3}{ac+2c} \rangle$	$(\frac{11}{ac+2c} \cdot s + \frac{3}{ac+2c}, \frac{5}{ac} \cdot s - \frac{1}{ac+2c}, s)$
$\langle y_2 + \frac{1}{ac+2c} \cdot y_3 - \frac{5}{ac}, y_1 - (\frac{3}{2}ac + 3c) \cdot y_3 + (\frac{1}{2}ac + 2c) \rangle$	$((\frac{3}{2}ac + 3c) \cdot s - (\frac{1}{2}ac + 2c), -\frac{1}{ac-2c} \cdot s + \frac{1}{2}ac, s)$
$\langle y_2 + \frac{1}{ac} \cdot y_3 - \frac{2}{ac-2c}, y_1 - \frac{11}{ac-4c} \cdot y_3 + \frac{3}{ac-2c} \rangle$	$(-\frac{11}{ac-4c} \cdot s - \frac{3}{ac-2c}, -\frac{5}{ac} \cdot s + \frac{2}{ac-2c}, s)$
$\langle y_2 + \frac{2}{ac-3c} \cdot y_3 - \frac{2}{ac-2c}, y_1 + (\frac{1}{4}ac + \frac{1}{4}c) \cdot y_3 + (\frac{1}{4}ac + \frac{1}{4}c) \rangle$	$(-(\frac{1}{4}ac + \frac{1}{4}c) \cdot s - (\frac{1}{4}ac + \frac{1}{4}c), -\frac{2}{ac-3c} \cdot s - (\frac{1}{4}ac + \frac{3}{4}c), s)$
$\langle y_2 + \frac{1}{ac-3c} \cdot y_3 - \frac{5}{ac}, y_1 + (\frac{3}{2}ac + 3c) \cdot y_3 - (\frac{1}{2}ac + 2c) \rangle$	$(-(\frac{3}{2}ac + 3c) \cdot s + (\frac{1}{2}ac + 2c), -\frac{1}{ac-3c} \cdot s + \frac{1}{2}ac, s)$
$\langle y_2 - \frac{2}{ac+3c} \cdot y_3 + \frac{2}{ac+3c}, y_1 - (\frac{1}{4}ac - \frac{1}{4}c) \cdot y_3 - (\frac{1}{4}ac - \frac{1}{4}c) \rangle$	$((\frac{1}{4}ac - \frac{1}{4}c) \cdot s + (\frac{1}{4}ac - \frac{1}{4}c), \frac{2}{ac+3c} \cdot s + (\frac{1}{4}ac - \frac{3}{4}c), s)$
$\langle y_2 + \frac{2}{ac+3c} \cdot y_3 - \frac{5}{ac}, y_1 - (\frac{3}{2}ac - 3c) \cdot y_3 - (\frac{1}{2}ac - 2c) \rangle$	$((\frac{3}{2}ac - 3c) \cdot s - (\frac{1}{2}ac - 2c), -\frac{1}{ac+2c} \cdot s - \frac{1}{2}ac, s)$
$\langle y_2 + \frac{2}{ac-3c} \cdot y_3 - \frac{1}{ac-2c}, y_1 - (\frac{3}{2}ac + \frac{1}{4}c) \cdot y_3 - (\frac{1}{4}ac + \frac{1}{4}c) \rangle$	$((\frac{1}{4}ac + \frac{1}{4}c) \cdot s + (\frac{1}{4}ac + \frac{1}{4}c), -\frac{2}{ac} \cdot s - (\frac{1}{4}ac + \frac{3}{4}c), s)$
$\langle y_2 + \frac{5}{ac} \cdot y_3 - \frac{1}{ac-2c}, y_1 - \frac{11}{ac-4c} \cdot y_3 - \frac{3}{ac-2c} \rangle$	$(\frac{11}{ac-4c} \cdot s + \frac{3}{ac-2c}, -\frac{5}{ac} \cdot s + \frac{1}{ac-2c}, s)$
$\langle y_2 - \frac{2}{ac+3c} \cdot y_3 + \frac{1}{ac+3c}, y_1 + (\frac{1}{4}ac - \frac{1}{4}c) \cdot y_3 + (\frac{1}{4}ac - \frac{1}{4}c) \rangle$	$(-(\frac{1}{4}ac - \frac{1}{4}c) \cdot s - (\frac{1}{4}ac - \frac{1}{4}c), \frac{2}{ac+3c} \cdot s + (\frac{1}{4}ac - \frac{3}{4}c), s)$
$\langle y_2 - \frac{5}{ac} \cdot y_3 + \frac{1}{ac+2c}, y_1 + \frac{11}{ac+4c} \cdot y_3 + \frac{3}{ac+2c} \rangle$	$(-\frac{11}{ac+4c} \cdot s - \frac{3}{ac+2c}, \frac{5}{ac} \cdot s - \frac{1}{ac+2c}, s)$

The 10 Eckardt points on C' have projective coordinates

$$(-1:1:0:0) \quad (-1:0:1:0) \quad (0:-1:1:0) \quad (-1:0:0:1) \quad (0:-1:0:1)$$
$$(0:0:-1:1) \quad (1:0:0:0) \quad (0:1:0:0) \quad (0:0:1:0) \quad (0:0:0:1).$$

Hence (after applying the transformation of Remark 11 and passing to the affine chart), the cubic C'' contains 7 of them with affine coordinates

$$(-\frac{1}{c}, -\frac{1}{c}, 0) \quad (\frac{1}{c}, -\frac{1}{c}, 0) \quad (0,0,1) \quad (-\frac{2}{c}, 0, -1) \quad (\frac{2}{c}, 0, -1) \quad (0, \frac{2}{c}, 1) \quad (0, -\frac{2}{c}, 1),$$

the remaining three of them lying at the plane at infinity $y_0 = 0$ with projective coordinates $(0:1:0:0)$, $(0:1:1:c)$, $(0:-1:1:c)$.

References

1. Böhm, J., Marais, M.S., van der Merwe, A.F.: `cubic.lib`, a SINGULAR library for contructing cubic surfaces and the lines thereon (2016)
2. Cayley, A.: On the triple tangent planes to a surface of the third order. Camb. Dublin Math. J. **IV**, 118–132 (1849)
3. Clebsch, A.: Die Geometrie auf den Flächen dritter Ordnung. J. für reine und angew. Math. **65**, 359–380 (1866)
4. Clebsch, A.: Ueber die Anwendung der quadratischen Substitution auf die Gleichungen 5-ten Grades und die geometrische Theorie des ebenen Fünfseits. Math. Ann. **4**(2), 284–345 (1871)
5. Coble, A.B.: Point sets and allied Cremona groups I. Trans. Am. Math. Soc. **16**(2), 155–198 (1915)
6. Cox, D., Little, J., O'Shea, D.: Ideals, Varieties, and Algorithms. An Introduction to Computational Algebraic Geometry and Commutative Algebra. Undergraduate Texts in Mathematics, 3rd edn. Springer, New York (2007)
7. Cremona, L.: Über die Polar-Hexaheder bei den Flächen dritter Ordnung. Math. Ann. **13**, 301–304 (1878). (Opere, t. 3, pp. 430–433)
8. Decker, W., Greuel, G.-M., Pfister, G., Schönemann, H.: Singular 4.0.2 – A computer algebra system for polynomial computations (2016)
9. Endrass, S.: Surf. A program for drawing curves and surfaces (2010). http://surf.sourceforge.net/
10. Greuel, G.M., Pfister, G.: A Singular Introduction to Commutative Algebra, extended edn. Springer, Berlin (2008). With contributions by Bachmann, O., Lossen, C., Schönemann, H., With 1 CD-ROM (Windows, Macintosh and UNIX)
11. Salmon, G.: On the triple tangent planes to a surface of the third order. Camb. Dublin Math. J. **IV**, 252–260 (1849)

Decomposing Solution Sets of Polynomial Systems Using Derivatives

Daniel A. Brake, Jonathan D. Hauenstein$^{(\boxtimes)}$, and Alan C. Liddell Jr.

Department of Applied and Computational Mathematics and Statistics,
University of Notre Dame, Notre Dame, IN, USA
{dbrake,hauenstein,aliddel1}@nd.edu

Abstract. A core computation in numerical algebraic geometry is the decomposition of the solution set of a system of polynomial equations into irreducible components, called the numerical irreducible decomposition. One approach to validate a decomposition is what has come to be known as the "trace test." This test, described by Sommese, Verschelde, and Wampler in 2002, relies upon path tracking and hence could be called the "tracking trace test." We present a new approach which replaces path tracking with local computations involving derivatives, called a "local trace test." We conclude by demonstrating this local approach with examples from kinematics and tensor decomposition.

Keywords: Numerical algebraic geometry · Trace test · Numerical irreducible decomposition

1 Introduction

Numerical algebraic geometry uses numerical methods to compute and manipulate the solution set to a given system of polynomial equations. Such a solution set can be decomposed into finitely many components yielding the irreducible decomposition. In numerical algebraic geometry, irreducible components are represented via a witness set with a numerical irreducible decomposition consisting of a witness set for each irreducible component. See [3,12] for a general overview of witness sets and computing a numerical irreducible decomposition.

The focus of this article is the pure-dimensional decomposition step in computing a numerical irreducible decomposition. Let X be a pure k-dimensional component of the solution set of f, namely $\mathcal{V}(f) = \{x \mid f(x) = 0\}$, and let $\mathcal{L}$ be a general linear space of codimension k. That is, X is a union of irreducible components of $\mathcal{V}(f)$ each having dimension k, say $X = X_1 \cup \cdots \cup X_m$. Given the finitely many points $W = X \cap \mathcal{L}$, called a witness point set for X, the pure-dimensional decomposition step partitions W into $X_1 \cap \mathcal{L}, \ldots, X_m \cap \mathcal{L}$ yielding witness point sets for the irreducible components of X.

All authors supported in part by NSF ACI 1460032, Sloan Research Fellowship, and Army Young Investigator Program (YIP).

G.-M. Greuel et al. (Eds.): ICMS 2016, LNCS 9725, pp. 127–135, 2016.
DOI: 10.1007/978-3-319-42432-3_16

There are two tools commonly used for pure-dimensional decomposition. First, random monodromy loops [10] aim to determine subsets of points in W contained in the same irreducible component. This relies on the fact that the set of smooth points of an irreducible algebraic set is connected.

Second, given $Z \subset W$, the trace test of [11] is used to verify that Z is a witness point set for some algebraic set, i.e., there exists $\mathcal{J} \subset \{1, \ldots, m\}$ such that

$$Z = \bigcup_{j \in \mathcal{J}} X_j \cap \mathcal{L}.$$

There are two ways to show that $|\mathcal{J}| = 1$, i.e., Z is a witness point set for an irreducible component. If Z was constructed as a result of using random monodromy loops, then each point in Z must lie on the same irreducible component. Another approach is to show that the trace test does not hold for any nonempty and proper subset of Z.

Since the trace test of [11] uses only path tracking, we will refer to this as the *tracking trace test*. We show that this test is the first in a family of three methods, which are based on the zeroth, first, and second derivatives, respectively. The third of these methods, which is built on computing second derivatives, computes these derivatives locally at each point in Z and hence we call it a *local trace test*.

The remainder is organized as follows. In Sect. 2, we describe linear traces in numerical algebraic geometry and present three computational approaches. Section 3 considers the extension to parameterized algebraic sets. We demonstrate the methods on two examples in Sect. 4 and conclude in Sect. 5.

2 Trace Test

Let $f : \mathbb{C}^N \to \mathbb{C}^n$ be a polynomial system with $X \subset \mathcal{V}(f) \subset \mathbb{C}^N$ a pure k-dimensional set. Let $\ell : \mathbb{C}^N \to \mathbb{C}^k$ be a general linear system with $\mathcal{L} = \mathcal{V}(\ell)$ and $W = X \cap \mathcal{L}$. If $X_1, \ldots, X_m$ are the irreducible components of X, the goal is to partition W into the m sets $W_1 = X_1 \cap \mathcal{L}, \ldots, W_m = X_m \cap \mathcal{L}$.

We first reduce to the multiplicity-one case as follows. Since the deflation sequence [9] with respect to f is the same for each $w \in W_i = X_i \cap \mathcal{L}$, we can first partition W based on deflation sequences. So, without loss of generality, we may assume that each point in W has the same deflation sequence. In particular, a byproduct of this computation is a polynomial system, which without loss of generality we call f, such that each irreducible component has multiplicity one.

We next reduce to the "square" case using Bertini's Theorem (see, e.g., [12, Theorem A.8.7] and [3, Theorem 9.3]). In particular, for a general $U \in \mathbb{C}^{(N-k)\times n}$, each X_i is an irreducible component of $\mathcal{V}(U \cdot f)$. Hence, without loss of generality, we may assume that $f : \mathbb{C}^N \to \mathbb{C}^{N-k}$ such that $X \subset \mathcal{V}(f)$ is pure k-dimensional and each irreducible component of X has multiplicity one with respect to f.

Suppose that $d = \deg X = |W|$ and $W = \{w_1, \ldots, w_d\}$. For a general $v \in \mathbb{C}^k$, consider the family of parallel slices $\mathcal{M}_t = \mathcal{V}(\ell - t \cdot v)$ for $t \in \mathbb{C}$, so that $\mathcal{M}_0 = \mathcal{L} = \mathcal{V}(\ell)$. For $i = 1, \ldots, d$, consider the paths $x_i(t)$ defined by

$$x_i(t) \in X \cap \mathcal{M}_t \text{ and } x_i(0) = w_i. \tag{1}$$

The following forms the basis of traces in numerical algebraic geometry.

Theorem 1 [11]. *With the setup as above, let $\mathcal{I} \subset \{1, \ldots, d\}$ be nonempty with $Z = \{w_i \mid i \in \mathcal{I}\} \subset W$. Then, there exists $\mathcal{J} \subset \{1, \ldots, m\}$ such that*

$$Z = \bigcup_{j \in \mathcal{J}} X_j \cap \mathcal{L}$$

if and only if

$$\mathrm{tr}_{\mathcal{I}}(t) = \sum_{i \in \mathcal{I}} x_i(t) \text{ is a vector of linear functions of } t. \tag{2}$$

This theorem can be used to create a trace test for images of algebraic sets using pseudowitness sets [8] (we consider a simple coordinate projection in Sect. 4.1), and for multihomogeneous witness sets [7].

Example 1. Consider the parabola $X = \mathcal{V}(f)$ where $f(\alpha, \beta) = \beta - \alpha^2$. For illustrative purposes, we consider the $\mathcal{L} = \mathcal{V}(\ell)$ where $\ell(\alpha, \beta) = 2\alpha + \beta - 3$ and take $v = \sqrt{-1}$. If $W = X \cap \mathcal{L} = \{w_1, w_2\} = \{(-3, 9), (1, 1)\}$, then

$$\mathrm{tr}_{\{1\}}(t) = \begin{bmatrix} -1 - \sqrt{4 + t\sqrt{-1}} \\ 5 + t\sqrt{-1} + 2\sqrt{4 + t\sqrt{-1}} \end{bmatrix} \text{ and } \mathrm{tr}_{\{2\}}(t) = \begin{bmatrix} -1 + \sqrt{4 + t\sqrt{-1}} \\ 5 + t\sqrt{-1} - 2\sqrt{4 + t\sqrt{-1}} \end{bmatrix}$$

are not linear in t, whereas

$$\mathrm{tr}_{\{1,2\}}(t) = \begin{bmatrix} 0 \\ 2\sqrt{-1} \end{bmatrix} \cdot t + \begin{bmatrix} -2 \\ 10 \end{bmatrix}$$

is indeed linear in t confirming that X is irreducible of degree 2.

Example 2. For each subsequent method, we will use the twisted cube curve

$$X = \{(s, s^2, s^3) \mid s \in \mathbb{C}\} \subset \mathbb{C}^3.$$

For illustrative purposes, we consider

$$f(\alpha, \beta, \gamma) = \begin{bmatrix} \beta - \alpha^2 \\ \gamma - \alpha^3 \end{bmatrix}, \quad \ell(\alpha, \beta, \gamma) = 2\alpha - 3\beta - \gamma + 2, \quad \text{and} \quad v = 1.$$

With $W = X \cap \mathcal{V}(\ell)$ where $|W| = 3$ and $\mathcal{I} = \{1, 2, 3\}$, Newton's identities yield

$$\mathrm{tr}_{\mathcal{I}}(t) = \begin{bmatrix} 0 \\ 0 \\ -3 \end{bmatrix} \cdot t + \begin{bmatrix} -3 \\ 13 \\ -39 \end{bmatrix} \tag{3}$$

which is linear in t.

The following three tests determine if $\mathrm{tr}_{\mathcal{I}}(t)$ is a linear function, i.e., there exists $a, b \in \mathbb{C}^N$ such that $\mathrm{tr}_{\mathcal{I}}(t) = a \cdot t + b$. They are derived using the fact that $\mathrm{tr}_{\mathcal{I}}(t)$ is linear if and only if $\dot{\mathrm{tr}}_{\mathcal{I}}(t)$ is constant if and only if $\ddot{\mathrm{tr}}_{\mathcal{I}}(t)$ is zero corresponding with the zeroth, first, and second derivatives of $\mathrm{tr}_{\mathcal{I}}(t)$. The zeroth derivative trace test is the tracking test of [11].

2.1   Zeroth Derivative Trace Test

The tracking trace test of [11] determines if $\mathrm{tr}_{\mathcal{I}}(t)$ is a linear function by evaluating it at 3 distinct sufficiently general values of t, say $t_1, t_2, t_3 \in \mathbb{C}$ facilitated by path tracking. Due to the genericity of $\mathcal{L}$ and v, one could take $t_1 = 0$, $t_2 = 1$, and $t_3 = -1$. That is, one needs to compute

$$\mathrm{tr}_{\mathcal{I}}(t_j) = \sum_{i \in \mathcal{I}} x_i(t_j)$$

where $x_i(t)$ defined in (1) are solution curves of $H : \mathbb{C}^N \times \mathbb{C} \to \mathbb{C}^N$ with

$$H(x,t) = \begin{bmatrix} f(x) \\ \ell(x) - t \cdot v \end{bmatrix} = 0. \tag{4}$$

With this setup, $\mathrm{tr}_{\mathcal{I}}(t)$ is linear in t if and only if

$$\frac{\mathrm{tr}_{\mathcal{I}}(t_2) - \mathrm{tr}_{\mathcal{I}}(t_1)}{t_2 - t_1} = \frac{\mathrm{tr}_{\mathcal{I}}(t_3) - \mathrm{tr}_{\mathcal{I}}(t_1)}{t_3 - t_1} = \frac{\mathrm{tr}_{\mathcal{I}}(t_3) - \mathrm{tr}_{\mathcal{I}}(t_2)}{t_3 - t_2}.$$

In the linear case, $\mathrm{tr}_{\mathcal{I}}(t) = a \cdot t + b$ where

$$a = \frac{\mathrm{tr}_{\mathcal{I}}(t_2) - \mathrm{tr}_{\mathcal{I}}(t_1)}{t_2 - t_1} \quad \text{and} \quad b = \mathrm{tr}_{\mathcal{I}}(t_1) - a \cdot t_1.$$

Example 3. With the setup from Example 2 and $t_1 = 0$, $t_2 = 1$, and $t_3 = -1$, the following table lists approximations of $x_i(t_j)$ for $i = 1, 2, 3$ and $j = 1, 2, 3$:

	$t_1 = 0$	$t_2 = 1$	$t_3 = -1$
$x_1(t_j)$	1.0000	0.8342	1.1284
	1.0000	0.6960	1.2733
	1.0000	0.5806	1.4368
$x_2(t_j)$	−0.5858	−0.3434	−0.7984
	0.3431	0.1179	0.6374
	−0.2010	−0.0405	−0.5089
$x_3(t_j)$	−3.4142	−3.4909	−3.3301
	11.6569	12.1861	11.0893
	−39.7990	−42.5401	−36.9280

so that

$$\mathrm{tr}_{\mathcal{I}}(0) = \begin{bmatrix} -3 \\ 13 \\ -39 \end{bmatrix}, \quad \mathrm{tr}_{\mathcal{I}}(1) = \begin{bmatrix} -3 \\ 13 \\ -42 \end{bmatrix}, \quad \mathrm{tr}_{\mathcal{I}}(-1) = \begin{bmatrix} -3 \\ 13 \\ -36 \end{bmatrix} \tag{5}$$

which one can use to easily recover (3).

2.2   First Derivative Trace Test

Since $\mathrm{tr}_{\mathcal{I}}(t)$ is linear if and only if $\dot{\mathrm{tr}}_{\mathcal{I}}(t)$ is constant, this can be decided by evaluating $\dot{\mathrm{tr}}_{\mathcal{I}}(t)$ at 2 distinct sufficiently general values of t, say $t_1, t_2 \in \mathbb{C}$, facilitated by path tracking and derivative computations. Due to the genericity of $\mathcal{L}$ and v, one could take $t_1 = 0$ and $t_2 = 1$. Due to the relationship between the paths $x_i(t)$ in (1) and the homotopy $H(x,t)$ in (4),

$$\dot{x}_i(t) = -J_x H(x_i(t), t)^{-1} \cdot J_t H(x_i(t), t) = \begin{bmatrix} Jf(x_i(t)) \\ J\ell(x_i(t)) \end{bmatrix}^{-1} \cdot \begin{bmatrix} 0 \\ v \end{bmatrix} \tag{6}$$

with corresponding Jacobian matrices $J_x H(x,t)$, $J_t H(x,t)$, $Jf(x)$, and $J\ell(x)$ so

$$\dot{\mathrm{tr}}_{\mathcal{I}}(t) = \sum_{i \in \mathcal{I}} \dot{x}_i(t) = \sum_{i \in \mathcal{I}} \begin{bmatrix} Jf(x_i(t)) \\ J\ell(x_i(t)) \end{bmatrix}^{-1} \cdot \begin{bmatrix} 0 \\ v \end{bmatrix}.$$

Therefore, $\mathrm{tr}_{\mathcal{I}}(t)$ is a linear function of t if and only if

$$\dot{\mathrm{tr}}_{\mathcal{I}}(t_1) = \dot{\mathrm{tr}}_{\mathcal{I}}(t_2) = \frac{\mathrm{tr}_{\mathcal{I}}(t_2) - \mathrm{tr}_{\mathcal{I}}(t_1)}{t_2 - t_1}.$$

In the linear case, $\mathrm{tr}_{\mathcal{I}}(t) = a \cdot t + b$ where

$$a = \dot{\mathrm{tr}}_{\mathcal{I}}(t_1) \ \text{ and } \ b = \mathrm{tr}_{\mathcal{I}}(t_1) - a \cdot t_1.$$

Thus, the first derivative trace test replaces evaluating $\mathrm{tr}_{\mathcal{I}}(t_3)$, a path tracking computation, with evaluating $\dot{\mathrm{tr}}_{\mathcal{I}}(t_1)$ and $\dot{\mathrm{tr}}_{\mathcal{I}}(t_2)$, a linear algebra computation. We emphasize here that finding $\dot{\mathrm{tr}}_{\mathcal{I}}(t_1)$ does involve path tracking, but the cost incurred due to tracking paths is half that of the zeroth derivative trace test.

Example 4. With the setup from Example 2, we consider $t_1 = 0$ and $t_2 = 1$ with the values of $x_i(t_j)$ listed in Example 3. The following table lists approximations of the six values of $\dot{x}_i(t_j)$ for $i = 1, 2, 3$ and $j = 1, 2$ computed using (6):

	$t_1 = 0$	$t_2 = 1$
	−0.1429	−0.1963
$\dot{x}_1(t_j)$	−0.2857	−0.3276
	−0.4286	−0.4099
	0.2230	0.2698
$\dot{x}_2(t_j)$	−0.2612	−0.1853
	0.2295	0.0954
	−0.0801	−0.0735
$\dot{x}_3(t_j)$	0.5469	0.5129
	−2.8009	−2.6855

so that

$$\dot{\mathrm{tr}}_{\mathcal{I}}(0) = \dot{\mathrm{tr}}_{\mathcal{I}}(1) = \begin{bmatrix} 0 \\ 0 \\ -3 \end{bmatrix} \tag{7}$$

which, together with $\mathrm{tr}_{\mathcal{I}}(0)$ in (5), one can easily recover (3).

2.3   Second Derivative Trace Test

Since $\mathrm{tr}_{\mathcal{I}}(t)$ is linear if and only if $\ddot{\mathrm{tr}}_{\mathcal{I}}(t) \equiv 0$, this can be decided by evaluating $\ddot{\mathrm{tr}}_{\mathcal{I}}(t)$ at a sufficiently general $t_1 \in \mathbb{C}$ facilitated by derivative computations. Due to the genericity of $\mathcal{L}$ and v, we take $t_1 = 0$. Hence, $x_i(0) = w_i$ by (1) and $\dot{x}_i(0)$ as in (6) so that

$$\mathrm{tr}_{\mathcal{I}}(0) = \sum_{i \in \mathcal{I}} x_i(0) = \sum_{i \in \mathcal{I}} w_i \ \text{ and } \ \dot{\mathrm{tr}}_{\mathcal{I}}(0) = \sum_{i \in \mathcal{I}} \dot{x}_i(0) = \sum_{i \in \mathcal{I}} \begin{bmatrix} Jf(w_i) \\ J\ell(w_i) \end{bmatrix}^{-1} \cdot \begin{bmatrix} 0 \\ v \end{bmatrix}.$$

Due to the structure of $H(x,t)$ in (4), $\frac{\partial^2 H}{\partial x \partial t} = 0$ and

$$\ddot{x}_i(0) = - \begin{bmatrix} Jf(w_i) \\ J\ell(w_i) \end{bmatrix}^{-1} \cdot \begin{bmatrix} \dot{x}_i(0)^T \cdot \mathrm{Hessian}(f_1)(w_i) \cdot \dot{x}_i(0) \\ \vdots \\ \dot{x}_i(0)^T \cdot \mathrm{Hessian}(f_n)(w_i) \cdot \dot{x}_i(0) \\ 0 \end{bmatrix} \tag{8}$$

where $\mathrm{Hessian}(f_j)(w_i)$ is the Hessian matrix of f_j evaluated at w_i. Hence, $\mathrm{tr}_{\mathcal{I}}(t)$ is a linear function of t if and only if

$$\ddot{\mathrm{tr}}_{\mathcal{I}}(0) = \sum_{i \in \mathcal{I}} \ddot{x}_i(0) = 0.$$

In the linear case, $\mathrm{tr}_{\mathcal{I}}(t) = a \cdot t + b$ where

$$a = \dot{\mathrm{tr}}_{\mathcal{I}}(0) \quad \text{and} \quad b = \mathrm{tr}_{\mathcal{I}}(0).$$

Thus, the second derivative trace test replaces all path tracking with second derivative computations performed locally and hence we call it a *local trace test*.

Example 5. With the setup from Example 2, we consider $t_1 = 0$ with the values of $x_i(0)$ and $\dot{x}_i(0)$ listed in Examples 3 and 4, respectively. Approximations of $\ddot{x}_i(0)$ for $i = 1, 2, 3$ computed using (8) are

$$\ddot{x}_1(0) = \begin{bmatrix} -0.0350 \\ -0.0292 \\ 0.0175 \end{bmatrix}, \quad \ddot{x}_2(0) = \begin{bmatrix} 0.0275 \\ 0.0671 \\ -0.1464 \end{bmatrix}, \quad \ddot{x}_3(0) = \begin{bmatrix} 0.0074 \\ -0.0380 \\ 0.1289 \end{bmatrix}$$

so that $\ddot{\mathrm{tr}}_{\mathcal{I}}(0) = 0$. Thus, $\mathrm{tr}_{\mathcal{I}}(0)$ and $\dot{\mathrm{tr}}_{\mathcal{I}}(0)$ computed in (5) and (7) yield (3).

3   Parameterizations

In Sect. 2, we considered pure-dimensional $X \subset \mathcal{V}(f)$, i.e., X was contained in the solution set of $f = 0$. In this section, we consider pure-dimensional sets which arise as the image of an algebraic set under an algebraic map. For simplicity, we only consider $X = \overline{\{p(y) \mid y \in \mathbb{C}^k\}} \subset \mathbb{C}^N$ where $p : \mathbb{C}^k \to \mathbb{C}^N$ has rank k, i.e., rank $Jp(y) = k$ for generic $y \in \mathbb{C}^k$, as more general situations follow using similar computations. With this setup, X is irreducible with $\dim X = k$ so that the main question is to determine $\deg X$ via a trace test. That is, given $Z \subset W = X \cap \mathcal{L}$ where $\ell : \mathbb{C}^N \to \mathbb{C}^k$ is a general linear system and $\mathcal{L} = \mathcal{V}(\ell)$, one aims to decide if $Z = W$ so that $d = \deg X = |W| = |Z|$.

Since we are given $Z \subset W$, let $Z = \{z_1, \ldots, z_q\}$. Since $X = \overline{p(\mathbb{C}^k)}$ and $\mathcal{L}$ is general, we know that there exists $y_1, \ldots, y_q \in \mathbb{C}^k$ such that $z_i = p(y_i)$. Let $v \in \mathbb{C}^k$ be general and $\mathcal{M}_t = \mathcal{V}(\ell - t \cdot v)$. For each $i = 1, \ldots, q$, we consider the paths $x_i(t) \in X$ and $u_i(t) \in \mathbb{C}^k$ defined by

$$x_i(t) = p(u_i(t)) \in X \cap \mathcal{M}_t \quad \text{and} \quad u_i(0) = y_i. \tag{9}$$

In particular, $u_i(t) \in \mathbb{C}^k$ satisfies the k equations $\ell(p(u_i(t))) = t \cdot v$.

With $\mathcal{I} = \{1, \ldots, q\}$, the trace tests in Sect. 2 involve the computation of $\mathrm{tr}_{\mathcal{I}}(t) = \sum_{i=1}^{q} x_i(t)$, $\dot{\mathrm{tr}}_{\mathcal{I}}(t) = \sum_{i=1}^{q} \dot{x}_i(t)$, and $\ddot{\mathrm{tr}}_{\mathcal{I}}(t) = \sum_{i=1}^{q} \ddot{x}_i(t)$. Thus, all that remains is to compute $\dot{x}_i(t)$ and $\ddot{x}_i(t)$, namely

$$\dot{u}_i(t) = (J\ell(x_i(t)) \cdot Jp(u_i(t)))^{-1} \cdot v, \quad \dot{x}_i(t) = Jp(u_i(t)) \cdot \dot{u}_i(t) \tag{10}$$

and

$$\ddot{x}_i(t) = \left(I - Jp(u_i(t)) \cdot (J\ell(x_i(t)) \cdot Jp(u_i(t)))^{-1} \cdot J\ell(x_i(t))\right) \begin{bmatrix} \dot{u}_i(t)^T \cdot \mathrm{Hessian}(p_1)(u_i(t)) \cdot \dot{u}_i(t) \\ \vdots \\ \dot{u}_i(t)^T \cdot \mathrm{Hessian}(p_N)(u_i(t)) \cdot \dot{u}_i(t) \end{bmatrix} \tag{11}$$

where $I \in \mathbb{C}^{N \times N}$ is the identity matrix.

Example 6. We again illustrate using the twisted cubic curve from Example 3 using the same ℓ and v. In this case, we have $x_i(t_j) = p(u_i(t_j))$ where $u_i(t_j) = (x_i(t_j))_1$, i.e., the first coordinate, and $p(s) = \begin{bmatrix} s, s^2, s^3 \end{bmatrix}^T$. Via (10) and (11), we have $\dot{u}_i(t_j) = (2 - 6u_i(t_j) - 3u_i(t_j)^2)^{-1}$,

$$\dot{x}_i(t_j) = \dot{u}_i(t_j) \cdot \begin{bmatrix} 1 \\ 2u_i(t_j) \\ 3u_i(t_j)^2 \end{bmatrix}, \quad \text{and} \quad \ddot{x}_i(t_j) = \dot{u}_i(t_j)^3 \cdot \begin{bmatrix} 6u_i(t_j) + 6 \\ 6u_i(t_j)^2 + 4 \\ 12u_i(t_j) - 18u_i(t_j)^2 \end{bmatrix}$$

which produces the values listed in the tables in Examples 4 and 5.

4 Examples

The following compares the zeroth, first, and second derivative trace tests on two large examples. These examples utilized `Bertini` [2] for the path tracking and used Python with NumPy [13] to perform the linear algebra computations. For simplicity in our comparison, we utilize serial computations for all three trace tests but note that all three tests could be easily parallelized.

4.1 A Curve from Kinematics

One problem solved in [4] is the so-called 8 path-point synthesis problem for four-bar linkages derived from classical work of [1,5]. That is, one aims to compute all four-bar planar linkages whose coupler curve passes through 8 given general points in the plane. Since one can freely set the orientation at one point, the 8 path-point synthesis problem is the one-pose and 7 path-point Alt-Burmester problem solved in [4] which defines a curve in $\mathbb{C}^8$ of degree 10,858. Following [4], we consider the system of 21 polynomials $f(x, y)$ where $x \in \mathbb{C}^8$ and $y \in \mathbb{C}^{14}$. Since the curve of interest is the natural projection of a solution curve in $\mathcal{V}(f)$ into $\mathbb{C}^8$, following [8], we take $\ell(x)$ as a random linear polynomial and $v \in \mathbb{C}$ random.

With this setup and $\mathcal{I} = \{1, \ldots, 10858\}$, we used the zeroth, first, and second derivative trace tests from Sect. 2 to verify that the degree of this curve in $\mathbb{C}^8$ is indeed 10,858 by showing that the first 8 coordinates of $\mathrm{tr}_{\mathcal{I}}(t)$ are linear in t. Using serial computations on an Intel Core i7, the zeroth derivative trace test took 21.6 min, the first derivative test took 9.5 min, and the second derivative test took 1.4 min.

4.2   A Secant Variety

In order to consider the border rank of the tensor corresponding to 2×2 matrix multiplication, the secant variety $X = \sigma_6(\mathbb{C}^4 \times \mathbb{C}^4 \times \mathbb{C}^4) \subset \mathbb{C}^{64}$ was considered in [6] which showed that $\dim X = 60$ and $\deg X = 15,456$. After selecting 60 random linear polynomials ℓ and random $v \in \mathbb{C}^{60}$, we performed the zeroth, first, and second derivative trace tests based on parameterizations in Sect. 3 which verified that the degree is indeed 15,456. Using serial computations on an AMD Opteron 6378 processor, the zeroth derivative test took 84.1 h, the first derivative test took 42.2 h, and the second derivative test took 0.2 h. The vast difference in computation time is due to the use of adaptive precision during tracking; larger systems such as this, having 60 variables, often require higher precision than hardware types provide.

5   Conclusion

Decomposition of a pure-dimensional algebraic set into its irreducible components is fundamental to computational algebraic geometry. In numerical algebraic geometry, the pure-dimensional decomposition is performed using random monodromy loops verified by a trace test. By replacing path tracking with local derivative computations, we have developed a local trace test which examples show is computationally advantageous. Due to these results, we are in the process of developing a robust, high-performance, and parallel implementation.

References

1. Alt, H.: Über die Erzeugung gegebener ebener Kurven mit Hilfe des Gelenkvierecks. ZAMM **3**(1), 13–19 (1923)
2. Bates, D.J., Hauenstein, J.D., Sommese, A.J., Wampler, C.W.: Bertini: software for numerical algebraic geometry. https://bertini.nd.edu
3. Bates, D.J., Hauenstein, J.D., Sommese, A.J., Wampler, C.W.: Numerically solving polynomial systems with Bertini. Software, Environments, and Tools, vol. 25. Society for Industrial and Applied Mathematics (SIAM), Philadelphia (2013)
4. Brake, D.A., Hauenstein, J.D., Murray, A.P., Myszka, D.H., Wampler, C.W.: The complete solution of Alt-Burmester synthesis problems for four-bar linkages. J. Mech. Robot. **8**(4), 041018 (2016)
5. Burmester, L.: Lehrbuch der Kinematic. Verlag Von Arthur Felix, Leipzig (1886)
6. Hauenstein, J.D., Ikenmeyer, C., Landsberg, J.M.: Equations for lower bounds on border rank. Exp. Math. **22**(4), 372–383 (2013)
7. Hauenstein, J.D., Rodriguez, J.I.: Numerical irreducible decomposition of multiprojective varieties (2015). arXiv:1507.07069
8. Hauenstein, J.D., Sommese, A.J.: Witness sets of projections. Appl. Math. Comput. **217**(7), 3349–3354 (2010)
9. Hauenstein, J.D., Wampler, C.W.: Isosingular sets and deflation. Found. Comput. Math. **13**(3), 371–403 (2013)

10. Sommese, A.J., Verschelde, J., Wampler, C.W.: Using monodromy to decompose solution sets of polynomial systems into irreducible components. In: Ciliberto, C., Hirzebruch, F., Miranda, R., Teicher, M. (eds.) Applications of Algebraic Geometry to Coding Theory, Physics and Computation, pp. 297–315. Springer, Netherlands (2001)
11. Sommese, A.J., Verschelde, J., Wampler, C.W.: Symmetric functions applied to decomposing solution sets of polynomial systems. SIAM J. Numer. Anal. **40**(6), 2026–2046 (2002)
12. Sommese, A.J., Wampler, C.W.: The Numerical Solution of Systems of Polynomials Arising in Engineering and Science. World Scientific Publishing Co. Pte. Ltd., Hackensack (2005)
13. van der Walt, S., Colbert, S.C., Varoquaux, G.: The NumPy array: a structure for efficient numerical computation. Comput. Sci. Eng. **13**(2), 22–30 (2011)

Calibration of Accelerometers and the Geometry of Quadrics

Laurent Evain[✉]

University of Angers, Angers, France
laurent.evain@univ-angers.fr
http://www.math.univ-angers.fr/~evain/

Abstract. We study a method of calibration of accelerometers usable on field. No tools are required except a computer. Since the method is purely mathematical, free from measurement tools errors, it is both precise and affordable. We prove that the calibration of an accelerometer with three axis is possible with 9 random measurements exactly when the sphere is the unique quadric containing the nine directions of measurements.

Keywords: Accelerometer · Calibration · Quadrics

1 Introduction

Calibration of accelerometers in laboratories are expensive when accuracy is needed. In contrast, on field methods are usually simple and affordable, at the price of precision. Mathematical methods, i.e. methods using only mathematical algorithms and well documented universal constants, fill the gap between the two approaches: Since a large number of decimals are computable on a small personal computer, mathematical methods are both very precise and usable on field.

The precision of mathematical formulas compared to methods involving physical devices is not new. It was known long ago before the development of computers. For, instance, Mohr in 1672 and Mascheroni in 1797 proved that points constructed with compass and straightedge can be constructed with compass only. In the preface of his book, Mascheroni explains how superior in precision is the compass compared to the straightedge. He hoped he could compute tables of sinus and cosinus using compass [7, p. 20]. However, in practice, a better precision for the tables can be obtained using rapidly converging series, for instance with the usual series expansion, a consequence of the Euler formula. Of course the advantage of mathematical methods is now even bigger with the advent of computers.

Here are some orders of magnitude for the readers not familiar with metrology. Consider the problem of measurement of the length of a perfect cylinder with a micrometer. We assume that the length of the cylinder is approximately 10^{-2} meters. To get an error of $0.01\,\%$, a micrometer with errors at most 10^{-6} m is required, whose price exceeds 1000 euros. Misalignment must be controlled, since the axis of the cylinder is not exactly parallel to the axis of measurement

© Springer International Publishing Switzerland 2016
G.-M. Greuel et al. (Eds.): ICMS 2016, LNCS 9725, pp. 136–141, 2016.
DOI: 10.1007/978-3-319-42432-3_17

of the micrometer. An error of one degree in the parallelism induces an error of 0.01523%. This means that an other investment to check the position the cylinder is needed.

By comparison, for the C99 format the rounding error on a number is about 10^{-16}%. Even with cumulative errors in the calculations, there are several orders of magnitude in favor of the computer for a lower price.

We propose a method of calibration of accelerometers free from any measurement tools or any physical device. It uses only computers and the well documented value of the gravitational constant g. To simplify the presentation and avoid normalization constants, we suppose that we have chosen units so that the gravitational constant is equal to 1.

The strategy is the following. We introduce a frame F that we call the calibrated frame. The calibration function C_F with respect to the frame F has to be compatible with the geometry of quadrics. This compatibility determines completely the calibration function C_F.

According to the semiconductor application notes [1], "the standard model used for the original factory calibration of a consumer grade digital accelerometer" assumes that the axis of the accelerometer are orthogonal *before* the calibration process. In this simpler context, pure mathematical methods have been proposed [4,5]. For the most general linear calibration presented here, the existing methods that we involve some external frame or external device and adjust relativly to that other frame/device using a large variety of distinct tools (GPS, static placement, dynamic placement...) [2,3,6]. The method proposed in the present note could easily be implemented in factory calibration.

Since the audience is composed mostly of mathematicians, the focus will be on the theoretical aspects, forgetting the practical noise considerations that spoil the theoretical model. These noise considerations are of course important, but they will be treated elsewhere.

2   Functionality and Applications

The output of an accelerometer at time t is by definition the vector $d(t)$ with three coordinates computed by the accelerometer. It may be displayed on a computer after a wifi connection, or processed in a computer to guide a robot for instance.

We put the accelerometer at rest on a table, in a random position. Let S be the unit sphere around the center p_0 of the accelerometer. The direction of measurement is a point $p \in S$ encoding the direction of the force exerted on the accelerometer. It is defined as follows.

Definition 1. *Let Δ be the half line starting at p_0 with vector $-g$, the opposite of the gravitational vector. The direction of measurement $p \in S$ in this random position is the intersection $S \cap \Delta$. The output when the direction of measurement is $p \in S$ is denoted $d(p)$.*

To calibrate an accelerometer, the steps followed by the user are the following:

- Put the accelerometer at rest in 9 different positions so that the sphere S is the unique quadric containing the 9 directions of measurement $p_1, \ldots, p_9$ (details below). Note the outputs $d(p_1), \ldots, d(p_9)$ of the accelerometer,
- Run the software with $d(p_1), \ldots, d(p_9)$ as input. The output of the software is an affine function C which is the calibration function (details below),
- Use the accelerometer to make the measurements: if the display of the accelerometer at time t is $d(t)$, the correct acceleration after calibration correction is $C(d(t))$.

In Step 1, the nine directions of measurements must be on a unique quadric. Usually, random positions are appropriate since the locus of bad positions is closed and has measure zero. However, one can prescribe a more detailed procedure to guarantee this step, according to the following proposition.

Proposition 2. *Let $p_1, \ldots, p_9$ be directions of measurements such that:*

- $p_1, \ldots, p_5$ *are coplanar in a plane P.*
- $p_6, \ldots, p_9$ *are not in the plane P.*
- $p_6, \ldots, p_9$ *are not coplanar.*

Then the sphere S is the unique quadric containing the points $p_1, \ldots, p_9$.

To compute the calibration function C in Step 2, the software applies the following algorithm:

- There exists a unique quadric Q containing the points $d(p_i)$ and Q is an ellipsoid,
- Using the standard Gauss reduction of quadratic forms, one finds a change of coordinates C that transforms the equation of the quadric to the equation of the sphere,
- The function C is the calibration function.

3   Background on Accelerometers

An accelerometer measures the proper acceleration of general relativity, or in other words, the non gravitational acceleration. This means that if the accelerometer has an acceleration (in the usual sense) R, decomposed formally as $R = g + a$, where g is the gravitational vector, the output of the accelerometer is a. In particular, at rest on a table, the usual acceleration $R = 0$ and the output of a perfectly calibrated accelerometer is the proper acceleration $-g$. This corresponds to our bodily sensations: sit on a chair, we experience the force below which pushes in a direction opposed to gravity.

Using an affine frame $F = (p_0, \ldots, p_3)$ composed of 4 points in $\mathbb{R}^3$, we may encode a proper acceleration a as a vector a_F. The thought experiment which transforms the acceleration a into a vector $a_F \in \mathbb{R}^3$ is the following. A small ball is attached using springs to the center p_0 of the frame. The acceleration pushes the ball to a position p. The three components of the vector $\boldsymbol{p_0 p}$ in the base $\boldsymbol{p_0 p_1}, \boldsymbol{p_0 p_2}, \boldsymbol{p_0 p_3}$ are the components of $\boldsymbol{a}_F$. The vector $\boldsymbol{p_0 p}$ is well defined up to a normalizing positive constant depending on the stiffness of the spring. To fix the normalization, we adjust the stiffness so that $||\boldsymbol{p_0 p}|| = ||\boldsymbol{a}||$ (the equality, hence the stiffness of the spring depends on the units used).

Definition 3. *Let F be an orthonormal frame. The calibration function C_F : $\mathbb{R}^3 \to \mathbb{R}^3$ of the frame F is defined by the relation $C_F(\boldsymbol{d}) = \boldsymbol{a}_F$, where d is the output of the accelerometer.*

An accelerometer is calibrated with respect to F if the output $\boldsymbol{d}$ displayed by the accelerometer is $\boldsymbol{a}_F$, or equivalently if the calibration function C_F is the Identity.

Suppose that the accelerometer is infinitely small. Then there is a canonical choice for the origin p_0 of the frame, namely the material point supporting the accelerometer. We may then vectorialize the affine situation by considering that p_0 is the zero of the vector space, and all base changes between orthonormal frames are elements of the orthogonal group. A frame change $F_1 \to F_2$ induces a change $C_{F_1} \to C_{F_2}$ for calibration functions, given by an action of the orthogonal group. To put the things more formally, we have the following proposition:

Proposition 4. *Consider the right action of the orthogonal group $O := O_3(\mathbb{R})$ on the space T of functions $f : \mathbb{R}^3 \to \mathbb{R}^3$ where an element $g \in O$ acts as $f.g = f \circ g$. Then the set of calibration functions is an orbit of the action of O on S.*

An easy corollary is that the linearity of the output of the accelerometer is equivalent to the linearity of the calibration function. To be more precise:

Corollary 5. *The following conditions are equivalent:*

- *There exists a calibration function C_F which is affine,*
- *All calibration functions are affine,*
- *There exists a frame F such that $\boldsymbol{a}_F$ is an affine function of the output d of the accelerometer,*
- *For every frame F, $\boldsymbol{a}_F$ is an affine function of d.*

4   Strategy of Calibration

We suppose that the hypothesis of the last corollary apply, i.e. that for every frame F, the display $\boldsymbol{a}_F$ is an affine function of d. For small values of the acceleration, this is a mild assumption since we may keep only the linear part in the Taylor development of the output. This is usually the data sheet of the accelerometer which indicates the range of utilization of the accelerometer.

The accelerometer is calibrated if the affine function is linear orthogonal. Obstructions to orthogonality may appear from the distortion caused during the welding process. The offset may be due to springs which do not follow the prescribed stiffness. The list of potential problems is a priori infinite, but this does not impact our approach. We suppose that the output is linear, whatever the reasons.

In this context, the accelerometer carries a special frame that we call the geometrical frame. We may define it mathematically or physically. The physical definition is the following. Put the accelerometer in a free fall. The ball attached to the springs is at the center p_0 and the display is (d_1, d_2, d_3). An acceleration is applied on the accelerometer so that the display is $(d_1 + 1, d_2, d_3)$ (resp. $(d_1, d_2 + 1, d_3)$ $(d_1, d_2, d_3 + 1)$). This acceleration moves the ball from p_0 to a position p_1 (resp. p_2, p_3). Then $F = (p_0, p_1, p_2, p_3)$ is the geometrical frame.

In mathematical terms, the geometrical frame of the accelerometer is the frame F such that the acceleration $\boldsymbol{a}_F$ coincides with the output d. The frame F is the result of the physical construction of the accelerometer. It is not orthonormal because of the defaults of the construction.

Definition 6. *The calibrated frame associated to an accelerometer is the Gram-Schmidt orthonormalization of the geometrical frame.*

The physical definition of the geometrical frame includes a thought experiment which is hardly practical. It follows that the geometrical frame, and its orthonormalization the calibrated frame, exist in theory but they are difficult to determine in practice.

We are going to compute the calibration function $C = C_F$ when F is the calibrated frame. To proceed, we remark that the calibration function C should respect the form of the base change given by Gram-Schmidt. Moreover, if we put the accelerometer at rest at any position p and if we read the display $d(p)$ in this position p, then $||C(d(p))|| = ||\boldsymbol{a}_F(p)||$. Since the accelerometer is at rest, the acceleration $\boldsymbol{a}_F(p)$ is opposite to the gravitational vector. It follows that $||C(d(p))|| = ||\boldsymbol{g}|| = 1$ according to our normalization. Mathematically, C may be determined by these conditions as we will see in the next section.

To summarize this argument in a precise statement, the calibration function C is compatible with the outputs in the sense of the following definition.

Definition 7. *Let $d_1, \ldots, d_n$ be outputs read on the accelerometer at n different positions. A calibration function C compatible with the outputs $d_1, \ldots, d_n$ is a function $C : \mathbb{R}^3 \to \mathbb{R}^3$, such that:*

- *C is affine*
- *the linear part $\boldsymbol{C}$ is represented by an upper triangular matrix with positive diagonal elements*
- *For every output d_i, then $C(d_i)$ is on the unitary sphere.*

Remark 8. *It may be surprising that it is possible to compute the calibration function with respect calibrated frame, whereas the calibrated frame is unknown. In fact, what we determine via the calibration function is the base change between the calibrated frame and the geometrical frame. Both frames are unknown, but the transfer matrix is known.*

5   The Results

Definition 9. *A set of points p_i on the unit sphere S is called affinely rigid if the only affine transformations ϕ of $\mathbb{R}^3$ satisfying $\phi(p_i) \in S$ are the orthogonal transformations.*

Theorem 10. *Let p_i be points on S. The following conditions are equivalent:*

– *The points p_i are affinely rigid*
– *S is the unique quadric containing the points p_i.*

The first condition of the theorem is the condition that arises naturally when we try to determine if the measurements determine the calibration. The reformulation given by the second condition is the key to transform a theoretical possibility of calibration into a practical calibration using the classification of quadrics.

As a corollary of the theorem, we obtain:

Corollary 11. *Let $\{p_i\}$ be an affinely rigid set on S and let $\{d_i = d(p_i)\}$ be the corresponding set of outputs of the accelerometer for measurements at rest with direction p_i. Then*

– *there exists a unique quadric Q containing the points d_i, namely $Q = d(S)$.*
– *An affine function $C : \mathbb{R}^3 \to \mathbb{R}^3$ with upper triangular linear part $\mathbf{C}$ is a calibration function compatible with the outputs if and only if $C(Q) \subset S$.*

Theorem 12. *Let $\{p_i\}$ be a set of directions on S and let $\{d_i = d(p_i)\}$ be the corresponding outputs of the accelerometer for measurements at rest with direction p_i. There is a unique calibration function C compatible with the outputs d_i if and only if the set of directions p_i is affinely rigid. When the calibration is unique, it is the unique affine function C, with $\mathbf{C}$ upper triangular such that $C(Q) = S$.*

References

1. http://cache.freescale.com/files/sensors/doc/app_note/AN4399.pdf
2. Hall, J.J., Williams II, R.L.: Inertial measurement unit calibration platform. J. Rob. Syst. **17**(11), 623–632 (2000)
3. Axelsson, P., Norrlf, M.: Method to estimate the position and orientation of a triaxial accelerometer mounted to an industrial manipulator. In: Proceedings of the 10th IFAC Symposium on Robot Control, pp. 283–288 (2012)
4. Won, S.P., Golnaraghi, F.: A triaxial accelerometer calibration method using a mathematical model. IEEE Trans. Instrum. Meas. **59**(8), 2144–2153 (2010). doi:10. 1109/TIM.2009.2031849
5. Grip, N., Sabourova, N.: Simple non-iterative calibration for triaxial accelerometers. Meas. Sci. Technol. **22**(12), 13 p. (2011)
6. Forsberg, T., Grip, N., Sabourova, N.: Non-iterative calibration for accelerometers with three non-orthogonal axes, reliable measurement setups and simple supplementary equipment. Meas. Sci. Technol. **24**(3), 14 p. (2013)
7. Mascheroni, L.: La Geometria del Compasso (1797)

On the Feasibility of Semi-algebraic Sets in Poisson Regression

Thomas Kahle$^{(\boxtimes)}$

Otto-von-Guericke Universität Magdeburg, Magdeburg, Germany
thomas.kahle@ovgu.de
https://www.thomas-kahle.de

Abstract. Designing experiments for generalized linear models is difficult because optimal designs depend on unknown parameters. The local optimality approach is to study the regions in parameter space where a given design is optimal. In many situations these regions are semi-algebraic. We investigate regions of optimality using computer tools such as YALMIP, QEPCAD, and MATHEMATICA.

Keywords: Algebraic statistics · Optimal experimental design · Poisson regression · Semi-algebraic sets

1 Introduction

Generalized linear models are a mainstay of statistics, but optimal experimental designs for them are hard to find, as they depend on unknown parameters of the model. A common approach to this problem is to study local optimality, that is, determine an optimal design for each fixed set of parameters. In practice, this means that appropriate parameters have to be guessed a priori, or fixed by other means. In [12] the authors approached this problem from a global perspective. They study the *regions of optimality* of fixed designs and demonstrate that these are often defined by semi-algebraic constraints. Their main tool is a general equivalence theorem due to Kiefer and Wolfowitz, which directly yields polynomial inequalities in the parameters. This makes these problems amenable to the toolbox of real algebraic geometry. In this extended abstract we pursue this direction for the Rasch Poisson counts model which is used in psychometry [6] in the design of mental speed tests. Analyzing saturated designs for this model amounts to studying the feasibility of polynomial inequality systems. We examine the state of computer algebra tools for this purpose and find that there is room for improvement.

2 Polynomial Inequality Systems in Statistics

For brevity we omit any details of statistical theory and focus on mathematical and computational problems. The interested reader should consult [12] and

© Springer International Publishing Switzerland 2016
G.-M. Greuel et al. (Eds.): ICMS 2016, LNCS 9725, pp. 142–147, 2016.
DOI: 10.1007/978-3-319-42432-3_18

its references. We also stick to that paper's notation. Throughout, fix a positive integer k, the *number of rules*, and another positive integer $d \leq k$, the *interaction order*. A *rule setting* is a binary string $x = (x_1, \ldots, x_k) \in \{0,1\}^k$. The *regression function of interaction order d* is the function $f : \{0,1\}^k \rightarrow \{0,1\}^p$ whose components are all square-free monomials of degree at most d in the indeterminates $x_1, \ldots, x_k$. The value p equals the number of square-free monomials of degree at most d and depends on d and k. For any $\beta \in \mathbb{R}^p$, the *intensity* of the rule setting $x \in \{0,1\}^k$ is

$$\lambda(x, \beta) = e^{f(x)^T \beta}.$$

The *information matrix* of x at β is the rank one matrix

$$M(x, \beta) = \lambda(\beta, x) f(x) f(x)^T.$$

The *information matrix polytope* is

$$P(\beta) = \mathrm{conv}\{M(x, \beta) : x \in \{0,1\}^k\}.$$

The case $d = 1$ and k arbitrary is known as the model with k *independent rules*. In this case $f(x) = (1, x_1, \ldots, x_k)$ and $p = 1 + k$. Then $P(0)$ is known as the *correlation polytope*, a well studied polytope in combinatorial optimization. This case is particularly well-behaved, well-studied, and relevant for practitioners. It was investigated in depth in [7–9,12].

The *pairwise interaction model* arises for $d = 2$, where

$$f(x) = (1, x_1, \ldots, x_k, x_1 x_2, x_1 x_3, \ldots, x_{k-1} x_k)$$

and $p = 1 + k + \binom{k}{2}$. This situation is already so intricate that neither an algebraic description of the model (the set of vectors $(\lambda(x, \beta))_{x \in \{0,1\}^k}$ parametrized by $\beta \in \mathbb{R}^p$) nor an explicit description of the polytope $P(\beta)$ are known.

An *approximate design* is a vector $(w_x)_{x \in \{0,1\}^k} \in [0,1]^{2^k}$ of non-negative weights with $\sum_x w_x = 1$. To each approximate design there is a matrix $M(w, \beta) = \sum_x w_x M(x, \beta) \in P(\beta)$. The main problem of classical design theory is to find designs w that are optimal with regard to some criterion. We limit ourselves to D-optimality, where the determinant ought to be maximized. To simplify the problem, we also only consider maximizing the determinant over $P(\beta)$, and not finding explicit weights w that realize an optimal matrix in $P(\beta)$. In non-linear regression, such as the Poisson regression considered here, this optimal solution depends on β (in linear regression it does not). Our approach is to consider the set of optimization problems for all β and subdivide them into regions where the optima are structurally similar. These regions of optimality are semi-algebraic.

In our setting, there are always matrices with positive determinant in $P(\beta)$. Since the vertices are rank one matrices, the optimum cannot be attained on any face that is the convex hull of fewer than p vertices. A design w is *saturated* if it achieves this lower bound, that is, $|\mathrm{supp}(w)| = p$.

As the logarithm of the determinant is concave, for each given β, the optimization problem can be treated with the tools of convex optimization. The design problem is to determine the changes in the optimal solution as β varies.

A special design, relevant for practitioners and studied in [12], is the *corner design* $w_{k,d}^*$. It is the saturated design with equal weights $w_x = 1/p$ for all $x \in \{0,1\}^k$ with $|x|_1 \le d$. For example, for $k = 3$ rules and interaction order $d = 2$ the regression function is $f(x_1, x_2, x_3) = (1, x_1, x_2, x_3, x_1 x_2, x_1 x_3, x_2 x_3)$ and there are $p = 7$ parameters. The corner design has weight $1/7$ on the seven binary 3-vectors different from $(1,1,1)$.

Saturated designs are mathematically attractive due to their combinatorial nature. It is reflected in the following classical theorem of Kiefer and Wolfowitz which is a main tool in the theory of optimal designs. See [15, Sect. 9.4] or [13] for details and proofs.

Theorem 1. *Let $X \subset \{0,1\}^k$ be of size p. There is a matrix with optimal determinant in the face* $\mathrm{conv}\{M(x,\beta) : x \in X\}$ *if and only if for all $x \in \{0,1\}^k$*

$$\lambda(x,\beta)(F^{-T}f(x))^T \psi^{-1}(\beta)(F^{-T}f(x)) \le 1.$$

where F is the $(p \times p)$-matrix with rows $f(x), x \in X$ and ψ is the diagonal matrix $\mathrm{diag}(e^{\beta_1}, \ldots, e^{\beta_p})$. *If this is the case, then the optimal point is $\frac{1}{p}\sum_{x \in X} M(x,\beta)$, the geometric center of the face.*

After changing the scale by the introduction of parameters $\mu_i = e^{\beta_i}$, Theorem 1 yields a system of rational polynomial inequalities in the μ_i. Together with the requirements $\mu_i > 0$, we find a semi-algebraic characterization of regions of optimality for saturated designs.

For example, the inequalities corresponding to the corner design are the topic of [12]. It can be seen that there always exist parameters $\beta_1, \ldots, \beta_p$ that satisfy the inequalities in Theorem 1. A good benchmark for our understanding of the semi-algebraic geometry of the Rasch Poisson counts model is to understand the other saturated designs, raised as [12, Question 3.7].

Question 1. When $\beta_i < 0$, for all $i = 1, \ldots, p$, is the corner design the only saturated design w that admits parameters β such w is D-optimal for β?

For $d = 1, k = 3$, Question 1 has been answered by Graßhoff et al. They have shown that, up to fractional factorial designs at $\beta = 0$, only the corner design yields a feasible system [9]. Using computer algebra, the case $d = 1, k = 4$ can be attacked.

3 Non-optimality of Saturated Designs for Four Predictors

Our benchmark problem for computational treatment of inequality systems is an extension of the content of [9] to the case $d = 1$ and $k = 4$. Together with Philipp Meissner, at the time of writing a master student, we have undertaken computational experiments. In this situation $p = 5$ and a saturated design is specified by a choice of its support $X \subset \{0,1\}^4$ with $|X| = 5$. A number of reductions applies. For example, if all 5 points lie in a three-dimensional cube, the determinant can be seen to be equal to zero throughout the face, so that optimality is precluded from the beginning. The hyperoctahedral symmetry acts on the designs and the inequalities. Therefore only one representative of each orbit has to be considered. After these reductions we are left with 17 systems of inequalities, one for each orbit of supports of saturated designs. One orbit corresponds to the corner design for which there always exist parameters at which it is optimal. It is conjectured that the remaining 16 saturated designs admit no parameters under which they are optimal. Theorem 1 translates this conjecture into the infeasibility of 16 inequality systems. The most complicated looking among them is the following.

$$4\mu_1\mu_2\mu_3\mu_4 + \mu_1\mu_3 + \mu_1\mu_2 + 4\mu_2\mu_3 + \mu_4 - 9\mu_2\mu_3\mu_4 \leq 0$$
$$4\mu_1\mu_2\mu_3\mu_4 + \mu_2\mu_3 + \mu_1\mu_2 + 4\mu_1\mu_3 + \mu_4 - 9\mu_1\mu_3\mu_4 \leq 0$$
$$4\mu_1\mu_2\mu_3\mu_4 + \mu_2\mu_3 + \mu_1\mu_3 + 4\mu_1\mu_2 + \mu_4 - 9\mu_1\mu_2\mu_4 \leq 0$$
$$\mu_1\mu_2\mu_3\mu_4 + \mu_2\mu_3 + \mu_1\mu_3 + \mu_1\mu_2 + \mu_4 - 9\mu_1\mu_2\mu_3 \leq 0$$
$$\mu_1\mu_2\mu_3\mu_4 + \mu_1\mu_3 + \mu_2\mu_3 + 4\mu_1\mu_2 + 4\mu_4 - 9\mu_3\mu_4 \leq 0$$
$$\mu_1\mu_2\mu_3\mu_4 + \mu_1\mu_2 + 4\mu_1\mu_3 + \mu_2\mu_3 + 4\mu_4 - 9\mu_2\mu_4 \leq 0$$
$$\mu_1\mu_2\mu_3\mu_4 + \mu_1\mu_2 + 4\mu_2\mu_3 + \mu_1\mu_3 + 4\mu_4 - 9\mu_1\mu_4 \leq 0$$
$$\mu_1\mu_2\mu_3\mu_4 + 4\mu_1\mu_3 + 4\mu_2\mu_3 + \mu_1\mu_2 + \mu_4 - 9\mu_3 \leq 0$$
$$\mu_1\mu_2\mu_3\mu_4 + 4\mu_1\mu_2 + \mu_1\mu_3 + 4\mu_2\mu_3 + \mu_4 - 9\mu_2 \leq 0$$
$$\mu_1\mu_2\mu_3\mu_4 + 4\mu_1\mu_2 + \mu_2\mu_3 + 4\mu_1\mu_3 + \mu_4 - 9\mu_1 \leq 0$$
$$4\mu_1\mu_2\mu_3\mu_4 + \mu_1\mu_2 + \mu_1\mu_3 + \mu_2\mu_3 + 4\mu_4 - 9 \leq 0$$
$$\mu_1 > 0, \quad \mu_2 > 0, \quad \mu_3 > 0, \quad \mu_4 > 0.$$

The interested reader is invited to try her favorite method of showing infeasibility of this system. We have first tried SDP methods. In the best situation, they would yield an Positivstellensatz infeasibility certificate (maybe for a relaxation). For this we used YALMIP [14] together with the MOSEK solver [2] to set up moment relaxations. While in general this method works and is reasonably easy to set up, it is not applicable here as the infeasibility of the system seems to depend on the strictness of the inequalities $\mu_i > 0$. Since spectrahedra are closed, the SDP method only works with closed sets. Tricks like introducing a new variables which represents the inverses of the μ_i lead to unbounded spectrahedra. Bounding these is equivalent to imposing an arbitrary bound $\mu_i \geq \epsilon$. With this the degrees of the Positivstellensatz certificate for infeasibility grow (quickly) when $\epsilon \to 0$. In total, the numerical method can give some intuition, but it is not feasible to yield proofs for the benchmark problem.

Our second attempt was to use QEPCAD [4], a somewhat dated open source implementation of quantifier elimination. The system is very easy to use, but unfortunately it seems to have problems already with small polynomial inequality systems due to a faulty memory management in the underlying library SACLIB. There have been attempts to rectify the situation [17], but their source code is unavailable and the authors are unreachable.

Finally, we tried the closed source implementation of quantifier elimination in MATHEMATICA [16] and were positively surprised about its power. Its function REDUCE quickly yields that FALSE is equivalent to the existence of $\mu_1, \ldots, \mu_4$ satisfying some of the 17 inequality systems. However, the benchmark system above seems out of reach. From here, the road is open to trying various semi-automatic tricks. For example, MATHEMATICA can confirm within a reasonable time frame that there is no solution to the above inequality system when $\mu_3 = \mu_4$ is also imposed. A summary of our findings will appear in the forthcoming master thesis of Philipp Meissner.

4   Outlook

Whoever takes an experimental stance towards mathematics will, from time to time, be faced with polynomial systems of equations and inequalities. We have shown one such a situation coming from statistics here and there are more to be found from the various equivalence theorems in design theory [15].

Deciding if such a system has a solution is a basic task. The technology to solve it should be developed to a degree that a practitioner can just work with off the shelf software to study their polynomial systems. For systems of equations this is a reality. There are several active open source systems that abstract Gröbner bases computations to a degree that one can simply work with ideals [1, 10, 11]. For systems of polynomial inequalities, the situation is not so nice. The method to exactly decide feasibility of general polynomial inequality systems is quantifier elimination [3, Chapter 14]. The only viable open source software for quantifier elimination is QEPCAD which appears unmaintained for about a decade. There do exist closed implementations that seem to work much better, for example in MATHEMATICA. Whether one accepts a proof by computation in a closed source system is a contentious matter.

Problem 1. Develop a fast and user-friendly open source tool to study the feasibility of polynomial inequality systems with quantifier elimination.

We shall not fear the complexity theory. The documentation and use cases of QEPCAD demonstrate that many interesting applications were in the reach of quantifier elimination already a decade ago. Gröbner bases were deemed impractical in view of their complexity theory, yet they are an indispensable tool now. We hope that in the future exact methods in semi-algebraic geometry can be developed to the same extend as exact methods in algebraic geometry are developed.

Finally, for experimentation one can always resort to numerical methods. Via the Nullstellensatz and the various Positivstellensätze the optimization community has developed very efficient methods to deal with polynomial systems of equations and inequalities [5].

Acknowledgement. The author is supported by the Research Focus Dynamical Systems (CDS) of the state Saxony-Anhalt.

References

1. Abbott, J., Bigatti, A.M., Lagorio, G.: CoCoA-5: a system for doing Computations in Commutative Algebra. http://cocoa.dima.unige.it
2. MOSEK ApS, The MOSEK optimization toolbox for MATLAB. Version 7.1 (revision 28) (2015)
3. Basu, S., Pollack, R.D., Roy, M.-F.: Algorithms in Real Algebraic Geometry, vol. 10. Springer, Heidelberg (2006)
4. Brown, C.W.: QEPCAD B: a program for computing with semi-algebraic sets using CADs. ACM SIGSAM Bull. **37**(4), 97–108 (2003)
5. De Loera, J.A., Malkin, P.N., Parrilo, P.A.: Computation with polynomial equations and inequalities arising in combinatorial optimization. In: Lee, J., Leyffer, S. (eds.) Mixed Integer Nonlinear Programming, pp. 447–481. Springer, New York (2012)
6. Doebler, A., Holling, H.: A processing speed test based on rule-based item generation: an analysis with the Rasch Poisson counts model. In: Learning and Individual Differences (2015). doi:10.1016/j.lindif.2015.01.013
7. Graßhoff, U., Holling, H., Schwabe, R.: Optimal design for count data with binary predictors in item response theory. In: Ucinski, D., Atkinson, A.C., Patan, M. (eds.) Advances in Model-Oriented Design and Analysis, pp. 117–124. Springer, Switzerland (2013)
8. Graßhoff, U., Holling, H., Schwabe, R.: Optimal design for the Rasch Poisson counts model with multiple binary predictors, Technical report (2014)
9. Graßhoff, U., Holling, H., Schwabe, R.: Poisson model with three binary predictors: when are saturated designs optimal? In: Steland, A., Rafajłowicz, E., Szajowski, K. (eds.) Stochastic Models, Statistics and Their Applications, pp. 75–81. Springer International Publishing, Switzerland (2015)
10. Grayson, D.R., Stillman, M.E.: Macaulay2, a software system for research in algebraic geometry. http://www.math.uiuc.edu/Macaulay2/
11. Greuel, G.-M., Pfister, G., Schönemann, H.: Singular 3.0 – a computer algebra system for polynomial computations. In: Kerber, M., Kohlhase, M. (eds.) Symbolic Computation and Automated Reasoning, The Calculemus-2000 Symposium, pp. 227–233. A. K. Peters Ltd., Natick (2001)
12. Kahle, T., Oelbermann, K.-F., Schwabe, R.: Algebraic geometry of Poisson regression. J. Algebraic Stat. (2015, to appear). arXiv:1510.05261
13. Kiefer, J., Wolfowitz, J.: The equivalence of two extremum problems. Can. J. Math. **12**(5), 363–365 (1960)
14. Löfberg, J.: YALMIP: a toolbox for modeling and optimization in MATLAB. In: Proceedings of the CACSD Conference, Taipei, Taiwan (2004)
15. Pukelsheim, F.: Optimal design of experiments. In: Classics in Applied Mathematics, vol. 50. SIAM (2006)
16. Wolfram Research, Mathematica 10.4.1 (2016)
17. Richardson, D.G., Krandick, W.: Compiler-enforced memory semantics in the SACLIB computer algebra library. In: Ganzha, V.G., Mayr, E.W., Vorozhtsov, E.V. (eds.) CASC 2005. LNCS, vol. 3718, pp. 330–343. Springer, Heidelberg (2005)

Combinatorial and Geometric View of the System Reliability Theory

Fatemeh Mohammadi[✉]

Institut für Mathematik, Technische Universität Berlin, 10623 Berlin, Germany
fatemeh.mohammadi@math.tu-berlin.de,
http://page.math.tu-berlin.de/~mohammad/

Abstract. Associated to every coherent system there is a canonical ideal whose Hilbert series encodes the reliability of the system. We study various ideals arising in the theory of system reliability. Using ideas from the theory of orientations, and matroids on graphs we associate a polyhedral complex to our system so that the non-cancelling terms in the reliability formula can be read from the labeled faces of this complex. Algebraically, this polyhedron resolves the minimal free resolution of these ideals. In each case, we give an explicit combinatorial description of non-cancelling terms in terms of acyclic orientations of graph and the number of regions in the graphic hyperplane arrangement. This resolves open questions posed by Giglio-Wynn and develops new connections between the theory of oriented matroid, the theory of divisors on graphs, and the theory of system reliability.

Keywords: System reliability · Betti numbers · Polyhedral cellular complex · Orientations

1 Introduction

Inspired by the work of Naiman-Wynn [NW92] and Giglio-Wynn [GW04] connecting system reliability to Hilbert functions of their associated ideals, we study reliability of networks through the lens of polyhedral geometry. We apply the *syzygy tool* from commutative algebra to encode the (non-cancelling) terms in the reliability formula for various systems. This gives a more clear insight into the structure of our systems.

The starting point of this paper is to study the following network flow reliability problem. Let $G = (V, E)$ be a graph. Assume that the vertices are reliable but each edge may fail (with the probability $1 - p_e$). A popular game in system reliability theory is to compute the probability of the union of certain events under various restrictions. The classical method to compute the system reliability is to apply the inclusion-exclusion principle of probability theory which is computationally expensive. On the other hand, the system reliability formula is equal to the numerator of Hilbert series of a certain ideal associated to the network. The special networks have been studied in [GW04], and the general

© Springer International Publishing Switzerland 2016
G.-M. Greuel et al. (Eds.): ICMS 2016, LNCS 9725, pp. 148–153, 2016.
DOI: 10.1007/978-3-319-42432-3_19

case was stated as an open problem. We recommend [Doh03, Sect. 6] and the survey articles [AB84] by Agrawal-Barlow, and [JMM88] by Johnson-Malek for an overview of the subject.

2 Source-to-Multiple-Terminal (SMT) System

A well-known example in the theory of system reliability is the SMT system. We fix a pointed graph (G, s) with the edge set $E(G)$, and the oriented edge set $\mathbb{E}(G)$. We study the probability that there exists at least one (oriented) path from s to every other vertex of G. We let $R = k[\mathbf{x}]$ be the polynomial ring in the variables $\{x_e : e \in E(G)\}$. The ideal corresponding to the SMT system is the spanning tree ideal of G. For each spanning tree T of G, let $\mathcal{O}_T$ denote the orientation of T with a unique source at s (i.e. the orientation obtained by orienting all paths away from s). Any spanning tree T of G gives rise to a monomial $\mathbf{x}^T = \prod_{e \in E(T)} x_e$. We define the *spanning tree ideal* $\mathfrak{I}_G$ as

$$\mathfrak{I}_G = \langle \mathbf{x}^T : T \text{ is a spanning tree of } G \rangle.$$

2.1 Reliability of the SMT System

Let $I \subset R$ be an ideal generated by monomials $I = \langle m_1, m_2, \ldots, m_\ell \rangle$. A graded free resolution of I is an exact sequence of the form

$$\mathcal{F} : 0 \to \cdots \to F_i \xrightarrow{\varphi_i} F_{i-1} \to \cdots \to F_0 \xrightarrow{\varphi_0} I \to 0$$

where all F_i's are free R-modules and all differential maps φ_i's are graded. The resolution is called *minimal free resolution* (MFR) if $\varphi_{i+1}(F_{i+1}) \subseteq \mathfrak{m}F_i$ for all $i \geq 0$, where $\mathfrak{m}$ is the ideal generated by all variables in R. The i-th *Betti number* $\beta_i(I)$ of I is the rank of F_i. The i-th *graded Betti number* in degree $\mathsf{j} \in \mathbb{Z}^m$, denoted by $\beta_{i,\mathsf{j}}(I)$, is the rank of the degree j part of F_i. These integers encode very subtle numerical information about the ideal (e.g. its Hilbert series). For the spanning tree ideal $\mathfrak{I}_G$, we express the numerator of its multigraded Hilbert function as

$$\mathcal{H}_{\mathfrak{I}_G}(\mathbf{x}) = 1 - \sum_{i=1}^{d} (-1)^{i+1} \Big(\sum_{\mathsf{j} \in \mathbb{N}^n} \beta_{i,\mathsf{j}}(\mathfrak{I}_G) \mathbf{x}^{\mathsf{j}} \Big) = 1 - \mathcal{R}_G(\mathbf{x}),$$

and we call the polynomial $\mathcal{R}_G(\mathbf{x})$ the *reliability polynomial* of its corresponding system. The evaluation of the reliability polynomial in p_e's gives us the probability that the system works and its evaluation in $1 - p_e$'s gives us the probability that the system fails. If $p_e = p$ for all edges, then the reliability formula of the SMT system can be expressed in terms of the Tutte polynomial $T(x, y)$ of G by

$$\mathcal{R}_G(p) = (1 - p)^{|E(G)| - |V(G)| + 1} p^{|V(G)| - 1} T\Big(1, \frac{1}{1 - p}\Big).$$

The following special class of acyclic partial orientations of G arises naturally in our setting, where g denotes the *genus* of the graph, i.e. $g = |E(G)| - |V(G)| + 1$.

Definition 2.1. *Fix a pointed graph* (G, s) *with the (oriented) edge set* $\mathbb{E}(G)$. *For each integer* $0 \leq k \leq g$, *an* oriented k-spanning tree $\mathcal{T}$ *of* (G, s) *is a connected subgraph of* G *on* $V(G)$ *with a unique source at* s *such that*

- $\mathbb{E}(\mathcal{T}) \subset \mathbb{E}(G)$ *with* $|\mathbb{E}(\mathcal{T})| = n - 1 + k$,
- $\mathcal{T}$ *is acyclic.*

The set of all oriented k-*spanning trees of* (G, s) *will be denoted by* $\mathfrak{S}_k(G, s)$. *The set* $\mathfrak{S}_0(G, s)$ *corresponds to the set* $\{O_T : T$ *is a spanning tree of* $G\}$.

Theorem 2.2. *There is a bijection between non-cancelling terms in the reliability polynomial* $\mathcal{R}_G(\mathbf{x})$, *and the set of multigraded Betti numbers of the spanning tree ideal, and the set of oriented* k-*spanning trees.*

2.2 Dual of the SMT System

For a system, its dual is defined such that a path set in a system is a cut set of its dual. In fact for the SMT system, the ideal corresponding to the dual system is the Alexander dual of $\mathfrak{T}_G$. We denote this ideal by $\mathcal{C}_G$ and we call it the cut ideal of graph. Hence by Alexander inversion formula [MS05, Theorem 5.14]

$$\mathcal{R}_G(\mathbf{x}) = \mathcal{H}_{\mathcal{C}_G}(1 - \mathbf{x}).$$

This connection enables us to obtain many numerical and intrinsic information about a network by looking instead at its dual arising in a different setting.

One natural way to describe a resolution of an ideal is through the construction of a polyhedral complex whose faces are labeled by monomials in such a way that the chain complex determining its cellular homology realizes a graded free resolution of the ideal. The study of cellular resolutions was initiated by Bayer-Sturmfels in [BS98].

We fix a pointed graph (G, s) on the vertex set $[n]$ with the edge set $E(G)$. Following [GZ83], we define the *graphic hyperplane arrangement* as follows. This arrangement lives in the Euclidean space $C^0(G, \mathbb{R})$, i.e. the vector space of all real-valued functions on $V(G)$ endowed with the bilinear form $\langle f_1, f_2 \rangle = \sum_{v \in V(G)} f_1(v) f_2(v)$. Let $C^1(G, \mathbb{R})$ be the vector space of all real-valued functions on $\mathbb{E}(G)$, and let $\partial^* : C^0(G, \mathbb{R}) \to C^1(G, \mathbb{R})$ denote the usual coboundary map. For each edge $e \in \mathbb{E}(G)$ let $\mathcal{H}_e \subset C^0(G, \mathbb{R})$ denote the hyperplane

$$\mathcal{H}_e = \{f \in C^0(G, \mathbb{R}) : (\partial^* f)(e) = 0\}.$$

Consider the arrangement $\mathcal{H}'_G = \{\mathcal{H}_e : e \in \mathbb{E}(G)\}$ in $C^0(G, \mathbb{R})$. Since G is connected, we know $\bigcap_{e \in \mathbb{E}(G)} \mathcal{H}_e$ is the 1-dimensional space of constant functions on $V(G)$. We define the *graphic arrangement* corresponding to G, denoted by $\mathcal{H}_G$, to be the restriction of $\mathcal{H}'_G$ to the hyperplane $\{f \in C^0(G, \mathbb{R}) : \sum_{v \in V(G)} f(v) = 0\}$.

There is a one-to-one correspondence between acyclic orientations of G and the regions of $\mathcal{H}_G$ (see [GZ83, Lemma 7.2]). In particular, the connected cuts of G are corresponding to the lowest dimensional regions of $\mathcal{H}_G$. We are mainly interested in acyclic orientations of G with a *unique source* at $s \in V(G)$. For this purpose, we define $\mathcal{H}_s = \{f \in C^0(G, \mathbb{R}) : f(s) = -1\}$. The restriction of the arrangement $\mathcal{H}_G$ to $\mathcal{H}_s$ will be denoted by $\mathcal{H}_G^s$. We denote the *bounded complex* (i.e. the polyhedral complex consisting of bounded cells) of $\mathcal{H}_G^s$ by $\mathcal{B}_G^s$. The restriction of $\mathcal{H}_G$ to $\mathcal{H}_s$ coincides with the restriction of $\mathcal{H}_G$ to

$$(\mathcal{H}_s)' = \{f \in C^0(G, \mathbb{R}) : \sum_{v \neq q} f(v) = 1\}.$$

The regions of $\mathcal{B}_G^s$ are corresponding to acyclic orientations with a unique source at s (see e.g., [GZ83, Theorem 7.3]). Fixing an orientation $\mathcal{O}$ of the graph G will fix the linear forms $(df)(e) = f(e_+) - f(e_-)$ for $e \in \mathcal{O}$ and gives an orientation to the hyperplane arrangement $\mathcal{H}_G^s$. The oriented matroid ideal associated to this oriented hyperplane arrangement $\mathcal{H}_G^s$ is called the *graphic oriented matroid ideal* (see [NPS02, MS15] for more details).

Theorem 2.3. *The polyhedral cell complex $\mathcal{B}_G^s$ supports a minimal free resolution for $\mathcal{C}_G$. In particular, the reliability polynomial $\mathcal{R}_G$ can be read from the faces of $\mathcal{B}_G^s$.*

2.3 Signature Analysis of the SMT System

While system reliability has been studied using Hilbert series of monomial ideals, this is not enough to understand in a deeper sense the behavior of the system under multiple simultaneous failures. In [MSdCW15] we introduce the lcm-filtration of a monomial ideal, and we study the Hilbert series of the ideals corresponding to multiple failures of the SMT system.

The Lcm-Filtration of Ideals. Let $I \subseteq R$ be a monomial ideal and $\{m_1, \ldots, m_r\}$ be a minimal monomial generating system of I. Let I_j be the ideal generated by the least common multiples of all sets of j distinct monomial generators of I,

$$I_j = \langle \operatorname{lcm}(\{m_i\}_{i \in \sigma}) : \sigma \subseteq \{1, \ldots, r\}, |\sigma| = j \rangle.$$

We call I_j the *j-fold lcm-ideal of I*. The ideals I_j form a descending filtration

$$I = I_1 \supseteq I_2 \supseteq \cdots \supseteq I_r,$$

which we call the *lcm-filtration of I*.

   Persistent homology (see e.g., [EH10, Wei11]) computes the topological features of a space presented as a simplicial complex. In this set-up there is a sequence of simplicial complexes as a filtration of the original space. The main objective is to compute the topological Betti numbers of the filtered space (a sequence of simplicial complexes). Based on Stanley-Reisner theory, in an ongoing work [MSdCW], we have defined a new filtration and we have computed all the topological Betti numbers of the space by algebraic methods. Since algebraic Betti numbers (of Stanley-Reisner ideals) contain topological Betti numbers as a subset, we look forward to pursue some applications in persistent homology.

2.4   Multiple Failures

Let S be a coherent system in which several minimal failures can occur at the same time. Let Y be the number of such simultaneous failures. The event $\{Y \geq 1\}$ is the event that at least one elementary failure event occurs, which is the same as the event that the system fails. If x^α and x^β are the monomials corresponding to two elementary failure events, then $\mathrm{lcm}(x^\alpha, x^\beta) = x^{\alpha \vee \beta}$ corresponds to the intersection of the two events and we have $Y \geq 2$. The corresponding ideal is $\langle x^\alpha \rangle \cap \langle x^\beta \rangle$. The full event $Y \geq 2$ corresponds to the ideal generated by all such pairs. The argument extends to $Y \geq k$ and to study the tail probabilities $\mathrm{prob}\{Y \geq k\}$.

Let K_n be the complete graph on n vertices. We are interested to compute the probability $\mathcal{R}_{K_n,j}(\mathbf{x})$ of having at least j failures in the network. Note that $\mathcal{R}_{K_n,1}(\mathbf{x})$ is the reliability polynomial of K_n. We let I_n be the cut ideal of K_n. For integer j, $1 \leq j \leq n$, we let $I_{n,j}$ be the j-fold lcm-ideal of I_n.

We denote by $\mathcal{P}_{n,k}$ the set of k-partitions of $[n]$. For any k-partition of $[n]$, we associate a monomial whose support is the set of edges between distinct blocks of the partition. For example for the partition $\sigma = 12|3|4$ of K_4 we associate the monomial $m_\sigma = x_{13}x_{14}x_{23}x_{24}x_{34}$. We let $P_{n,k}$ be the ideal generated by the monomials associated to the partitions in $\mathcal{P}_{n,k}$.

Theorem 2.4. *For all integers k and n, $1 \leq k \leq n$ we have*

$$I_{n,2^{k-1}} = I_{n,2^{k-1}+1} = \cdots = I_{n,2^k-1} = P_{n,k+1}.$$

In particular $\mathcal{R}_{K_n,j}(\mathbf{x}) = 1 - \mathcal{H}_{P_{n,k+1}}(\mathbf{x})$ for all indices $2^{k-1} \leq j \leq 2^k - 1$, where $\mathcal{H}_{P_{n,k+1}}$ denotes the numerator of Hilbert series of $P_{n,k+1}$.

2.5   Failure Distributions and Signatures

We finish this section by applying the above considerations on the lcm-filtration lof system ideals to probability analysis. We have a tool for computing moments of the probability distribution of the number of elementary cuts (failures) of a given system.

Lemma 2.5. *Let C be the set of elementary cuts for a network reliability problem, and let Y be the number of elementary cuts. Then under the Erdös-Rényi independence model with probability p, the expectation of Y considered as a random variable, is given by*

$$E(Y) = \sum_{\alpha \in C} p^{|\alpha|},$$

where $|\alpha|$ is the degree of α.

For the complete networks, the computation of the moment of the distribution of Y, the number of elementary cuts, is straightforward.

Theorem 2.6. *For the complete graph K_{2r+1} the mean value is:*

$$\mu_{2r+1} = \sum_{k=1}^{r} \binom{2r}{k} p^{k(2r-k)}$$

and for K_{2r} the mean value is

$$\mu_{2r} = \sum_{k=1}^{r-1} \binom{2r}{k} p^{k(2r-k)} + \frac{1}{2} \binom{2r}{r} p^{r^2}.$$

References

[AB84] Agrawal, A., Barlow, R.E.: A survey of network reliability and domination theory. Oper. Res. **32**(3), 478–492 (1984)

[BS98] Bayer, D., Sturmfels, B.: Cellular resolutions of monomial modules. J. Reine Angew. Math. **502**, 123–140 (1998)

[Doh03] Dohmen, K.: Improved Bonferroni Inequalities via Abstract Tubes: Inequalities and Identities of Inclusion-Exclusion Type. Lecture Notes in Mathematics, vol. 1826. Springer, Berlin (2003)

[EH10] Edelsbrunner, H., Harer, J.: Computational topology: an introduction. Am. Math. Soc. (2010)

[GW04] Giglio, B., Wynn, H.P.: Monomial ideals and the Scarf complex for coherent systems in reliability theory. Ann. Statist. **32**, 1289–1311 (2004)

[GZ83] Greene, C., Zaslavsky, T.: On the interpretation of Whitney numbers through arrangements of hyperplanes, zonotopes, non-Radon partitions, and orientations of graphs. Trans. Amer. Math. Soc. **280**(1), 97–126 (1983)

[JMM88] Johnson Jr., A.M., Malek, M.: Survey of software tools for evaluating reliability, availability, and serviceability. ACM Comput. Surv. (CSUR) **20**(4), 227–269 (1988)

[MS05] Miller, E., Sturmfels, B.: Combinatorial Commutative Algebra. Graduate Texts in Mathematics, vol. 227. Springer-Verlag, New York (2005)

[MS15] Mohammadi, F., Shokrieh, F.: Divisors on graphs, binomial and monomial ideals, and cellular resolutions. Mathematische Zeitschrift, pp. 1–44 (2015)

[MSdCW] Mohammadi, F., Sáenz-de Cabezón, E., Wynn, H.P.: Persistent homology based on lcm-filtration for monomial ideals. in preparation

[MSdCW15] Mohammadi, F., Sáenz-de Cabezón, E., Wynn, H.P.: Types of signature analysis in reliability based on Hilbert series. arXiv preprint 2015. arXiv:1510.04427

[NPS02] Novik, I., Postnikov, A., Sturmfels, B.: Syzygies of oriented matroids. Duke Math. J. **111**(2), 287–317 (2002)

[NW92] Naiman, D.Q., Wynn, H.P.: Inclusion-exclusion-Bonferroni identities and inequalities for discrete tube-like problems via Euler characteristics. Ann. Statist. **20**, 43–76 (1992)

[Wei11] Weinberger, S.: What is. persistent homology? Not. AMS **58**(1), 36–39 (2011)

Software of Polynomial Systems

Need Polynomial Systems
Be Doubly-Exponential?

James H. Davenport[1($\boxtimes$)] and Matthew England[2($\boxtimes$)]

[1] Department of Computer Science, University of Bath, Bath BA2 7AY, UK
J.H.Davenport@bath.ac.uk
[2] Faculty of Engineering, Environment and Computing, School of Computing,
Electronics and Maths, Coventry University, Coventry CV1 5FB, UK
Matthew.England@coventry.ac.uk
http://people.bath.ac.uk/masjhd/,
http://computing.coventry.ac.uk/~mengland/

Abstract. Polynomial Systems, or at least their algorithms, have the reputation of being doubly-exponential in the number of variables (see the classic papers of Mayr & Mayer from 1982 and Davenport & Heintz from 1988). Nevertheless, the Bezout bound tells us that number of zeros of a zero-dimensional system is singly-exponential in the number of variables. How should this contradiction be reconciled?

We first note that Mayr and Ritscher in 2013 showed the doubly exponential nature of Gröbner bases is with respect to the dimension of the ideal, not the number of variables. This inspires us to consider what can be done for Cylindrical Algebraic Decomposition which produces a doubly-exponential number of polynomials of doubly-exponential degree.

We review work from ISSAC 2015 which showed the number of polynomials could be restricted to doubly-exponential in the (complex) dimension using McCallum's theory of reduced projection in the presence of equational constraints. We then discuss preliminary results showing the same for the degree of those polynomials. The results are under primitivity assumptions whose importance we illustrate.

Keywords: Computer algebra · Cylindrical algebraic decomposition · Equational constraint · Gröbner bases, Quantifier elimination

1 Introduction

We consider the title question for two of the main tools for polynomial systems: *Gröbner Bases* (GB) and *Cylindrical Algebraic Decomposition* (CAD). For both the common claims of "doubly exponential", refers to "doubly exponential in the number of variables n". All other dependencies, on polynomial degrees d, polynomial coefficient length l, or number of polynomials m, are themselves polynomial in these quantities (albeit with the exponent of d and m possibly exponential in n).

© Springer International Publishing Switzerland 2016
G.-M. Greuel et al. (Eds.): ICMS 2016, LNCS 9725, pp. 157–164, 2016.
DOI: 10.1007/978-3-319-42432-3_20

In Sect. 2 we recall recent improvements to the analysis for GB which inspires us to revisit the complexity of CAD in Sect. 3. Here we describe how recent work for CAD in the presence of equational constraints (equations logically implied by the input) allows for a more subtle analysis. The progress is under the assumption of primitive equational constraints and in Sect. 4 we elaborate on the importance of this.

2   Gröbner Bases

A *Gröbner Basis* (GB) is a particular generating set of an ideal I (defined with respect to a monomial ordering). One definition is that the ideal generated by the leading terms of I is generated by the leading terms of the GB. GB theory allows properties of the ideal to be deduced such as dimension and number of zeros and so are one of the main practical tools for working with polynomial systems. Introduced by Buchberger in his PhD thesis of 1965 [10]; there has been much research to improve and optimise GB calculation, with the F_5 algorithm [21] perhaps the most used approach currently.

It is common (and the authors have done this, to write) "[31] shows that the computation of Gröbner bases is doubly exponential in the number of variables". It is unfortunately also common simply to write "[31] shows that the computation of Gröbner bases is doubly exponential", which while strictly correct if one counts the number of bits in a suitable encoding, is not particularly helpful.

However, we have known for a long time that the complexity of a Gröbner base of a zero-dimensional ideal is "only" singly-exponential in n [27]. These days, a much better reference is [32], which establishes both upper and lover bounds which are *singly* exponential in n, but *doubly* exponential in r, the actual dimension of the ideal. Clearly $r \leq n$, and only in the worst case is $r = n$.

Though we are currently unable to capitalise on the fact, we note that the examples of [31,32] are of non-radical ideals. The effective Nullstellensatz of [26] is only singly-exponential in the number of variables for membership in the *radical of* an ideal, giving us reason to believe it may be possible to prove a singly-exponential bound for radical ideals. GB technology is also needed to realise similar improvements to the complexity bound of CAD, as discussed next.

3   Cylindrical Algebraic Decomposition

3.1   Background

A *cylindrical algebraic decomposition* (CAD) is a *decomposition* of $\mathbb{R}^n$ into cells. The cells are arranged *cylindrically*, meaning the projections of any pair with respect to the given ordering are either equal or disjoint. We assume variables labelled according to their ordering (so the projections considered are $(x_1, \ldots, x_\ell) \rightarrow (x_1, \ldots, x_k)$ for $k < \ell$) with the highest ordered variable present said to be the *main variable*. Finally, by *algebraic* we mean semi-algebraic: each cell can be described with a finite sequence of polynomial constraints.

A CAD is produced to be invariant for input; originally *sign-invariant* for a set of input polynomials (so on each cell each polynomial is positive, zero or negative), and more recently *truth-invariant* for input Boolean-valued formulae built from the polynomials (so on each cell each formula is either true or false). Unlike Gröbner Bases we may now consider general polynomial systems instead of just equations.

CAD usually involves two phases. The first *projection*, applies operators recursively on polynomials, each time producing a set with one less variable which together define the *projection polynomials*. These are used in the second phase, *lifting*, to build CADs incrementally by dimension. First a CAD of the real line is built according to the real roots of the univariate polynomials. Next, a CAD of $\mathbb{R}^2$ is built by repeating the process over each cell in $\mathbb{R}^1$ with the bivariate polynomials evaluated at a sample point of the cell in $\mathbb{R}^1$. We call the cells where a polynomial vanishes *sections* and those regions in-between *sectors*, which together form the *stack* over the cell. Taking the union of these stacks gives the CAD of $\mathbb{R}^2$. The process is repeated until a CAD of $\mathbb{R}^n$ is produced. In each lift we extrapolate the conclusions drawn from working at a sample point to the whole cell requiring validity theorems for the projection operator used.

CAD was originally introduced by Collins for quantifier elimination (QE) in real closed fields [1] with applications since ranging from parametric optimisation [22] and epidemic modelling [9], to reasoning with multi-valued functions [15] and the derivation of optimal numerical schemes [20]. There has been much work on improving Collins' original approach most notably refinements to the projection operator [6,23,28]; early termination of lifting [14,35]; and symbolic-numeric schemes [25,33]. Some recent advances include dealing with multiple formulae [3,4]; local projection [7,34]; decompositions via complex space [2,12]; and the development of heuristics for CAD problem formulation [5,17,36] including machine learning approaches [24].

3.2   Complexity

CAD has long been known to have worst case complexity doubly exponential [8,16]. Suppose the input consists of m polynomials (perhaps derived from formulae) in n variables of maximum degree d in any one variable. Section 2.3 of [4] describes in detail how the complexity of CAD algorithms may be measured in terms of a bound on the total number of cells produced (closely correlated to the timings but allowing for simpler implementation independent comparisons) based on improvements to techniques introduced by McCallum's thesis. In particular, the dominant term in that bound for a sign-invariant CAD produced using the algorithm of [28] is

$$(2d)^{2^n-1}m^{2^n-1}2^{2^{n-1}-1}. \tag{1}$$

I.e. the CAD grows doubly exponentially with the number of variables n. The analysis shows that by the end of the projection stage we have M polynomials in $\mathbb{R}^1$, each of degree D, where $D = d^{2^{O(n)}}$ and $M = m^{2^{O(n)}}$. However, [8,16] respectively find lower bounds with $D = d^{2^{\Omega(n)}}$ and $M = m^{2^{\Omega(n)}}$ with the underlying polynomials all simple, showing that the doubly-exponential difficulty of CAD resides in *the complicated number of ways simple polynomials can interact.*

So Need CAD Be Doubly Exponential? Given the previous discussion the answer is yes, but as with GB we need not settle for "doubly exponential *in the number of variables n*". We might hope for "doubly exponential *in the dimension*", but this is thwarted by the fact that the examples of [8,16] are in fact zero-dimensional. Nevertheless, we can take advantage of certain dimensional reductions when made explicit through the identification of *equational constraints* (ECs), polynomial equations logically implied by formulae.

The presence of an EC restricts the dimension of the solution space and so we may expect the CAD to be doubly exponential in $n - \ell$ where ℓ is the number of ECs taken advantage of. Of course, we would no longer be building CADs sign-invariant for polynomials but ones truth-invariant for formulae. The present authors have demonstrated this first for the part of the bound dependent on m (number of polynomials) in [18] and then for the part dependent on d (maximum degree) in [19] (work recently accepted for publication).

3.3   CAD with Multiple ECs

Collins noticed that in the presence of an EC a truth-invariant CAD need only be sign-invariant throughout for the defining polynomial of the EC with other polynomials sign-invariant only on the sections of that polynomial [13]. This led McCallum to develop restrictions to his projection operator from [28] in [29] (for the first projection) and [30] (for subsequent projections). See [18, Section 2.1] for a more detailed summary of this theory. These operators work with a single EC and so the CAD algorithm may take advantage of only one in each main variable. However, [30] also introduced a process to derive ECs in lower main variables based on the observation that the resultant of the polynomials defining two ECs itself defines an EC.

In [18] the present authors reviewed the theory of reduced projection operators. In particular we introduced two refinements to the lifting phase of CAD which follow from McCallum's theory of reduced projection operators:

1. Minimising lifting polynomials: When lifting *to* $\mathbb{R}^k$ if there exists an EC with main variable k then we need only lift with respect to (isolate roots of) this.
2. Minimising real root isolation: When lifting *over* $\mathbb{R}^k$ if there exists an EC with main variable k then we need only isolate real roots over sections (allowing sectors to be trivially lifted to a cylinder).

These refinements require us to discard two embedded principles of CAD:

– That the projection polynomials are a fixed set: we now differ the polynomials used in projection from lifting and keep track of which relate to ECs.

- That the invariance structure of the final CAD can be expressed in terms of sign-invariance of polynomials: The final CAD may not no longer be sign-invariant for any one polynomial polynomials, even ECs, but is still guaranteed to be truth invariant for the formula.

In [18, Section 5] we used the complexity analysis techniques of [4] to show that a CAD in which the first ℓ projections had a designated EC had dominant term complexity bound of the form $(2d)^{\mathcal{O}(2^n)}(2m)^{\mathcal{O}(2^{n-\ell})}$. I.e. we have reduced the number of polynomials involved accordingly but not their degree.

The present authors considered what could be done with respect to the degree recently in [19]. The theory of iterated resultants as considered by Busé and Mourrain [11] suggested that the iterated univariate resultants produced by CAD (and in particular in the identification of ECs for subsequent projections) were more complicated than the information they needed to encode. The true multivariate resultants were contained as a factor and grow in degree exponentially rather than doubly exponentially. The key result had to be adapted from [11] to change the arguments from total degree in all variables to the degree in at most one variable required for bounding the number of CAD cells produced.

The authors proposed using GB technology for the generation of the ECs in subsequent projections to realise this limit in degree growth. This leads to the other projection polynomials growing exponentially in $O(\ell^2)$ but remember that these are not used during lifting ([18] improvement (1) from above) and thus not counted towards the cell count bounds (although they do boost the degree of polynomials involved in projections without ECs). The outcome of this approach is a dominant term complexity bound of the form $(\ell d)^{\mathcal{O}(2^{n-\ell})}(2m)^{\mathcal{O}(2^{n-\ell})}$.

Restrictions. There are some restrictions to the work as acknowledged in [18, 19]. First, the analysis assumes the designated ECs are in strict succession at the start of projection. This restrictions was made to ease the complexity analysis (with the formal algorithm specification and implementations not adhering).

The substantial restriction is that the theory of CAD with multiple ECs is only developed for primitive ECs. Possibilities to remove this restriction are discussed in [18] and could involve leveraging the TTICAD theory of [3, 4]. A TTICAD (truth-table invariant CAD) allows for savings from ECs when building a CAD for multiple formulae at once. Currently the theory is only developed for ECs in the main variable of the system and so an analogous extension to subsequent projections is first required for TTICAD itself.

4 The Primitivity Restriction

We finish by considering the classic complexity results of [8, 16] in light of the above recent progress. We see the importance of the aforementioned primitivity restriction.

The examples in both [8,16] rest on the following construction. Let $P_k(x_k, y_k)$ be the statement $x_k = f(y_k)$ and then define recursively

$$P_{k-1}(x_{k-1}, y_{k-1}) := \tag{2}$$

$$\underbrace{\exists z_k \forall x_k \forall y_k}_{Q_k} \underbrace{\left((y_{k-1} = y_k \wedge x_k = z_k) \vee (y_k = z_k \wedge x_{k-1} = x_k)\right)}_{L_k} \Rightarrow P_k(x_k, y_k).$$

This is $\exists z_k (z_k = f(y_{k-1}) \wedge x_{k-1} = f(z_k))$, i.e. $x_{k-1} = f(f(y_{k-1}))$. It is repeated nesting of this procedure that builds the doubly-exponential growth, so that

$$Q_{k-1} L_{k-1} \Rightarrow (Q_k L_k \Rightarrow P_k(x_k, y_k)), \tag{3}$$

gives $x_{k-2} = f(f(f(f(y_{k-2}))))$ etc. Rewriting (3) in prenex form gives

$$Q_{k-1} Q_k \neg L_{k-1} \vee \neg L_k \vee P_k(x_k, y_k). \tag{4}$$

The negation of (4) is therefore

$$\overline{Q}_{k-1} \overline{Q}_k L_{k-1} \wedge L_k \wedge \neg P_k(x_k, y_k), \tag{5}$$

where the $\overline{}$ operator interchanges $\forall$ and $\exists$.

Now, L_k can be rewritten as

$$L_k = (y_{k-1} = y_k \vee y_k = z_k) \wedge (y_{k-1} = y_k \vee x_{k-1} = x_k)$$
$$\wedge (x_k = z_k \vee y_k = z_k) \wedge (x_k = z_k \vee x_{k-1} = x_k) \tag{6}$$

and further

$$L_k = (y_{k-1} - y_k)(y_k - z_k) = 0 \wedge (y_{k-1} - y_k)(x_{k-1} - x_k) = 0$$
$$\wedge (x_k - z_k)(y_k - z_k) = 0 \wedge (x_k - z_k)(x_{k-1} - x_k) = 0, \tag{7}$$

which shows L_k to be a conjunction of (imprimitive) equational constraints. This is true for any L_i, hence the propositional part of (5) is a conjunction of eight equalities, mostly imprimitive, and $\neg P_k(x_k, y_k)$. Furthermore there are equalities whose main variables are the first variables to be projected if we try to produce a quantifier-free form of (5). But the quantifier-free form of (5) describes the complement of the semi-algebraic varieties in [8] or [16] (depending which P_k we take) and these have doubly-exponential complexity in n.

The discussion of this section shows the relevance of the primitivity restriction discussed at the end of the previous section and imposed in the work of [18,19]. It may be more than a technicality to remove it.

Acknowledgements. This work was originally supported by EPSRC grant: EP/J003247/1 and is now supported by EU H2020-FETOPEN-2016-2017-CSA project SC^2 (712689). We are also grateful to Professor Buchberger for reminding JHD that Gröbner Bases were applicable to CAD complexity.

References

1. Arnon, D., Collins, G.E., McCallum, S.: Cylindrical algebraic decomposition I: The basic algorithm. SIAM J. Comput. **13**, 865–877 (1984)
2. Bradford, R., Chen, C., Davenport, J.H., England, M., Moreno Maza, M., Wilson, D.: Truth table invariant cylindrical algebraic decomposition by regular chains. In: Gerdt, V.P., Koepf, W., Seiler, W.M., Vorozhtsov, E.V. (eds.) CASC 2014. LNCS, vol. 8660, pp. 44–58. Springer, Heidelberg (2014)
3. Bradford, R., Davenport, J.H., England, M., McCallum, S., Wilson, D.: Cylindrical algebraic decompositions for boolean combinations. In: Proceedings of the ISSAC 2013, pp. 125–132. ACM (2013)
4. Bradford, R., Davenport, J.H., England, M., McCallum, S., Wilson, D.: Truth table invariant cylindrical algebraic decomposition. J. Symbolic Comput. **76**, 1–35 (2015)
5. Bradford, R., Davenport, J.H., England, M., Wilson, D.: Optimising problem formulation for cylindrical algebraic decomposition. In: Carette, J., Aspinall, D., Lange, C., Sojka, P., Windsteiger, W. (eds.) CICM 2013. LNCS, vol. 7961, pp. 19–34. Springer, Heidelberg (2013)
6. Brown, C.W.: Improved projection for cylindrical algebraic decomposition. J. Symbolic Comput. **32**(5), 447–465 (2001)
7. Brown, C.W.: Constructing a single open cell in a cylindrical algebraic decomposition. In: Proceedings of the ISSAC 2013, pp. 133–140. ACM (2013)
8. Brown, C.W., Davenport, J.H.: The complexity of quantifier elimination and cylindrical algebraic decomposition. In: Proceedings of the ISSAC 2007, pp. 54–60. ACM (2007)
9. Brown, C.W., El Kahoui, M., Novotni, D., Weber, A.: Algorithmic methods for investigating equilibria in epidemic modelling. J. Symbolic Comput. **41**, 1157–1173 (2006)
10. Buchberger, B.: Bruno Buchberger's PhD thesis (1965): An algorithm for finding the basis elements of the residue class ring of a zero dimensional polynomial ideal. J. Symbolic Comput. **41**(3–4), 475–511 (2006)
11. Busé, L., Mourrain, B.: Explicit factors of some iterated resultants and discriminants. Math. Comput. **78**, 345–386 (2009)
12. Chen, C., Maza, M.M., Xia, B., Yang, L.: Computing cylindrical algebraic decomposition via triangular decomposition. In: Proceedings of the ISSAC 2009, pp. 95–102. ACM (2009)
13. Collins, G.E.: Quantifier elimination by cylindrical algebraic decomposition - 20 years of progress. In: Caviness, B., Johnson, J. (eds.) Quantifier Elimination and Cylindrical Algebraic Decomposition. Texts and Monographs in Symbolic Computation, pp. 8–23. Springer, Heidelberg (1998)
14. Collins, G.E., Hong, H.: Partial cylindrical algebraic decomposition for quantifier elimination. J. Symbolic Comput. **12**, 299–328 (1991)
15. Davenport, J.H., Bradford, R., England, M., Wilson, D.: Program verification in the presence of complex numbers, functions with branch cuts etc. In: Proceedings of the SYNASC 2012, pp. 83–88. IEEE (2012)
16. Davenport, J.H., Heintz, J.: Real quantifier elimination is doubly exponential. J. Symbolic Comput. **5**(1–2), 29–35 (1988)
17. England, M., Bradford, R., Chen, C., Davenport, J.H., Maza, M.M., Wilson, D.: Problem formulation for truth-table invariant cylindrical algebraic decomposition by incremental triangular decomposition. In: Watt, S.M., Davenport, J.H., Sexton, A.P., Sojka, P., Urban, J. (eds.) CICM 2014. LNCS, vol. 8543, pp. 45–60. Springer, Heidelberg (2014)

18. England, M., Bradford, R., Davenport, J.H.: Improving the use of equational constraints in cylindrical algebraic decomposition. In: Proceedings of the ISSAC 2015, pp. 165–172. ACM (2015)
19. England, M., Davenport, J.H.:The complexity of cylindrical algebraic decomposition with respect to polynomial degree. In: Proceedings of CASC 2016. Springer (2016, to appear). Preprint available at http://arxiv.org/abs/1605.02494
20. Erascu, M., Hong, H.: Synthesis of optimal numerical algorithms using real quantifier elimination (Case Study: Square root computation). In: Proceedings of the ISSAC 2014, pp. 162–169. ACM (2014)
21. Faugère, J.C.: A new efficient algorithm for computing Gröbner bases without reduction to zero (F5). In: Proceedings of the ISSAC 2002, pp. 75–83. ACM (2002)
22. Fotiou, I.A., Parrilo, P.A., Morari, M.: Nonlinear parametric optimization using cylindrical algebraic decomposition. In: 2005 European Control Conference Decision and Control, CDC-ECC 2005, pp. 3735–3740 (2005)
23. Han, J., Dai, L., Xia, B.: Constructing fewer open cells by GCD computation in CAD projection. In: Proceedings of the ISSAC 2014, pp. 240–247. ACM (2014)
24. Huang, Z., England, M., Wilson, D., Davenport, J.H., Paulson, L.C., Bridge, J.: Applying machine learning to the problem of choosing a heuristic to select the variable ordering for cylindrical algebraic decomposition. In: Watt, S.M., Davenport, J.H., Sexton, A.P., Sojka, P., Urban, J. (eds.) CICM 2014. LNCS, vol. 8543, pp. 92–107. Springer, Heidelberg (2014)
25. Iwane, H., Yanami, H., Anai, H., Yokoyama, K.: An effective implementation of a symbolic-numeric cylindrical algebraic decomposition for quantifier elimination. In: Proceedings of the SNC 2009, pp. 55–64 (2009)
26. Kollár, J.: Sharp effective Nullstellensatz. J. Am. Math. Soc. **1**, 963–975 (1988)
27. Lazard, D.: Gröbner Bases, Gaussian elimination and resolution of systems of algebraic equations. In: van Hulzen, J.A. (ed.) ISSAC 1983 and EUROCAL 1983. LNCS, vol. 162, pp. 146–156. Springer, Heidelberg (1983)
28. McCallum, S.: An improved projection operation for cylindrical algebraic decomposition. In: Caviness, B., Johnson, J. (eds.) Quantifier Elimination and Cylindrical Algebraic Decomposition. Texts and Monographs in Symbolic Computation, pp. 242–268. Springer, Heidelberg (1998)
29. McCallum, S.: Factors of iterated resultants and discriminants. J. Symbolic Comput. **27**(4), 367–385 (1999)
30. McCallum, S.: On propagation of equational constraints in CAD-based quantifier elimination. In: Proceedings of the ISSAC 2001, pp. 223–231. ACM (2001)
31. Mayr, E.W., Meyer, A.R.: The complexity of the word problems for commutative semigroups and polynomial ideals. Adv. Math. **46**(3), 305–329 (1982)
32. Mayr, E.W., Ritscher, S.: Dimension-dependent bounds for Gröbner bases of polynomial ideals. J. Symbolic Comput. **49**, 78–94 (2013)
33. Strzeboński, A.: Cylindrical algebraic decomposition using validated numerics. J. Symbolic Comput. **41**(9), 1021–1038 (2006)
34. Strzeboński, A.: Cylindrical algebraic decomposition using local projections. In: Proceedings of the ISSAC 2014, pp. 389–396. ACM (2014)
35. Wilson, D., Bradford, R., Davenport, J.H., England, M.: Cylindrical algebraic sub-decompositions. Math. Comput. Sci. **8**, 263–288 (2014)
36. Wilson, D., England, M., Davenport, J.H., Bradford, R.: Using the distribution of cells by dimension in a cylindrical algebraic decomposition. In: Proceedings of the SYNASC 2014, pp. 53–60. IEEE (2014)

On the Implementation of CGS Real QE

Ryoya Fukasaku[1(✉)], Hidenao Iwane[2], and Yosuke Sato[3]

[1] Tokyo University of Science, Tokyo, Japan
fukasaku@rs.tus.ac.jp
[2] Fujitsu Laboratories LTD/National Institute of Informatics, Tokyo, Japan
iwane@jp.fujitsu.com
[3] Tokyo University of Science, Tokyo, Japan
ysato@rs.kagu.tus.ac.jp

Abstract. A CGS real QE method is a real quantifier elimination (QE) method which is composed of the computation of comprehensive Gröbner systems (CGSs) based on the theory of real root counting. Its fundamental algorithm was first introduced by Weispfenning in 1998. We further improved the algorithm in 2015 so that we can make a satisfactorily practical implementation. For its efficient implementation, there are several key issues we have to take into account. In this extended abstract we introduce them together with some important techniques for making an efficient CGS real QE implementation.

Keywords: Comprehensive Gröbner systems · Real quantifier elimination

1 Introduction

Study of real quantifier elimination (QE) is an important research topic of many areas such as mathematics, computer science, engineering, etc. The cylindrical algebraic decomposition (CAD) algorithm introduced in [5] has been the most efficient real QE method up to the present date together with the improvements by many successive works. For a special type of real QE problems, however, we may have a more practical method. For a quantified formula containing many equalities, we have an alternative real QE method composed of the computations of comprehensive Gröbner systems (CGSs) based on the theory of real root counting, which is called a *CGS real QE* method in this extended abstract. The first CGS real QE algorithm was introduced by V. Weispfenning in [15]. It is implemented in Redlog [12] as the command rlhqe [6]. Unfortunately, however, we cannot say that it achieves pre-eminent performance over CAD based real QE implementations even for quantified formulas with many equalities. In [7] a simpler and more intuitive CGS real QE algorithm was introduced by us. Our algorithm enables us to have a satisfactorily practical implementation of CGS real QE. We have implemented our algorithm in the computer algebra system Maple as a subcommand of SyNRAC [14] and released as free software in [2]. Though our program is a prototype, according to our computation experiments,

© Springer International Publishing Switzerland 2016
G.-M. Greuel et al. (Eds.): ICMS 2016, LNCS 9725, pp. 165–172, 2016.
DOI: 10.1007/978-3-319-42432-3_21

it is superior to other implementations for many examples which contain many equalities. For an efficient implementation of our algorithm there are several key issues we have to take into account. In this extended abstract we introduce them together with some important techniques for making an efficient CGS real QE implementation such as the simplification method reported in [8]. Some of them are embedded in our latest CGS real QE program [3] which achieves much better performance than the previous one.

The extended abstract is organized as follows. In Sect. 2, we give a minimum description of CGS real QE. The reader is referred to [7] for more detailed description. In Sect. 3, we describe the simplification method introduced in [8] in more detail together with some computation data we have obtained by our new implementation [3]. In Sect. 4, we introduce our new theoretical result which is an easy consequence of the main result of [10] but enables us to have a more efficient method to deal with strict inequalities than the original method given in [7], although it has not been implemented in [3] yet.

2   Preliminary

In the rest of the extended abstract $\mathbb{Q}$, $\mathbb{R}$ and $\mathbb{C}$ denote the field of rational numbers, real numbers and complex numbers respectively. $\bar{X}$ denotes some variables $X_1, \ldots, X_n$. $T(\bar{X})$ denotes a set of terms in $\bar{X}$. For an ideal $I \subset \mathbb{R}[\bar{X}]$, $V_{\mathbb{R}}(I)$ and $V_{\mathbb{C}}(I)$ denote the varieties of I in $\mathbb{R}$ and $\mathbb{C}$ respectively (i.e., $V_{\mathbb{R}}(I) = \{\bar{c} \in \mathbb{R}^n | \forall f \in I \ f(\bar{c}) = 0\}$ and $V_{\mathbb{C}}(I) = \{\bar{c} \in \mathbb{C}^n | \forall f \in I \ f(\bar{c}) = 0\}$). Let I be a zero dimensional ideal in a polynomial ring $\mathbb{R}[\bar{X}]$. Considering the residue class ring $\mathbb{R}[\bar{X}]/I$ as a vector space over $\mathbb{R}$, let $t_1, \ldots, t_d$ be its basis. For an arbitrary $h \in \mathbb{R}[\bar{X}]/I$ and each i, j $(1 \leq i, j \leq d)$ we define a linear map $\theta_{h,i,j}$ from $\mathbb{R}[\bar{X}]/I$ to $\mathbb{R}[\bar{X}]/I$ by $\theta_{h,i,j}(f) = ht_i t_j f$ for $f \in \mathbb{R}[\bar{X}]/I$. Let $q_{h,i,j}$ be the trace of $\theta_{h,i,j}$ and M_h^I be a symmetric matrix such that the (i, j)-th component is given by $q_{h,i,j}$. The characteristic polynomial of M_h^I is denoted by $\chi_h^I(x)$. The dimension of $\mathbb{R}[\bar{X}]/I$ is denoted by $\dim(\mathbb{R}[\bar{X}]/I)$. For a polynomial $f(x) \in \mathbb{R}[x]$, the signature of $f(x)$, denoted $\sigma(f(x))$, is an integer which is equal to 'the number of positive real roots of $f(x) = 0$' - 'the number of negative real roots of $f(x) = 0$', that is, $\sigma(f(x)) = \#(\{c \in \mathbb{R} | f(c) = 0, c > 0\}) - \#(\{c \in \mathbb{R} | f(c) = 0, c < 0\})$. The signature of M_h^I, denoted $\sigma(M_h^I)$, is defined as the signature $\sigma(\chi_h^I(x))$ of its characteristic polynomial. The real root counting theorem found independently in [1,10] is the following assertion.

Theorem 1. $\sigma(M_h^I) = \#(\{\bar{c} \in V_{\mathbb{R}}(I) | h(\bar{c}) > 0\}) - \#(\{\bar{c} \in V_{\mathbb{R}}(I) | h(\bar{c}) < 0\})$.

We have the following corollary.

Corollary 2. $\sigma(M_1^I) = \#(V_{\mathbb{R}}(I))$.

Using an obvious relation $h \geq 0 \Leftrightarrow \exists z \ z^2 = h$, we have the following fact.

Lemma 3. Let $h_1, \ldots, h_l$ be polynomials in $\mathbb{R}[\bar{X}]$ and $\bar{Z} = Z_1, \ldots, Z_l$ be new variables. Using the same notations as above, let J be an ideal in $\mathbb{R}[\bar{X}, \bar{Z}]$ defined by $J = I + \langle Z_1^2 - h_1, \ldots, Z_l^2 - h_l \rangle$. Then the following equation holds for some l' $(0 \leq l' \leq l)$:

$$\#(V_\mathbb{R}(J)) = 2^{l'} \#(\{\bar{c} \in V_\mathbb{R}(I) | h_1(\bar{c}) \geq 0, \ldots, h_l(\bar{c}) \geq 0\}).$$

The next theorem plays an important role in our CGS real QE method of [7].

Theorem 4. Let I be a zero dimensional ideal of $\mathbb{R}[\bar{X}]$ and $J = I + \langle Z_1^2 - h_1, \ldots, Z_l^2 - h_l \rangle$ be an ideal of $\mathbb{R}[\bar{X}, \bar{Z}]$ with polynomials $h_1, \ldots, h_l \in \mathbb{R}[\bar{X}]$. Let k be a dimension of $\mathbb{R}[\bar{X}]/I$ and $\{t_1, \ldots, t_k\} \subset T(\bar{X})$ be a basis of the vector space $\mathbb{R}[\bar{X}]/I$, then $\{t_1 Z_1^{e_1} Z_2^{e_2} \cdots Z_l^{e_l}, \ldots, t_k Z_1^{e_1} Z_2^{e_2} \cdots Z_l^{e_l} | (e_1, e_2, \ldots, e_l) \in \{0,1\}^l\}$ forms a basis of the vector space $\mathbb{R}[\bar{X}, \bar{Z}]/J$. Let M_g^J denote a symmetric matrix and χ_g^J denote its characteristic polynomial for a polynomial $g \in \mathbb{R}[\bar{X}]$ induced by the above basis of $\mathbb{R}[\bar{X}, \bar{Z}]/J$. Let M_g^I denote a symmetric matrix and χ_g^I denote its characteristic polynomial for a polynomial $g \in \mathbb{R}[\bar{X}]$ induced by the above basis of $\mathbb{R}[\bar{X}]/I$. Then we have the following equation for some non-zero constant c:

$$\chi_g^J(2^l x) = c \Pi_{(e_1,e_2,\ldots,e_l) \in \{0,1\}^l} \chi_{g h_1^{e_1} h_2^{e_2} \cdots h_l^{e_l}}^I(x).$$

As an easy consequence of this theorem, we have the following fact.

Theorem 5. $\{\bar{c} \in V_\mathbb{R}(I) | h_1(\bar{c}) \geq 0, \ldots, h_l(\bar{c}) \geq 0\} \neq \emptyset \Leftrightarrow \sigma(\chi_1^J(x)) > 0 \Leftrightarrow \sigma(\Pi_{(e_1,e_2,\ldots,e_l) \in \{0,1\}^l} \chi_{h_1^{e_1} h_2^{e_2} \cdots h_l^{e_l}}^I(x)) > 0.$

Based on this theorem we can eliminate quantifiers from the following basic formula by the computation of a CGS:

$$\phi(\bar{Y}) \wedge \exists \bar{X}(f_1(\bar{X},\bar{Y})=0 \wedge \cdots \wedge f_k(\bar{X},\bar{Y})=0 \wedge h_1(\bar{X},\bar{Y}) \geq 0 \wedge \cdots \wedge h_l(\bar{X},\bar{Y}) \geq 0) \quad (1)$$

where $f_i, h_j \in \mathbb{Q}[\bar{X}, \bar{Y}]$ for each i, j and $\phi(\bar{Y})$ is a quantifier free formula with free variables $\bar{Y} = Y_1, \ldots, Y_m$ such that the ideal $I = \langle f_1(\bar{X}, \bar{a}), \ldots, f_k(\bar{X}, \bar{a}) \rangle$ in $\mathbb{R}[\bar{X}]$ is zero dimensional for each $\bar{a} \in \mathbb{R}^m$ satisfying $\phi(\bar{a})$.

Let $\mathcal{S}$ be the subset of $\mathbb{R}^m$ defined by $\mathcal{S} = \{\bar{a} \in \mathbb{R}^m | \phi(\bar{a})\}$. Regarding $\bar{Y}$ as parameters, compute a minimal CGS $\mathcal{G} = \{(\mathcal{S}_1, G_1), \ldots, (\mathcal{S}_s, G_s)\}$ of $\{f_1(\bar{X}, \bar{Y}), \ldots, f_k(\bar{X}, \bar{Y})\}$ over $\mathcal{S}$. For values $\bar{a} \in \mathcal{S}$, let $I = \langle f_1(\bar{X}, \bar{a}), \ldots, f_k(\bar{X}, \bar{a}) \rangle$ be an ideal in $\mathbb{R}[\bar{X}]$ and $h_1, \ldots, h_l$ be polynomials $h_1(\bar{X}, \bar{a}), \ldots, h_l(\bar{X}, \bar{a})$ in $\mathbb{R}[\bar{X}]$. The ideal J and the characteristic polynomial $\chi_1^J(x)$ are defined as in the above theorems. By the properties of a minimal CGS we can have a uniform representation form of $\chi_1^J(x)$ for $\bar{a}$ in a segment $\mathcal{S}_i$. That is, it has a uniform representation $\chi_1^J(x) = x^t + \frac{p_{t-1}(\bar{a})}{q_{t-1}(\bar{a})} x^{t-1} + \cdots + \frac{p_1(\bar{a})}{q_1(\bar{a})} x + \frac{p_0(\bar{a})}{q_0(\bar{a})}$ for every $\bar{a} \in \mathcal{S}_i$ where $p_{t-1}(\bar{Y}), \ldots, p_0(\bar{Y}), q_{t-1}(\bar{Y}), \ldots, q_0(\bar{Y})$ are polynomials in $\mathbb{Q}[\bar{Y}]$ such that $q_{t-1}(\bar{a}) \neq 0, \ldots, q_0(\bar{a}) \neq 0$ for any $\bar{a} \in \mathcal{S}_i$. The above theorem together with Descartes' rule of signs and this representation enables us to compute a quantifier free formula $\psi_i(\bar{Y})$ such that it is equivalent to (1) for each specialization $\bar{Y} = \bar{a}$ for any $\bar{a} \in \mathcal{S}_i$. Using the defining formula $\theta_i(\bar{Y})$ of $\mathcal{S}_i$, we can have a quantifier free formula $(\theta_1(\bar{Y}) \wedge \psi_1(\bar{Y})) \vee \cdots \vee (\theta_s(\bar{Y}) \wedge \psi_s(\bar{Y}))$ equivalent to (1).

Our CGS real QE algorithm introduced in [7] eliminates all quantifiers from an arbitrary quantified formula by applying the above computation recursively.

3   Simplification

Consider the following first order formula:

$$\phi(\bar{Y}) \equiv \exists X (X = f(\bar{Y}) \wedge h_1(X, \bar{Y}) \geq 0 \wedge h_2(X, \bar{Y}) \geq 0).$$

Considering $\bar{Y}$ as parameters, let $I = \langle X - f(\bar{Y}) \rangle$ be the zero dimensional ideal of $\mathbb{R}[X]$ and J be the ideal $I + \langle Z_1^2 - h_1(X, \bar{Y}), Z_2^2 - h_2(X, \bar{Y}) \rangle$ of $\mathbb{R}[X, Z_1, Z_2]$ with new variables Z_1, Z_2. The corresponding characteristic polynomial $\chi_1^J(x)$ has the following form with $H_1 = h_1(f(\bar{Y}), \bar{Y})$ and $H_2 = h_2(f(\bar{Y}), \bar{Y})$:

$$\chi_1^J(x) = (x - 1)(x - H_1)(x - H_2)(x - H_1 H_2).$$

By the first assertion of Theorem 5, the following relation holds:

$$\phi(\bar{Y}) \Leftrightarrow \sigma(\chi_1^J(x)) > 0.$$

If we directly apply Descartes' rule of signs to the expanded quartic polynomial of $\chi_1^J(x)$ we have the following quantifier free formula equivalent to $\phi(\bar{Y})$:

$$(A_0 \leq 0 \wedge A_1 \neq 0) \vee (0 \leq A_1 \wedge 0 \leq A_2 \wedge 0 < A_3) \vee (A_1 \leq 0 \wedge 0 \leq A_2 \wedge A_3 < 0)$$

where A_3, A_2, A_1, A_0 are the coefficients of $\chi_1^J(x)$, i.e., $\chi_1^J(x) = x^4 + A_3 x^3 + A_2 x^2 + A_1 x + A_0$. So $A_3 = -(H_1 H_2 + H_1 + H_2 + 1)$, $A_2 = H_1^2 H_2 + H_1 H_2^2 + 2H_1 H_2 + H_1 + H_2$, $A_1 = -(H_1^2 H_2^2 + H_1^2 H_2 + H_1 H_2^2 + H_1 H_2)$ and $A_0 = H_1^2 H_2^2$.

Since the given formula $\phi(\bar{Y})$ is obviously equivalent to $H_1 \geq 0 \wedge H_2 \geq 0$, the method is certainly inefficient even though we can get the above formula by some simplification technique. This unpleasant situation can be resolved by the following equation derived from the second assertion of Theorem 5:

$$\sigma(\chi_1^J(x)) = \sigma(x - 1) + \sigma(x - H_1) + \sigma(x - H_2) + \sigma(x - H_1 H_2).$$

Using this equation we can easily have the following relation:

$$\sigma(\chi_1^J(x)) > 0 \Leftrightarrow \sigma(x - H_1) \geq 0 \wedge \sigma(x - H_2) \geq 0 \Leftrightarrow H_1 \geq 0 \wedge H_2 \geq 0.$$

By the second assertion of Theorem 5, we also have the following relation:

$$\sigma(\chi_1^J(x)) > 0 \Leftrightarrow \sum_{(e_1, e_2, \ldots, e_l) \in \{0,1\}^l} \sigma(\chi_{h_1^{e_1} h_2^{e_2} \cdots h_l^{e_l}}^I(x)) > 0.$$

Using it we can obtain much simpler representation formula of $\sigma(\chi_1^J(x)) > 0$ for arbitrary $m = \dim(\mathbb{R}[\bar{X}]/I)$ and l than the one introduced in [7] obtained directly applying Descartes' rule of signs to the expanded polynomial of $\chi_1^J(x)$.

Example 6. *Let $m = 2, l = 2$. Note first that $\sigma(\chi_1^J(x)) \geq 0$ and $\sigma(\chi_1^I(x)) \geq 0$ hold by Corollary 2. Secondly, $\sigma(\chi_1^J(x)) = 0$, $\sigma(\chi_1^J(x)) = 4$ or $\sigma(\chi_1^J(x)) = 8$ by Lemma 3. Thirdly, $2 \geq \sigma(\chi_1^I(x)), \sigma(\chi_{h_1}^I(x)), \sigma(\chi_{h_2}^I(x)), \sigma(\chi_{h_1 h_2}^I(x)) \geq -2$ because all of them are quadratic polynomials. So the following holds:*

$$\sigma(\chi_1^J(x)) > 0 \Leftrightarrow \sigma(\chi_{h_1}^I(x)) + \sigma(\chi_{h_2}^I(x)) + \sigma(\chi_{h_1 h_2}^I(x)) > 0.$$

For arbitrary m and l we can similarly compute much simpler representation formula of $\sigma(\chi_1^J(x)) > 0$ than the ones reported in [7]. Those representation formulas are employed in our new CGS real QE implementation [3].

We have computed many real QE problems. In the following, we show some computation data of 5 problems among them. We also give several data of our computation experiments using other real QE implementations for each problem.

Table 1 indicates the computation time of a problem measured in seconds (truncated after the decimal point) for each implementation. '0' means that the computation time is within 1 second, '>' means that the computation does not terminate within 1 hour, 'E' means the computation was crashed with some error. Tables 2 and 3 are for measuring the simplicity of output quantifier free formulas. Table 2 indicates the number of atomic formulas contained in the output formula which can be considered as a barometer of the complexity of its logical structure. Table 3 indicates the number of occurrence of the symbols '+' and '−' in the output formula which can be considered as a barometer of the size of all polynomials contained in the output formula. **cgs** is our new implementation of [3], **cgs-15** is our previous implementation of [2], **syn** is a CAD based QE program of SyNRAC in Maple 18 [14], **rc** is a real QE program in RegularChains package (Version 2015-10-27) in Maple 18 [4], **red** and **res** are Resolve and Reduce of Mathematica 10.3 [9], **rl** and **rlh** are the regular QE rlqe of Redlog and the CGS real QE rlhqe of Redlog [12], **qep** is QEPCAD (Version B-1.69) [11]. All the computations were done by the same computer environment with an Intel CORE i7 CPU 2.60 GHz with 8 GB memory OS Ubuntu 14.10.

Problem 1. $\exists v_5 \, ((-v_1)v_5^2 + v_5^3 - 1 = 0 \wedge v_4 v_5^2 + v_5^3 + v_3 v_5 + v_2 = 0 \wedge v_1 \le v_5 \wedge 0 < v_3)$.

Problem 2. $\exists x \, (ax^4 + x^5 + bx^3 + cx^2 + x + 1 = 0 \wedge 4ax^3 + 5x^4 + 3bx^2 + 2cx + 1 = 0 \wedge 12ax^2 + 20x^3 + 6bx + 2c = 0 \wedge 0 \le x)$.

Problem 3. $\exists x \exists y \exists z \, (xy + axz + yz - 1 = 0 \wedge xyz + xz + xy = a \wedge xz + yz - az - x - y - 1 = 0 \wedge axy = byz \wedge ayz = bzx \wedge c < x + y + z)$.

Problem 4. $\exists x \exists y \exists z \, (axyz + yz - 1 = 0 \wedge xyz + x + 1 = 0 \wedge xz + yz - z - x - b = 0 \wedge 0 \le ay)$.

Problem 5. $\exists c_2 \exists s_2 \exists c_1 \exists s_1 \, (r - c_1 + l(s_1 s_2 + c_1 c_2) = 0 \wedge z - s_1 - l(s_1 c_2 - s_2 c_1) = c_1 \wedge s_1^2 + c_1^2 = 1 \wedge s_2^2 + c_2^2 = 1 \wedge 0 < 4c_1 r + 2c_1 z + 2c_2 l + 5s_1^2)$.

Our previous implementation **cgs-15** based on the old representation may produce a quantifier free formula slightly faster than our new one. However, the obtained quantifier free formula is more complicated in general. Problem 4 and 5 are among such typical examples.

4   How to Deal with Strict Inequalities

In order to eliminate quantifiers from an arbitrary first order formula, we also need to deal with the following basic formula with strict inequalities:

$$\phi(\bar{Y}) \wedge \exists \bar{X}(f_1(\bar{X}, \bar{Y}) = 0 \wedge \cdots \wedge f_k(\bar{X}, \bar{Y}) = 0 \wedge h_1(\bar{X}, \bar{Y}) > 0 \wedge \cdots \wedge h_l(\bar{X}, \bar{Y}) > 0) \quad (2)$$

Table 1. Computation time

	1	2	3	4	5
cgs	0	0	0	2	1
cgs-15	0	0	10	1	1
syn	>	>	>	49	E
rc	3	>	>	6	>
red	>	>	>	1120	>
res	>	>	>	0	>
rl	>	>	>	0	>
rlh	1	1	E	E	>
qep	E	E	>	604	>

Table 2. Number of atomic formulas

	1	2	3	4	5
cgs	35	26	25	50	30
cgs-15	98	51	256	93	144
syn	–	–	–	10091	–
rc	74	–	–	273	–
red	–	–	–	13	–
res	–	–	–	230	–
rl	–	–	–	76	–
rlh	32	108	–	–	–
qep	–	–	–	13	–

Table 3. Number of +, −

	1	2	3	4	5
cgs	224	148	146	320	459
cgs-15	666	452	1585	1245	31697
syn	–	–	–	103024	–
rc	588	–	–	308	–
red	–	–	–	39	–
res	–	–	–	1384	–
rl	–	–	–	481	–
rlh	2651	5310	–	–	–
qep	–	–	–	27	–

with the same conditions of (1). Our algorithm introduced in [7] computes the CGS of $\{f_1(\bar{X}, \bar{Y}), \ldots, f_k(\bar{X}, \bar{Y}), Z_1^2 h_1(\bar{X}, \bar{Y}) - 1, \ldots, Z_l^2 h_l(\bar{X}, \bar{Y}) - 1\}$ regarding $\bar{Y}$ as parameters with new variables $\bar{Z} = Z_1, \ldots, Z_l$ w.r.t. a term order such that $\bar{X} \gg \bar{Z}$, i.e. each variable X_i is lexicographically greater than any variable Z_j. For a segment $\mathcal{S}_i$ such that $G_i \neq \{1\}$, G_i has the following form:

$$G_i = \{g_1(\bar{X}, \bar{Y}), \ldots, g_t(\bar{X}, \bar{Y}), Z_1^2 - p_1(\bar{X}, \bar{Y}), \ldots, Z_l^2 - p_l(\bar{X}, \bar{Y})\}$$

where $\langle g_1, \ldots, g_t \rangle$ is the saturation ideal $\langle f_1, \ldots, f_k \rangle : (h_1 \cdots h_l)^\infty$ in $\mathbb{R}[\bar{X}]$ and each p_j is a inverse of h_j in the residue class ring $\mathbb{R}[\bar{X}]/\langle g_1, \ldots, g_t \rangle$ for any specialization of $\bar{Y}$ by elements $\bar{a}$ in $\mathcal{S}_i$. Since we have the following relation:

$$f_1(\bar{X}, \bar{a}) = 0 \wedge \cdots \wedge f_k(\bar{X}, \bar{a}) = 0 \wedge h_1(\bar{X}, \bar{a}) > 0 \wedge \cdots \wedge h_l(\bar{X}, \bar{a}) > 0$$
$$\Leftrightarrow g_1(\bar{X}, \bar{a}) = 0 \wedge \cdots \wedge g_t(\bar{X}, \bar{a}) = 0 \wedge p_1(\bar{X}, \bar{a}) > 0 \wedge \cdots \wedge p_l(\bar{X}, \bar{a}) > 0$$
$$\Leftrightarrow g_1(\bar{X}, \bar{a}) = 0 \wedge \cdots \wedge g_t(\bar{X}, \bar{a}) = 0 \wedge p_1(\bar{X}, \bar{a}) \geq 0 \wedge \cdots \wedge p_l(\bar{X}, \bar{a}) \geq 0.$$

We can apply the method described in Sect. 2 to eliminate quantifiers.

Computation of a saturation ideal, however, is very heavy in general. This is another reason that our previous program can not deal with many inequalities. If we somehow know that the saturation ideal $\langle f_1, \ldots, f_k \rangle : (h_1 \cdots h_l)^\infty$ is equal to the original ideal $\langle f_1, \ldots, f_k \rangle$, which is equivalent to that $h_1 \neq 0 \wedge \cdots \wedge h_l \neq 0$ for any zero of $\langle f_1, \ldots, f_k \rangle$, before the computation of the saturation ideal, we can avoid this heavy computation. The following theorem which is an easy consequence of the main result of [10] is useful for this purpose.

Theorem 7. Let I be a zero dimensional ideal in $\mathbb{R}[\bar{X}]$. Let $\dim(\mathbb{R}[\bar{X}]/I) = m$. Hence, $\deg(\chi_h^I(x)) = m$ for any $h \in \mathbb{R}[\bar{X}]$. Let $\dim(\mathbb{R}[\bar{X}]/\sqrt{I}) = m'$, then each characteristic polynomial $\chi_h^I(x)$ has a factor $x^{m-m'}$ and $\sigma(\chi_h^{\sqrt{I}}(x)) = \sigma(\chi_h^I(x)/x^{m-m'})$. Furthermore, $h(\bar{a}) \neq 0$ for any $\bar{a} \in \mathbb{V}_{\mathbb{C}}(I)$ if and only if the factor of x in $\chi_h^I(x)$ is exactly $x^{m-m'}$.

Example 8. Let us consider $\exists X(X^2 + AX + B = 0 \wedge X > 0)$. For the ideal $I = \langle X^2 + AX + B \rangle \subset \mathbb{Q}[X]$ with parameters A and B, we have the characteristic polynomials $\chi_1^I(x) = x^2 + (-A^2 + 2B - 2)x + A^2 - 4B$ and $\chi_X^I(x) = x^2 + (A^3 + A - 3AB)x + A^2B - 4B^2$. When $B \neq 0$, both of them has an exactly same factor of x and we have

$$\exists X(X^2 + AX + B = 0 \wedge X > 0) \Leftrightarrow \exists X(X^2 + AX + B = 0 \wedge X \geq 0).$$

We need to compute the saturation ideal $I : X^\infty$ only when $B = 0$. It is much lighter computation than the one for arbitrary A and B.

5 Conclusion and Remarks

We have improved our previous CGS real QE software using the technique described in Sect. 3. While the previous software achieves better performance than the other real QE implementations for many formulas containing many equalities, it has a weak point for formulas containing many inequalities. Our new software achieves much better performance for such formulas than the previous one. The another technique described in Sect. 4 has not been implemented yet. We expect that we can further improve the software.

References

1. Becker, E., Wörmann, T.: On the trace formula for quadratic forms. In: Recent Advances in Real Algebraic Geometry and Quadratic Forms, Berkeley, CA, 1990/1991; San Francisco, CA, 1991, pp. 271–291. Contemporary Mathematics, vol. 155, American Mathematical Society Providence, RI (1994)
2. The first prototype CGS real QE program released (2015). http://www.mi.kagu.tus.ac.jp/~fukasaku/issac2015
3. The latest prototype CGS real QE program released (2016). http://www.mi.kagu.tus.ac.jp/~fukasaku/CGSQE2016

4. Chen, C., Maza, M.M.: Quantifier elimination by cylindrical algebraic decomposition based on regular chains. In: Proceedings of International Symposium on Symbolic and Algebraic Computation (ISSAC 2014), pp. 91–98. ACM (2014)
5. Collins, G.E.: Quantifier elimination for real closed fields by cylindrical algebraic decomposition. In: Brakhage, H. (ed.) Automata Theory and Formal Languages. LNCS, vol. 33, pp. 134–183. Springer, Heidelberg (1975)
6. Dolzmann, A., Gilch, L.A.: Generic hermitian quantifier elimination. In: Buchberger, B., Campbell, J. (eds.) AISC 2004. LNCS (LNAI), vol. 3249, pp. 80–93. Springer, Heidelberg (2004)
7. Fukasaku, R., Iwane, H., Sato, Y.: Real quantifier elimination by computation of comprehensive Gröbner systems. In: Proceedings of International Symposium on Symbolic and Algebraic Computation (ISSAC 2015), pp. 173–180. ACM (2015)
8. Fukasaku, R., Iwane, H., Sato, Y.: Improving a CGS-QE algorithm. In: Kotsireas, I.S., Rump, S.M., Yap, C.K. (eds.) MACIS 2015. LNCS, vol. 9582, pp. 231–235. Springer, Heidelberg (2016). doi:10.1007/978-3-319-32859-1_20
9. Mathematica Tutorial (RealPolynomialSystems). http://reference.wolfram.com/language/tutorial/RealPolynomialSystems.html
10. Pedersen, P., Roy, M.-F., Szpirglas, A.: Counting real zeroes in the multivariate case. In: Proceedings of the Effective Methods in Algebraic Geometry, pp. 203–224 (1993)
11. QEPCAD-Quantifier elimination by partial cylindrical algebraic decomposition. http://www.usna.edu/CS/qepcadweb/B/QEPCAD.html
12. Redlog: an integral part of the interactive computer algebra system reduce. http://www.redlog.eu/
13. Suzuki, A., Sato, Y.: A simple algorithm to compute comprehensive Gröbner bases using Gröbner bases. In: Proceedings of International Symposium on Symbolic and Algebraic Computation (ISSAC 2006), pp. 326–331. ACM (2006)
14. SyNRAC: a software package for quantifier elimination. http://www.fujitsu.com/jp/group/labs/en/resources/tech/freeware/synrac/
15. Weispfenning, V.: A new approach to quantifier elimination for real algebra. In: Caviness, B.F., Johnson, J.R. (eds.) Quantifier Elimination and Cylindrical Algebraic Decomposition, pp. 376–392. Springer, Vienna (1998)

Common Divisors of Solvable Polynomials in JAS

Heinz Kredel[(✉)]

University of Mannheim, Mannheim, Germany
kredel@rz.uni-mannheim.de
http://www.uni-mannheim.de

Abstract. We present generic, type safe (non-unique) common divisors of solvable polynomials software. The solvable polynomial rings are defined with non-commuting variables, moreover, in case of parametric (solvable) coefficients the main variables may not commute with the coefficients. The interface, class organization is described in the object-oriented programming environment of the Java Algebra System (JAS). The implemented algorithms can be applied, for example, in solvable extension field and root construction. We show the design and feasibility of the implementation in the mentioned applications.

Keywords: Generic multivariate solvable polynomials · Common divisors

1 Introduction

We are interested in computations in solvable polynomial rings with coefficients from a (solvable) skew field. For example (notation see Sect. 2)

$$\mathbb{Q}(x, y, z, t; Q_x)/_{\mathcal{I}}\{r; Q_r\},$$

where $Q_x = \{z * y = yz + x, t * y = yt + y, t * z = zt - z\}$ with $\mathcal{I} = (t^2 + z^2 + y^2 + x^2 + 1)$, Q_r is eventually empty. Elements from the quotient skew field $\mathbb{Q}(x, y, z, t; Q_x)$ are constructed as fractions, that is, pairs of (nominator, denominator), of solvable polynomials from the solvable polynomial ring $\mathbb{Q}\{x, y, z, t; Q_x\}$. The nominator and denominator are then checked if they reduce to zero modulo the ideal $\mathcal{I}$, and if they do, the respective number is replaced by zero. Note, it is not possible to replace the respective polynomial by the reduced normal form by the ideal, since this is a different fraction.

The implementation of this example in the Java Algebra System (JAS), [7], requires the generic classes `GenSolvablePolynomial<BigRational>` for the solvable polynomials over a coefficient field, here the rational numbers `BigRational`, next the class `SolvableIdeal<BigRational>` for a twosided ideal in the solvable polynomial ring, and class `SolvableLocalResidue<BigRational>` for elements from the solvable quotient ring.

© Springer International Publishing Switzerland 2016
G.-M. Greuel et al. (Eds.): ICMS 2016, LNCS 9725, pp. 173–180, 2016.
DOI: 10.1007/978-3-319-42432-3_22

These elements can be created with the help of a corresponding 'factory', a means to construct them as Java objects. In the example, first, a solvable polynomial ring mfac is constructed, then the relations from Q_x in rel are added. The factory mfac provides a method parse() to read a polynomial p from a string. Then a a twosided ideal id is created from the polynomial p. With this ideal a solvable local residue ring efac is constructed.

```
String[] vars = new String[] { "x", "y", "z", "t" };
BigRational cfac = new BigRational(1);
GenSolvablePolynomialRing<BigRational> mfac;
mfac = new GenSolvablePolynomialRing<BigRational>(cfac,
                              TermOrderByName.INVLEX, vars);
GenSolvablePolynomial<BigRational> p;
List<GenSolvablePolynomial<BigRational>> rel;
rel = new ArrayList<GenSolvablePolynomial<BigRational>>();
// add relations to Q_x
//z, y,  y * z + x,
p = mfac.parse("z"); rel.add(p);
p = mfac.parse("y"); rel.add(p);
p = mfac.parse("y * z + x"); rel.add(p);
//t, y,  y * t + y,
//t, z,  z * t - z
...
mfac.addSolvRelations(rel);
p = mfac.parse("t^2 + z^2 + y^2 + x^2 + 1"); F.add(p);
SolvableIdeal<BigRational> id;
id = new SolvableIdeal<BigRational>(mfac, F,
                              SolvableIdeal.Side.twosided);
id.doGB(); // compute twosided GB
SolvableLocalResidueRing<BigRational> efac;
efac = new SolvableLocalResidueRing<BigRational>(id);
```

The construction of elements a and b of the solvable local residue ring is by reading the polynomial with method parse() and then constructing it using factory efac. Now, in this ring, a is invertible and its inverse is computed as c. Next f is computed as b.multiply(c).multiply(a).

```
SolvableLocalResidue<BigRational> a, b, c, d, e, f;
p = mfac.parse("t + x + y + 1");
a = new SolvableLocalResidue<BigRational>(efac, p);
p = mfac.parse("z^2+x+1");
b = new SolvableLocalResidue<BigRational>(efac, p);
c = a.inverse();
f = b.multiply(c).multiply(a);
b.equals(f); // --> true, since (b * 1/a) * a == b
```

Now b equals $(b * a^{-1}) * a$ and in the sample program b equals f, but f is a fraction consisting of 150 polynomial terms in the nominator and denominator, whereas b consists only of 4 polynomial terms. This exemplifies the importance of simplification of such fractions.

In case of commutative polynomial rings, which are unique factorization domains, one can compute greatest common divisors of any two polynomials. This can then be used in quotient fields to divide the nominator and the denominator to reduce the fraction to lowest terms. In the non-commutative (solvable) case, solvable polynomial rings are also factorization domains, however, the factorization may not be unique. Moreover, we must distinguish the order of the factors as multiplication is no more commutative, and consequently distinguish left common divisors and right common divisors. The developed algorithms are of utmost importance for the performance of computations in quotients and localizations of solvable polynomial rings.

In this paper we present as work in progress an experimental first step in the computation of common divisors. The implementation is discussed in the Java computer algebra system (called JAS). It provides generic multivariate solvable polynomials in an object oriented, type safe and thread safe approach to computer algebra, see [7,8,10]. It provides a well designed software library implemented in Java, thus leveraging software and hardware improvements over time. For an introduction to JAS see the cited articles.

Other computer algebra systems with implementations of non-commutative polynomials rings are FELIX [1], Plural in Singular [11,12] or NCPOLY in Reduce [3], to name a few. For more related work see the discussion in [9], for example [4,13] or [2].

Outline. In the next section we summarize some definitions from [9] and in Sect. 3 we will sketch some generic left and right common divisor implementations. In the last section we will draw some conclusions.

2   Solvable Polynomial Rings

Recall from [5] the concept of solvable polynomial rings with commutator relations between variables and from [6] the concept of solvable polynomial rings with additional commutator relations between variables and coefficients. Solvable polynomials rings S are associative rings $(S, 0, 1, +, -, *)$, with

$$S = \mathbf{K}\{X_1, \ldots, X_n; Q; Q'\}, \tag{1}$$

characterized as polynomial rings over skew fields $\mathbf{K}$ in variables $X_1, \ldots, X_n, n \geq 0$, together with a new non-commutative product '$*$', defined by means of commutator relations $Q = \{X_j * X_i = c_{ij} X_i X_j + p_{ij} : 0 \neq c_{ij} \in \mathbf{K}, X_i X_j > p_{ij} \in S, 1 \leq i < j \leq n\}$ between the variables with respect to a $*$-compatible term order $>$ on $S \times S$ (extended from the order $>$ on the set of terms), and commutator relations $Q' = \{X_i * a = c_{ai} a X_i + p_{ai} : 0 \neq c_{ai} \in \mathbf{K}, p_{ai} \in \mathbf{K}, 1 \leq i \leq n, a \in \mathbf{K}\}$ between the variables and the coefficients. In case, the commutator relations Q or Q' are empty, the respective relations are treated as commutative, and if $\mathbf{K}$ is commutative, S is a commutative polynomial ring $S = \mathbf{K}[X_1, \ldots, X_n]$.

2.1   Parametric Solvable Polynomial Coefficient Rings

Recall from [9] the definition of parametric solvable polynomial rings with coefficients from solvable polynomials and commutator relations between variables of the 'main' ring and the coefficient ring.

$$S = \mathbf{R}\{U_1, \ldots, U_m; Q_u\}\{X_1, \ldots, X_n; Q_x; Q'_{ux}\} \tag{2}$$

The coefficients are from a solvable polynomial ring $R = \mathbf{R}\{U_1, \ldots, U_m; Q_u\}$ in the variables $U_1, \ldots, U_m$, together with commutator relations Q_u between the U variables, $Q_u = \{U_j * U_i = c_{uij} U_i U_j + p_{uij} : 0 \neq c_{uij} \in \mathbf{R}, U_i U_j > p_{uij} \in R, 1 \leq i < j \leq m\}$. Q'_u is assumed to be empty, i.e. the elements of $\mathbf{R}$ commute with the U variables. For the main solvable polynomial ring $S = R\{X_1, \ldots, X_n; Q_x; Q'_{ux}\}$ in the variables $X_1, \ldots, X_n$ over R, there are commutator relations between the X variables Q_x and between the U and X variables Q'_{ux}, $Q_x = \{X_j * X_i = c_{xij} X_i X_j + p_{xij} : 0 \neq c_{xij} \in R, X_i X_j > p_{xij} \in S, 1 \leq i < j \leq n\}$ and $Q'_{ux} = \{X_j * U_i = c_{ij} U_i X_j + p_{ij} : 0 \neq c_{ij} \in \mathbf{R}, U_i X_j > p_{ij} \in S, 1 \leq i \leq m, 1 \leq j \leq n\}$. The p_{ij} are allowed to lie in S, and not only in R, provided that $p_{ij} < U_i X_j$. It is assumed that the elements of $\mathbf{R}$ commute with the U and X variables. The term orders $<$ are assumed to be $*$-compatible in the respective rings.

2.2   Recursive Solvable Polynomial Rings

The solvable polynomial ring from equation (2) does not completely match the situation of the ring from (1) as it still does not allow commutator relations between arbitrary base coefficients and main variables. However, it models *recursive* solvable polynomial rings, where main variables can be shifted to coefficient variables and vice versa as desired by an application.

$$S_k = \mathbf{R}\{X_1, \ldots, X_k; Q_k\}\{X_{k+1}, \ldots, X_n; Q_n; Q'_{kn}\}, \quad 0 \leq k \leq n \tag{3}$$

The cases $k = 0$ and $k = n$ recover the usual non-parametric cases: $S_0 = S_n = \mathbf{R}\{X_1, \ldots, X_n; Q\}$.

The implementation of the recursive solvable polynomials is discussed in [9]. They are implemented in classes `RecSolvablePolynomial<C>` and `RecSolvablePolynomialRing<C>`. This classes extend the recursive solvable polynomials `GenSolvablePolynomial<GenPolynomial<C>>` and inherit most methods, except for the new multiplication according to Q'_{ux}. The factory class extends the recursive solvable polynomial ring `GenSolvablePolynomialRing<GenPolynomial<C>>`. It contains the two commutator tables for the variable relations Q_x and Q_u, and additionally there is a relation table `coeffTable`, which contains the commutator relations of Q'_{ux}. The implementation of the $*$-multiplication is described in detail in [9].

These polynomials are used in the next session to compute common divisors.

3   Generic Common Divisors

In this section we explain the implementation of the generic common divisors.

3.1   Recursive Algorithm

With the help of the recursive solvable polynomials it is possible to design recursive common divisor algorithms. From a generic multivariate solvable polynomial from $S' = \mathbf{R}\{X_1, \ldots, X_n; Q_n; Q'_{kn}\}$, in class `GenSolvablePolynomial<C>` one constructs a univariate solvable polynomial with coefficients from multivariate solvable polynomials in $S = \mathbf{R}\{X_1, \ldots, X_{n-1}; Q_k\}\{X_n; Q_n; Q'_{kn}\}$ in class `GenSolvablePolynomial<GenPolynomial<C>>`. In this class the relations Q'_{kn} are empty as they can not be expressed. So we convert the ring and the polynomials to class `RecSolvablePolynomial` and add the missing relations between the main variable and the coefficient variables to the `coeffTable` relation table so that finally $S' = S$.

The main idea for the recursive common divisor algorithm is as in the commutative case (see e.g. [8]): for a univariate solvable polynomial with (multivariate) solvable polynomial coefficients compute the (left, right) content of the coefficients by recursion. Remove the content by division to obtain a primitive (univariate) polynomial. For such polynomials compute a greatest common divisor by taking successive pseudo remainders until a zero remainder appears. Note, the computation of pseudo remainders require the computation of (left, right) Ore conditions to match the coefficients to be eliminated.

3.2   Class Design

Part of the algorithm relations and the interface and class layout is depicted in Fig. 1. It shows an interface `GreatestCommonDivisor` which defines the methods `leftGcd()` and `rightGcd()` to compute left or right common divisors. Further methods are `leftContent()` and `rightContent()` to compute a left or right common divisor of the coefficients, as well as methods `leftPrimitivePart()` and `rightPrimitivePart()` to divide a polynomial by the respective content. Moreover there are methods to compute common multiples and construct lists of polynomials with mutually common divisor one `leftCoPrime()`.

The interface is parametrized by a type C, which is restricted to implement the `GcdRingElem` interface. The `GcdRingElem` interface is itself parametrized by the type C. This allows for recursive coefficient types. The `GcdRingElem` interface defines all methods needed for ring arithmetic. It includes also a method `inverse()` to compute inverses of ring elements if they exist.

The class `GreatestCommonDivisorAbstract` implements all methods of the interface `GreatestCommonDivisor` and defines abstract methods for univariate polynomials `leftBaseGcd()` and `rightBaseGcd()`, and for univariate recursive polynomials `leftRecursiveUnivariateGcd()` and `rightRecursiveUnivariateGcd()`. These are to be provided by the concrete classes, currently only `GreatestCommonDivisorSimple` and `GreatestCommonDivisorPrimitive`.

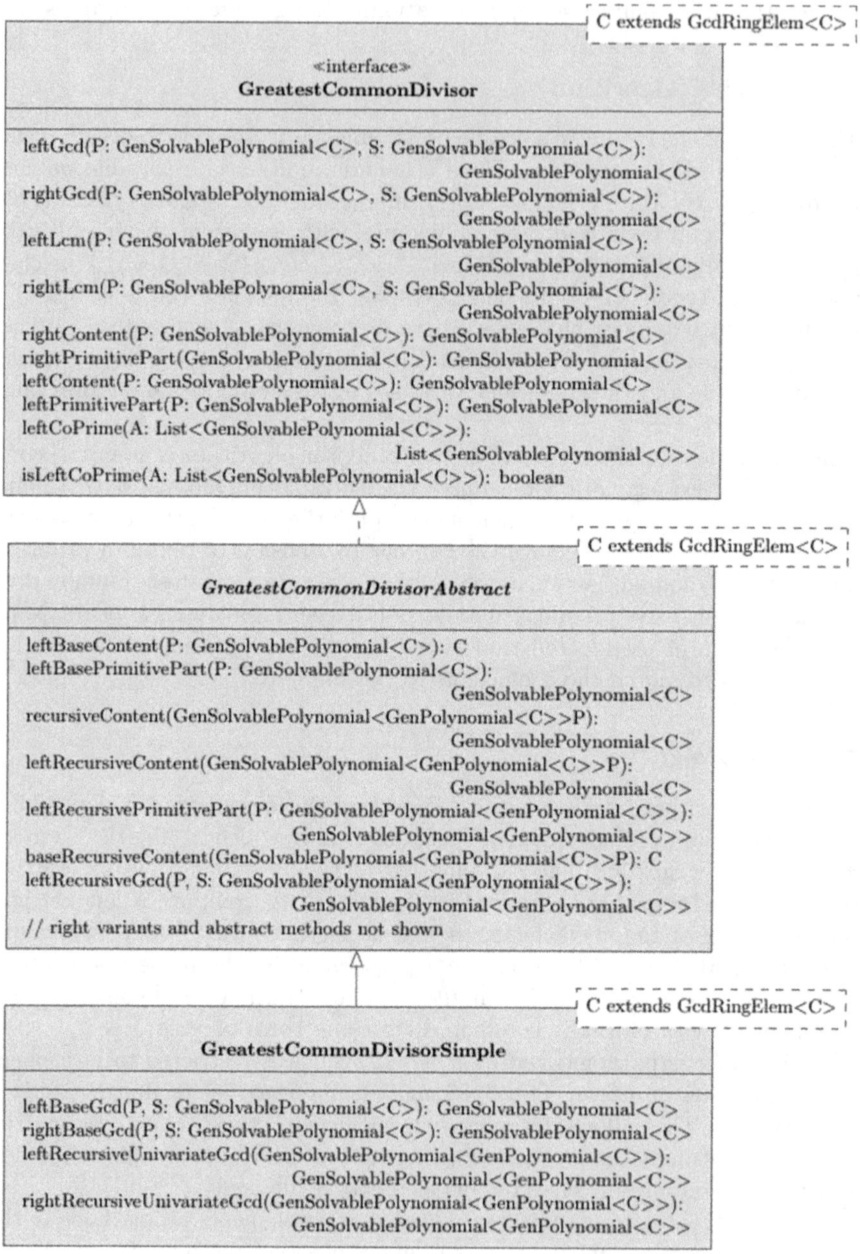

C extends GcdRingElem<C>

«interface»
GreatestCommonDivisor

leftGcd(P: GenSolvablePolynomial<C>, S: GenSolvablePolynomial<C>):
                                    GenSolvablePolynomial<C>
rightGcd(P: GenSolvablePolynomial<C>, S: GenSolvablePolynomial<C>):
                                    GenSolvablePolynomial<C>
leftLcm(P: GenSolvablePolynomial<C>, S: GenSolvablePolynomial<C>):
                                    GenSolvablePolynomial<C>
rightLcm(P: GenSolvablePolynomial<C>, S: GenSolvablePolynomial<C>):
                                    GenSolvablePolynomial<C>
rightContent(P: GenSolvablePolynomial<C>): GenSolvablePolynomial<C>
rightPrimitivePart(GenSolvablePolynomial<C>): GenSolvablePolynomial<C>
leftContent(P: GenSolvablePolynomial<C>): GenSolvablePolynomial<C>
leftPrimitivePart(P: GenSolvablePolynomial<C>): GenSolvablePolynomial<C>
leftCoPrime(A: List<GenSolvablePolynomial<C>>):
                                    List<GenSolvablePolynomial<C>>
isLeftCoPrime(A: List<GenSolvablePolynomial<C>>): boolean

C extends GcdRingElem<C>

GreatestCommonDivisorAbstract

leftBaseContent(P: GenSolvablePolynomial<C>): C
leftBasePrimitivePart(P: GenSolvablePolynomial<C>):
                                    GenSolvablePolynomial<C>
recursiveContent(GenSolvablePolynomial<GenPolynomial<C>>P):
                                    GenSolvablePolynomial<C>
leftRecursiveContent(GenSolvablePolynomial<GenPolynomial<C>>P):
                                    GenSolvablePolynomial<C>
leftRecursivePrimitivePart(P: GenSolvablePolynomial<GenPolynomial<C>>):
                            GenSolvablePolynomial<GenPolynomial<C>>
baseRecursiveContent(GenSolvablePolynomial<GenPolynomial<C>>P): C
leftRecursiveGcd(P, S: GenSolvablePolynomial<GenPolynomial<C>>):
                            GenSolvablePolynomial<GenPolynomial<C>>
// right variants and abstract methods not shown

C extends GcdRingElem<C>

GreatestCommonDivisorSimple

leftBaseGcd(P, S: GenSolvablePolynomial<C>): GenSolvablePolynomial<C>
rightBaseGcd(P, S: GenSolvablePolynomial<C>): GenSolvablePolynomial<C>
leftRecursiveUnivariateGcd(GenSolvablePolynomial<GenPolynomial<C>>):
                            GenSolvablePolynomial<GenPolynomial<C>>
rightRecursiveUnivariateGcd(GenSolvablePolynomial<GenPolynomial<C>>):
                            GenSolvablePolynomial<GenPolynomial<C>>

Fig. 1. UML diagram of common divisor classes.

As in the commutative case we have a class `SGCDFactory` with static methods `getImplementation()` or `getProxy()` to obtain suitable implementations of `GreatestCommonDivisor` based on the type of the coefficients [8]. Class `SGCDParallelProxy` will run two implementations in parallel and return the result of the fastest computation (using `invokeAny()` of `ExecutorService` in `java.util.concurrent`).

3.3   Example Continued

We are now able to use these algorithms in various settings and explore the performance and feasibility. In the example from Sect. 1, we can compute a left common divisor p and right divide the nominator and the denominator by it, to construct the new fraction e, which is z**2 + x + 1.

```
GreatestCommonDivisorAbstract<BigRational> engine;
engine = new GreatestCommonDivisorSimple<BigRational>(cfac);
p = engine.leftGcd(f.num,f.den);
// p = ( x**2 * z * t**2 + 3 * x * z * t**2 + 2 * z * t**2 + x**2 *
// ..
// + 26 * x * y + 11 * y + 4 * x**4 + 19 * x**3 + 36 * x**2 + 31 * x + 7)
GenSolvablePolynomial<BigRational>[] qr;
qr = FDUtil.<BigRational> rightBasePseudoQuotientRemainder(f.num, p);
fn = qr[0]; // ( z**2 + x + 1 ), qr[1] == 0
qr = FDUtil.<BigRational> rightBasePseudoQuotientRemainder(f.den, p);
fd = qr[0]; // 1, qr[1] == 0
e = new SolvableLocalResidue<BigRational>(efac, fn, fd);
// e = ( z**2 + x + 1 )
e.equals(b); // --> true
```

In a next step we make use of the common divisor in the constructor of class `SolvableLocalResidue` to reduce the fraction to lower terms. In a first step with left gcd computation and right division and in a second step by right gcd computation and left division. There are two utility methods `leftGcdCofactors` and `rightGcdCofactors` in class `FDUtil` to do this in a combined operation. Similar adjustments are incorporated in the constructors of classes `SolvableLocal` and `SolvableQuotient`. This shows the feasibility of the common divisor computation for the improvement of the computation in solvable quotient rings. It reduces the size of the fractions and so improves the performance of such computations.

4   Summary and Conclusions

With the recursive solvable polynomial rings, the definition of commutator relations between polynomial variables and coefficient variables is possible. This enabled the construction of recursive algorithms, like the computation of (left, right) common divisors. With the insight gained from this implementation, the classes need a redesign and need to be reimplemented. The common divisors improve the computation in quotient and other local solvable rings by the ability to reduce fractions to lower terms. Using these more efficient implementations as coefficient rings of solvable polynomial rings makes computations of roots, common divisors and ideal constructions over skew fields more feasible. The algorithms have been implemented in JAS in a type-safe, object oriented way with generic coefficients. This gives more efficient simplifiers to reduce intermediate expression swell in such fields. Still, the solvable multiplication has high complexity and only small examples are practical to compute.

Acknowledgments. We thank Thomas Becker for discussions on the implementation of a generic polynomial library and Raphael Jolly for the fruitful cooperation. We thank moreover our colleagues Wolfgang K. Seiler, Thomas Sturm, Axel Kramer, Victor Levandovskyy, Joachim Apel, Markus Aleksy and others for various discussions on the design and the requirements for JAS and its mathematical foundations. Thanks also for helpful suggestions from the reviewers.

References

1. Apel, J., Klaus, U.: FELIX - an assistant for algebraists. In: Proceedings of the 1991 International Symposium on Symbolic and Algebraic Computation, ISSAC 1991, Bonn, Germany, 15–17 July 1991, pp. 382–389 (1991)
2. Apel, J., Lassner, W.: Computation and simplification in Lie fields. In: EUROCAL 1987, pp. 468–478 (1987)
3. Apel, J., Melenk, H.: NCPOLY: computation in non-commutative polynomial ideals. Technical report (2004). http://www.reduce-algebra.com/docs/ncpoly.pdf
4. Bueso, J.L., Gómez-Torrecillas, J., Verschoren, A.: Algorithmic Methods in Non-Commutative Algebra: Applications to Quantum Groups. Kluwer Academic Publishers, Dordrecht (2003)
5. Kandri Rody, A., Weispfennning, V.: Non-commutative Gröbner bases in algebras of solvable type. J. Symbol Comput. **9**(1), 1–26 (1990)
6. Kredel, H.: Solvable Polynomial Rings. Dissertation, Universität Passau, Passau (1992)
7. Kredel, H.: On a Java computer algebra system, its performance and applications. Sci. Comput. Program. **70**(2–3), 185–207 (2008)
8. Kredel, H.: Unique factorization domains in the Java computer algebra system. In: Sturm, T., Zengler, C. (eds.) ADG 2008. LNCS, vol. 6301, pp. 86–115. Springer, Heidelberg (2011)
9. Kredel, H.: Parametric solvable polynomial rings and applications. In: Gerdt, V.P., Koepf, W., Seiler, W.M., Vorozhtsov, E.V. (eds.) CASC 2015. LNCS, vol. 9301, pp. 275–291. Springer, Heidelberg (2015)
10. Kredel, H.: The Java Algebra System (JAS). Technical report, since 2000. http://krum.rz.unimannheim.de/jas/
11. Levandovskyy, V.: PLURAL, a non–commutative extension of singular: past, present and future. In: Iglesias, A., Takayama, N. (eds.) ICMS 2006. LNCS, vol. 4151, pp. 144–157. Springer, Heidelberg (2006)
12. Levandovskyy, V., Schönemann, H.: Plural: a computer algebra system for non-commutative polynomial algebras. In: Proceedings of the Symbolic and Algebraic Computation, International Symposium ISSAC 2003, Philadelphia, USA, pp. 176–183 (2003)
13. Mora, T.: An introduction to commutative and noncommutative Gröbner bases. Theor. Comput. Sci. **134**(1), 131–173 (1994)

An Online Computing and Knowledge Platform for Differential Equations

Yinping Liu[1]($\boxtimes$), Ruoxia Yao[2]($\boxtimes$), Zhibin Li[1], Le Yang[3], and Zhian Zhang[3]

[1] Institute of Systems Science, East China Normal University,
Shanghai, People's Republic of China
ypliu@cs.ecnu.edu.cn
[2] Department of Computer Science, Shaanxi Normal University,
Xi'an, People's Republic of China
rxyao2@hotmail.com
[3] Department of Computer Science, East China Normal University,
Shanghai, People's Republic of China

Abstract. A Web-based knowledge database and computing platform for nonlinear differential equations is presented, which could provide computing and graphing based on symbolic computing system Maple and some of its built-in packages. Users can not only calculate specific types of analytical solutions of nonlinear differential systems by calling the packages, but also carry out any symbolic computations associated with equations and other kinds of simple computations in an interactive mode with visual output. The knowledge database of differential equations has all functions of the general database. Furthermore, each equation has a web page to show its properties and research results. In addition, each mathematica formula is stored in its infix form in the knowledge database and can be displayed visually.

Keywords: Nonlinear differential equation · Symbolic computation · Online · Computing platform · Knowledge database

1 Introduction

The expansion of Internet led to high development of online computing, teaching and learning. It is extremely valuable to make mathematical computing and online education easier over the Internet. In recent years, online developing of computing software is becoming a public focus, and some symbolic computation software companies have released more and more online computing softwares or education applications. Related online scientific computing softwares or applications include Web-enabled systems [1–7], Parallel and Distributed CAS (Computer Aided Systems) computing projects [8–10], Grid-enabled systems [11–17], Scientific computing in the cloud [18,19] and online computing apps for mobile devices [20,21]. The related online mathematics education softwares or applications include Maple T.A. [22], computational knowledge engine Wolfram|Alpha [2], DMAS [23], Math-Pass [24], Mathway.com [25,26], MEGA [27], E-GEMS [28], etc.

© Springer International Publishing Switzerland 2016
G.-M. Greuel et al. (Eds.): ICMS 2016, LNCS 9725, pp. 181–188, 2016.
DOI: 10.1007/978-3-319-42432-3_23

Differential equation is one of the most widely applied branches in modern mathematics, and it is also one of the most active branches in the field of mathematical physics. From natural science to social science almost all subject areas, more and more people prefer to solve problems using differential equation methods. Many algorithms and techniques have been born to solve differential equations and to analyze the properties of differential equations. As a consequence, more new research results are emerging constantly. However, Many researchers, engineers and technicians can not make good use of those new results due to the lack of knowledge of mathematics and physics. There are already two specialized equation websites [29], where the equation is stored in the format of picture. To our knowledge, there is not any online differential equation database or knowledge database so far. Therefore, developing an online knowledge database for differential equations is really meaningful and valuable.

In the last 10 years, the research interests of our group are mainly concentrated on symbolic computation and related software development of differential equations. Around solving and integrability analysis of mathematical physics equations, we have developed a dozen different offline related softwares based on Maple. However, the utilization rate of these offline softwares is rather low. In the past two years, an online symbolic computation platform for nonlinear differential equations is proposed and implemented. Furthermore, we are developing an online knowledge database of differential equations which is connected with this computing platform.

2   The Computing Platform

The computing platform named NDEmathema is an online B/S computing platform. Users could use NDEmethema via a Web browser (Chrome, Firefox, IE,

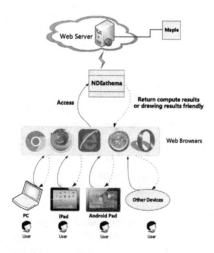

Fig. 1. The functional diagram of NDEmathema

Safari, Opera etc.). Figure 1 illustrates the functional diagram of NDEmethema. NDEmathema could be accessed not only from desktop/laptop computers or supercomputers, but also by tablet PCs, smart phones and other mobile devices. Most of our programs have been embedded in the platform. Therefore, users could use it to solve different types of nonlinear differential systems. For the convenience of usage, each embedded package not only has an Independent User Interface, but also has been integrated in the Integrated User Interface of NDE-mathema. All results are displayed visually in this platform. Furthermore, users can export visual outputs as Maple Worksheet file or HTML file. In this way, the exported Maple Worksheet file could be downloaded and to be used later. In addition, in each Independent User Interface, users could use the corresponding Maple package easily without needing to learn any syntax of any symbolic computing software; in the Integrated User Interface, users could input the commands according to the APIs of the related Maple package.

In the following a sample of the embedded package is given to show the Independent User Interface and the Integrated User Interface of the computing platform, respectively.

The Elliptic equation method is widely applied because of its simplicity. The basic idea of the method applied to a single PDE in $u(x,t)$ works as follows: In a travelling frame of reference, $\xi = kx + ct$, one transforms a PDE into an ODE in the new independent variable T. Since T is a solution of the Elliptic equation $T' = \sqrt{a_0 + a_1 T + a_2 T^2 + a_3 T^3 + a_4 T^4}$, all the derivatives of T are polynomials or radical expression of itself. Therefore, via a chain rule, a polynomial PDE in $u(x,t)$ is transformed into an ODE in $U(T)$, which has polynomial coefficients in T or $\sqrt{a_0 + a_1 T + a_2 T^2 + a_3 T^3 + a_4 T^4}$. One then seeks for polynomial solutions of the ODE, and thus generate a subset of the set of all solutions for the original PDE. As the Elliptic equation has a series of different function solutions for different parameters $(a_i, i = 1, 2, 3, 4)$ constraints. The program RAEEM [30] is a complete implementation of the Elliptic equation method. RAEEM can automatically deliver a series of possible different types of function expansion solutions, which include polynomial, rational function, exponential function, trigonometric

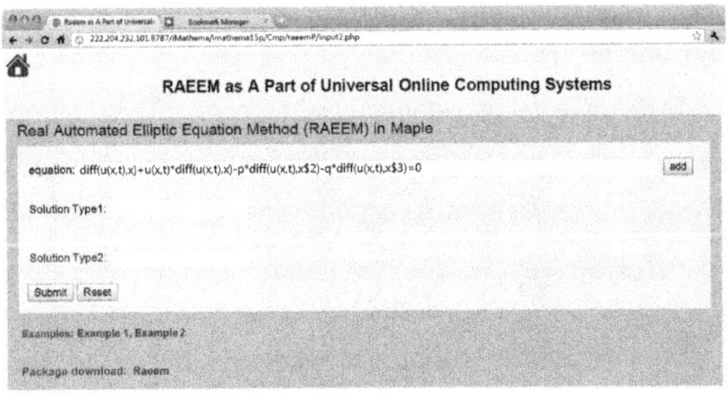

Fig. 2. The independent user interface of the package *RAEEM*

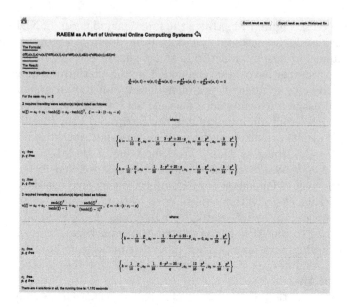

Fig. 3. Output web page of the example

function, hyperbolic function, elliptic function types of solutions, etc. RAEEM has been embedded in the platform NDEmathema (Fig. 2).

The Independent User Interface of RAEEM is shown as follows:

The related visual outputs on the Web are shown in Fig. 3.

By clicking the buttons "Export as Maple Worksheet" in the upper right corner of the resulting page. The related Maple Worksheet is shown in Fig. 4.

For this example, the corresponding Integrated Input Interface is shown in Fig. 5. More details of the platform will be reported at the conference.

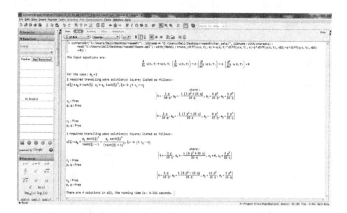

Fig. 4. Maple worksheet file of the example

Fig. 5. The Integrated user interface of the NDEmathema

3   The Knowledge Database of Differential Equation

The knowledge database of differential equations is an online and open system. This facilitates the distributed development by the advanced users in different regions. The homepage of the database is shown below. It should be noted that we just have Chinese version now, the English version will be developed next year. The following pictures are only for the conference demonstration (Fig. 6).

For each equation, there is an equation page to show its properties and research results. Some achievements are shown by connecting as many as possible relevant documents or web sites (the maximum numbers of documents and web sites are both 50). In particular, a Maple worksheet is established and hyperlinked to demonstrate step by step some of the main results of the equation. The equation page of the KdV equation is shown in Fig. 7. All of the different types of equations are stored in a database. Then users can establish a sub database by

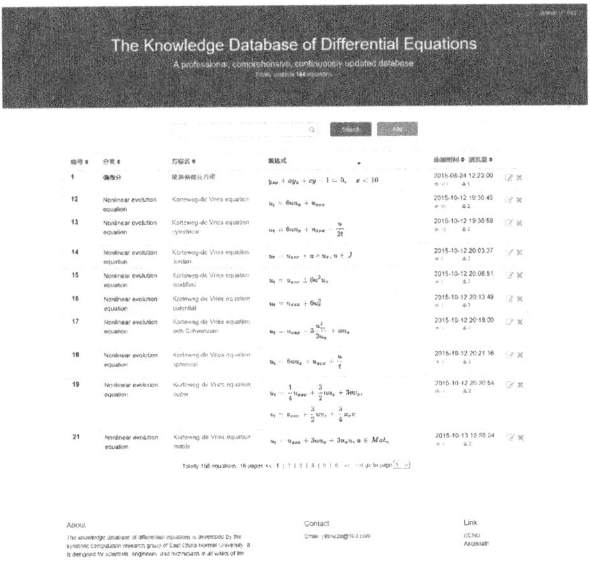

Fig. 6. The homepage of the knowledge database

Fig. 7. The equation page of the KdV equation

Fig. 8. A sample of sub database

a keyword query. Users can carry out any database operations on a sub database. A sample sub database is shown in Fig. 8.

Each mathematica formula is stored in its infix form in the database and can be displayed visually in the form of two-dimensional mathematical formula. More details will be shown at the conference.

References

1. Wang, P., Gray, S., Kajler, N., Lin, D., Liao, W., Zou, X.: IAMC architecture andprototyping: a progress report. In: Proceedings of the 2001 International Symposium on Symbolic and Algebraic Computation, pp. 337–344. ACM (2001)
2. Hoy, M.B.: Wolfram|Alpha: a brief introduction. Med. Ref. Serv. Q. **29**, 67–74 (2010)
3. Wolfram Research Inc. webMathematica. http://www.wolfram.com/products/webmathematica
4. Perez, F., Granger, B.E.: IPython: a system for interactive scientific computing. Comput. Sci. Eng. **9**, 21–29 (2007)
5. Zimmer, J., Kohlhase, M.: System description: the MathWeb software bus for distributed mathematical reasoning. In: Voronkov, A. (ed.) CADE 2002. LNCS (LNAI), vol. 2392, pp. 139–143. Springer, Heidelberg (2002)
6. Maxima-online.org. Maxima-online. http://maxima-online.org
7. McGettrick, M.: Online Gröbner basis [OGB]. ACM SIGSAM Bull. **38**, 19–21 (2004)
8. Schreiner, W., Mittermaier, C., Bosa, K.: Distributed maple: parallel computer algebra in networked environments. J. Symbolic Comput. **35**, 305–347 (2003)
9. Pau, C., Schreiner, W.: Distributed mathematica-User and Reference Manual. RISC Report 00–25, RISC, JK University, Linz, Australia (2000)
10. Almasi, G., Cascaval, C., Padua, D.A.: Mat marks: a shared memory environment for matlab programming. In: Proceedings of the 8th IEEE International Symposium on High Performance Distributed Computing, p. 21 (1999)
11. Agrawal, S., Dongarra, J., Seymour, K., Vadhiyar, S.: Netsolve: past, present, and future-a look at a grid enabled server. In: Grid Computing: Making the Global Infrastructure a Reality, pp. 615–624 (2003)
12. Wu, Y., Liao, W., Wang, P., Lin, D., Yang, G.: An internet accessible gridcomputing system: Grid-elimino. In: Proceedings of IAMC, pp. 1–8 (2003)
13. Tanaka, Y., Nakada, H., Sekiguchi, S., Suzumura, T., Matsuoka, S.: Ninf-G: a reference implementation of RPC-based programming middleware for grid computing. J. Grid Comput. **1**, 41–51 (2003)
14. Cox, S., Keane, A.: Grid enabled optimisation and design search for engineering (geodise). In: NeSC Workshop on Applications and Testbeds on the Grid, pp. 20–33 (2002)
15. Pound, G.E., Eres, M.H., Wason, J.L., Jiao, Z., Keane, A.J., Cox, S.J.: A grid-enabled problem solving environment (PSE) for design optimisation within matlab. In: Proceedings of Parallel and Distributed Processing Symposium. International, pp. 50–57 (2003)
16. Petcu, D., Dubu, D., Paprzycki, M.: Extending maple to the grid: design and implementation. In: Third International Symposium on Algorithms, Models and Tools for Parallel Computing on Heterogeneous Networks, pp. 209–216 (2011)

17. Amestoy, P., Pantel, M.: Grid-TLSE: a web expertise site for sparse linear algebra. In: Sparse Days and Grid Computing at St. Girons Workshop, p. 192 (2003)
18. Srirama, S., Batrashev, O., Vainikko, E.: Scicloud: scientific computing on the cloud. In: Proceedings of the 2010 10th IEEE/ACM International Conference on Cluster, Cloud and Grid Computing, pp. 579–580 (2010)
19. Rehr, J.J., Vila, F.D., Gardner, J.P., Svec, L., Prange, M.: Scientific computing in the cloud. Comput. Sci. Eng. **12**, 34–43 (2010)
20. MathWorks, Inc. Matlab mobile. http://www.mathworks.com/products/matlab-mobile
21. Wolfram Research Inc. Wolframalpha mobile. http://products.wolframalpha.com/mobile
22. Maplesoft Inc. Mapleta. http://www.maplesoft.com.cn/products/mapleta
23. Al-shomrani, S., Wang, P.: DMAS: a web-based distributed mathematics assessment system. In: International Conference on Learning, pp. 3–6 (2008)
24. Su, W., Wang, P.S., Li, L.: Mathpass: a remedial mathematics system with concept checking. In: CICM (Conferences on Intelligent Computer Mathematics), pp. 11–12 (2010)
25. Mathway. https://mathway.com
26. Siekmann, J., Benzmller, C., Autexier, S.: Computer supported mathematics with mega. J. Appl. Logic **4**, 533–559 (2006)
27. Team Members of E-GEMS Project, E-gems. http://www.cs.ubc.ca/nest/egems/index.html
28. Lopez-Morteo, G., Lpez, G.: Computer support for learning mathematics: a learning environment based on recreational learning objects. Comput. Educ. **48**, 618–641 (2007)
29. Polyanin, A.D.: EqWorld (The World of Mathematical Equations). http://eqworld.ipmnet.ru
30. Li, Z.-B., Liu, Y.-P.: RAEEM: a Maple package for finding a series of exact traveling wave solutions for nonlinear evolution equation. Comput. Phys. Commun. **163**, 191–201 (2004)

Software for Numerically Solving
Polynomial Systems

SIROCCO: A Library for Certified Polynomial Root Continuation

Miguel Ángel Marco-Buzunariz[1,3(✉)] and Marcos Rodríguez[2,3]

[1] Universidad of Zaragoza, Zaragoza, Spain
mmarco@unizar.es
[2] Centro Universitario de la Defensa de Zaragoza, Zaragoza, Spain
marcos@unizar.es
[3] IUMA. Instituto Universitario de Matemáticas y Aplicaciones, Zaragoza, Spain
https://riemann.unizar.es/~mmarco,
http://www.imark.es

Abstract. The classical problem of studying the topology of a plane algebraic curve is typically handled by the computation of braid monodromies. The existence of arithmetic Zariski pairs implies that purely algebraic methods cannot provide those braids, so we need numerical methods at some point. However, numerical methods usually have the problem that floating point arithmetic introduces rounding errors that must be controlled to ensure certified results. We present SIROCCO (The source code and documentation is available in: https://github.com/miguelmarco/sirocco), a library for certified polynomial root continuation, specially suited for this task. It computes piecewise linear approximations of the paths followed by the roots. The library ensures that there exist disjoint tubular neighborhoods that contain both the actual path and the computed approximation. This fact proves that the braids corresponding to the approximation are equal to the ones corresponding to the actual curve. The validation is based on interval floating point arithmetic, the Interval Newton Criterion and auxiliary lemmas. We also provide a SageMath interface and auxiliary routines that perform all the needed pre and post-processing tasks. Together this is an "out of the box" solution to compute, for instance, the fundamental group of the complement of an affine complex curve.

Keywords: Validated numerics · Interval arithmetic · Homotopy · Algebraic curves

1 Introduction

The problem of studying the topology of the embedding of a curve $\mathcal{C}$ in the complex projective plane $\mathbb{CP}^2$ is a classical one in algebraic geometry. One of the most important tools in this theory is the braid monodromy. It is defined as follows. Consider $\mathcal{C}$ be a degree d curve in $\mathbb{CP}^2$. Fix a point $p \in \mathbb{CP}^2 \setminus \mathcal{C}$. The set of lines going through p forms a $\mathbb{CP}^1$. So we have a fibration

© Springer International Publishing Switzerland 2016
G.-M. Greuel et al. (Eds.): ICMS 2016, LNCS 9725, pp. 191–197, 2016.
DOI: 10.1007/978-3-319-42432-3_24

$$\mathbb{C} \lhook\joinrel\longrightarrow \mathbb{CP}^2 \setminus \{p\}$$
$$\downarrow \pi$$
$$\mathbb{CP}^1$$

Generically, the fibres intersect the curve $\mathcal{C}$ in d distinct points. However, there is a finite set Δ of points of $\mathbb{CP}^1$ such that their preimages contains less than d points. This set Δ will be called the *discriminant* of π. Consider $Sym^d(\mathbb{C})$ the d'th symmetric product of $\mathbb{C}$. This is the configuration space of d different points in $\mathbb{C}$. It is well known that its fundamental group $\pi_1(Sym^d(\mathbb{C}))$ is the braid group in d strands B_d. As we have seen before, we have a well defined map

$$\mathbb{CP}^1 \setminus \Delta \to Sym^d(\mathbb{C})$$

which induces a map of fundamental groups

$$\pi_1(\mathbb{CP}^1 \setminus \Delta) \to B_d$$

This map is called the *braid monodromy* of the curve with respect to the projection π. It is usually presented as a list of braids, corresponding to the images of a good system of generators of $\pi_1(\mathbb{CP}^1 \setminus \Delta)$. VanKampen gave in [6] a method to compute the fundamental group of $\mathbb{CP}^2 \setminus \mathcal{C}$ from the braid monodromy. Moreover, Carmona [3] proved that the braid monodromy itself determines the topology of the pair $(\mathbb{CP}^2, \mathcal{C})$. The previous paragraphs show the interest of computing the braid monodromy. However, the existence of pairs of curves, defined by polynomials whose coefficients are Galois conjugated in some number field, but with nonhomeomorphic embeddings (the so-called Zariski pairs, see [1] for a survey on the subject), shows us that this braids cannot be computed by purely algebraic methods. In the following sections we will present a numerical (yet certified) method to compute them. Our approach will consist of computing a piecewise linear approximation of the strands of each braid. This approximation must produce the same braid as the original strands. In order to ensure this, we propose a method that certifies that the approximations live inside disjoint tubular neigbourhoods of the actual strands (See Fig. 2). The method is based on a homotopic continuation of the roots via a predictor–corrector scheme, plus a validation step using interval arithmetic and interval Newton Operator. In fact the final scheme is "predictor–validator–corrector". The predictor uses implicit differentiation and linear extrapolation. The corrector is just the classical Newton Method.

2   Validated Numerics

In the following we will need the concepts of complex interval, interval polynomial and interval evaluation.

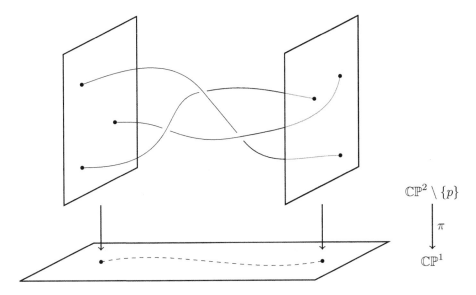

Fig. 1. Braid over a path

Definition 1. *A complex interval is a set of the form $\{x + i \cdot y \mid a \le x \le b, c \le y \le d\}$ for some $a, b, c, d \in \mathbb{R}$, $a \le b$, $c \le d$. The set of complex intervals will be denoted as $I\mathbb{C}$.*

Given $S \subseteq \mathbb{C}$, its interval closure (that is, the smallest element of $I\mathbb{C}$ that contains S) will be denoted by $[S]$. Note that complex numbers are a particular case of complex intervals. $I\mathbb{C}$ is not a ring, however, by abuse of notation, we will talk about polynomials over this set. For example, if $If = Ia_0 + Ia_1x + \cdot Ia_nx^n \in I\mathbb{C}[x]$, it can be thought of as a way to represent the set $S(If) = \{a_0 + a_1x + \cdots a_nx^n \in \mathbb{C}[x] \mid a_i \in Ia_i \forall i\}$.

Definition 2. *An evaluation scheme is a map*

$$E : I\mathbb{C}[x] \times I\mathbb{C} \mapsto I\mathbb{C}$$

such that for every $f \in S(If)$, and every $z \in Iz$, the evaluation $f(z)$ is in $E(If, Iz)$.

Analogously, we can define interval polynomials of two variables and their corresponding evaluation schemes. For example, the usual interval arithmetic is an evaluation scheme. From now on we will assume that we have fixed evaluation schemes E_1, E_2 for univariate and bivariate interval polynomials respectively. By abuse of notation, $E_1(If, Iz)$ will be denoted by $If(Iz)$, and $E_2(If, Ix, Iy)$ will be denoted by $If(Ix, Iy)$.

2.1 Newton Method

A basic tool to prove statements with a computer which cannot be proved in a symbolic way is the Interval Newton Method [7,9]. Among all its possible

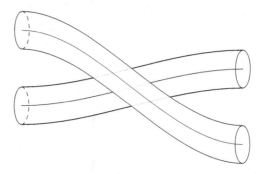

Fig. 2. Tubular neigbourhoods of the strands. The piecewise linear approximations live inside them.

formulations, we present here the one for complex univariate polynomials, since it is the one we need.

Theorem 1. *Let $f : \Omega \to \mathbb{C}$, $\Omega \subseteq \mathbb{C}$ an open set, $f \in \mathcal{C}^\infty(\Omega)$. Let $Y \in I\mathbb{C}$, $y_0 \in Y \subset \Omega$. Let us assume that $0 \notin [f'(Y)]$. We call the Interval Newton Operator:*

$$N(y_0, Y, f) = y_0 + f(y_0)/[f'(Y)].$$

Then:

- *If $y_1, y_2 \in Y$ such that $f(y_1) = f(y_2)$, then $y_1 = y_2$.*
- *If $N(y_0, Y, f) \subseteq Y$, then $\exists | y^* \in Y$ such that $f(y^*) = 0$.*
- *If $y_1 \in Y$ such that $f(y_1) = 0$, then $y \in N(y_0, Y, f)$.*
- *If $N(y_0, Y, f) \cap Y = \emptyset$, then $f(y) \neq 0$, $\forall y \in Y$.*

A detailed proof of this Theorem can be found in reference [10]. Further details on validation methods can be found in [7]. Note that, in the same way we defined an evaluation scheme for polynomials, we can also define an evaluation scheme for the Newton operator. Again, let us assume that we have fixed such an evaluation scheme.

Corollary 1 (Newton method for interval polynomials). *Consider $If \in I\mathbb{C}[x,y]$ a complex intervalar polinomial. Let $y_0 \in \mathbb{C}$, $Ix, Iy \in I\mathbb{C}$ such that $y_0 \in Iy$. Consider If_{Ix} the univariate intervalar polynomial resulting from evaluating If at $x = Ix$. If $N(y_0, Iy, If_{Ix}) \subseteq Iy$, then for every $f \in If$ and every $x \in Ix$, there exists a unique root (counted with multiplicity) of f_x in Iy. Moreover, this root lies in $N(y_0, Iy, If_{Ix})$.*

3   The SIROCCO Library

In this section we present a C library developed for the purpose described in the Introduction. The core function provided by the library is called `homotopyPath`.

It takes as input a polynomial $f(x, y)$ (as a list of its coefficients) and an approximation y_0 of a root $f(0, y)$. Its output is a list with the points that determine a good (as explained in the Introduction) approximation of the path followed by y_0 as x moves from 0 to 1. A simple change of variable allows us to translate any other linear path to this one. In the following of this section we will briefly present the method implemented in SIROCCO. First we set a trivial Lemma to ensure that the neighbourhoods we use will be disjoint. The description of the method will follow the notation set in this lemma.

Lemma 1. *Let C_1, C_2 be two concentric squares with horizontal and vertical sides, being C_2 three times bigger than C_1. Let C_1', C_2' another pair of squares with the same properties. If there exists points $x \in C_1 \setminus C_2'$ and $y \in C_1' \setminus C_2$, then C_1 and C_1' are disjoint.*

3.1   The Validated Continuation Algorithm

The basic outline of the method consists on the following: Start with a polynomial $f_0(x, y) \in \mathbb{C}[x, y]$, an interval polynomial $If(x, y) \in I\mathbb{C}[x, y]$ that contains it (can be f_0 itself), and $y_{inp} \in \mathbb{C}$ an approximate root of $f(0, y)$. Set $x_0 = 0$, $y_0 = y_{inp}$ For each step we want to compute two boxes $Ix \times IY_1$ and $Ix \times IY_2$ such that $\forall x \in Ix, \forall f \in If$, $f_x(y)$ has a unique root in IY_1 and in IY_2. IY_1 and IY_2 are both centered in y_0. IY_2 will be three times wider than IY_1 (and they will play the roles of C_1 and C_2 in Lemma 1). Then:

1. Estimate[1] an initial value $\delta > 0$ to be the radius of IY_1.
2. Apply Corollary 1 to If with $Ix = [x_0]$, $Iy = C_i$ as in Lemma 1 ($i = 1, 2$). Keep reducing δ until Corollary 1 is satisfied.
3. Estimate[2] $h > 0$ to be the stepsize for the predictor.
4. Apply Corollary 1 to If with $Ix = [x_0, x_0 + h]$, $Iy = C_i$. Keep reducing h until the Corollary is satisfied.
5. Apply classical Newton method to correct y_0 for polynomial f_{x_0+h}. Use corrected value to update y_0. Set x_0 to $x_0 + h$.

Stages 1 to 5 shall be repeated until x_0 reaches 1. In order to obtain longer steps in the validation, in stages 3 and 4 we use the following trick. We define an auxiliar polynomial.

$$g(x, y) = f(x + x_0, y + a(x + x_0))$$

This change of variables sends the point (x_0, y_0) to $(0, y_0)$ and transform the implicit curve given by $f(x, y) = 0$ into a curve whose implicit derivative at the translated point vanishes (see Fig. 3). The validation of this new polynomial in a rectangular interval box implies the validation of the original polynomial in the box transformed by the change of variables. In practice, the longer stepsizes that this trick allows compensates the computation effort of the change of variables. Experimental evidence shows an important speedup.

---

[1] This estimation is derived from the degree 2 Taylor expansion of the polynomial.
[2] This estimation is derived from the degree 2 Taylor expansion of the implicit function defined by $f(x, y) = 0$.

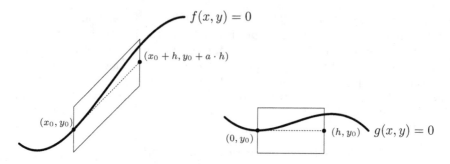

Fig. 3. Neighborhoods related through the change of variables ϕ.

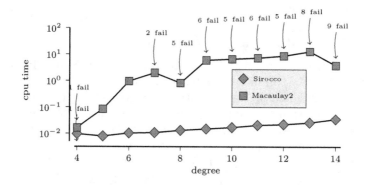

Fig. 4. Timing comparison. CPU-time vs degree of validated polynomial using softwares SIROCCO and Macaulay2

4   Comparison and Timimgs

Up to the authors's knowledge, there are several software packages able to perform validated homotopy continuation, such as pss5 [8], Cadenza [5] or the NumericalAlgebraicGeometry package [2] of Macaulay2 [4]. However, they have different objectives. We will now compare the performance of our implementation with the one by Leykin in the NumericalAlgebraicGeometry package. We remark again that the purpose of NumericalAlgebraicGeometry package is not to compute braid monodromies, but to find solutions of polynomial systems. In that sense this comparison is not fully fair. Our comparison consisted in timing the computation of the strand starting at one root y_0 of a polynomial $f(x, y)$ from $x = 0$ to $x = 1$ for several polynomials. The polynomials were chosen randomly among the polynomials of degree 4 to 14 (10 polynomials of each degree). All measurements were made by averaging 5 runs of the program on the same input. The test platform has an Intel Core i5-4570 CPU running at 3.20 GHz, with 8GB of RAM. Both softwares are configured to throw an error message when they are not able to validate a step (those timings are not taken into account in the average) All timings are expressed in seconds. In Fig. 4 we can see that our implementation is consistently faster (as we could expect) than the one in Macaulay2. Moreover, it is also more robust, since it gives the right answer in cases where Macaulay2 could not guarantee the correctness. The difference in timings varies greatly, but, on average, our implementation is about an order of magnitude faster.

References

1. Bartolo, E.A., Cogolludo, J.I., Tokunaga, H.: A survey Zariski pairs. Adv. Stud. Pure Math. **50**, 1–100 (2008)
2. Beltrán, C., Leykin, A.: Robust certified numerical homotopy tracking. Found. Comput. Math. **13**(2), 253–295 (2013)
3. Ruber, J.C.: Monodroma de trenzas de curvas algebraicas planas. Ph.D. thesis, Universidad de Zaragoza (2003)
4. Grayson, D.R., Stillman, M.E.: Macaulay2, a software system for research in algebraic geometry. http://www.math.uiuc.edu/Macaulay2/
5. Hauenstein, J.D., Haywood, I., Liddell Jr., A.C., Cadenza: certifying homotopy paths for polynomial systems. http://www.nd.edu/aliddel1/research/cadenza
6. Van Kampen, E.R.: On the fundamental group of an algebraic curve. Am. J. Math. **55**(1–4), 255–260 (1933)
7. Krawczyk, R., Neumaier, A.: An improved interval newton operator. J. Math. Anal. Appl. **118**(1), 194–207 (1986)
8. Malajovich, G.: Polynomial System Solver. https://sourceforge.net/projects/pss5/
9. Moore, R.E.: Interval Analysis, vol. 4. Prentice-Hall, Englewood Cliffs (1966)
10. Zgliczyński, P.: Interval krawczyk and newton method, February 2007, Lecture notes. http://ww2.ii.uj.edu.pl/~zgliczyn/cap07/krawczyk.pdf

An Implementation of Exact Mixed Volume Computation

Anders Nedergaard Jensen$^{(\boxtimes)}$

Technische Universität Kaiserslautern, Kaiserslautern, Germany
jensen@math.au.dk
http://home.math.au.dk/jensen/

Abstract. Mixed volumes of lattice polytopes play a central role in numerical and tropical algebraic geometry. We present an implementation of a new algorithm for their computation based on tropical homotopy continuation, which is a combinatorial procedure using ideas from numerical algebraic geometry. While the mathematical aspects of the algorithm are presented elsewhere, here we mainly address technical details of the implementation, in particular how it was made fast and reliable. The implementation is distributed as part of the library gfanlib.

Keywords: Tropical geometry · Mixed volumes · Numerical algebraic geometry

1 Introduction

Given n convex bounded sets $P_1, \ldots, P_n \subseteq \mathbb{R}^n$ the function $f : \mathbb{R}^n_{\geq 0} \to \mathbb{R}_{\geq 0}$ given by $(\lambda_1, \ldots, \lambda_n) \mapsto \text{Volume}_n(\lambda_1 P_1 + \cdots + \lambda_n P_n)$ turns out to be a polynomial function and the coefficient of $\lambda_1 \cdots \lambda_n$ in the corresponding polynomial is called the *mixed volume* of $P_1, \ldots, P_n$. Mixed volumes were studied by Minkowski, but due to their appearance in the BKK Theorem, they have in recent years played important roles in enumerative, tropical and numerical algebraic geometry.

Theorem 1 (Bernstein, Khovanskii, Kushnirenko, 1975). *Let $f_1, \ldots, f_n \in \mathbb{C}[x_1, \ldots, x_n]$ be polynomials. Then the number of isolated solutions to the polynomial system $f_1(x) = \cdots = f_n(x) = 0$ with $(x_1, \ldots, x_n) \in (\mathbb{C} \setminus \{0\})^n$ is (counting multiplicities) bounded by the mixed volume of the Newton polytopes of $f_1, \ldots, f_n$.*

There have been many attempts to implement effective[1] algorithms for mixed volume computation, i.e. [6–8]. Indeed the available software is capable of computing mixed volumes of systems whose numerical solutions cannot be determined — simply because there are too many. Thus for the immediate purpose of solving, faster algorithms may not seem particularly important. However, for the human interaction process, being able to quickly determine mixed volumes

---

[1] Mixed volume computation, just as volume computation, is #P-hard [2].

© Springer International Publishing Switzerland 2016
G.-M. Greuel et al. (Eds.): ICMS 2016, LNCS 9725, pp. 198–205, 2016.
DOI: 10.1007/978-3-319-42432-3_25

```
***************** CLING *****************
* Type C++ code and press enter to run it *
*             Type .q to exit              *
*******************************************
[cling]$ #include "gfanlib_circuittableint.cpp"
[cling]$ #include "gfanlib_paralleltraverser.cpp"
[cling]$ #include "gfanlib_mixedvolume.cpp"
[cling]$ using namespace std;
[cling]$ using namespace gfan;
[cling]$ using namespace gfan::MixedVolumeExamples;
[cling]$ auto s={cyclic(2),cyclic(3),cyclic(5),cyclic(7)};
[cling]$ for(auto v:s)cout<<mixedVolume(v)<<" ";cout<<endl;
2 6 70 924
[cling]$ .q
```

Fig. 1. A session with the C++ "interpreter" CLING demonstrating how we can turn gfanlib into a simple interactive computer algebra system capable of computing mixed volume. We ran CLING with options `-l /usr/lib/libgmp.so.3.5.2 -std=c++0x`.

will allow choosing the right approach more quickly. There are other cases where speed of mixed volume computation is important. For example, when we don't need one big, but rather many small mixed volumes, such as when we wish to determine the multiplicities in a stable intersection of a set of tropical hypersurfaces.

Correctness is another issue. In [6] the authors write:

"However, for many polynomial systems in real applications, listed in Table B1 and Table B2 in Sect. 4, MixedVol-2.0 produced the same mixed volume for all different sets of random liftings, whereas DEMiCs failed to provide a unique mixed volume with respect to different liftings."

comparing the implementation in [6] to that of [8]. However, it seems that there has been no serious attempt in recent implementations to eliminate the possibility of round-off errors. This may in practise not be a problem for many systems, but it is a problem if the software is used in a mathematical proof.

Finally, memory usage is an issue – not so much because we will run out of RAM on conventional machines, but rather because memory accesses slow down computation and limits what exotic architectures our software may run on.

We address these three issues with our new implementation. Distributed as part of the C++ library gfanlib [4], other systems can benefit from it. How to use the library is demonstrated in Figs. 1 and 2. The algorithm itself, which uses no linear programming, was proposed at MEGA 2015 and is a variant of the algorithm in [7]. The mathematical details are described in [5], while we here address some software engineering aspects. We hope that a general audience will find the discussion interesting.

```cpp
#include <iostream>
#include "gfanlib.h"
using namespace std;
using namespace gfan;

int main(){
    try{
        cout << mixedVolume(MixedVolumeExamples::cyclic(12),16) << endl;
        cout << mixedVolume(MixedVolumeExamples::gaukwa(7),8) << endl;
    }
    catch (...){
        cerr << "Error - most likely an integer overflow." << endl;
        return 1;
    }
    return 0;
}
```

Fig. 2. A sample C++ program demonstrating how to run the mixed volume computation with 16 or 8 threads and catch possible exceptions.

2 A Brief Description of the Algorithm

We briefly describe the tropical homotopy approach to finding mixed volumes. The main object of study is overlays of tropical hypersurfaces.

Definition 1. *Given a matrix $A \in \mathbb{Z}^{n \times m}$ and a vector $\omega \in \mathbb{R}^m$, the tropical hypersurface $T(A, \omega)$ is defined as*

$$T(A, \omega) := \{x \in \mathbb{R}^n : \max_i(\omega_i + A_{1i}x_1 + \cdots + A_{ni}x_n) \text{ is attained at least twice}\}.$$

A tropical hypersurface is a polyhedral complex of codimension 1. We will consider n such hypersurfaces with a total of $m = m_1 + \cdots + m_n$ terms. We let $\mathcal{A} = (A_1, \ldots, A_n)$ be the tuple of matrices with $A_i \in \mathbb{Z}^{n \times m_i}$ and $\omega \in \mathbb{R}^m = \mathbb{R}^{m_1} \times \cdots \times \mathbb{R}^{m_n}$ be the concatenated vector of coefficients.

Example 1. The tropical hypersurfaces defined by $\omega = ((0, -1, 0, 0), (0, 0, 1, 0))$ and

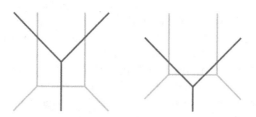

Fig. 3. Different overlays of tropical hypersurfaces defined by the same pair of matrices. As coefficients change in the tropical homotopy one intersection point splits into two.

$$\mathcal{A} = \left(\begin{pmatrix} 0 & 1 & 2 & 1 \\ 0 & 0 & 0 & 1 \end{pmatrix}, \begin{pmatrix} 1 & 0 & 1 & 2 \\ 0 & 1 & 1 & 1 \end{pmatrix} \right)$$

are shown in Fig. 3 (left). The right picture is for $\omega' = ((0, -1, 0, \frac{3}{2}), (0, 0, 1, 0))$.

In the case of generic coefficients, given an intersection point, for each hypersurface, the maximum in Definition 1 is attained exactly twice. That is, to each intersection point p we may associate a tuple $M = ((a_1, b_1), \ldots, (a_n, b_n)) \in \{1, \ldots, m_1\}^2 \times \cdots \times \{1, \ldots, m_n\}^2$ containing the indices to the columns of the A_i where the maximum was attained. We call M the mixed cell dual to p.

The purpose of the tropical homotopy algorithm is to keep track of the mixed cells as ω changes along a straight line ℓ towards some $\omega' \in \mathbb{R}^m$. To prevent ℓ from passing through low dimensional cones, ω is perturbed symbolically, by letting $\omega \in (\mathbb{R}(\varepsilon))^m$ with $\varepsilon > 0$ symbolic and small. We give the specifications:

Algorithm 2 (Tropical homotopy).
Input: *A tuple $\mathcal{A}$, coefficients ω and ω' and the mixed cells dual to $\cap_i T(A_i, \omega_i)$.*
Output: *The mixed cells dual to $\cap_i T(A_i, \omega'_i + \varepsilon \omega_i)$.*

2.1 A Mixed Cell Cone

The tropical homotopy attempts to treat each mixed cell independently. Therefore it is relevant to know the set of ω's for which a particular cell appears.

Theorem 3. *The set C_M of all $\omega \in \mathbb{R}^m$ giving rise to a particular M appearing as a dual mixed cell is an m-dimensional cone with $m - 2n$ facets.*

The facet normals are in the null space of the *Cayley matrix* of $\mathcal{A}$. In Example 1

$$\text{Cayley}(A_1, A_2) = \begin{pmatrix} 0 & 1 & 2 & 1 & 1 & 0 & 1 & 2 \\ 0 & 0 & 0 & 1 & 0 & 1 & 1 & 1 \\ 1 & 1 & 1 & 1 & 0 & 0 & 0 & 0 \\ 0 & 0 & 0 & 0 & 1 & 1 & 1 & 1 \end{pmatrix}$$

and the four marked columns constitute the center mixed cell in Fig. 3 (left). Each of the other columns together with those indexed by M forms a matrix of rank $2n$. Each non-zero vector v in the null space is a facet normal of C_M. In the indicated case $v = (1, 0, 1, -2, 0, -2, 0, 2, 0)$, and we may check that v separates ω and ω'. What happens to M after ω crosses the facet is encoded in v.

We may assign multiplicities to each intersection point, namely the absolute value of the determinant of the submatrix of the Cayley matrix indexed by the mixed cell. The multiplicities happen to sum to $\text{MixVol}(\text{New}(f_1), \ldots, \text{New}(f_n))$. Consequently, the mixed volume can be found by finding any generic tropical intersection $\cap_i T(A_i, \omega_i)$. Next we discuss how to build up such intersection.

2.2 Tropical Regeneration

In numerical regeneration [3], to solve $f_1 = \cdots = f_n = 0$, one starts with the solution to a linear system consisting of n equations in n unknowns. Successively each linear equation is first replaced by a product of linear equations and thereafter the desired f_i. During this process, the solutions to the intermediate systems are tracked. The tropical regeneration process is its *tropicalisation*.

We start with the configuration $(L_1, \ldots, L_n)$ where the L_i denote $n \times (n+1)$ matrices with a zero vector and the standard basis vectors as columns. Then we replace L_1 by the configuration B_1 which is just L_1 scaled so that the columns of A_1 are in the convex hull of the columns of B. Solutions to a system with matrices

$$((A_1, B_1), L_2, \ldots, L_n)$$

are known for coordinates of the coefficients associated to A_1 being very small. As we instead make coefficients of L_1 small via tropical homotopy, we obtain a set of mixed cells for the tuple (after deleting solutions going to infinity):

$$(A_1, L_2, \ldots, L_n).$$

Repeating this process for the other A_i we eventually reach the set of solutions to a system with exponent matrices $(A_1, \ldots, A_n)$. The mixed volume is obtained by summing the multiplicities, i.e. the absolute values of the determinants.

Starting out from one solution to the linear system, the solution point splits up into many that may later join, go to infinity or split up further as the regeneration proceeds. Using reverse search [1], the graph of all these homotopy paths may be turned into a tree, allowing easy organisation of a memoryless traversal.

3   Technical Implementation Details

A comparison to other implementations can be found in Fig. 4. Note that those are running on different hardware and using different algorithms, with the algorithm of [7] begin closest to ours. Our timings are roughly 50 times faster than the first arXiv.org version of [5]. Moreover, the new version achieves this while still catching overflows, should any arise. In this section we will describe how this performance gain was obtained and address a few other technical details.

Example	n	Mixed volume	1 thread	16 threads	Ratio	[7] 8 threads	[6] 1 thread
Cyclic-15	15	35243520	461.3	35.8	12.9	4070	36428
Noon-20	20	3486784361	59.0	4.8	12.4	6460	1109
Chandra-21	21	1048576	151.6	11.5	13.2	7580	1067
Katsura-17	18	131070	4.5	0.5	9.5	5310	75619
Eco-22	22	1048576	102.7	8.5	12.1	8750	

Fig. 4. Timings in seconds for our implementation compared to timings reported in [6,7]. Our timings are for a dual Intel Xeon E2670 CPU system, while a "2.4 GHz Intel Core 2 Quad CPU" was used in [6] and a "SGI Altix ICE 8400" system in [7]. For Gaukwa-7 our implementation had integer overflows, while [6,7] had no problems.

Profiling. To improve performance of software it is important to measure any possible progress. We found the Linux command line tool `perf` convenient, alternatingly doing `perf record ./a.out` and `perf report` to find the hotspots.

3.1   The Circuit Table

Given a tuple $(A_1, \ldots, A_n)$ and a mixed cell $M = ((a_1, b_1), \ldots, (a_n, b_n))$, we may use Theorem 3 to find out which circuits of the Cayley matrix give the irredundant inequalities of C_M. The tuple M selects columns of the Cayley matrix that form an invertible $(2n) \times (2n)$ matrix D. Our first attempt was to keep D^{-1} around to easily produce the desired circuits. The drawback of this approach is that D need not be invertible over $\mathbb{Z}$ and we had to resolve to floating point arithmetic, cast results to integers and verify them in exact arithmetic – a costly procedure.

The solution is to not store D^{-1}, but instead update the circuits as M changes.

Definition 2. *Given $\mathcal{A}$ and M, their circuit table is an $m \times m$ matrix, with each row indexed by $a \notin M$ containing a non-zero kernel vector of the Cayley matrix of $\mathcal{A}$ with support contained in $M \cup \{a\}$.*

This leaves $2n$ rows of the circuit matrix undefined. We will fix the scaling of each row by letting its coordinates be the maximal minors of the $2n \times (2n + 1)$ submatrix of $\text{Cayley}(A_1, \ldots, A_n)$ indexed by $M \cup \{a\}$. Therefore our rows are not circuits in the strict sense, as they need not be primitive.

The most important feature of the circuit matrix is that if we substitute an entry of M, it is possible to keep the matrix updated using integer row operations, i.e. cancelling out a particular coordinate common between two circuits.

Cache-Friendly Memory Layout. While matrices on which we want to do row operations are conveniently stored row-wise in memory, the circuit matrix is quite sparse. In a typical situation with $n = 10$ it may be of size 100×100, while having at most $2n + 1 = 21$ non-zero entries per row. Moreover, most of the entries appear in pairs with opposite sign due to the last n rows of the Cayley matrix. As all but one entry of a row appears in a column indexed by M, we may therefore store the matrix compactly as a 100×10 matrix and an additional vector of entries. It however turns out that many of the operations required to update the circuit matrix can conveniently and cache-friendly be performed as multiply-add row operations of the packed 10×100 transposed matrix of type

$$\text{row}_i := \frac{1}{\gamma}(\alpha\,\text{row}_i + \beta\,\text{row}_j) \tag{1}$$

where α, β and γ are integers. While our implementation treat generic integer implementations, we should imagine that the entries above are 32 bit ints.

Eliminating Integer Divisions. The update (1) above is done roughly for every row of the packed circuit matrix and requires an exact division for each entry. This is costly compared to the addition and multiplications. A common trick for low-level optimisation is to replace the division by multiplication. If γ is odd, then

$\gcd(\gamma, 2^{32}) = 1$ and therefore $[\gamma]$ has a multiplicative inverse in $\mathbb{Z}/2^{32}\mathbb{Z}$. On the other hand, if γ is even, we can reduce to the odd case by binary right shifts. The γ remains constant for the update of the full matrix as it happens to be the volume of the mixed cell in question, i.e. the determinant of D.

Vectorisation. In the update (1) each entry of a row gets the same treatment. This allows the use of the by now common 128 bit SIMD vector instructions. They can do 4 integer operations simultaneously. While we use a highlevel C++ class to encapsulate matrix data, to use the vector instructions we need to pass the restrict keyword, telling for example that the data pointed to by a, respectively b, can only be accessed through a, respectively b:

```
void muladd(int* __restrict__ a, int* __restrict__ b, int alph, int len);
```

Passing compiler options -O3 -mavx -msse2 -finline-limit=1000, gcc will emit the desired SIMD instructions. It is possible that the code would be more efficient if we forced data to be aligned at 128 bit boundaries and len to be divisible by 4. However, as we would like the code to automatically exploit 256 bit instructions in the future we decided not to make the code 128 bit specific.

Overflow Checking. If we know bounds on the entries of the ith and jth rows in (1) then we may be able to conclude that the operation does not overflow. However, a bound would have to be computed again for the next update. Therefore we compute the minimum and maximum entry of the new row$_i$ in the same loop that performs (1) and store the results for later. The code will translate into two vector instructions (max and min) per loop iteration.

In the implementation there are actually four different loops for doing (1) depending on the bounds on the arguments and whether a shift is needed. In the worst case the operation needs to be done in 64 bit to catch the overflow. To actually perform this checking without overflows, we put further restrictions on the entries of $\mathcal{A}$ and ω' (must fit in 16 bit) and on m ($m < 2^{15}$). These restrictions are checked automatically in the initialisation step of the algorithm.

3.2   Templates

Our description above has mainly considered the case where the integers are of 32 bit. With the use of C++ templates, we have written the code for updating the packed circuit table generically, so that the type can be replaced at compile time. For example one may choose to make an integer type with arbitrary precision, which should be called in case of an overflow.

3.3   Parallelisation

The key to easy parallelisation on multicore architectures is abstraction. We make a class representing the concept of an enumeration tree traverser. It will be located at a vertex of the tree and has a simple interface for asking for the number of children and moving up and down in the tree. With this class at hand we make, say 16, instances of such objects and hand them and the actual parallelisation over to an abstract tree traversal library. Such library was contributed to the gfan project by Bjarne Knudsen. It uses the C++11 standard for dealing with threads.

The only other issue to take care of for good performance is not stressing the memory allocator, which means that matrices whose sizes are known at initialisation are never reallocated. With our 16 core machine we obtain speedups of a factor 13.5 in the best cases. This is a reasonable speedup as the machine clocks down by roughly 10 percent when many cores are in use.

3.4 Exceptions

We use the C++ exceptions to handle issues with overflows. If a subroutine called by the enumeration graph traverser overflows, it will throw an exception. We do not count on C++ handling exceptions across threads and therefore catch the exception in the abstract traverser, which will then make sure that the computation is aborted in all threads. The library then throws a new exception indicating that an overflow occurred. We then have the option to redo the computation with higher precision.

For this strategy to work it is important that our code is exception safe. To be that, the most important feature is that we design our classes so that allocated memory is freed in destructors following the RAII principle (Resource Acquisition Is Initialization). In particular we avoid **malloc** and **new** but use STL containers instead (Standard Template Library).

References

1. Avis, D., Fukuda, K.: Reverse search for enumeration. Discrete Appl. Math. **65**(1–3), 21–46 (1996). First International Colloquium on Graphs and Optimization (GOI) (1992) (Grimentz)
2. Dyer, M., Gritzmann, P., Hufnagel, A.: On the complexity of computing mixed volumes. SIAM J. Comput. **27**(2), 356–400 (1998)
3. Hauenstein, J.D., Sommese, A.J., Wampler, C.W.: Regeneration homotopies for solving systems of polynomials. Math. Comput. **80**(273), 345–377 (2011)
4. Jensen, A.N.: Gfan, a software system for Gröbner fans. http://home.math.au.dk/jensen/software/gfan/gfan.html
5. Jensen, A.N.: Tropical homotopy continuation (2016). arXiv:1601.02818
6. Lee, T.-L., Li, T.-Y.: Mixed volume computation in solving polynomial systems. Contemp. Math. **556**, 97–112 (2011)
7. Malajovich, G.: Computing mixed volume and all mixed cells in quermassintegral time. Found. Comput. Math. 1–42 (2014). arXiv:1412.0480. http://link.springer.com/article/10.1007%2Fs10208-016-9320-1
8. Mizutani, T., Takeda, A., Kojima, M.: Dynamic enumeration of all mixed cells. Discrete Comput. Geom. **37**(3), 351–367 (2007)

Primary Decomposition in SINGULAR

Hans Schönemann[✉]

Department of Mathematics, University of Kaiserslautern, Kaiserslautern, Germany
hannes@mathematik.uni-kl.de
http://www.singular.uni-kl.de/

Abstract. Singular is a comprehensive and steadily growing computer algebra system, with particular emphasis on applications in algebraic geometry, commutative algebra, and singularity theory.

Singular provides highly efficient core algorithms, a multitude of advanced algorithms in the above fields, an intuitive, C-like programming language, easy ways to make it user-extendable through libraries, and a comprehensive online manual and help function.

Singular's core algorithms handle Gröbner resp. standard bases and free resolutions, polynomial factorization, resultants, characteristic sets, and numerical root finding. Symbolic-numeric solving in Singular starts with a decomposition to a triangular system or the primary decomposition of (the radical of) an ideal. New developments for primary decomposition will be presented in this paper: identifying sub problems allows an early split of the radical. A primary decomposition of these sub problems can be lifted to (not necessary primary) decomposition of the original problem: the subsequent primary decomposition will be faster.

After the symbolic preprocessing numerical solving of these smaller and easier to solve systems can be achieved by Singular's implementation of Laguerre's algorithm or by integrating other systems.

1 Introduction

Symbolic-numeric solving in Singular [DGPS] starts with a decomposition of the radical, which is the first step in the computation of the primary decomposition. Further decomposition into triangular systems provide a suitable starting point for numeric solving for each component of the solution space. New developments for primary decomposition act as a preprocessing step and allow an initial splitting of the problem into smaller sub-problems: factorizing Gröbner and identifying sub problems.

A proper ideal Q of a ring R is said to be **primary** if $f, g \in R$, $fg \in Q$ and $f \notin Q$ implies $g \in \sqrt{Q}$. In this case, $P = \sqrt{Q}$ is a prime ideal, and Q is also said to be a **P-primary ideal**. Given any ideal I of R, a **primary decomposition** of I is an expression of I as an intersection of finitely many primary ideals.

Now suppose that R is Noetherian. Then every proper ideal I of R has a primary decomposition. We can always achieve that such a decomposition $I = \bigcap_{i=1}^{r} Q_i$ is **minimal**. That is, the prime ideals $P_i = \sqrt{Q_i}$ are all distinct and none of the Q_i can be left out. In this case, the P_i are uniquely determined

© Springer International Publishing Switzerland 2016
G.-M. Greuel et al. (Eds.): ICMS 2016, LNCS 9725, pp. 206–211, 2016.
DOI: 10.1007/978-3-319-42432-3_26

by I and are referred to as the **associated primes** of I. If P_i is minimal among $P_1, \ldots, P_r$ with respect to inclusion, it is called a **minimal associated prime** of I.

The minimal associated primes of I are precisely the minimal prime ideals containing I. Their intersection is equal to $\sqrt{I}$. Every primary ideal occurring in a minimal primary decomposition of I is called a **primary component** of I. The component is said to be **isolated** if its radical is a minimal associated prime of I. Otherwise, it is said to be **embedded**. The isolated components are uniquely determined by I, the others are far from being unique.

From the definitions, it is clear that there is a number of different tasks coming with primary decomposition. These range from computing radicals via computing the minimal associated primes to computing a full primary decomposition. A variety of corresponding algorithms is implemented in the SINGULAR library `primdec.lib`. The two main algorithms for computing a full primary decomposition are `primdecSY` and `primdecGTZ`. A detailed description is given in [DGP].

The various Gröbner base computations dominate the computation. Therefore, a splitting into simpler parts renders some problems possible. In this paper I will present two such methods: the factorizing Gröbner (decreasing the degree) and a new method: identifying sub-problems with less variables (decreasing the number of variables).

2   Primary Decomposition by the Algorithm of Gianni, Trager, Zacharias

2.1   Splitting Tool

The starting point of the GTZ-algorithm is the following simple observation:

Lemma (Splitting Tool). If $I \subset R$ is an ideal, if $h \in R$ is a polynomial, and if $m \geq 1$ is an integer such that $I : \langle h \rangle^\infty = I : \langle h \rangle^m$, then

$$I = \left(I : \langle h \rangle^m \right) \cap \langle I, h^m \rangle.$$

The key result on which the algorithm is based specifies which polynomials h are considered:

If $I : f = I : f^2$ for some f, then $I = (I : f) \cap (I, f)$.
If $fg \in I$ and $(f, g) = R$, then $I = (I, f) \cap (I, g)$.
If $fg \in I$, then $\sqrt{I} = \sqrt{(I, f)} \cap \sqrt{(I, g)}$.
If $f^r \in I$, then $\sqrt{I} = \sqrt{(I, f)}$.
If $J \subseteq R$ an ideal, then $\sqrt{I} = \sqrt{I : J} \cap \sqrt{I + J} = \sqrt{I : J} \cap \sqrt{I : (I : J)}$.

2.2   Primary Decomposition: Reduction to Dimension 0

Proposition. Let $I \subsetneq K[\underline{x}] = R$ be a proper ideal, and let $\underline{u} \subset \underline{x}$ be a subset of maximal cardinality such that $I \cap K[\underline{u}] = \{0\}$. Then:

- The ideal $I\,K(\underline{u})[\underline{x}\setminus\underline{u}] \subset K(\underline{u})[\underline{x}\setminus\underline{u}]$ is zero-dimensional.
- Let $> = (>_{\underline{x}\setminus\underline{u}}, >_{\underline{u}})$ be a global product ordering on $K[\underline{x}]$, and let G be a Gröbner basis for I with respect to $>$. Then G is a Gröbner basis for $I\,K(\underline{u})[\underline{x}\setminus\underline{u}]$ with respect to the monomial ordering obtained by restricting $>$ to the monomials in $K[\underline{x}\setminus\underline{u}]$. Further, if $h \in K[\underline{u}]$ is the least common multiple of the leading coefficients of the elements of G (regarded as polynomials in $K(\underline{u})[\underline{x}\setminus\underline{u}]$), then

$$I\,K(\underline{u})[\underline{x}\setminus\underline{u}] \cap K[\underline{x}] = I : \langle h \rangle^{\infty}.$$

- All primary components of the ideal $I\,K(\underline{u})[\underline{x}\setminus\underline{u}] \cap K[\underline{x}]$ have the same dimension, namely $\dim I$.
  Further, if $I\,K(\underline{u})[\underline{x}\setminus\underline{u}] = Q_1 \cap \ldots \cap Q_r$ is the minimal primary decomposition, then

$$I\,K(\underline{u})[\underline{x}\setminus\underline{u}] \cap K[\underline{x}] = (Q_1 \cap K[\underline{x}]) \cap \ldots \cap (Q_r \cap K[\underline{x}])$$

is the minimal primary decomposition, too.

If $>$ is a global monomial ordering on $K[\underline{x}]$, then every subset $\underline{u} \subset \underline{x}$ of maximal cardinality satisfying $L_>(I) \cap K[\underline{u}] = \{0\}$ is also a subset of maximal cardinality such that $I \cap K[\underline{u}] = \{0\}$. By recursion, the proposition allows us to reduce the general case of primary decomposition to the zero-dimensional case. In turn, if $I \subset K[\underline{x}]$ is a zero-dimensional ideal "in general position" (with respect to the lexicographic order satisfying $x_1 > \cdots > x_n$), and if h_n is a generator for $I \cap K[x_n]$, the minimal primary decomposition of I is obtained by factorizing h_n. In characteristic zero, the condition that I is in general position can be achieved by means of a generic linear coordinate transformation.

2.3   Zero-Dimensional Primary Decomposition

The lexicographical Gröbner basis of a zero-dimensional ideal I contains one polynomial f of only the last variable. Let $f_1^{\alpha_1}\ldots f_r^{\alpha_r} = f$ the decomposition of f in irreducible factors.

Then the minimal primary decomposition of I is given by

$$I = \cap_{k=1}^{r}(I, f_k^{\alpha_k}).$$

3   Preprocessing: Factorizing Buchberger Algorithm

The factorizing Buchberger algorithm is the combination of Buchberger algorithm with factorization: each new element for the Gröbner basis will be factorized, and, if reducible, used to split the computation into several branches corresponding to the factors. Applied to an ideal $I = (f_1, ..., f_s)$ it computes a list of Gröbner bases $G_1, ..., G_r$ such that

$$V(I) = V(G_1) \cup \ldots \cup V(G_r)$$

The $V(G_i)$ need not be irreducible, so this algorithm is mainly used as a preprocessing step, substituting the initial Gröbner base computation of the GTZ algorithm, see [C].

This preprocessing steps lowers the degrees of the input polynomials to subsequent Gröbner base computations, decreasing their difficulty.

4   Preprocessing: Identifying Sub-Problems

Polynomial systems constructed from practical problems often include, even in the starting polynomials, subsystems (resp. subsets) which contain less than the full set of variables.

A simple search for such subsystems in the input polynomials to the initial Gröbner basis algorithm of GTZ (and also during the (lexicographic, i.e. variables eliminating) Gröbner basis computations.

Lemma (Subsystem Lemma). If the ideal s given as $I = (f_1, ..., f_s)$ with $J = (f_1, .., f_l)$ being the subsystem $(l < s)$, and $V(J) = V(J_1) \cup ... \cup V(J_k)$ a decomposition of J then

$$V(J_1 + I) \cup ... \cup V(J_k + I)$$

is a decomposition of I.

If a subsytem J splits into several components, the degree of each component will be lower than the degree of the original system I. The algorithm proceeds as follows:

(1.) find a subset $g_1, ..., g_l$ of $f_1, ..., f_s$ which involves only NN variables
(2.) try to decompose $(g_1, ..., g_l)$
(3.) if it does split, return the set of the resulting sub-problems $J_k + I$
(4.) if $(g_1, ..., g_l)$ does not split, increase NN
(5.) if $NN < N$ continue at step 1
(6.) the search for smaller subsystem was not successful: return I

This preprocessing, if successful, can be very helpful: it allows the decomposition of ideals which are not tractable otherwise.

5   Examples

The library `primdec.lib` implements the primary decomposition algorithms in SINGULAR: the main routines are `primdecGTZ` for the algorithm of Gianni, Trager and Zacharias resp. `primdecSY` for the algorithm of Shimoyama and Yokoyama.

The following ideal will be decomposed into 2 parts, each given as a pair of the primary component and the prime component:

```
LIB "primdec.lib";
ring  r = 0,(x,y,z),lp;
poly  p = z2+1;
poly  q = z3+2;
ideal I = p*q^2,y-z2;
primdecGTZ(i);
[1]:
   [1]:
      _[1]=z6+4z3+4
      _[2]=y-z2
   [2]:
      _[1]=z3+2
      _[2]=y-z2
  [2]:
   [1]:
      _[1]=z2+1
      _[2]=y-z2
   [2]:
      _[1]=z2+1
      _[2]=y-z2
```

In order to modify the main algorithm (GTZ resp. SY), additional arguments (of type string or int) can be passed to primdecGTZ resp. primdecSY.

Some parts of these algorithms are available separately: minAssGTZ for minimal associated primes (resp. minAssChar).

In the following example the search for subsystems gives 5 equations in 5 variables which decomposes in 20 parts. The computation time for minAssGTZ is 1.7 s while the unmodified algorithm needs 3.6 s:

```
LIB "primdec.lib";
ring rs= 0,(a,b,c,d,e,g,h),dp;
poly f0= a + b + c + d + e + 1;
poly f1= a + b + c + d + e;
poly f2= a*b + b*c + c*d + a*e + d*e;
poly f3= a*b*c + b*c*d + a*b*e + a*d*e + c*d*e;
poly f4= a*b*c*d + a*b*c*e + a*b*d*e + a*c*d*e + b*c*d*e;
poly f5= a*b*c*d*e - 1;
poly f6=g2+h2+e2;
poly f7=gha;
ideal gls= f1,f2,f3,f4,f5,f6,f7;
int ti=timer;
list L=minAssGTZ(gls,"subsystem"); timer-ti;
1700
ti=timer;
L=minAssGTZ(gls); timer-ti;
3600
```

6   Alternative Solutions

There are other algorithms to compute the primary decomposition of an ideal which can also include these preprocessing steps. As they are not starting with an initial Gröbner basis computation, the integration of these preprocessing is not straightforward.

The algorithm by Eisenbud, Huneke, Vasconcelos avoids the time consuming elimination and decomposes into equidimensional parts. Its main splitting tool is (cf. [EHV])

Proposition. If $I \subseteq R = K[x_1, ..x_n]$ is an ideal, then the equidimensional hull of I $E(I)$ is $AnnExt_R^{n-d}(R/I, R)$, where $d = dim(I)$.

This works well if the ideal have components of many different dimensions.

The other algorithm is the decomposition by characteristic series (cf. [SY]) which requires a different approach there this preprocessing does not fit very well.

References

[C]   Czapor, S.R.: Solving algebraic equations: combining Buchberger's algorithm with multivariate factorization. J. Symb. Comp. **7**(1), 49–53 (1998)

[EHV]   Eisenbud, D., Huneke, C., Vasconcelos, W.: Direct methods for primary decomposition. Invent. Math. **110**, 207–235 (1992)

[DGP]   Decker, W., Greuel, G.-M., Pfister, G.: Primary decomposition: algorithms and comparisons. In: Greuel, G.-M., Matzat, B.H., Hiss, G. (eds.) Algorithmic Algebra and Number Theory, pp. 187–220. Springer, Heidelberg (1998)

[DGPS] Decker, W., Greuel, G.-M., Pfister, G., Schönemann, H.: Singular 4-0-3 – A computer algebra system for polynomial computations

[SY]   Shimoyama, T., Yokoyama, K.: Localization and primary decomposition of polynomial ideals. J. Symb. Comp. **22**, 247–277 (1996)

Border Basis for Polynomial System Solving and Optimization

Philippe Trébuchet[1], Bernard Mourrain[2(✉)], and Marta Abril Bucero[2]

[1] ANSSI, Paris, France
Philippe.Trebuchet@lip6.fr
[2] Inria Sophia Antipolis Méditerranée, AROMATH, Valbonne, France
{Bernard.Mourrain,Marta.Abril_Bucero}@inria.fr

Abstract. We describe the software package BORDERBASIX dedicated to the computation of border bases and the solutions of polynomial equations. We present the main ingredients of the border basis algorithm and the other methods implemented in this package: numerical solutions from multiplication matrices, real radical computation, polynomial optimization. The implementation parameterized by the coefficient type and the choice function provides a versatile family of tools for polynomial computation with modular arithmetic, floating point arithmetic or rational arithmetic. It relies on linear algebra solvers for dense and sparse matrices for these various types of coefficients. A connection with SDP solvers has been integrated for the combination of relaxation approaches with border basis computation. Extensive benchmarks on typical polynomial systems are reported, which show the very good performance of the tool.

1 Border Basis Algorithms

In this section, we briefly describe the border basis algorithms and the algebraic solvers available in the package BORDERBASIX. Let $R = \mathbb{K}[x_1, \ldots, x_n]$ be the ring of polynomials in the variables $x_1, \ldots, x_n$ with coefficients in a field $\mathbb{K}$. Let $f_1, \ldots, f_m$ be the equations to be solved and $I = (f_1, \ldots, f_n)$ the ideal of R generated by these equations. The algebraic approach implemented in this package to solve the set of equations $\{f_1, \ldots, f_m\}$ proceeds in two steps:

(a) Compute the quotient algebra structure $\mathcal{A} = R/I$ represented by a (monomial) basis and the operators of multiplication by the variables.
(b) Compute the roots of the system from the operators of multiplication by the variables, when $\dim \mathcal{A} < \infty$.

The main algorithm of the package BORDERBASIXis the computation of border bases which provides the algebra structure of $\mathcal{A}$.

A border basis is defined with respect to a set B of monomials, connected to 1 (if $m \in B$ either $m = 1$ or $\exists i_0 \in [1, n]$ and $m' \in B$ such that $m = x_{i_0} m'$). Let $B^+ := B \cup x_1 B \cup \ldots \cup x_n B$ and $\partial B = B^+ \setminus B$. The computation of a border basis goes through the construction of a family F of polynomials of the form

© Springer International Publishing Switzerland 2016
G.-M. Greuel et al. (Eds.): ICMS 2016, LNCS 9725, pp. 212–220, 2016.
DOI: 10.1007/978-3-319-42432-3_27

$$f_\alpha = \mathbf{x}^\alpha - \sum_{\beta \in B} c_{\alpha,\beta} \mathbf{x}^\beta$$

which are in $\langle B^+ \rangle$ with only one term denoted $\gamma(f_\alpha) = \mathbf{x}^\alpha$ in ∂B and the other monomials of its support in B. A family of polynomials of this form is called a *rewriting* family. The family is *graded* if $\deg(\gamma(f)) = \deg(f)$ for all $f \in F$.

Let $\mathcal{M}$ be the set of monomials in the variables $x_1, \dots, x_n$. For $S \subset R, t \in \mathbb{N}$, S_t is the set of elements of S of degree $\leq t$. We denote by $\langle S \,|\, t \rangle$ the vector space spanned by the elements sm such that $\deg(sm) \leq t$ for $s \in S$ and $m \in \mathcal{M}$.

A rewriting family F is said to be *complete* in degree t if it is graded and satisfies $(\partial B)_t \subseteq \gamma(F)$; that is, each monomial $m \in \partial B$ of degree at most t is the leading monomial of some (necessarily unique) $f \in F$.

Given a rewriting family F which is complete in degree t, we define recursively the projection $\pi_{F,B}$ on B along F in the following way: $\forall m \in \mathcal{M}$,

- if $m \in B_t$, then $\pi_{F,B}(m) = m$,
- if $m \in (\partial B)_t \ (= (B^{[1]} \setminus B^{[0]})_t)$, then $\pi_{F,B}(m) = m - f$, where f is the (unique) polynomial in F for which $\gamma(f) = m$,
- if $m \in (B^{[k]} \setminus B^{[k-1]})_t$ for some integer $k \geq 2$, write $m = x_{i_0} m'$, where $m' \in B^{[k-1]}$ and $i_0 \in [1, n]$ is the smallest possible variable index for which such a decomposition exists, then $\pi_{F,B}(m) = \pi_{F,B}(x_{i_0} \pi_{F,B}(m'))$.

The map $\pi_{F,B}$ extends by linearity to a linear map from $\mathbb{K}[\mathbf{x}]_t$ onto $\langle B \rangle_t$. By construction, $f = \gamma(f) - \pi_{F,B}(\gamma(f))$ and $\pi_{F,B}(f) = 0$ for all $f \in F_t$. The next theorem shows that, under some natural commutativity condition, the map $\pi_{F,B}$ coincides with the linear projection from $\mathbb{K}[\mathbf{x}]_t$ onto $\langle B \rangle_t$ along the vector space $\langle F \,|\, t \rangle$ (see [16]):

Theorem 1. *Assume that B is connected to 1 and let F be a rewriting family for B, complete in degree $t \in \mathbb{N}$. Suppose that, for all $m \in \mathcal{M}_{t-2}$,*

$$\pi_{F,B}(x_i \pi_{F,B}(x_j m)) = \pi_{F,B}(x_j \pi_{F,B}(x_i m)) \text{ for all } i, j \in [1, n]. \tag{1}$$

Then $\pi_{F,B}$ coincides with the linear projection of $\mathbb{K}[\mathbf{x}]_t$ on $\langle B \rangle_t$ along the vector space $\langle F \,|\, t \rangle$; that is, $\mathbb{K}[\mathbf{x}]_t = \langle B \rangle_t \oplus \langle F \,|\, t \rangle$.

The commutation polynomials or C-polynomials are the polynomials:
$$\pi_{F,B}(x_i \pi_{F,B}(x_j m)) - \pi_{F,B}(x_j \pi_{F,B}(x_i m))$$
for $m \in B$, $i, j \in [1, n]$ such that $x_i m \in \partial B$ or $x_j m \in \partial B$.

1.1 Border Basis Computation

The border basis algorithm computes a rewriting family F which satisfies the relation (1) for any degree t. It proceeds incrementally degree by degree with a candidate monomial set B for the basis of $\mathcal{A}$ and the rewriting family F for B at a given degree t. At each degree, the non-zero polynomials deduced from the relations (1) are added to F.

In the main loop of the algorithm, the following operations are performed:

1. **prolongation**: determine the new elements of B in degree $t + 1$ and the elements $\tilde{F}$ of F^+ which are in $\langle B^+ \rangle$;
2. **matrix construction**: compute the coefficient matrix M of $\tilde{F}$ with respect to B^+;
3. **linear reduction**: compute a (sparse) LU decomposition of the matrix M and update the rewriting family $\tilde{F}$;
4. **commutation reduction**: reduce the C-polynomials with respect to $\tilde{F}$ and update B, F and t;

This loop is iterated until a complete rewriting family for B which satisfies (1) is obtained in the case of a zero-dimensional ideal [16] or until the Gotzmann regularity criterion is satisfied [17]. The computation is controlled by a choice function, which select "leading" monomials for the construction of rewriting families.

1.2   Real Radical Computation

The border basis algorithm has been extended to compute the real radical of an ideal [14], by integrating in the main loop a new operation:

5. **moment kernel**: compute a linear functional, which is positive on the sum of squares and orthogonal to $\langle F \mid t \rangle$, by solving a Semi-Definite Program; compute a basis of the kernel of its moment matrix in degree $\frac{t}{2}$ and add it to F.

1.3   Polynomial Optimisation

The border basis algorithm is also extended to compute the minimum of a polynomial $f(\mathbf{x})$ on a basic closed semi-algebraic set S defined by a set of constraints $\mathbf{g}^0 = 0$, $\mathbf{g}^+ \geq 0$ and the points where this minimum is reached (if they exist)[2]. The following operations are inserted in the main loop:

5.1. **optimal moment kernel**: compute a linear functional, which minimizes f on the preordering or quadratic module generated by the set of constraints, by solving a Semi-Definite Program, for moments associated to monomials in $B_{\frac{t}{2}} \cdot B_{\frac{t}{2}}$;
5.2. **flat extension test**: check if the moment sequence satisfies a flat extension property, and compute a basis and the border associated to the moment matrix kernel in degree $\frac{t}{2}$.

1.4   Root Finding

In the case of a zero-dimensional ideal, the last step in the resolution process consists in computing the roots from one or several operators of multiplication [5,11].

The eigenspaces associated to the transposed operator M_1^t of multiplication by the variable x_1 in $\mathcal{A}$ are computed. The first coordinates of the roots are given by the eigenvalues of M_1^t. If the eigenspaces are one-dimensional the other coordinates are deduced. Otherwise the transposed operator of multiplication by x_2 restricted to these eigenspaces is computed as well as its eigenspaces. It determines the second coordinates associated to a given first coordinate. This process is repeated until all coordinates of the roots are determined.

2   Software

The package BORDERBASIX is implemented in C++. It contains classes of multivariate polynomials represented as lists of monomials; classes for the border basis algorithms and for the solution of polynomial systems by eigenvector computation and linear algebra tools for dense and sparse matrices;

All the implementations are parameterized (templated) by the coefficient type. So that it is possible to use several number types for the computation of border bases. The set of number types effectively used in this computation includes modular arithmetic, multi-modular arithmetic, double, long double, double double, extended arithmetic based on GMP such as rational numbers (mpq), floating point numbers based on the GMP type mpf or on the library mpfr.

For linear algebra on dense matrices, a templated version of BLAS and LAPACK [3] libraries has been developed and is available in the linalg sub-package of BORDERBASIX. It includes the specialization of some of the arithmetic functions needed to control the precision of the computation. The main functionalities on dense matrices used in this library are Singular Value Decomposition, eigenvalue and eigenvector computation.

For sparse matrices, a templated version of SUPERLU [7] is also available, so that it can be used with general arithmetic number type. The main functionality used in this library is the solution of sparse linear systems by a direct method, for the computation of a rewriting family from the coefficient matrix of polynomials in the main loop of the border basis algorithm.

The connection with SDP solvers is developed in two ways. For the solver CSDP, a connection with a templated version has been implemented. The solver MOSEK [18] is also linked directly with the border basis implementation. For the solvers SDPA, SDPA_GMP [9], a file interface is used to describe the SDP problem to be solved and the SDP solver is called in an external process. The result is output in a file and read for the next step of the border basis computation. Since the SDP part is the most expensive part of whole computation, using files is not too penalizing.

The package contains approximatively 250 000 lines[1] of C++ code. It is accessible from http://www-sop.inria.fr/teams/galaad/software/bbx. It is also part of the software project MATHEMAGIX (www.mathemagix.org).

3   Benchmarks

In this section we analyse the behavior of our software on characteristic inputs:

---

[1] counted with cloc.

- Generic zero dimensional systems, because they provide the simplest case, with no degree drop, trivial syzygies and sharp zero dimensional detection;
- Cyclic-n test suite because they lead to very sparse bases and because the computation of the latter involves finding many non trivial syzygies.

The benchmarks have been performed on an `Intel Corei7-3610QM CPU@ 2.30 GHz` with 6Go of DDR3 1600 MHz.

First we compare the basis computation, we emphasize here that once the border basis is computed all the multiplication matrices are available. This is not the case for a Gröbner basis and the computation of just one multiplication matrix can be costly as shown in [12].

We have indicated by – test cases that have failed for unexplained reason and by `mem` cases that failed because of lack of memory.

`mac` choice function is the choice function that returns one monomial of highest degree and highest partial degree.

3.1 Katsura-n

The systems Katsura are zero dimensional complete intersection. Resultant theory give a characterization of a basis that is canonical but is not a Gröbner basis. First of all we present timings using modular arithmetic. The reason is that the behavior exposed here is also the one obtained using multi-modular computations. We then present numerical computations and show two different algorithms to recover the roots from the border basis, the first one described in [5], and the second in [11,21]. The system katsura-n has 2^n solutions. In the following table, we show *raw* timing using our method and other Gröbner engines.

n	Bbx mac	Bbx grl	Magma [4]	Sing. (std) [6]	Sing. (SlimGB) [6]	Giac [20]	Fgb [8]
7	0.06	0.09	0.018	0.01	0.01	0.05	0.1
8	0.23	0.43	0.14	2	3	0.37	0.5
9	1.05	2.41	0.84	13	35	2.15	2.7
10	5.43	18.23	5.4	100	333	11.6	22.5
11	33.41	127.23	37.26	1043	3408	87.2	172.6
12	240.69	1029.15	602.02	>15000	>15000	715.55	mem
13	1985.35	10432.12	4700.1	-	-	mem	-
14	13121.62	>15000	mem	-	-	-	-

Most of the time difference between *Bbx mac* order and *Bbx grl* order is due to the time spent to perform reduction of the C-polynomials, this operation has not been made scalable yet and also would greatly benefit from adaptation of the signature based criterions.

3.2 Cyclic-n

This family of polynomial systems comes from permutation theory. These systems are very far from being a complete intersection and have a complicated first syzygy module making them a standard benchmark case for Gröbner bases computations. It is noticeable here that the Gröbner basis computed for each system is very sparse, i.e. the multiplication matrix is costly to compute from the basis. This partly explains the difference of timings between bbx and the other softwares.

n	Bbx mac	Bbx grl	Magma grl [4]	Singular [6]	Giac [20]	Fgb [8]
5	0.12 s	0.05 s	0.01 s	(0.01)	(0.16 s)	(0.01 s)
6	1.09 s	0.18 s	0.10 (0.01) s	(0.01)	(0.15 s)	(6.63 s)
7	65.46 s	7.24 s	12.78 (0.28) s	(3)	(0.61 s)	(2.6 s)

The total time (in seconds) for computing the multiplication matrix by a variable is reported in this table. The time for computing Gröbner bases is given between parentheses. We did not consider higher systems of Cyclic-n which are not zero-dimensional.

What is also striking here is the timing difference between BORDERBASIX and the classical Gröbner engines. The explanation is that for Cyclic-n systems the Gröbner basis is very far from giving at least one multiplication matrix. As shown in [12] in such a case the computation of the multiplication matrix is the bottleneck of the resolution process. For instance it took 12.5 s with Magma to compute the multiplication matrix by x_0, the first variable, for the Cyclic-7 problem.

3.3   Floating Point Computation

In this section we present the floating point computation that are available inside BORDERBASIX. We show the time needed and the accuracy of the computed basis (the error is computed from the rational basis).

We use the Katsura-6 system as support benchmark, for it has only 64 solutions that are suitable for double precision treatment. The error estimates and approximations are performed, computing a certification as exposed in [11,21].

arith	Time basis	Time solve	Time cert	Error on basis	Error on sol
MPQ	22.5	-	-	0	-
double	0.058	0.06	0.60	10^{-10}	10
long double	0.069	0.32	3.34	10^{-30}	10^{-14}
MPF 128	0.13	3.38	17.64	10^{-38}	10^{-36}
MPF 256	0.25	3.57	56.9	10^{-76}	10^{-75}

We present here the same comparison for katsura-7:

arith	Time basis	Time solve	Time cert	Error on basis	Error on sol
MPQ	22.5	-	-	0	-
double	0.058	-	-	10^{-10}	-
long double	0.68	0.32	3.35	10^{-30}	10^{-14}
MPF 128	0.95	31.57	217.9	10^{-38}	10^{-35}
MPF 256	0.98	32.4	220.24	10^{-76}	10^{-75}

Problem	v	c	d	sol	o	p	s	t
⋄ Robinson	2	0	6	8	4	21	15	0.10
⋄ Motzkin	2	0	6	4	4	26	15	0.08
⋄ Motzkin perturbed	3	1	6	1	5	167	56	0.90
⋄ L'01, Ex. 1	2	0	4	1	2	8	6	0.022
⋄ L'01, Ex. 2	2	0	4	1	2	8	6	0.022
⋄ L'01, Ex. 3	2	0	6	4	4	25	15	0.075
L'01, Ex. 5	2	3	2	3	2	14	6	0.037
F, Ex. 4.1.4	1	2	4	2	2	4	3	0.023
F, Ex. 4.1.6	1	2	6	2	3	6	4	0.023
F, Ex. 4.1.7	1	2	4	1	2	4	3	0.022
F, Ex. 4.1.8	2	5	4	1	2	13	6	0.031
F, Ex. 4.1.9	2	6	4	1	4	44	15	0.11
F, Ex. 2.1.1	5	11	2	1	3	461	56	4.61
F, Ex. 2.1.2	6	13	2	1	2	209	26	0.46
F, Ex. 2.1.3	13	35	2	1	2	2379	78	34.55
F, Ex. 2.1.4	6	15	2	1	2	209	26	0.43
F, Ex. 2.1.5	10	31	2	1	2	1000	66	12.31
F, Ex. 2.1.6	10	25	2	1	2	1000	66	6.05
F, Ex. 2.1.7(1)	20	30	2	1	2	10625	231	1083.60
F, Ex. 2.1.7(5)	20	30	2	1	2	10625	231	1117.33
F, Ex. 2.1.8	24	58	2	1	2	3875	136	311.54
F, Ex. 2.1.9	10	11	2	1	2	714	44	1.98
F, Ex. 3.1.3	6	16	2	1	2	209	26	0.61
L'09 cbms1	3	3	3	5	3	26	17	0.14
L'09 rediff3	3	3	2	2	2	7	7	0.06
L'09 quadfor2	4	12	4	2	3	48	19	0.45
Simplex	15	16	2	1	2	3059	120	65.73
Tensor Ex. 4.2	6	0	8	4	8	2340	210	59.38

We emphasize here that most of the numerical computation timing is spent for getting a numerical certificate not for performing the actual computation of the roots!

3.4 Polynomial Optimization

In the following examples, the border basis computation is combined with Semi-Definite Programming to compute the optimum of a polynomial function over a basic semi-algebraic set [2].

The minimizer ideal is zero-dimensional and the algorithm outputs a numerical approximation of the minimizer points and generators of the minimizer ideal, after a finite number of relaxations. The SDP solver used in this computation is MOSEK.

The table records the number of variables (v), the number of inequality and equality constraints (c), the maximum degree of the constraints and of the polynomial to minimize (d), the number of minimizer points (sol), the maximal order (o), the maximal number of parameters (p), the maximal size of the moment matrices (s) in the SDP problems, and the total CPU time in seconds (t).

The examples L'09 are from [13], L'01 from [15] and F are from [10]. New equality constraints are added, following [1], to the initial problems in the examples marked with ⋄. The example "Simplex" is the optimization of a quadratic polynomial over the simplex. "Tensor" is an example from best rank-2 approximation of a tensor from [19]. Experiments are made on an Intel Core i5 2.40 GHz with 8 Gb of RAM.

References

1. Bucero, M.A., Mourrain, B.: Exact relaxation for polynomial optimization on semi-algebraic sets (2013). http://hal.inria.fr/hal-00846977
2. Bucero, M.A., Mourrain, B.: Border basis relaxation for polynomial optimization. J. Symb. Comput. **74**, 378–399 (2015)
3. Anderson, E., Bai, Z., Bischof, C., Demmel, J., Dongarra, J., Du Croz, J., Greenbaum, A., Hammarling, S., McKenney, A., Ostrouchov, S., Sorensen, D.: LAPACK Users' Guide. SIAM, Philadelphia (1992). http://www.netlib.org/lapack/
4. Bosma, W., Cannon, J., Playoust, C.: The magma algebra system I. The user language. J. Symb. Comput. **24**(3–4), 235–265 (1997)
5. Corless, R.M., Gianni, P.M., Trager, B.M.: A reordered Schur factorization method for zero-dimensional polynomial systems with multiple roots. In: Küchlin, W.W. (ed.) Proceedings of ISSAC, pp. 133–140 (1997)
6. Decker, W., Greuel, G.-M., Pfister, G., Schönemann, H.: Singular 4-0-2 – A computer algebra system for polynomial computations (2015). www.singular.uni-kl.de
7. Demmel, J.W., Eisenstat, S.C., Gilbert, J.R., Liu, J.W.H., Li, X.S.: A supernodal approach to sparse partial pivoting. SIAM J. Matrix Anal. Appl. **20**, 720–755 (1999)
8. Faugère, J.-C.: FGb: a library for computing Gröbner bases. In: Fukuda, K., Hoeven, J., Joswig, M., Takayama, N. (eds.) ICMS 2010. LNCS, vol. 6327, pp. 84–87. Springer, Heidelberg (2010)

9. Fujisawa, K., Fukuda, M., Kobayashi, K., Kojima, M., Nakata, K., Nakata, M., Yamashita, M.: SDPA (SemiDefinite Programming Algorithm) (2008)
10. Floudas, C.A., Pardalos, P.M., Adjiman, C.S., Esposito, W.R., Gumus, Z.H., Harding, S.T., Klepeis, J.L., Meyer, C.A., Schweiger, C.A.: Handbook of Test Problems in Local and Global Optimization. Kluwer Academic Publishers, Dordrecht (1999)
11. Graillat, S., Trébuchet, P.: A new algorithm for computing certified numerical approximations of the roots of a zero-dimensional system. In: ISSAC 2009, pp. 167–173 (2009)
12. Huot, L.: Polynomial systems solving and elliptic curve cryptography. Ph.D. thesis, Université Pierre et Marie Curie (UPMC) (2013)
13. Lasserre, J.-B.: Moments, Positive Polynomials and Their Applications. Imperial College Press, London (2009)
14. Lasserre, J.-B., Laurent, M., Mourrain, B., Rostalski, P., Trébuchet, P.: Moment matrices, border bases and real radical computation. J. Symb. Comput. **51**, 63–85 (2012)
15. Lasserre, J.B.: Global optimization with polynomials and the problem of moments. SIAM J. Optim. **11**, 796–817 (2001)
16. Mourrain, B., Trébuchet, P.: Generalized normal forms and polynomials system solving. In: Kauers, M. (ed.) ISSAC 2005, pp. 253–260 (2005)
17. Mourrain, B., Trébuchet, P.: Border basis representation of a general quotient algebra. In: van der Hoeven, J. (ed.) ISSAC 2012, pp. 265–272 (2012)
18. MOSEK ApS. The MOSEK optimization library (2015). www.mosek.com
19. Ottaviani, G., Spaenlehauer, P.-J., Sturmfels, B.: Exact solutions in structured low-rank approximation. SIAM J. Matrix Anal. Appl. **35**(4), 1521–1542 (2014)
20. Parisse, B.: Giac/XCas, a free computer algebra system. Technical report, University of Grenoble (2008)
21. Trébuchet, P.: A new certified numerical algorithm for solving polynomial systems. In: SCAN 2010, pp. 1–8 (2010)

High-Precision Arithmetic, Effective
Analysis and Special Functions

Recursive Double-Size Fixed Precision Arithmetic

Alexis Breust[1], Christophe Chabot[1], Jean-Guillaume Dumas[1],
Laurent Fousse[1], and Pascal Giorgi[2(⊠)]

[1] Laboratoire J. Kuntzmann, Université Grenoble Alpes, Grenoble, France
alexis.breust,christophe.chabot,jean-guillaume.dumas,
laurent.fousse@imag.fr
[2] LIRMM, CNRS, Université Montpellier, Montpellier, France
pascal.giorgi@lirmm.fr

Abstract. We propose a new fixed precision arithmetic package called
RECINT. It uses a recursive double-size data-structure. Contrary to arbitrary precision packages like GMP, that create vectors of words on the
heap, RECINT large integers are created on the stack. The space allocated
for these integers is a power of two and arithmetic is performed modulo that power. Operations are thus easily implemented recursively by a
divide and conquer strategy. Among those, we show that this packages
is particularly well adapted to Newton-Raphson like iterations or Montgomery reduction. Recursivity is implemented via doubling algorithms
on templated data-types. The idea is to extend machine word functionality to any power of two and to use template partial specialization to
adapt the implemented routines to some specific sizes and thresholds.
The main target precision is for cryptographic sizes, that is up to several
tens of machine words. Preliminary experiments show that good performance can be attained when comparing to the state of art GMP library:
it can be several order of magnitude faster when used with very few
machine words. This package is now integrated within the Givaro C++
library and has been used for efficient exact linear algebra computations.

1 Introduction

Mathematical computations that needs integers above machine word precision
are compelled to rely on a third party library. Among arbitrary precision libraries
for integers, GMP [11] (or its fork MPIR - www.mpir.org) is the most renown and
efficient. The underlying structure of GMP/MPIR integers is based on an array
of machine word integers that are accessed through a pointer. For instance, a
256-bits integers a can been represented by a dimension four array $[a_0, a_1, a_2, a_3]$
such that $a = a_0 + a_1 2^{64} + a_2 2^{128} + a_1 2^{192}$. Additionally to the pointer, GMP/
MPIR stores two integers that represent respectively the number of allocated
words and the number of used words, ensuring the dynamic of the precision.

This material is based on work supported in part by the Agence Nationale pour la
Recherche under Grant ANR-11-BS02-013 HPAC.

© Springer International Publishing Switzerland 2016
G.-M. Greuel et al. (Eds.): ICMS 2016, LNCS 9725, pp. 223–231, 2016.
DOI: 10.1007/978-3-319-42432-3_28

Indeed, such a structure is well designed to efficiently handle arbitrary precision but can be too costly when one knows in advance the targeted precision. Furthermore, when dealing with multiple integers at the same time, i.e. in matrix computation, access through pointers breaks cache mechanisms and penalizes performances. Hopefully, one can access to a low level API (mpn level) to both handle fixed precision and multiple integer without suffering from an heavy interface. However, this approach let the memory management to the user and cannot be incorporated to code that have been designed at a higher level. The purpose of our work is to provide an alternative to the GMP library for fixed precision integers that allow the flexibility of a high level library while still being efficient. In particular, our approach is to design very simple codes that extend naturally machine word to other powers of two.

A common and efficient way to compute over prime finite fields is to first perform the operation over the integers and map the result back to the field via modular reduction. In this specific case, the use of arbitrary precision is not relevant since precision is fixed by the cardinality of the finite field. Indeed, doubling the precision is often sufficient. To further optimize this approach, one can use precompiled code to tackle specific range of modulus. This approach has been proposed in the MPFQ library for cryptographic size moduli [10]. While being very efficient to perform scalar computation over finite field, i.e. modular exponentiation, the design of this library does not allow to use lazy modular reduction that proved very efficient for linear algebra [7].

Besides fixed precision integers, our package also provides prime fields support that can be easily embedded in high level code, while offering very good performance, in particular for matrix computations.

2 Functionality

2.1 Fixed Precision Arithmetic : Extending the Word Size

With fixed precision integers, the maximum number of bit is given in advance and all arithmetic operations are done within this precision. In particular, if K is the maximum bitsize (i.e. the precision) then all computations are done modulo 2^K. This corresponds to forgetting the carry for addition/subtraction or to getting the lowest K-bits part of the integer products.

Fixing K in advance facilitates the storage of integers through static array of size $\lceil K/64 \rceil$ on 64-bits architecture. In order to mimic word-size integers and to fully use the memory, we focus only on supporting integers of size a power of two: $K = 2^k$, with $k >= 6$. This assumption will allow us to provide a simple recursive data structure that eases manipulation and implementation of most arithmetic operations as explained in Sect. 3.1.

2.2 Fixed Precision Types

Our C++ package is devoted to provide functionality for fixed integer types. As for native integers, we provide signed and unsigned types. Furthermore, we extend this by providing a modular integer type that allows, e.g., computation over prime finite fields as a standard type. We denote by `ruint<k>` the type of our unsigned integer at precision 2^k (for instance we also define: `using ruint128 = ruint<7>; using ruint256 = ruint<8>;` etc.). Then the other types (signed and modular) are implemented with this structure. Our fixed precision integers are integrated in the Givaro library and available at https://github.com/linbox-team/givaro.

Signed Type. The recursive signed integer is defined as a `rint<k>` type. The data structure is just a `ruint<k>` and operations are performed via unsigned operations thanks to a two's complement representation.

Modular Integer Type. A special type `rmint<k>` is provided for modular operations. The data structure is composed of an unsigned value, `ruint<k>`, as well as a global `ruint<k>` for the modulus that is shared by all modular elements. Modular elements modulo p are represented in the range $[0, p-1]$ and all operations guarantee that values are always reduced to this range.

2.3 Integer Operations

Our package provides basic integer operations defined both as mathematical operator $(\times, +, -, /)$ or with their functional variants (mul, add, sub, div). A simple example of using our package for doing vector dot product with 256-bit integers is given below:

```
ruint<8> *A,*B,res=0; ... for(size_t i=0;i<n;i++) res+=A[i]*B[i];
```

Elements of type `ruint<8>` can be easily converted to `rint<8>` for computing with signed integer, or to `rmint<8>` for computing modulo a 256-bit prime. Several other functions are available (gcd, axpy, exponentiation, comparison, shift) as well as some internal functions that eases algorithm implementation, e.g. addition with carry, internal access to the structure, etc.

Two Types of Modular Reduction. One can easily switch between two type of modular reduction. In particular, one can choose between classic modular reduction using Euclidean division or the one using Montgomery's method [13], described in Sect. 3.2. The choice of representation is done at compile time, by a template constant: `rmint<k>`, or `rmint<K, MG_INACTIVE>` gives you the classical one, while `rmint<K, MG_ACTIVE>` uses Montgomery's reduction.

3 Underlying Theory and Technical Contribution

One drawback of arbitrary precision integers is to use a dynamic structure to handle the variation of integers value. There, defining contiguous structure of such integers leads to break cache prefetching when accessing data. Generally, one alternative is then to use a low level API where dynamics of the integers is delegated to the programmer. Unfortunately, this often becomes incompatible with the design of code at a high level, and then may not allow to re-use existing code. In the following section, we present our prototype designing a simple data structure preserving good performances.

3.1 Template Recursive Data Structure

Using specificities of fixed precision integers, we can represent an integer of 2^k bits as two integers of 2^{k-1} bits and therefore use a recursive representation. `ruint<k>` is our recursive unsigned integer of 2^k bits. The base case `ruint<6>` is called a limb and it corresponds to a native 64-bits integer. With these conventions, a `ruint<k>` can store any value between 0 and $2^{2^k} - 1$. In order to ensure a static contiguous storage, we chose to use a template recursive data structure with partial specialization. Note that the compiler will always completely unroll this structure at compiled time.

Our C++ template structure `ruint<>` is given hereafter together with an example for 256-bits integers, where `High` and `Low` correspond respectively to the leading and trailing 2^{k-1}-bits of the 2^k-bit integer.

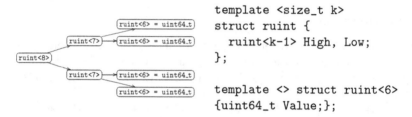

```
template <size_t k>
struct ruint {
    ruint<k-1> High, Low;
};

template <> struct ruint<6>
{uint64_t Value;};
```

3.2 Arithmetic Operations

Thanks to the recursive structure of `ruint<k>`, the implementation of arithmetic operations can be done by a recursive approach. For instance, addition can be done by two recursive calls with a carry propagation in between. Of course, the base case is mapped to the corresponding word-size integer operation. One major interest of such approach is that the compiler will unroll all the recursive calls leading to reduced control flow overhead and better instruction scheduling. As shown in Fig. 1 (left), such approach leads to better performance against the GMP library for the addition of two integers (when the result is stored in one of the operand) up to 1024 bits.

Multiplication is handled via a naive approach (or via Karatsuba's method), still using a recursive implementation, as shown in the following code:

```
template <size_t K> inline void lmul_naive
(ruint<K>& ah, ruint<K>& al, const ruint<K>& b, const ruint<K>& c) {
        bool rmid, rlow; ruint<K> bcmid, blcl;
        lmul_naive(blcl, b.Low, c.Low);              // Low part
        lmul_naive(bcmid, b.High, c.Low);            // Middle part
        laddmul(rmid, bcmid, b.Low, c.High, bcmid);  // Middle part
```

```
    laddmul(ah, b.High, c.High, bcmid.High);    // High part
    copy(al.Low, blcl.Low);
    add(rlow, al.High, blcl.High, bcmid.Low);
    if (rlow)  add_1(ah);
    if (rmid)  add_1(rmid, ah.High);
}
```

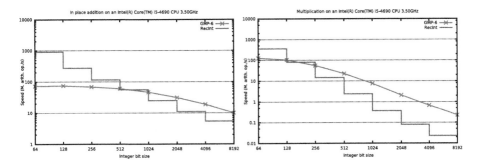

Fig. 1. Comparing integer addition (left) and multiplication (right) with GMP library

Still, without using any specific assembly code, contrary to GMP, our approach allows to have good speed-up for small integers up to 256-bits, as shown in Fig. 1 (right). Remark: We use Karatsuba's method only for more than 1024 bits.

Montgomery's Modular Reduction. Classical modular reduction uses integer division to map the result back into the desired integer range. Besides good theoretical complexity [2], this approach is not efficient in practice since hardware divisions are costly. If there exists a radix β such that divisions with β are inexpensive and $gcd(\beta, M) = 1$, Montgomery gives in [13] a method for modular reduction without trial division. The idea of Montgomery is to compute $C\beta^{-1} \bmod M$ instead of computing $C \bmod M$. As shown in [2, Algorithm 2.7], the fast version of this reduction requires to compute integer products modulo β and integer division by β. Assuming β to be a power of 2, this provides an algorithm without any division.

When the integer C to be reduced has a precision 2^{k+1} and the modulo has a precision of 2^k, i.e. $\beta = 2^{2^k}$, this boils down to two multiplications at precision 2^k. The design of our recursive integers is naturally compliant to such method and implementation is almost straightforward. From our first experiment, this straightforward implementation gives 1.5 speedup against GMP for 128-bits integer but becomes not competitive for larger modulus.

Inversion Modulo 2^{2^k}. On of the main operation in the setup of Montgomery reduction is the computation of the inverse of M modulo $\beta = 2^{2^k}$. The classical

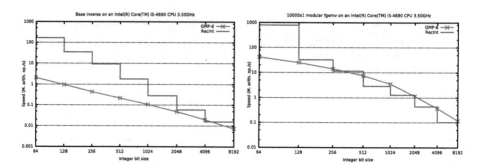

Fig. 2. Newton-Raphson iteration for the inverse modulo a prime power (left) and Modular matrix-vector multiplication (right)

algorithm for this is to use the extended Euclidean algorithm. But with the special structure of B, it is better to use a Newton-Raphson iteration [1,5], doubling the precision at each iteration. Thus here also our recursive structure is perfectly suited to this kind of algorithm, as shown in Fig. 2 (left). Note that GMP uses an extended gcd.

Recursive Division. For the integer division, our structure is also well suited to a recursive algorithm. Thus, we use that of [3] which uses two sub-algorithms dividing respectively 2 digits by 1 digit and 3 halves by 2. They allow then a recursive division of an s-digits integer by a r-digits integer with complexity $O(rs^{\log(3)-1} + r\log(s))$.

Recursive Shift. One operation is slightly more complicated on a recursive structure: shifting. At each recursive level, five cases have to be explored: no shift; shifting the lower part on the high part, in parts or completely; shifting more than half the word; shifting more than the whole word. For instance, if the shift d is such that $0 < d < 2^{2^{k-1}}$, then $b = a << d$ satisfies: $b.High = (a.High \ll d) \oplus (a.Low \gg (2^{k-1} - d))$ and $b.Low = (a.Low \ll d)$.

4 Application

4.1 Dense Linear Algebra: Freivalds Certificate

For dense linear algebra, a basic building block is the matrix-vector multiplication. We show in Fig. 2 (right) a multi-precision version of a modular matrix-vector multi-precision. There, we use the FFLAS-FFPACK package [7], version 2.2.1 (http://linalg.org/projects/fflas-ffpack) and just change the underlying representation from GMP to RECINT.

This possibility of easily changing the base representation can also be extremely important when certifying the results. In verified computing, a client (the Verifier) will check the results provided by a server (typically a cloud, the Prover). In dense linear algebra the basic tools for this is Freivalds' certificate for matrix-multiplication [9]. It uses matrix-vector multiplications with random vectors to check that $C = AB$, via $Cv - A(Bv) \overset{?}{=} 0$ on a random projection. With this certificate it is then possible to check any fast dense linear algebra computations with any precision [12, Sect. 5]. These certificates have a double goal. First this is a way to improve the confidence in a result, and this is even more the case if the underlying data structure is different for the computation and for the certificate. Second it provides a way to check outsourced computations. This is economically viable only if the Verifier's time is faster than the Prover's time. We show in Fig. 2 (right) that RecInt makes it possible to gain on all dense linear algebra verification, when operations require a few machine words.

4.2 Sparse Linear Algebra: Prime Power Rank

The Smith normal form is useful, in topology for instance, for computing the homology of a simplicial complex over the integers. There, the involved matrices are quite often sparse and computations reduce to computing the rank of these matrices modulo some prime powers [8, Sect. 5.1]. We report in Table 1 on timings using the LinBox library [6], version 1.4.1 (https://github.com/linbox-team/linbox). The last invariant factor of the matrix S16.231 × 231 is 2^{63}, while that of S22.1002 × 1002 is 2^{85}, so both need some extra precision to be computed. But as this factor is not known in advance, a strategy is to double the precision, until the rank modulo the prime power reaches the integer rank of the matrix. For the considered matrices we would stop at 256 bits, but we see in Table 1 that the strategy can be faster until 512 bits. As a comparison we also give timings modulo 3^{40} to show that this is not specific to the characteristic 2.

Table 1. Rank modulo a prime power http://hpac.imag.fr/Matrices/Tsuchioka

Matrix	mod	int64_t	RecInt					GMP
			6	7	8	9	10	
S16.231 × 231	3^{40}	-	**0.01 s**	0.05 s	0.12 s	0.37 s	1.28 s	0.19 s
S16.231 × 231	2^{63}	**0.03 s**	0.03 s	0.09 s	0.29 s	1.15 s	4.64 s	1.85 s
S16.231 × 231	2^{64}	-	-	**0.09 s**	0.30 s	1.15 s	4.64 s	1.85 s
S22.1002 × 1002	3^{40}	-	**0.34 s**	1.04 s	2.60 s	7.67 s	25.84 s	6.32 s
S22.1002 × 1002	2^{63}	-	**2.60 s**	7.00 s	23.10 s	88.24 s	356.79 s	154.22 s
S22.1002 × 1002	2^{86}	-	-	**7.37 s**	24.17 s	90.30 s	357.44 s	190.36 s

4.3 Towards an FPGA Implementation

We present here the first attempts towards an implementation on a FPGA. To build a programmable hardware (http://shiva.minalogic.net/), we had to provide basic arithmetic libraries to be used in an Elliptic curve based encryption scheme. We chose to use a dedicated software transforming C++ source into VHDL called GAUT [4]. The creation of a VHDL program can be split in the following steps: compilation of the C++ source and creation of the corresponding graph; compilation of the library containing the required operations; synthesis of the VHDL program and estimation of performance.

Table 2. FPGA area of modular exponentiation for different output flows, with RecInt integers from 128 to 512 bits

128 bits	op/s	7812	15625	31250	62500	125000	250000
	Flipflops	3040	6100	7008	9538	16617	32199
256 bits	op/s	976	1953	3906	7812	15625	31250
	Flipflops	3618	4391	6923	7654	8776	14542
512 bits	op/s	61	122	244	488	976	1953
	Flipflops	3553	3553	3553	4704	5729	7458

Table 2 shows some simulations of a modular exponentiation on a Virtex 5. We made the output flow vary in order to check the effect on the required size on the FPGA. We notice that required size can be significantly reduced if we accept a lower output flow. One nice point is that, due to the simple recursive structure of RecInt, these results have been obtained without significant modifications on the C++ source and automatically transformed. They are not optimal but rather promising, with no significant work load.

5 Conclusion and Perspective

Our RecInt package is a first attempt to provide high level API for fixed precision integers that are usable out of the box. Thus it is easy to switch from native integer types and to still provide good performance when compared to standard libraries. To further improve the performance, we need to introduce SIMD vectorized instructions in the design of our specializations and remove as much as possible conditional jumps. Modular multiplication should also benefit from either Barret's method or Montgomery-Svoboda's algorithm [2, Sect. 2.4.2].

References

1. Arazi, O., Qi, H.: On calculating multiplicative inverses modulo 2^m. IEEE Trans. Comput. **57**(10), 1435–1438 (2008). http://dx.org/10.1109/TC.2008.54

2. Brent, R., Zimmermann, P.: Modern Computer Arithmetic. Cambridge University Press, New York (2010)

3. Burnikel, C., Ziegler, J.: Fast recursive division. Technical Report MPI-I-98-1-022, Max Planck Institute fr Informatik, October 1998. http://www.mpi-sb.mpg.de/ziegler/TechRep.ps.gz

4. Coussy, P., Chavet, C., Bomel, P., Heller, D., Senn, E., Martin, E.: GAUT: a high-level synthesis tool for DSP applications. From algorithm to digital circuit. In: Coussy, P., Morawiec, A. (eds.) High-Level Synthesis, pp. 147–169. Springer, Netherlands (2008). http://dx.org/10.1007/978-1-4020-8588-8_9

5. Dumas, J.-G.: On Newton-Raphson iteration for multiplicative inverses modulo prime powers. IEEE Trans. Comput. **63**(8), 2106–2109 (2014). http://dx.org/10.1109/TC.2013.94

6. Dumas, J.-G., Gautier, T., Giesbrecht, M., Giorgi, P., Hovinen, B., Kaltofen, E., Saunders, B.D., Turner, W.J., Villard, G.: LinBox: a generic library for exact linear algebra. In: ICMS 2002, Beijing, China, pp. 40–50, August 2002. http://ljk.imag.fr/membres/Jean-Guillaume.Dumas/Publications/icms.pdf

7. Dumas, J.-G., Giorgi, P., Pernet, C.: Dense linear algebra over prime fields. ACM Trans. Math. Softw. **35**(3), 1–42 (2008). http://dx.org/10.1145/1391989.1391992

8. Dumas, J.-G., Saunders, B.D., Villard, G.: On efficient sparse integer matrix Smith normal form computations. J. Symbol Comput. **32**(1/2), 71–99 (2001). http://dx.org/10.1006/jsco.2001.0451

9. Freivalds, R.: Fast probabilistic algorithms. In: Bečvář, J. (ed.) MFCS 1979. LNCS, vol. 74, pp. 57–69. Springer, Heidelberg (1979)

10. Gaudry, P., Thomé, E.: The mpFq library and implementing curve-based key exchanges. In: SPEED: Software Performance Enhancement for Encryption and Decryption, Amsterdam, Netherlands, pp. 49–64. ECRYPT Network, June 2007. http://hal.inria.fr/inria-00168429

11. Granlund, T.: The GNU multiple precision arithmetic library, v6.1, November 2015. http://gmplib.org

12. Kaltofen, E.L., Nehring, M., Saunders, B.D.: Quadratic-time certificates in linear algebra. In: ISSAC 2011, San Jose, USA, pp. 171–176, June 2011. http://www.math.ncsu.edu/kaltofen/bibliography/11/KNS11.pdf

13. Montgomery, P.L.: Modular multiplication without trial division. Math. Comput. **44**(170), 519–521 (1985). http://dx.org/10.1090/S0025-5718-1985-0777282-X

CAMPARY: Cuda Multiple Precision Arithmetic Library and Applications

Mioara Joldes[1]([⊠]), Jean-Michel Muller[2], Valentina Popescu[2],
and Warwick Tucker[3]

[1] LAAS-CNRS, 7 Avenue du Colonel Roche, 31077 Toulouse, France
`joldes@laas.fr`
[2] LIP Laboratory, ENS Lyon, 46 Allée d'Italie, 69364 Lyon Cedex 07, France
`{jean-michel.muller,valentina.popescu}@ens-lyon.fr`
[3] Department of Mathematics, Uppsala University, Box 480, 75106 Uppsala, Sweden
`warwick@math.uu.se`

Abstract. Many scientific computing applications demand massive numerical computations on parallel architectures such as Graphics Processing Units (GPUs). Usually, either floating-point single or double precision arithmetic is used. Higher precision is generally not available in hardware, and software extended precision libraries are much slower and rarely supported on GPUs. We develop CAMPARY: a multiple-precision arithmetic library, using the CUDA programming language for the NVidia GPU platform. In our approach, the precision is extended by representing real numbers as the unevaluated sum of several standard machine precision floating-point numbers. We make use of error-free transforms algorithms, which are based only on native precision operations, but keep track of all rounding errors that occur when performing a sequence of additions and multiplications. This offers the simplicity of using hardware highly optimized floating-point operations, while also allowing for rigorously proven rounding error bounds. This also allows for easy implementation of an interval arithmetic. Currently, all basic multiple-precision arithmetic operations are supported. Our target applications are in chaotic dynamical systems or automatic control.

Keywords: Floating-point arithmetic · Multiple precision library · GPGPU computing · Error-free transform · Floating-point expansions · Dynamical systems · Hénon map, Semi-definite programming

1 Introduction

CAMPARY is a multiple-precision arithmetic library which targets mainly applications deployed on NVIDIA GPU platforms (compute capability 2.0 or greater). Both a CPU version (in C++ language) and a GPU version (written in CUDA C programming language [1]) are freely available at http://homepages.laas.fr/mmjoldes/campary/. Our library provides extended precisions on the order of

© Springer International Publishing Switzerland 2016
G.-M. Greuel et al. (Eds.): ICMS 2016, LNCS 9725, pp. 232–240, 2016.
DOI: 10.1007/978-3-319-42432-3_29

a few hundreds of bits. Currently, all basic multiple-precision arithmetic operations $(+, -, *, /, \sqrt{})$ are supported. Our implementation is very flexible: we provide templated precision sizes and overloaded operators. The library contains two levels of algorithms: *(i)* certified algorithms with rigorous error bounds and output constraints; *(ii)* "quick-and-dirty" algorithms that perform well for the average case, but do not consider the corner cases (i.e. cancellation prone computations). The later one also comes with a code generation module that allows for optimal performance.

The library was initially developed and tuned for long time iteration of chaotic dynamical systems in extended precision. At present, we aim at handling applications which require both extended precision and high performance computing.

2   Context and Related Software

Currently most floating-point (FP) calculations are done using single-precision (also called binary32) or double-precision (also called binary64) arithmetic. The majority of today's available processors (including GPUs) offer very fast implementations of FP arithmetic in these two formats, and comply with the IEEE 754-2008 standard for FP arithmetic [2]. This standard defines five *rounding functions* (round downwards, upwards, towards zero, to the nearest "ties to even", and to the nearest "ties to away"). An arithmetic operation should return the result as if computed using *infinite precision* and then applying the rounding function. Such an operation is said to be *correctly rounded*. The IEEE 754-2008 standard enforces the correct rounding of all basic arithmetic operations (addition, multiplication, division and square root). This requirement improves the portability of numerical software and also makes it possible and relatively easy to build an *interval arithmetic* (i.e., we get sure lower and upper bounds on the exact result).

However, several high-performance computing (HPC) problems require higher precision (also called *multiple precision*), up to a few hundred bits. For instance, in the field of chaotic dynamical systems, such problems appear in both mathematical questions (e.g., the study of strange attractors such as the Hénon attractor [3], in bifurcation analysis and stability of periodic orbits) and in applications to celestial mechanics (e.g., long-term stability of the solar system [4]). Multiple precision is also used in computational geometry (several techniques we use were initially developed for this domain) [5]. An example in *experimental mathematics* is the high-accuracy computation of *kissing numbers*, i.e. the maximal number of non-overlapping unit spheres that simultaneously can touch a central unit sphere [8]. That approach is based on very accurate solving of numerically sensitive semi-definite optimization problems (SDP). A recent increased interest in high precision SDP libraries comes also from ill-conditioned problems in quantum chemistry or control theory [9].

As of today, higher precision, such as *quad-precision* (binary128) or more has not yet been implemented in hardware on widely distributed processors, and the most common solution is software emulation. Arbitrary precision, i.e., the user's ability to choose the precision for each calculation, is now available in most computer algebra systems such as Mathematica, Maple or Sage. Furthermore, GNU MPFR [6] is an open source library written in C, that provides, besides arbitrary precision, correct rounding for each atomic basic operation. However, the versatility of such multiple precision libraries, which are able to manipulate numbers with tens of thousands –or even more– of bits, can sometimes be a quite heavy alternative to use, when a precision up to a few hundred bits is sufficient and we have strong performance requirements. Moreover, these libraries are rather difficult to port to recent highly parallel architectures, such as GPUs, since they implement very complex arithmetic algorithms, and they employ non-trivial memory management. Their complexity also makes them very difficult to prove formally.

In order to take advantage of the availability and efficiency of standard floating-point operations, our approach consists in representing higher precision numbers as unevaluated sums of several FP numbers (of different magnitudes). This representation is called double-double when the numbers are made up with two double-precision numbers, triple-double for three double-precision numbers, etc., and *floating-point expansion* in the general case (an arbitrary number of terms). The arithmetic operations on such representations are based on the use of *error-free transforms*, namely algorithms that allow one to compute the error of a FP addition or multiplication exactly. For instance, the sum of two FP numbers can be represented *exactly* as a FP number which is the correct rounding of the sum, plus a second FP number corresponding to the rounding error. Under certain assumptions, this decomposition can be computed at a very low cost by a simple sequence of standard precision FP operations. For instance, assuming that a and b are two FP numbers and that the rounding function, denoted RN, is one of the two round-to-nearest functions defined by IEEE 754-2008, a simple algorithm (called *2Sum* and due to Knuth [10]) computes the decomposition of $a + b$ using only 6 FP operations (see Algorithm 1). Similarly, if a FMA operator[1] is available, *2ProdFMA* returns π and the error e (namely $ab - \pi$) in 2 FP operations (see Algorithm 2). Algorithms like this can be extended to be used with arbitrary precision computations by chaining, resulting into the so called *distillation* algorithms [11].

Algorithm 1. 2Sum(a, b)

$s \leftarrow \text{RN}(a + b)$
$a' \leftarrow \text{RN}(s - b)$
$b' \leftarrow \text{RN}(s - a')$
$\delta_a \leftarrow \text{RN}(a - a')$
$\delta_b \leftarrow \text{RN}(b - b')$
$t \leftarrow \text{RN}(\delta_a + \delta_b)$
return s, t

Algorithm 2. 2ProdFMA(a, b)

$\pi \leftarrow \text{RN}(ab)$
$e \leftarrow \text{RN}(ab - \pi)$
return π, e

---

[1] A FMA (Fused Multiply-Add) operator evaluates an expression of the form $xy + t$ with one final rounding only.

It is thus possible to compute very accurate values even when rounding occurs at the intermediate operation's level. However, proving correctness and computing error bounds for this kind of algorithms is quite tricky and often their formal proof is necessary. Currently, the only available and easily portable code for manipulating such floating-point expansions is Bailey's QD library [7]. It provides double-double (DD) and quad-double (QD) arithmetic. It is known that most operations implemented in this library do not come with proven error bounds and correct or directed rounding is not supported. It is thus usually impossible to assess the final accuracy of these operations and no interval arithmetic can be constructed based on this library. However, the performance results of QD are very good on tested problems (e.g. on SDP instances [9]).

We generalize or modify this kind of algorithms in order to prove their correctness and keep good performances. We provide intermediary formats (such as triple-double) and also we generalize the use of expansions to those based on single-precision (for some processors which support only this format).

3   Key Features

CAMPARY is fully supported on and suitable for GPUs. This is because most available GPUs are compliant with the IEEE 754-2008 standard for FP arithmetic for both single and double precision; all rounding modes are provided and dynamic rounding mode change is supported without penalties. The `fma` instruction is supported in all devices with CUDA compute capability ≥ 2.0.

We implemented and proved new algorithms for normalizing, adding, multiplying, dividing and square rooting FP expansions. The method we use for computing the reciprocal and the square root of a FP expansion is based on an adapted Newton-Raphson iteration, where the intermediate calculations are done using "truncated" operations (additions, multiplications) involving FP expansions. We gave a thorough error analysis showing that it allows for very accurate computations (see [12]). We also introduced a new multiplication algorithm for FP expansions with arbitrary precision, up to the order of tens of FP elements in mind. The main feature consists in the partial products being accumulated in a specially designed data structure that has the regularity of a fixed-point representation while allowing the computation to be naturally carried out using native FP types. This allows us to easily avoid unnecessary computation and to obtain a rigorous accuracy analysis. The correctness and accuracy proofs of the algorithm and performance comparisons with existing libraries are presented in [13].

Fully certified algorithms like the aforementioned usually come with a performance cost. Thus, we chose to offer besides these, some so-called "quick-and-dirty" algorithms. (i) The certified ones[2] come with correctness proofs and they ensure the resulted expansion to be non-overlapping. Roughly speaking this means that an FP expansion carries sufficient information by ensuring that the

---

[2] Certified algorithms are available in *multi_prec_certified.h* file.

each two consecutive terms, say u_i and u_{i+1} are sufficiently far apart; for example, $|u_{i+1}| \leq \text{ulp}(u_i)$, where ulp is the unit in the last place [10, Chap. 2]. This is achieved by using different re-normalization algorithms, depending on the methods used for computing. Moreover, these algorithms offer a very tight error bound on the result. *(ii)* The "quick-and-dirty"[3] use faster versions of re-normalization algorithms. In most cases the result is going to be the same as obtained when computing with the certified level, even the non-overlapping condition can be achieved. The result may be uncertain if cancellation happens during intermediate computations; this can generate intermediate 0s or even non-monotonic expansions in the result. Also the worst case error bounds that we are able to prove are not as tight as in the certified level. We recommend the use of the "quick-and-dirty" if the performance requirements are strong, especially if there is a possibility to a posteriori check the correctness of the numerical result.

4   Applications

In what follows we briefly describe two applications (one achieved and one ongoing work) for CAMPARY.

4.1   Hénon Map Iteration

In [3], we studied the behavior of the Hénon map, a classical two-parameter, invertible map $h(x, y) = (1 + y - ax^2, bx)$. Depending on the two parameters a and b, this map can be chaotic, regular (the attractor of the map is a stable periodic orbit, also called sink), or a combination of these. We were interested in observing whether near the classical parameters $a = 1.4$ and $b = 0.3$, the Hénon map is chaotic and supports a strange attractor. This property has been observed numerically, but the question whether the Hénon attractor is indeed chaotic (trajectories belonging to the attractor are aperiodic and sensitive to initial conditions) or not remains open. In order to disprove this conjecture and find sinks for parameters close to the classical ones, we need to compute very long orbits for a large amount of initial points and parameters. Iterating the map for various initial points is a classical SIMD parallel problem, so a GPU implementation was done. For a double-precision implementation we obtain a significant speed-up of 21.5x compared to a multi-threaded CPU implementation. In order to tackle the conjecture, we used CAMPARY. The strategy for locating sinks is briefly the following: *(1)* We long term iterate the map ($10^6 \sim 10^9$ iterations) for various combinations of parameters and initial points in order to identify (using some additional tricks) some "numerical periodic orbits". A GPU code snippet for iterating the map is in Fig. 1. This very computationally intensive search process is parallelized; *(2)* At the end of the search, we rigorously prove the existence of periodic orbits using methods from interval analysis. This part is checked "on-line" on a CPU architecture. A performance results comparison with QD and MPFR is given in Table 1 (the "quick-and-dirty" version of the algorithms is used for CAMPARY).

---

[3] "Quick-and-dirty" algorithms are available in *multi_prec.h* file.

with QD and MPFR is given in Table 1 (the "quick-and-dirty" version of the algorithms is used for CAMPARY).

```
#define prec 4
/*device fct to be run using prec*doubles precision*/
__host__ __device__ void henon_iterate(double x0, double y0,
      double a, double b, long int ITER) {
      /*init multi_prec template vars*/
      multi_prec<prec,double> x_i(x0);
      multi_prec<prec,double> y_i(y0);
      multi_prec<prec,double> x_old;
      for (long int i=1; i <= ITER; i++) {
          /*Compute iterates*/
          x_i = y_i + 1.0 - a*x_i*x_i;
          y_i = b*x_old;
      }
}
```

Fig. 1. Example of usage of template `multi_prec` types and operations with 4-doubles precision in a host or device code that performs Hénon map iterations

Table 1. Peak number of Hénon map orbits/second for double vs. extended precision obtained using 10^6 iterations/orbit: (left) CAMPARY vs. QD library on a Tesla GPU[C2075]; (right) CAMPARY vs. MPFR (both parallelized with OpenMP on 8 threads) on Intel i7-3820 @3.60 GHz.

Prec	CAMPARY	QD
double	102398	
2-d	7608	4539
4-d	1788	618

Prec	CAMPARY	MPFR
2 doubles (106 bits)	227	11.8
3 doubles (159 bits)	76	10.6
4 doubles (212 bits)	37	10.1
6 doubles (318 bits)	15	8.9
8 doubles (424 bits)	8	7.9

4.2 SDP Programming

We currently consider the large-scale numerically sensitive semi-definite programs (SDP) on linear matrix inequalities (LMI). SDP can be seen as an extension of linear programming to the cone of symmetric matrices with positive eigenvalues, and where the linear vector inequalities are replaced by LMI. LMI are an important modeling tool in various areas of signal processing or automatic control. Currently, SDPA [9] is the leading multiple precision HPC SDP solver. Versions of SDPA (SDPA-GMP, SDPA-DD, SDPA-QD) use different multiple-precision libraries for performing accurate computations. Among those, SDPA-DD and SDPA-QD are reported to be the fastest on the market. In our present study, we replaced QD with CAMPARY at the compilation step of SDPA (no other tuning performed). We considered test problems from [14] where the performances of the previously mentioned libraries are compared. Preliminary results given in Table 2 show that CAMPARY is competitive for this kind of application also.

Table 2. The optimal value, relative gaps, primal/dual feasible errors, iterations and time for solving some ill-posed problems from SDPLIB by SDPA-QD, -DD, -CAMPARY

Problem	SDPA-DD	SDPA-QD	SDPA-CAMPARY		
			(2D)	(3D)	(4D)
gpp124-1	optimal: -7.3430762652465377				
Relative gap	$7.72e-04$	$1.91e-18$	$7.46e-04$	$6.72e-12$	$1.43e-18$
p.feas.error	$5.42e-20$	$2.86e-41$	$2.71e-20$	$2.72e-29$	$6.88e-41$
d.feas.error	$4.40e-14$	$3.48e-21$	$1.25e-14$	$2.72e-16$	$6.41e-21$
Iteration	24	57	24	39	66
Time (s)	3.580	94.58	13.25	55.57	127.52
gpp250-1	optimal: $-1.5444916882934067e+01$				
Relative gap	$5.29e-04$	$4.75e-18$	$5.22e-04$	$5.42e-12$	$5.03e-18$
p.feas.error	$3.89e-20$	$2.58e-41$	$1.35e-20$	$1.18e-30$	$6.43e-42$
d.feas.error	$9.78e-14$	$1.64e-21$	$3.52e-14$	$5.92e-16$	$1.14e-21$
Iteration	25	58	25	46	56
Time (s)	28.93	762.89	132.1	527.33	856.16
gpp500-1	optimal: $-2.5320543879075787e+01$				
Relative gap	$1.008e-03$	$2.13e-18$	$3.67e-04$	$5.78e-12$	$8.52e-18$
p.feas.error	$1.01e-20$	$5.73e-39$	$1.35e-20$	$2.01e-28$	$3.76e-42$
d.feas.error	$5.29e-14$	$1.70e-21$	$1.47e-13$	$1.03e-16$	$3.81e-21$
Iteration	25	55	26	42	58
Time (s)	230.05	5738.42	1027	3759.72	7146.72
qap10	optimal: $-1.0926074684462389e+03$				
Relative gap	$3.84e-05$	$2.06e-14$	$9.82e-05$	$2.40e-10$	$3.86e-14$
p.feas.error	$2.54e-21$	$1.09e-46$	$8.27e-22$	$2.64e-34$	$9.85e-47$
d.feas.error	$4.91e-14$	$2.97e-30$	$2.62e-13$	$1.98e-22$	$1.18e-29$
Iteration	20	36	19	29	36
Time (s)	30.46	645.28	115.3	371.88	762.53
hinf3	optimal: $5.6940778009669388e+01$				
Relative gap	$1.35e-08$	$5.30e-31$	$2.59e-06$	$2.47e-24$	$1.98e-31$
p.feas.error	$2.75e-24$	$1.18e-54$	$1.65e-23$	$7.10e-39$	$2.37e-55$
d.feas.error	$3.82e-14$	$1.74e-38$	$3.66e-14$	$1.79e-29$	$7.89e-42$
Iteration	30	47	24	46	48
Time (s)	0.02	0.11	0.03	0.07	0.12

5   Conclusion and Future Developments

Although initially used as a prototype for extended precision iterations of dynamical systems, CAMPARY has become a self-contained multiple precision arithmetic library mainly tuned for NVidia GPUs. We provide support for all arithmetic operations, so the first extension is to use and test it in the context of programs that make use of linear algebra, like SDP programming. Preliminary CPU implementations show good results. On short term, we intend to provide a GPU implementation for SDPA-CAMPARY. Concerning the certified part, a current ongoing work aims at formally proving our arithmetic algorithms using the Coq proof assistant [15]. A first proof of the renormalization algorithm is almost completed. A long term goal is to provide elementary functions also.

References

1. NVIDIA, NVIDIA CUDA Programming Guide 5.5 (2013)
2. IEEE Computer Society, IEEE Standard for Floating-Point Arithmetic, IEEE Standard 754-2008, August 2008. http://ieeexplore.ieee.org/servlet/opac?punumber=4610933
3. Joldes, M., Popescu, V., Tucker, W.: Searching for sinks for the Hnon map using a multiple-precision GPU arithmetic library. SIGARCH Comput. Archit. News **42**(4), 63–68 (2014)
4. Laskar, J., Gastineau, M.: Existence of collisional trajectories of Mercury, Mars and Venus with the Earth. Nature **459**(7248), 817–819 (2009)
5. Priest, D.H.: Algorithms for arbitrary precision floating point arithmetic. In: Kornerup, P., Matula, D.W. (eds.) Proceedings of the 10th IEEE Symposium on Computer Arithmetic (Arith-10), pp. 132–144. IEEE Computer Society Press, Los Alamitos (1991)
6. Fousse, L., Hanrot, G., Lefèvre, V., Pélissier, P., Zimmermann, P.: MPFR: a multiple-precision binary floating-point library with correct rounding. ACM Trans. Math. Softw. **33**(2) (2007). Art. 13, 115
7. Hida, Y., Li, X.S., Bailey, D.H.: Algorithms for quad-double precision floating-point arithmetic. In: Burgess, N., Ciminiera, L. (eds.) Proceedings of the 15th IEEE Symposium on Computer Arithmetic (ARITH-16), Vail, CO, pp. 155–162, June 2001
8. Mittelmann, H.D., Vallentin, F.: High-accuracy semidefinite programming bounds for kissing numbers. Exp. Math. **19**(2), 175–179 (2010)
9. Yamashita, M., Fujisawa, K., Fukuda, M., Kobayashi, K., Nakata, K., Nakata, M.: Latest Developments in the SDPA Family for Solving Large-Scale SDPs. In: Anjos, M.F., Lasserre, J.B. (eds.) Handbook on Semidefinite, Conic and Polynomial Optimization. International Series in Operations Research & Management Science, vol. 166, pp. 687–713. Springer, US (2012)
10. Muller, J.-M., Brisebarre, N., de Dinechin, F., Jeannerod, C.-P., Lefèvre, V., Melquiond, G., Revol, N., Stehlé, D., Torres, S.: Handbook of Floating-Point Arithmetic, Birkhäuser Boston, 2010, ACM G.1.0; G.1.2; G.4; B.2.0; B.2.4; F.2.1. ISBN 978-0-8176-4704-9
11. Rump, S.M., Ogita, T., Oishi, S.: Accurate floating-point summation part I: faithful rounding. SIAM J. Sci. Comput. **31**(1), 189–224 (2008)

12. Joldes, M., Marty, O., Muller, J.-M., Popescu, V.: Arithmetic algorithms for extended precision using floating-point expansions. IEEE Trans. Comput. **65**(4), 1197–1210 (2016)
13. Muller, J.-M., Popescu, V., Tak, P., Tang, P.: A new multiplication algorithm for extended precision using floating-point expansions. In: Proceedings of the 23rd Symposium on Computer Arithmetic (ARITH-23), July 2016, to appear
14. Nakata, M.: A numerical evaluation of highly accurate multiple-precision arithmetic version of semidefinite programming solver: SDPA-GMP, -QD and -DD. In: 2010 IEEE International Symposium on Computer-Aided Control System Design (CACSD). IEEE (2010)
15. The Coq Development team. The Coq proof assistant referencemanual, Inria, Version 8.0 (2004). http://coq.inria.fr

On the Computation of Confluent Hypergeometric Functions for Large Imaginary Part of Parameters b and z

Guillermo Navas-Palencia[1,2]([⊠]) and Argimiro Arratia[2]

[1] Numerical Algorithms Group Ltd, Oxford, UK
guillermo.navas@nag.co.uk
[2] Department of Computer Science,
Universitat Politècnica de Catalunya, Barcelona, Spain
argimiro@cs.upc.edu

Abstract. We present an efficient algorithm for the confluent hypergeometric functions when the imaginary part of b and z is large. The algorithm is based on the steepest descent method, applied to a suitable representation of the confluent hypergeometric functions as a highly oscillatory integral, which is then integrated by using various quadrature methods. The performance of the algorithm is compared with open-source and commercial software solutions with arbitrary precision, and for many cases the algorithm achieves high accuracy in both the real and imaginary parts. Our motivation comes from the need for accurate computation of the characteristic function of the Arcsine distribution or the Beta distribution; the latter being required in several financial applications, for example, modeling the loss given default in the context of portfolio credit risk.

Keywords: Confluent hypergeometric function · Complex numbers · Steepest descent

1 Introduction

The confluent hypergeometric function of the first kind $_1F_1(a;b;z)$ or Kummer's function $M(a,b,z)$ arises as one of the solutions of the limiting form of the hypergeometric differential equation, $z\frac{d^2w}{dz^2}+(b-z)\frac{dw}{dz}-aw=0$, for $b \notin \mathbb{Z}^-\cup\{0\}$, see [1, Sect. 13.2.1]. Another standard solution is $U(a,b,z)$, which is defined by the property $U(a,b,z) \sim z^{-a}$, $z \to \infty$, $|\mathrm{ph}\,z| \leq (3/2)\pi - \delta$, where δ is an arbitrary small positive constant such that $0 < \delta \ll 1$. Different methods have been devised for evaluating the confluent hypergeometric functions, although we are mainly interested in methods involving their integral representations. As stated in [1, Sects. 13.4.1 and 13.4.4], the functions $_1F_1(a;b;z)$ and $U(a,b,z)$ have the following integral representations, respectively

$$_1F_1(a;b;z) = \frac{\Gamma(b)}{\Gamma(a)\Gamma(b-a)} \int_0^1 e^{zt}t^{a-1}(1-t)^{b-a-1}\,dt, \quad \Re(b) > \Re(a) > 0 \quad (1)$$

© Springer International Publishing Switzerland 2016
G.-M. Greuel et al. (Eds.): ICMS 2016, LNCS 9725, pp. 241–248, 2016.
DOI: 10.1007/978-3-319-42432-3_30

$$U(a,b,z) = \frac{1}{\Gamma(a)} \int_0^\infty e^{-zt} t^{a-1}(1+t)^{b-a-1}\, dt, \quad \Re(a) > 0, |\mathrm{phz}| < \frac{1}{2}\pi \quad (2)$$

Furthermore, Kummer's transformations (cf. [1, Sect. 13.2 (vii)]), can be applied in situations where the parameters are not valid for some methods, or when the regime of parameters causes numerical instability,

$$_1F_1(a;b;z) = e^z\, _1F_1(b-a;b;-z), \quad U(a,b,z) = z^{1-b}U(a-b+1,2-b,z) \quad (3)$$

In other cases, recurrences relations can be applied (cf. [1, Sect. 13.3]). A recent survey of numerical methods for computing the confluent hypergeometric function can be found in [9,10], where the authors provide *roadmaps* with recommendations for which methods should be used in each situation.

2   Algorithm

The presented method for the computation of the confluent hypergeometric functions is based on the application of suitable transformations to highly oscillatory integrals and posterior numerical evaluation by means of quadrature methods. Some direct methods for $_1F_1(a;b;iz)$ can be applied for moderate values of $|\Im(z)|$, however a more general approach is the use of the numerical steepest descent method, which turns out to be very effective for the regime of parameters of interest. First, we briefly explain the path of steepest descent. Subsequently, we introduce the steepest descent integrals for those cases where $|\Im(b)|, |\Im(z)| \to \infty$.

2.1   Path of Steepest Descent

For this work we consider the ideal case for analytic integrand with no stationary points. We follow closely the theory developed in [3]. Let us consider the oscillatory integral

$$I = \int_\alpha^\beta f(x)e^{i\omega g(x)}\, dx \quad (4)$$

where $f(x)$ and $g(x)$ are smooth functions. By applying the steepest descent method, the interval of integration is substituted by a union of contours on the complex plane, such that along these contours the integrand is non-oscillatory and exponentially decaying. Given a point $x \in [\alpha, \beta]$, we define the path of steepest descent $h_x(p)$, parametrized by $p \in [0, \infty)$, such that the real part of the phase function $g(x)$ remains constant along the path. This is achieved by solving the equation $g(h_x(p)) = g(x) + ip$. If $g(x)$ is easily invertible, then $h_x(p) = g^{-1}(g(x) + ip)$, otherwise root-finding methods are employed, see [3, Sect. 5.2]. Along this path of steepest descent, integral (4) is transformed to

$$I[f; h_x] = e^{i\omega g(x)} \int_0^\infty f(h_x(p))h_x'(p)e^{-\omega p}\, dp$$

$$= \frac{e^{i\omega g(x)}}{\omega} \int_0^\infty f\left(h_x\left(\frac{q}{\omega}\right)\right)h_x'\left(\frac{q}{\omega}\right)e^{-q}\, dq \quad (5)$$

and $I = I[f; h_\alpha] - I[f; h_\beta]$ with both integrals well behaved. In the cases where $\beta = \infty$, this parametrization gives $I = I[f; h_\alpha] - 0$.

A particular case of interest is when $g(x) = x$. Then the path of steepest descent can be taken as $h_x(p) = x + ip$, and along this path (4) is written as

$$\int_\alpha^\beta f(x)e^{i\omega x}\, dx = \frac{ie^{i\omega\alpha}}{\omega}\int_0^\infty f\left(\alpha + i\frac{q}{\omega}\right)e^{-q}\, dq - \frac{ie^{i\omega\beta}}{\omega}\int_0^\infty f\left(\beta + i\frac{q}{\omega}\right)e^{-q}\, dq \quad (6)$$

2.2 $U(a, b, z)$, $\Im(z) \to \infty$

Integral representation (2) can be transformed into a highly oscillatory integral

$$U(a, b, z) = \frac{1}{\Gamma(a)}\int_0^\infty e^{-\Re(z)t}t^{a-1}(1+t)^{b-a-1}e^{-i\Im(z)t}\, dt \quad (7)$$

Taking $g(t) = t$, $g'(t) = 1 \neq 0$ and there are no stationary points. Therefore, in this case we only have one endpoint and the steepest descent integral obtained by (6) is reduced to a single line integral,

$$U(a, b, z) = \frac{i}{\omega\Gamma(a)}\int_0^\infty e^{-\Re(z)i\frac{q}{\omega}}\left(i\frac{q}{\omega}\right)^{a-1}\left(1 + i\frac{q}{\omega}\right)^{b-a-1}e^{-q}\, dq \quad (8)$$

2.3 $U(a, b, z)$, $\Im(b) \to \infty$

In this case, the path of integration is modified to avoid a singularity at $t = 0$, as can be seen after performing the transformation to a highly oscillatory integral,

$$U(a, b, z) = \frac{1}{\Gamma(a)}\int_0^\infty e^{-zt}t^{a-1}(1+t)^{b-a-1}\, dt$$

$$= \frac{e^z}{\Gamma(a)}\int_1^\infty e^{-zt}(t-1)^{a-1}t^{\Re(b)-a-1}e^{i\Im(b)\log(t)}\, dt \quad (9)$$

Now, we solve the path of steepest descent at $t = 1$ with $g(t) = \log(t)$, which in this case results trivial, $h_1(p) = e^{\log(1)+ip} = e^{ip}$ and $h_1'(p) = ie^{ip}$.

Likewise, no stationary points besides $t = \infty$ are present, and therefore there are no further contributions. The steepest descent integral is given by

$$U(a, b, z) = \frac{ie^z}{\omega\Gamma(a)}\int_0^\infty e^{\phi(q,\omega)}(\mu(q,\omega) - 1)^{a-1}\mu(q,\omega)^{\Re(b)-a-1}e^{-q}\, dq \quad (10)$$

where $\mu(q, \omega) = i\frac{q}{\omega}$ and $\phi(q, \omega) = -ze^{\mu(q,\omega)} + \mu(q,\omega)$.

2.4 $_1F_1(a, b, z)$, $\Im(z) \to \infty$

Similarly, we transform integral (1) into a highly oscillatory integral

$$_1F_1(a; b; z) = \frac{\Gamma(b)}{\Gamma(a)\Gamma(b-a)}\int_0^1 e^{\Re(z)t}t^{a-1}(1-t)^{b-a-1}e^{i\Im(z)t}\, dt \quad (11)$$

Again with $g(t) = t$ and the transformation stated in (6), we obtain, after some calculations, the steepest descent integrals given by

$$
\begin{aligned}
{}_1F_1(a;b;z) = \tfrac{\Gamma(b)}{\Gamma(a)\Gamma(b-a)} \tfrac{i}{\omega} \Bigg[&\int_0^\infty e^{\Re(z)i\frac{q}{\omega}} \left(i\tfrac{q}{\omega}\right)^{a-1} \left(1 - i\tfrac{q}{\omega}\right)^{b-a-1} e^{-q}\, dq \\
&- e^{i\omega} \int_0^\infty e^{\Re(z)(1+i\frac{q}{\omega})} \left(1 + i\tfrac{q}{\omega}\right)^{a-1} \left(-i\tfrac{q}{\omega}\right)^{b-a-1} e^{-q}\, dq \Bigg]
\end{aligned} \quad (12)
$$

2.5 ${}_1F_1(a, b, z)$, $\Im(b) \to \infty$

For this case we can use the following connection formula [1, Sect. 13.2.41], valid for all $z \neq 0$,

$$
\frac{1}{\Gamma(b)} {}_1F_1(a;b;z) = \frac{e^{\mp\pi i a}}{\Gamma(b-a)} U(a,b,z) + \frac{e^{\pm\pi i(b-a)}}{\Gamma(a)} e^z U(b-a,b,ze^{\pm\pi i}) \quad (13)
$$

2.6   Numerical Quadrature Schemes

Adaptive Quadrature for Oscillatory Integrals. The integrand in (11) can be rewritten in terms of its real and imaginary parts to obtain two separate integrals with trigonometric weight functions, the oscillatory factor, given the property,

$$
\int_0^1 f(t)e^{i\omega t}\, dt = \int_0^1 f(t)\cos(\omega t)\, dt + i \int_0^1 \sin(\omega t)\, dt \quad (14)
$$

Thus, we obtain the following integral representation for ${}_1F_1(a;b;z)$ when $|\Im(z)| \to \infty$,

$$
\begin{aligned}
{}_1F_1(a;b;z) = \frac{\Gamma(b)}{\Gamma(a)\Gamma(b-a)} \Bigg[&\int_0^1 e^{\Re(z)t} t^{a-1}(1-t)^{b-a-1} \cos(\Im(z)t)\, dt \\
&+ i \int_0^1 e^{\Re(z)t} t^{a-1}(1-t)^{b-a-1} \sin(\Im(z)t)\, dt \Bigg]
\end{aligned} \quad (15)
$$

These type of integrals can be solved using specialized adaptive routines, such as the routine `gsl_integration_qawo` from the GNU Scientific Library [2]. This routine combines Clenshaw-Curtis quadrature with Gauss-Kronrod integration. Numerical examples can be found in [8], which show that this method works reasonably well for moderate values of $|\Im(z)|$. Unfortunately, this method cannot be directly applied to $U(a, b, z)$, and Kummer's transformation [1, Sect. 13.2.42], valid for $b \notin \mathbb{Z}$, is needed.

Gauss-Laguerre Quadrature. An efficient approach for infinite integrals with an exponentially decaying integrand is classical Gauss-Laguerre quadrature. Laguerre polynomials are orthogonal with respect to e^{-x} on $[0, \infty)$. Hence, using n-point quadrature yields an approximation,

$$
I[f; h_x] \approx Q[f; h_x] := \frac{e^{i\omega g(x)}}{\omega} \sum_{k=1}^n w_k f\left(h_x\left(\frac{x_k}{\omega}\right)\right) h_x'\left(\frac{x_k}{\omega}\right) \quad (16)
$$

As stated in [3], the approximation error by the quadrature rule behaves asymptotically as $\mathcal{O}(\omega^{-2n-1})$ as $\omega \to \infty$. As an illustrative example, let us consider the asymptotic expansion for $U(a, b, z)$ when $|z| \to \infty$, which can be deduced by applying Watson's lemma [11] to (8),

$$U(a, b, z) \sim z^{-a} \sum_{n=0}^{\infty} \frac{(a)_n (a - b + 1)_n}{n! (-z)^n}, \quad |\mathrm{ph}\, z| \le \frac{3}{2}\pi - \delta \tag{17}$$

The error behaves asymptotically as $\mathcal{O}(z^{-n-1})$, as notice by truncating the asymptotic expansion after n terms. Therefore, the asymptotic order of the Gauss-Laguerre quadrature is practically double using the same number of terms. A formula for the error of the n-point quadrature approximation (16) is

$$E = \frac{(n!)^2}{(2n)!} f^{2n}(\zeta), \quad 0 < \zeta < \infty \tag{18}$$

According to this formula and under the general assumption that $a, b \in \mathbb{R} \setminus \mathbb{N}$, f is infinitely differentiable on $[0, \infty)$, we can use the general Leibniz rule for the higher derivatives of a product of m factors to obtain the derivative of order $2n$,

$$((f \cdot g) \cdot h)^{(2n)} = \sum_{j=0}^{2n} \sum_{k=0}^{2n-j} \frac{(2n)! \cdot f^{(j)} g^{(k)} h^{(2n-k-j)}}{j! k! (2n - k - j)!} \tag{19}$$

where

$$f(x) = e^{-\Re(z) i x / \omega}, \quad g(x) = \left(1 + i\frac{x}{\omega}\right)^{b-a-1}, \quad h(x) = \left(i\frac{x}{\omega}\right)^{a-1} \tag{20}$$

and the $2n$ derivatives are given by

$$\sum_{j=0}^{2n} \sum_{k=0}^{2n-j} \frac{(2n)!(-1)^j}{j! k! (2n-k-j)!} \left(\frac{\Re(z) i}{\omega}\right)^j e^{-\Re(z) i x / \omega} \frac{\left(\frac{i}{\omega}\right)^k}{(b-a-1)_{-k}} \left(1 + i\frac{x}{\omega}\right)^{b-a-1-k}$$
$$\times \frac{\left(\frac{i}{\omega}\right)^{2n-k-j}}{(a-1)_{-2n+k+j}} x^{a-1-2n+k+j} \tag{21}$$

where $(a)_n$ is the Pochhammer symbol or rising factorial. An error bound in terms of a, b and z might be obtained from (21). Ideally, the error bound shall be tight enough without increasing the total computation time excessively. However, as can be seen below, numerical experiments indicate that the number of terms n rarely exceeds 50 for moderate values of the remaining parameters, typically if $|a|, |b| \cdot 10 < |\omega|$, for the case $U(a, b, iz)$ or $_1F_1(a, b, iz)$. Finally, for large parameters we apply logarithmic properties to the integrand in order to avoid overflow or underflow.

2.7 Numerical Examples

In this section, we compare our algorithm (NSD) with other routines in double precision floating-point arithmetic in terms of accuracy and computation time[1].

---

[1] Intel(R) Core(TM) i5-3317U CPU at 1.70 GHz.

Note that just a few packages in double precision allow the evaluation of the confluent hypergeometric function with complex argument. For this study we use Algorithm 707: CONHYP, described in [6,7] and Zhang and Jin implementation (ZJ) in [12]. Both codes are written in Fortran 90 and were compiled using `gfortran 4.9.3` without optimization flags. We implemented a simple prototype of the described methods using `Python 3.5.1` and the package `SciPy` [5], therefore there is plenty of room for improvement, and is part of ongoing work. Nevertheless, as shown in Table 1, our algorithm clearly outperforms aforementioned codes, being more noticeable as z increases. In order to test the accuracy, we use `mpmath` [4] with 20 digits of precision to compute the relative errors.

Table 1. Relative errors for routines computing the confluent hypergeometric function for complex argument. N: number of Gauss-Laguerre quadratures. (∗): precision in mpmath increased to 30 digits. (E): convergence to incorrect value. ($-$): overflow.

$_1F_1(a,b,z)$	CONHYP	ZJ	NSD	N
$(1,4,50i)$	3.96e−13/4.29e−18i	1.50e−15/4.28e−18i	1.15e−16/1.11e−16i	2
$(3,10,30+100i)$	1.27e−13/1.28e−13i	6.83e−17/1.07e−14i	2.48e−17/1.24e−14i	25
$(15,20,200i)$	9.20e−13/9.20e−13i	E	8.43e−16/7.93e−16i	25
$(400,450,1000i)$	8.32e−12/1.00e−11i	−	1.37e−12/1.02e−13i	50
$(2,20,50-2500i)$	1.35e−11/1.35e−11i	7.30e−11/2.10e−09i	4.75e−16/6.41e−16i	20
$(500,510,100-1000i)$	4.10e−13/3.68e−12i	−	4.71e−13/3.11e−16i	50
$(2,20,-20000i)$	−	5.79e−10/7.99e−07i	5.92e−16/3.62e−14i	10
$(900,930,-10^{10}i)$	−	−	6.78e−13/6.77e−13i	20
$(4000,4200,50000i)^*$	−	−	6.04e−12/5.99e−12i	80

Table 2 and Fig. 1 summarize the testing results and general performance of the algorithm for $U(a,b,z)$. As can be observed, 13–14 digits of precision in real and imaginary part are typically achieved. A similar precision for $_1F_1(a;ib;z)$ is expected. In terms of computational time, we compare our implementation in Python with MATLAB R2013a. As shown in Table 3, the MATLAB routine `hypergeom` is significantly slow for large imaginary parameters.

Table 2. Error statistics for $U(a,b,iz)$ and $U(a,ib,z)$ using $N=100$ quadratures.

Function	Min	Max	Mean
$U(a,b,iz)$	1.97e−18/2.04e−17i	9.97e−13/2.50e−11i	1.34e−14/6.94e−14i
$U(a,ib,z)$	6.57e−18/6.17e−18i	1.49e−11/8.55e−12i	1.38e−13/1.43e−13i

3   Applications

Besides the necessity of accurate and reliable methods for the regime of parameters and argument considered, confluent hypergeometric functions can be encountered in several scientific applications. In this paper, we focus on applications in statistics, more precisely on the evaluation of characteristic functions,

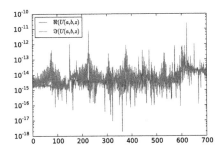 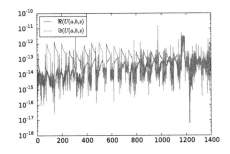

Fig. 1. Relative error in computing $U(a, b, z)$. Error in $U(a, b, iz)$ for $a \in [2, 400], b \in [-500, 500], z \in [10^3, 10^6]$ (left) and $U(a, ib, z)$ for $a \in [10, 100], b \in [10^3, 10^4], z \in [10, 100]$ (right). 700 and 1400 tests, respectively. (Color figure online)

Table 3. Comparison in terms of cpu time. MATLAB second evaluation in parenthesis.

$_1F_1(a; b; z)$	MATLAB	NSD
$(2, 20, -20000i)$	1.509 (0.068)	0.033
$(900, 930, -10^{10}i)$	5.594 (0.739)	0.035
$(4000, 4200, 50000i)$	488.384(18.127)	0.043

which can be defined in terms of confluent hypergeometric functions. Characteristic functions appear in many financial econometric models, for example modelling a beta-distributed loss given default in portfolio credit risk models (see [8, Sect. 4.4.2]). Let us consider three statistical distributions:

- Characteristic function of the Beta distribution.

$$\phi_X(t) = {}_1F_1(\alpha; \alpha + \beta; it) \tag{22}$$

where $\alpha, \beta > 0$. Thereby, the regime of parameters holds for the integral representation in (12).

- The standard Arcsine distribution is a special case of the Beta distribution with $\alpha = \beta = 1/2$, therefore we obtain a similar characteristic function, which can be identically computed.

$$\phi_X(t) = {}_1F_1\left(\frac{1}{2}; 1; it\right) \tag{23}$$

- The characteristic function for the F–distribution is defined in terms of the confluent hypergeometric function of the second kind,

$$\phi_X(t) = \frac{\Gamma((p+q)/2)}{\Gamma(q/2)} U\left(\frac{p}{2}, 1 - \frac{q}{2}, -\frac{q}{p} it\right) \tag{24}$$

where $p, q > 0$, are the degrees of freedom. In this case we can use the integral representation in (8).

4   Conclusions

We have presented an efficient algorithm for computing the confluent hypergeometric functions with large imaginary parameter and argument, which emerges as an alternative to asymptotic expansions. The numerical experiments show promising results and fast convergence as the imaginary part increases. Throughout this paper we have been considering real values for the remaining parameters, otherwise the function f becomes oscillatory. The numerical steepest descent method is not insensitive to oscillations in f, although in some cases this can be treated by applying other transformations. In cases where that is not possible, other methods have to be considered. Finally, a suitable integral representation for $|\Im(a)| \to \infty$ carry more complications and is part of future work.

References

1. NIST digital library of mathematical functions. Release 1.0.10 of 07 August 2015. In: Olver, F. W.J., et al. (ed.) NIST Handbook of Mathematical Functions. Cambridge University Press, NY (2010). http://dlmf.nist.gov/
2. Galassi, M., et al.: GNU Scientific Library Reference Manual, 3rd edn. ISBN 0954612078. http://www.gnu.org/software/gsl/
3. Huybrechs, D., Vandewalle, S.: On the evaluation of highly oscillatory integrals by analytic continuation. SIAM J. Numer. Anal. 44(3), 1026–1048 (2006)
4. Johansson, F., others.: mpmath: a Python library for arbitrary-precision floating-point arithmetic (version 0.19) (2013). http://mpmath.org/
5. Jones, E., Oliphant, E., Peterson, P., et al.: SciPy: open source scientific tools for Python (2001). http://www.scipy.org/
6. Nardin, M., Perger, W.F., Bhalla, A.: Algorithm 707. CONHYP: a numerical evaluator of the confluent hypergeometric function for complex arguments of large magnitudes. ACM Trans. Math. Softw. **18**, 345–349 (1992)
7. Nardin, M., Perger, W.F., Bhalla, A.: Numerical evaluation of the confluent hypergeometric function for complex arguments of large magnitudes. J. Comput. Appl. Math. **39**, 193–200 (1992)
8. Navas-Palencia, G.: Portfolio credit risk: models and numerical methods. MSc in Statistics and Operations Research Dissertation, Universitat Politècnica de Catalunya (2016). http://upcommons.upc.edu/bitstream/handle/2117/82265/memoria.pdf
9. Pearson, J.W.: Computation of hypergeometric functions. MSc in Mathematical Modelling and Scientific Computing Dissertation, University of Oxford (2009). http://people.maths.ox.ac.uk/~porterm/research/pearson_final.pdf
10. Pearson, J.W., Olver, S., Porter, M.A.: Numerical methods for the computation of the confluent and gauss hypergeometric functions (2015). arXiv:1407.7786
11. Watson, G.N.: The harmonic functions associated with the parabolic cylinder. Proc. Lond. Math. Soc. **2**, 116–148 (1918)
12. Zhang, S., Jin, J.: Computation of Special Functions. Wiley, New York (1996)

Mathematical Optimization

Parallelization of the FICO Xpress-Optimizer

Timo Berthold[1(✉)], James Farmer[2], Stefan Heinz[1], and Michael Perregaard[2]

[1] FICO (Fair Isaac Corporation), Berlin, Germany
{timoberthold,stefanheinz}@fico.com
[2] FICO, Birmingham, UK
{jamesfarmer,michaelperregaard}@fico.com

Abstract. Many optimization problems arising in practice can be modeled as mixed integer programs (MIPs). In this paper, we present the new parallelization concept for the state-of-the-art MIP solver FICO Xpress-Optimizer. A natural precondition to achieving reasonabling speedups from parallelization is maintaining a high workload of the available computational resources. At the same time, reproducibility and reliability are key requirements for mathematical optimization software; solvers like the FICO Xpress-Optimizer are expected to be deterministic. The resulting synchronization latencies render the goal of a satisfying workload a challenge in itself.

We address this challenge by following a partial information approach and separating the concepts of simultaneous tasks and independent threads from each other. Our computational results indicate that this leads to a much higher CPU workload and thereby to an improved scaling on modern high-performance CPUs. As an added value, the solution path that the FICO Xpress-Optimizer takes is not only deterministic in a fixed environment, but, to a certain extent, thread-independent.

Keywords: Mathematical optimization · Mixed integer programming · Parallelization

1 Introduction

Mixed integer programming (MIP) has become one of the most important techniques in Operations Research and Discrete Optimization. This paper deals with solving mixed integer programming (MIP) problems in parallel. Throughout this paper without loss of generality, we assume that the MIP is given in the following, general form:

$$\min\{\langle c, x \rangle : Ax \leq b, \ x_I \in \mathbb{Z}^{|I|}\}, \tag{1}$$

with matrix $A \in \mathbb{R}^{m \times n}$, vectors $b \in \mathbb{R}^m$ and $c \in \mathbb{R}^n$, and a subset $I \subseteq \{1, \ldots, n\}$.

State-of-the-art MIP solvers are based on *LP-based branch-and-bound* [8] in which the problem is recursively split into smaller subproblems, thereby creating a so-called *branching tree*. The basic branch-and-bound tree search typically is enhanced by a large number of sophisticated algorithms to keep the enumeration effort small. These include numerous heuristic methods to devise primal feasible

© Springer International Publishing Switzerland 2016
G.-M. Greuel et al. (Eds.): ICMS 2016, LNCS 9725, pp. 251–258, 2016.
DOI: 10.1007/978-3-319-42432-3_31

solutions, and a number of cutting plane separation algorithms to increase the lower bound value obtained by the Linear Programming (LP) relaxation, see, e.g., [10]. In practice, this allows for a dramatic reduction in the size of the branching tree. Typically, problems with ten thousand variables and constraints (i.e., approximately $2^{10\,000}$ potential solutions) can be solved by investigating a few hundred thousand branch-and-bound nodes.

Intuitively, tree search algorithms appear easy to parallelize. However, an efficient parallelization of LP-based branch-and-bound is a challenging task, see e.g., [11]. This is partially due to fact that the decisions involved depend upon each other (e.g., branching decision are based on history information) and partially due to the fact that the effort of solving a single branching node can vary tremendously. What makes a scalable parallelization of LP-based branch-and-bound most challenging, however, is to require determinism, i.e., making the algorithm always take the same solution path for identical input. A well-designed dynamic load balancing mechanism is an essential part of parallelizing LP-based branch-and-bound. In this paper, we present the new parallelization concept for the FICO Xpress-Optimizer.

The FICO Xpress Optimization Suite is a toolbox for mathematical optimization [3,9]. It features software tools used to model and solve linear, integer, quadratic, nonlinear, and robust optimization problems. The core solver of this suite is the FICO Xpress-Optimizer, a state-of-the-art MIP solver which combines ease of use with speed and flexibility. All implementations described in this paper have been conducted within the FICO Xpress-Optimizer.

2   Tasks Instead of Threads: A New Parallel Framework for the FICO Xpress-Optimizer

When designing a parallel framework for a MIP solver, we need to take into consideration the various algorithmic components that might exploit parallelization and the different means of doing so.

- **Tree search:** The ability to process individual nodes and subtrees in parallel is a major driver since for many MIP instances, a huge portion of the solving time goes into tree search.
- **Strong branching:** A natural candidate for parallelization, given that for early nodes it is a major cost factor and that individual strong branches are hopefully very similar in timewise effort.
- **LP solving:** The barrier algorithm for solving the root LP can be parallelized effectively. Additionally, the FICO Xpress-Optimizer is the first MIP solver using a parallel dual simplex algorithm [6], e.g., for LP solving during the root node cutting loop.
- **Heuristics:** Heuristics rarely depend on other tasks or vice versa and can always be run, mostly independent of the state of the search. Hence, they can be used to fill up any available time.

- **Cutting:** During cutting plane separation, similar algorithms are applied in serial with identical or at least similar input data, making it a natural candidate for parallelization. One could run different separators for the same LP solution in parallel or the same separators for different LP solutions.
- **Concurrent MIP starts:** A brute-force strategy for parallelization is to concurrently start single-threaded solves of slightly perturbed copies of the same MIP problem. Recently it has been shown that this is surprisingly effective, see [4].
- **Decomposition:** If a problem naturally decomposes, the MIPs should be solved in parallel. Decomposition methods such as Dantzig and Wolfe [2] also lend themselves naturally to parallelism.

So far, only some of these aspects were addressed in the FICO Xpress-Optimizer. Each of them used to have its own parallelization. The existing parallel framework was originally implemented as a job queue for parallel strong branching. It assumed all threads to be symmetric and the set of threads to remain constant. Determinism was accomplished by having all threads pass through synchronization barriers in thread order. A thread was only allowed through the a barrier when all threads had passed through the previous barrier. The synchronization was based on counts of dual simplex iterations, with some adjustments for heuristics. Data synchronization was very costly and therefore performed rarely. To synchronize data, all threads were halted at a synchronization barrier, and only when all threads were waiting, was data synchronized across all workers. All but one thread would therefore be idle while the synchronization was performed.

2.1   Considerations Behind a New Parallel Framework

The new parallel framework comes in two parts:

- A general task scheduler which is independent of the concrete MIP solving application. It can handle the execution of interdepending tasks in a deterministic fashion, with a focus on tasks being of different levels of complexity
- A parallel MIP implementation that makes use of the task scheduler using callbacks

The rationale of this separation is to develop a parallel task scheduler that is generic enough that it can be applied for all multi-threading purposes within the FICO Xpress-Optimizer. This includes the applications listed above, but also additional applications such as multi-starts in FICO Xpress-Nonlinear.

A core aspect of the new design is the capability to handle *asymmetric* tasks that might have different levels of complexity. It should not only be possible to have, e.g., cutting, heuristics and branch-and-bound dives parallelized individually, but to be able to run tasks of each type at the same time.

A problem with barrier based synchronization is that it does not scale well on large numbers of cores. One consequence is to break with the one-to-one association between threads and *tasks* to be performed. There should always be new tasks available for a thread when it completes a previous task, without needing to wait for all (or: any) other thread. The tasks therefore need to be independent of which threads executes them.

As an important consequence, by breaking the link between threads and tasks, it is possible to make the solution path independent of the number of threads used – it only depends on the number of tasks created. By ensuring that there are sufficient tasks in the work queue for waiting workers and by making the exchange of information more dynamic, it becomes less important to have an accurate measure for assigning deterministic stamps. Note that it is an important goal of our design to make the exchange of information less costly, mostly by a tendency to use "out-dated" information. As an extreme case of addressing ups and downs of data synchronization, Fischetti et al. recently presented a deterministic parallelization framework for constraint programming that does not make use of any information exchange [5].

2.2   Implementation of a New Parallel Framework

The new scheduler is a callback driven multi-threaded scheduler similar to an operating system kernel. Each thread of the scheduler should be able to execute any task in the queue.

A task consists of three stages: creation, execution and collection. Note that there typically are delays between creation and execution as well as between the end of execution and collection. During creation, a task is defined by the callback functions to call for execution and collection and by private data associated with the task. Additionally, a task might have dependencies on other tasks or create locks on global data. Finally, to improve cache performance, a task might have an affinity towards a certain thread.

Threads do not share data at designated synchronization points. At the time when a task is created, it gets a *deterministic stamp*. The task may only use information which is itself tagged with a smaller stamp. By this, the task uses only a subset of information that could be available if a synchronization had been triggered when the task was created. We refer to this concept as *partial information*. The idea is that the potential performance loss from using slightly "outdated" information will be easily made up with the performance gained by dropping the need for regular complete synchronization.

Like in the old parallelization, the computation of deterministic stamps is based on dual simplex iterations. Adjustments are made for simplex iterations inside heuristics. Other time-intensive procedures, e.g., calls to the barrier algorithm, are approximately converted into equivalents of dual simplex iterations.

When information is collected, all data that is transferred back to global data receives a deterministic stamp. All tasks that have a stamp which is greater than this, will be allowed to use that information. Those stamps serve two purposes: they provide a deterministic ordering of events, and they let each task broadcast its current local information so the scheduler knows which data can be made available to which tasks. Note that it might happen that tasks are created at a later point in real time but with a smaller deterministic stamp. When a thread performs a long-running task, e.g., solving a whole sub-tree, it might update its local copy of global data. Likewise, it might send its local data back to make it globally available. E.g., new incumbent solutions will be made globally available shortly after they have been found.

There are two fundamentally different types of global data: pooled data and updated data. With pooled data, we refer to a set of independent pieces of data, such as feasible solutions, cuts, or conflict constraints. The only difference between a static synchronization at certain synchronization points and a *dynamic synchronization* with deterministic stamps is that each of the information pieces needs to receive an individual stamp. For updated data, i.e. scalar statistics such as counters, averages, and so forth, the situation is more complex. This requires being able to present different snapshots of the data for different deterministic stamps.

Occasionally, collected data will be deleted. This holds in particular for pooled data, such as cuts or conflicts; but updated data might be reset, too. Data deletion can be handled the same way as data creation: The point of deletion corresponds to a deterministic stamp, and tasks might only consider this data deleted when they themselves operate on a larger stamp. This implies that data is not deleted instantly, but kept around until all tasks with a smaller stamp have been processed, which creates a memory overhead.

There is one major benefit in switching from a static synchronization system to a dynamic synchronization based on deterministic stamps. It significantly lowers the need to have the deterministic counters used for stamps to be an accurate approximation to real elapsed time. Rough proportionality is sufficient, since having a certain stamp "off" first of all implies that particular information will be used later than it could, but it will not completely stop all other threads from continuing their work. However, synchronization delays might still appear in this new framework, when a task requests data with a deterministic stamp, for which other, data-providing, tasks have not yet reached that stamped point. The more the deterministic stamps are off, the more likely such delays are. This could be pre-empted by using more "outdated" information to avoid delays.

Finally, it is the scheduler's responsibility to ensure scalability and proper CPU utilization. The MIP solver "only" needs to provide a sufficient number of tasks at every time. Compare the issue of ramp-up and ramp-down phases as, e.g., described in [12]. In the FICO Xpress-Optimizer 8.0, the solution path will only depend on the number of tasks that are present at any point in time, not on the number of threads. The maximum number of tasks can be set by the control MAXMIPTASKS. By default, it is computed automatically and lies in between two and four times the number of threads. By setting MAXMIPTASKS accordingly, it is possible to get the same solution path as for a run on a machine with a different number of threads, no matter whether the actual machine has more or less threads.

3 Computational Results

Our computational experiments compare the FICO Xpress-Optimizer 7.9 [3] with an internal beta version of the FICO Xpress-Optimizer 8.0. The experiments were conducted on a cluster of 40 core Intel Xeon E5-2690 CPUs at 3 GHz with 24 MB cache and 256 GB main memory, running Windows Server 2012 R2. More precisely, these are 40 virtual cores on 20 physical cores and hence, running times for more than 20 threads will be slower than they would which 40 actual physical cores. Both, the old and the new parallelization framework, use the Windows synchronization API, or the Pthreads library as parallel programming API when running on Unix.

As a test set, we chose the benchmark set from MIPLIB2010 [7]. We ran each FICO Xpress-Optimizer version with 1 thread, 4 threads, 12 threads, and 40 threads (the maximum on the underlying hardware). A summary of the results is given in Table 1. Column "threads" gives the number of threads with which the respective FICO Xpress-Optimizer version was run. Column "CPU load" presents the average workload, as a number between 0 % and 100 %. More precisely, it gives the average of the measured CPU time divided by the elapsed wall clock times the number of requested threads. Column "time in s" shows the shifted geometric mean of the running (wall clock) time for each of the solvers. The shifted geometric mean of values $t_1, \ldots, t_n$ with shift s is defined as $\sqrt[n]{\prod(t_i + s)} - s$, see, e.g., [1]. We use a shift of $s = 10$. Column "speedup" gives the speedup gained through parallelization w.r.t. running a single-threaded FICO Xpress-Optimizer. We used a time limit of one hour.

Table 2 presents the same statistics for the same solvers, but restricted to those 29 instances from the MIPLIB2010 benchmark set, for which at least one of the eight different solvers took at least 100 000 branch-and-bound nodes.[1]

Table 1. Comparing the FICO Xpress-Optimizer 7.9 and the FICO Xpress-Optimizer 8.0 w.r.t. different numbers of threads: average CPU workload, shifted geometric mean of solving time, and the speedup factor w.r.t. single-threaded run; complete MIPLIB2010 benchmark set (87 instances).

Threads	CPU load		Time in s		Speedup	
	Xpr7.9	Xpr8.0	Xpr7.9	Xpr8.0	Xpr7.9	Xpr8.0
1	99.4	99.7	121.4	117.4	–	–
4	63.9	67.5	57.2	54.4	2.1	2.2
12	43.2	53.8	42.1	34.7	2.9	3.4
40	25.8	33.1	35.2	28.6	3.4	4.1

---

[1] Those were the instances rocII-4-11, ns1766074, aflow40b, bnatt350, csched010, danoint, dfn-gwin-UUM, gmu-35-40, iis-100-0-cov, m100n500k4r1, n3div36, neos-1337307, neos18, neos-849702, neos-916792, newdano, noswot, ns1830653, pg5_34, pigeon-10, ran16x16, reblock67, rmine6, sp98ic, timtab1, vpphard, bab5, glass4, iis-bupa-cov.

Table 2. Comparing the FICO Xpress-Optimizer 7.9 and the FICO Xpress-Optimizer 8.0 w.r.t. different numbers of threads: comparing average CPU workload, shifted geometric mean of solving time, and the speedup factor w.r.t. single-threaded run; MIPLIB2010 benchmark set, at least one solver took at least 100 000 nodes (29 instances).

Threads	CPU load		Time in s		Speedup	
	Xpr7.9	Xpr8.0	Xpr7.9	Xpr8.0	Xpr7.9	Xpr8.0
1	99.5	99.9	477.3	480.2	–	–
4	78.4	90.8	168.7	151.5	2.8	3.2
12	56.6	83.5	104.6	68.4	4.6	7.0
40	36.6	57.1	83.2	49.4	5.7	9.7

Looking at the results in Table 1 in more detail, we see that with the FICO Xpress-Optimizer 8.0, we indeed observe a higher CPU load on 4, 12 and 40 threads. That the one-thread solve records slightly below 100 % usage is due to system calls. On all four different thread numbers, the FICO Xpress-Optimizer 8.0 improved its performance over the FICO Xpress-Optimizer 7.9. Of course, this is not only due to the new parallelization framework (in particular the improvement for single thread). Thus, it makes sense to compare the speedup factors w.r.t. the single-threaded run to get an impression of the impact of our new parallelization framework. We again observe a better performance of the FICO Xpress-Optimizer 8.0 on 4, 12 and 40 threads. In particular, the FICO Xpress-Optimizer 8.0 improved on the larger thread numbers.

The FICO Xpress-Optimizer typically solves 10 out of 87 MIPLIB2010 benchmark instances at the root node, and there are a couple more that require really small search trees. Naturally, the effects of parallelization come most into play for instances with large search trees. Therefore, we present the same numbers for the subset of instances for which at least one of the FICO Xpress-Optimizer versions took at least 100 000 search nodes. The differences between the FICO Xpress-Optimizer 7.9 and the FICO Xpress-Optimizer 8.0 become more obvious here. For 12 threads and 40 threads, the managed workload is a roughly a factor of 1.5 larger with the FICO Xpress-Optimizer 8.0: 83.5 % compared to 56.6 % and 57.1 % compared to 36.6 %, respectively. If we look at instances with a particularly good workload, we also notice a big difference. For the FICO Xpress-Optimizer 7.9 with 4, 12, and 40 threads, not a single instance had a workload of more than 90 %. For the FICO Xpress-Optimizer 8.0, running with 4 threads, 19 out of 29 instances achieved a workload of more than 90 %. For 12 threads, it were ten instances, and four instances for 40 threads.

Not surprisingly, the speedup factors are much more significant, when only looking at those instances that heavily rely on parallelization. We see that for 4 threads, the new parallel framework gives a nice improvement of 14 % in the parallel speedup: factor 3.2 versus 2.8. It is excelling, however, on larger thread numbers. For a typical 12 core desktop computer, this difference already grows to 50 % (speedup factor 7.0 versus 4.6). For a high-end 40 core machine, the speedup factor of the new parallel framework is more than 70 % better than with the old parallelization. Please note again, that the numbers for 40 threads are based on results from running on 40 virtual, not physical, cores and thereby underestimating the full potential.

4   Conclusion

We presented a parallelization framework that uses a dynamic synchronization scheme to exchange data. An important part of our implementation is the concept of partial information: we prefer starting a task with "deprecated" data to waiting for updated information. Furthermore, we separated the concepts of simultaneous tasks and independent threads from each other. Our computational results indicate that this leads to a much higher CPU workload and thereby to a significantly improved scaling on modern high-performance CPUs.

References

1. Achterberg, T.: Constraint integer programming. Technische Universität Berlin (2007)
2. Dantzig, G., Wolfe, P.: Decomposition principle for linear programs. Oper. Res. **8**(1), 101–111 (1960)
3. FICO Xpress-Optimizer, Reference Manual. http://www.fico.com/xpress
4. Fischetti, M., Lodi, A., Monaci, M., Salvagnin, D., Tramontani, A.: Improving branch-and-cut performance by random sampling. Math. Programm. Comput. 1–20 (2015)
5. Fischetti, M., Monaci, M., Salvagnin, D.: Self-splitting of workload in parallel computation. In: Simonis, H. (ed.) CPAIOR 2014. LNCS, vol. 8451, pp. 394–404. Springer, Heidelberg (2014)
6. Huangfu, Q., Hall, J.: Parallelizing the dual revised simplex method. Technical report 1503.01889, ArXiv e-prints (2015)
7. Koch, T., Achterberg, T., Andersen, E., Bastert, O., Berthold, T., Bixby, R.E., Danna, E., Gamrath, G., Gleixner, A.M., Heinz, S., Lodi, A., Mittelmann, H., Ralphs, T., Salvagnin, D., Steffy, D.E., Wolter, K.: MIPLIB 2010. Math. Progam. Comput. **3**, 103–163 (2011)
8. Land, A.H., Doig, A.G.: An automatic method of solving discrete programming problems. Econometrica **28**(3), 497–520 (1960)
9. Laundy, R., Perregaard, M., Tavares, G., Tipi, H., Vazacopoulos, A.: Solving hard mixed-integer programming problems with Xpress-MP: a MIPLIB 2003 case study. Inf. J. Comput. **21**(2), 304–313 (2009)
10. Nemhauser, G.L., Wolsey, L.A.: Integer and Combinatorial Optimization. Wiley, New York (1988)
11. Ralphs, T.K., Ladányi, L., Saltzman, M.J.: Parallel branch, cut and price for large-scale discrete optimization. Math. Program. B **98**(1–3), 253–280 (2003)
12. Shinano, Y., Achterberg, T., Berthold, T., Heinz, S., Koch, T.: ParaSCIP: a parallel extension of SCIP. Competence in High Performance Computing 2010, pp. 135–148. Springer, Heidelberg (2011)

PolySCIP

Ralf Borndörfer[1]([⊠]), Sebastian Schenker[1,2], Martin Skutella[2]([⊠]),
and Timo Strunk[1]([⊠])

[1] Zuse Institute Berlin, Berlin, Germany
{borndoerfer,schenker,strunk}@zib.de
[2] TU Berlin, Berlin, Germany
{schenker,skutella}@math.tu-berlin.de
https://www.zib.de,
https://www.coga.tu-berlin.de

Abstract. PolySCIP [1] is a new solver for multi-criteria integer and multi-criteria linear programs handling an arbitrary number of objectives. It is available as an official part of the non-commercial constraint integer programming framework SCIP. It utilizes a lifted weight space approach to compute the set of supported extreme non-dominated points and unbounded non-dominated rays, respectively. The algorithmic approach can be summarized as follows: At the beginning an arbitrary non-dominated point is computed (or it is determined that there is none) and a weight space polyhedron created. In every next iteration a vertex of the weight space polyhedron is selected whose entries give rise to a single-objective optimization problem via a combination of the original objectives. If the optimization of this single-objective problem yields a new non-dominated point, the weight space polyhedron is updated. Otherwise another vertex of the weight space polyhedron is investigated. The algorithm finishes when all vertices of the weight space polyhedron have been investigated. The file format of PolySCIP is based on the widely used MPS format and allows a simple generation of multi-criteria models via an algebraic modelling language.

Keywords: Multi-criteria optimization · Multi-objective optimization · Efficient solutions · Pareto-optimal solutions · Non-dominated points · Weight space partition · Weight set decomposition

1 Introduction and Motivation

Multi-objective integer programs (MOIPs) and multi-objective linear programs (MOLPs) can be considered as a generalization of single-objective IPs and single-objective LPs. Despite many applications, the wide availability and the (commercial) success of single-objective solvers (see e.g. [7–10]), there are almost no solvers for MOIPs and MOLPs: until recently only Symphony [3], a solver for bi-objective mixed IPs, and Bensolve [4], a solver for vector optimization problems (including MOLPs), were available. PolySCIP is a new solver for MOIPs and MOLPs handling an arbitrary number of objectives. Its name is composed of the

© Springer International Publishing Switzerland 2016
G.-M. Greuel et al. (Eds.): ICMS 2016, LNCS 9725, pp. 259–264, 2016.
DOI: 10.1007/978-3-319-42432-3_32

greek word πολύς meaning "many" and SCIP [5,6], a non-commercial constraint integer programming framework. One of the main drivers for the development of PolySCIP was the need for a multi-objective solver for problems in sustainable manufacturing [11] where the three dimensions of sustainability, i.e., the economic, environmental and social dimension, can be considered as different objective functions which need to be optimized simultaneously (Fig. 1).

2   Problem Formulation and Basic Definitions

PolySCIP solves problems of the form:

$$\min / \max \ (c_1 \cdot x, \dots, c_k \cdot x)$$
$$\text{s.t. } Ax \leq b,$$
$$x \in \mathbb{Z}^n \vee \mathbb{Q}^n,$$

where $k \geq 2$, $c_i \in \mathbb{Q}^n$ for $i \in [k]$, $A \in \mathbb{Q}^{m \times n}$ and $b \in \mathbb{Q}^m$.

Let $\mathcal{X}$ denote the feasible space of the given problem and let $\mathcal{Y} = \{(c_1 \cdot x, \dots, c_k \cdot x) : x \in \mathcal{X}\}$ be the corresponding image in objective space. In the rest of this paper we will consider minimization problems and state definitions with respect to the latter. A point $y^* \in \mathcal{Y}$ is called *non-dominated* if there is no $y \in \mathcal{Y}$ such that $y_i \leq y_i^*$ for $i \in [k]$ with $y_i < y_i^*$ for at least one i. A solution

Fig. 1. Feasible space of a bi-criteria integer maximization problem and corresponding set in objective space with dominated points (blue), supported non-dominated points (red) and unsupported non-dominated point (green) for a maximization problem. (Color figure online)

Fig. 2. Feasible space of a bi-criteria linear maximization problem and corresponding set in objective space with extreme non-dominated points (red) for a maximization problem. (Color figure online)

$x^* \in \mathcal{X}$ corresponding to a non-dominated point $y^* = (c_1 \cdot x^*, \ldots, c_k \cdot x^*)$ is called *efficient*. A non-dominated point $y \in \mathcal{Y}$ that lies in the interior of the convex hull of $\mathcal{Y}$ is called *unsupported* whereas a non-dominated point $y \in \mathcal{Y}$ that lies on the boundary of the convex hull of $\mathcal{Y}$ is called *supported*. A supported non-dominated point $y \in \mathcal{Y}$ that is an extreme point of the convex hull of $\mathcal{Y}$ is called a *supported extreme non-dominated (SEN)* point. It is an important and well-known theorem (see [12,13] for more details) that for a supported efficient solution $x^* \in \mathcal{X}$ there is a positive weight vector $(w_1, \ldots, w_k) \in \mathbb{R}_+^k$ such that x^* is the optimal solution to the single-objective problem $\min_{x \in \mathcal{X}} \sum_{i=1}^{k} w_i c_i \cdot x$. Due to algorithmic and complexity issues it is clear that we cannot just try every possible weight vector and we need to avoid unnecessary iterations that would yield the same non-dominated point over and over again (Fig. 2).

3   Lifted Weight Space Approach

Taking an objective space perspective, for a non-dominated point $y \in \mathcal{Y}$ we can consider the corresponding weight set $W(y) = \{w \in \mathbb{R}_+^k : w \cdot y \leq w \cdot y' \ \forall y' \in \mathcal{Y}\}$. Benson and Sun [14] were the first to investigate the structure of these weight sets. Przybylski et al. [15] and Özpeynirci and Köksalan [16] utilized them directly in an algorithmic framework. Our approach lifts the *original* weight sets by one dimension also taking into account (all values less than) the achieved weighted objective value. Let $Y \subset \mathcal{Y}$ be the set of all SEN points and let P be the lifted weight space polyhedron defined by $P = \{(a, w) \in \mathbb{R} \times \Lambda : a \leq w \cdot y \ \forall y \in Y\}$ where $\Lambda = \{w \in \mathbb{R}^k : w_i \geq 0, \ \sum_{i=1}^{k} w_i = 1\}$. For a subset $\bar{Y} \subset Y$, the partial weight space polyhedron $\bar{P} = \{(a, w) \in \mathbb{R} \times \Lambda : a \leq w \cdot y \ \forall y \in \bar{Y}\}$ contains P since it is defined by a subset of the constraints of P. The basic idea is to find in each iteration a cutting plane (given by a new non-dominated point $y \in Y \setminus \bar{Y}$) that cuts off parts of $\bar{P}$ that do not belong to P. Thus, the algorithm can be interpreted as a cutting plane algorithm with respect to the weight space polyhedron. The difference to Benson's outer approximation algorithm [17] or its dual variant [18] is that our approach operates on the (lifted) weight space whereas the latter two work directly in the objective space.

The algorithm begins with an initial SEN point y^1, which can be found by optimizing the objectives lexicographically, and initializes the partial weight space polyhedron with y^1, i.e., $\bar{P} = \{(a, w) \in \mathbb{R} \times \Lambda : a \leq w \cdot y \ \forall y \in \{y^1\}\}$. At this stage all vertices of $\bar{P}$ are unmarked. After the initialization phase, in every iteration an unmarked vertex $(a, w) \in \bar{P}$ is chosen and the weighted optimization problem $\min_w = \min_{x \in \mathcal{X}} \sum_{i=1}^{k} w_i c_i \cdot x$ is solved. Let $\bar{Y}$ be the set of SEN points found so far, let $\tilde{x} \in \mathcal{X}$ be the computed solution of $\min_w$ and let $\tilde{y} \in \mathcal{Y}$ be the corresponding point in objective space. If $a > w \cdot \tilde{y}$, then $(a, w) \in \bar{P}$ is cut off and $\tilde{y}$ is a new non-dominated point which was not found so far. In this case, the partial weight space polyhedron $\bar{P}$ is updated to $\bar{P} = \{(a, w) \in \mathbb{R} \times \Lambda : a \leq w \cdot y \ \forall y \in \bar{Y} \cup \{\tilde{y}\}\}$. On the other hand, if $a \leq w \cdot \tilde{y}$, then no new non-dominated point was found by considering $\min_w$ and the vertex $(a, w) \in \bar{P}$ is marked. The algorithm stops when all vertices of $\bar{P}$ are marked which implies that $\bar{P} = P$. Note that for some weights $w \in \Lambda$, the problem $\min_w$ might be unbounded yielding unbounded non-dominated rays.

4   Implementation

PolySCIP is written in C++11. The optimization of the weighted single objective problems $\min_{x \in \mathcal{X}} \sum_{i=1}^{k} w_i c_i \cdot x$ is done via interface calls to SCIP.

In order to keep track of the partial weight space polyhedron $\bar{P}$ and its vertices, the 1-skeleton of $\bar{P}$ is stored via an undirected graph data structure from the LEMON Graph Library [19]. Nodes in the 1-skeleton represent vertices of $\bar{P}$ and edges between nodes in the 1-skeleton represent adjacent vertices. The initial partial weight space polyhedron is given by

$$y_1^1 w_1 + \ldots + y_k^1 w_k \geq a, \qquad \text{(H-rep)}$$
$$w_1 + \ldots + w_k = 1,$$
$$w_1, \ldots, w_k \geq 0,$$

where $y^1 = (y_1^1, \ldots, y_k^1) \in \mathcal{Y}$ is the first non-dominated point found by optimizing the objectives lexicographically. $\bar{P}$ is initialized with the k vertices of H-rep and the undirected graph data structure representing the 1-skeleton is initialized as a complete graph with k nodes each representing a distinct vertex. In the course of the algorithm, if a vertex of $\bar{P}$ is cut off by a cutting plane given by a newly found non-dominated point, then the graph data structure is updated: nodes representing vertices which are cut off by the cutting plane are deleted from the graph, a new node representing the new vertex is added and the adjacency relationships to the remaining nodes are updated. Unmarked nodes, i.e., vertices of $\bar{P}$ that need to be investigated, are kept in a queue.

5   File Format

The MOP (multi-objective problem) file format is based on the widely used MPS file format (see [20] for more details). MPS is column-oriented and all model components (variables, rows, etc.) receive a name. An objective in MPS is indicated by an N followed by the name in the ROWS section. Similarly, in MOP the objectives are indicated by an N followed by the name in the ROWS section. In general, MPS might not be as human readable as other formats. However, one of the main reasons to base the file format of PolySCIP on it is its easy extension towards several objectives and its wide availability in other linear and integer programming software packages which (hopefully) minimizes the effort to extend other available MPS parsers to parse MOP as well. Furthermore, no user is expected to write MOP files by hand, but to use an algebraic modelling language that does the job (see [2] for an example).

The following simple equation-based bi-criteria integer problem

$$\text{minimize} \quad \text{Obj1: } 3x_1 + 2x_2 - 4x_3$$
$$\text{Obj2: } \quad x_1 + x_2 + 2x_3$$

subject to

$$\text{Eqn:} \quad x_1 + x_2 + x_3 \quad = 2$$
$$\text{Lower: } x_1 + 0.4x_2 \quad \le 1.5$$
$$x_1, \ x_2, \ x_3 \quad \ge 0$$
$$x_1, \ x_2, \ x_3 \quad \in \mathbb{Z}$$

could be written in MOP format as follows:

```
NAME            BICRIT
OBJSENSE
 MIN
ROWS
 N  Obj1
 N  Obj2
 E  Eqn
 L  Lower
COLUMNS
      x#1       Lower           1
      x#1       Eqn             1
      x#1       Obj2            1
      x#1       Obj1            3
      x#2       Lower           0.4
      x#2       Eqn             1
      x#2       Obj2            1
      x#2       Obj1            2
      x#3       Eqn             1
      x#3       Obj2            2
      x#3       Obj1           -4
RHS
      RHS       Eqn             2
      RHS       Lower           1.5
BOUNDS
 LI BOUND       x#1             0
 LI BOUND       x#2             0
 LI BOUND       x#3             0
ENDATA
```

References

1. PolySCIP website: http://polyscip.zib.de
2. PolySCIP user guide: http://polyscip.zib.de/download/userguide.pdf

3. Ralphs, T., Guzelsoy, M., Mahajan, A.: SYMPHONY Version 5.5 User's Manual. Lehigh University (2013). https://projects.coin-or.org/SYMPHONY
4. Löhne, A.: Vector Optimization with Infimum and Supremum. Springer, Heidelberg (2011). www.bensolve.org
5. Achterberg, T.: SCIP: solving constraint integer programs. Math. Program. Comput. **1**(1), 1–41 (2009)
6. Gamrath, G., Fischer, T., Gally, T., et al.: The SCIP Optimization Suite 3.2. ZIB-Report 15–60 (2016)
7. SCIP Optimization Suite. http://scip.zib.de
8. Gurobi Optimization. www.gurobi.com
9. FICO Xpress Optimization Suite. www.fico.com/en/products/fico-xpress-optimization-suite
10. ILOG Cplex Optimization Studio. www-03.ibm.com/software/products/en/ibmilogcpleoptistud
11. CRC (2011) Observation of strains: Sustainable Manufacturing. www.sustainable-manufacturing.net
12. Isermann, H.: Proper efficiency and the linear vector maximum problem. Oper. Res. **22**(1), 189–191 (1974)
13. Ehrgott, M.: Multicriteria Optimization. Springer, Heidelberg (2005)
14. Benson, H.P., Sun, E.: Outcome space partition of the weight set in multiobjective linear programming. J. Optim. Theory Appl. **105**(1), 17–36 (2000)
15. Przybylski, A., Gandibleux, X., Ehrgott, M.: A recursive algorithm for finding all nondominated extreme points in the outcome set of a multiobjective integer programme. INFORMS J. Comput. **22**(3), 371–386 (2010)
16. Özpeynirci, Ö., Köksalan, M.: An exact algorithm for finding extreme supported nondominated points of multiobjective mixed integer programs. Manage. Sci. **56**(12), 2302–2315 (2010)
17. Benson, H.P.: An outer approximation algorithm for generating all efficient extreme points in the outcome set of a multiple objective linear programming problem. J. Global Optim. **13**(1), 1–24 (1998)
18. Ehrgott, M., Löhne, A., Shao, L.: A dual variant of Benson's "outer approximation algorithm" for multiple objective linear programming. J. Global Optim. **52**(4), 757–778 (2012)
19. LEMON Graph Library. https://lemon.cs.elte.hu/trac/lemon
20. MPS format: http://lpsolve.sourceforge.net/5.5/mps-format.htm

Advanced Computing and Optimization Infrastructure for Extremely Large-Scale Graphs on Post Peta-Scale Supercomputers

Katsuki Fujisawa[1(✉)], Toshio Endo[2], and Yuichiro Yasui[3]

[1] The Institute of Mathematics for Industry,
Kyushu University & JST CREST, Fukuoka, Japan
fujisawa@imi.kyushu-u.ac.jp
[2] Global Scientific Information and Computing Center,
Tokyo Institute of Technology & JST CREST, Tokyo, Japan
endo@is.titech.ac.jp
[3] Center for Co-Evolutional Social Systems,
Kyushu University & JST COI, Fukuoka, Japan
y-yasui@imi.kyushu-u.ac.jp

Abstract. In this talk, we present our ongoing research project. The objective of this project is to develop advanced computing and optimization infrastructures for extremely large-scale graphs on post peta-scale supercomputers. We explain our challenge to Graph 500 and Green Graph 500 benchmarks that are designed to measure the performance of a computer system for applications that require irregular memory and network access patterns. The 1st Graph500 list was released in November 2010. The Graph500 benchmark measures the performance of any supercomputer performing a BFS (Breadth-First Search) in terms of traversed edges per second (TEPS). In 2014 and 2015, our project team was a winner of the 8th, 10th, and 11th Graph500 and the 3rd to 6th Green Graph500 benchmarks, respectively. We also present our parallel implementation for large-scale SDP (SemiDefinite Programming) problem. The semidefinite programming (SDP) problem is a predominant problem in mathematical optimization. The primal-dual interior-point method (PDIPM) is one of the most powerful algorithms for solving SDP problems, and many research groups have employed it for developing software packages. We solved the largest SDP problem (which has over 2.33 million constraints), thereby creating a new world record. Our implementation also achieved 1.774 PFlops in double precision for large-scale Cholesky factorization using 2,720 CPUs and 4,080 GPUs on the TSUBAME 2.5 supercomputer.

Keywords: Graph analysis · Breadth-first search · Optimization problem · High performance computing · Supercomputer · Big data

1 Introduction

The objective of our ongoing research projects (which we call GraphCREST) promoted by JST (Japan Science and Technology Agency) is to develop an

© Springer International Publishing Switzerland 2016
G.-M. Greuel et al. (Eds.): ICMS 2016, LNCS 9725, pp. 265–274, 2016.
DOI: 10.1007/978-3-319-42432-3_33

advanced computing and optimization infrastructure for extremely large-scale graphs on the peta-scale and/or exa-scale supercomputers. The extremely large-scale graphs that have recently emerged in various application fields, such as big-data analysis, transportation, social networks, cyber-security, and bioinformatics, require fast and scalable analysis. The number of vertices in the graph networks has grown from billions to trillions and that of the edges from hundreds of billions to tens of trillions. For example, a graph that represents the interconnections of all the neurons of the human brain has over 89 billion vertices and over 100 trillion edges. To analyze these extremely large-scale graphs, we require a new generation exa-scale supercomputer, which will not appear until 2018 to 2020, and therefore, we need a new framework of software stacks for extremely large-scale graph analysis systems, such as parallel graph analysis and optimization libraries on multiple CPUs and GPUs, hierarchal graph stores using non-volatile memory (NVM) devices, and graph processing and visualization systems. In this talk, we explain our ongoing research project and show its remarkable results.

2   Graph500 and Green Graph500 Benchmarks

The Graph500[1] and Green Graph 500[2] benchmarks are designed to measure the performance of a computer system for applications that require irregular memory and network access patterns. The detailed instructions of the Graph500 benchmark are described as follows:

1. **Step1: Edge List Generation.** First, the benchmark generates an edge list of an undirected graph with $n(= 2^{SCALE})$ vertices and $m(= n \cdot edge\_factor)$ edges;
2. **Step2: Graph Construction.** The benchmark constructs a suitable data structure, such as CSR (Compressed Sparse Row) graph format, for performing BFS from the generated edge list;
3. **Step3: BFS.** The benchmark performs BFS to the constructed data structure to create a BFS tree. Graph500 employs TEPS (Traversed Edges Per Second) as a performance metric. Thus, the elapsed time of a BFS execution and the total number of processed edges determine the performance of the benchmark;
4. **Step4: Validation.** Finally, the benchmark verifies the results of the BFS tree. Note that the benchmark iterates Step3 and Step4 64 times from randomly selected start points, and the median value of the results is adopted as the score of the benchmark.

Previous studies [1,2] have proposed hybrid approaches that combine a well-known *top-down* algorithm and an efficient *bottom-up* algorithm for large frontiers at **Step3** above. This reduces some unnecessary searching of outgoing edges

---

[1] http://www.graph500.org.
[2] http://green.graph500.org.

Our achievements in Graph500

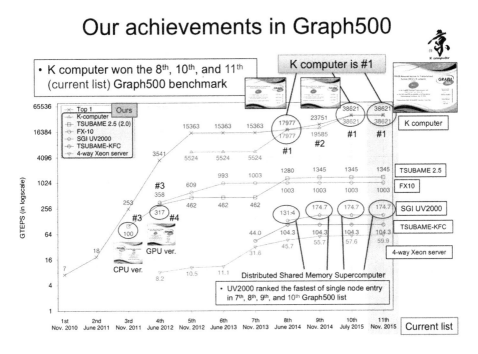

Fig. 1. Our major achievements in Graph500 benchmark

in the BFS traversal of a small-world graph, such as a Kronecker graph. BFS is an essential kernel for graph traversals, it is widely applied to graph analysis applications. BFS has been tuned for both single-node systems and large-scale multi-node distributed systems [3–6]. In 2013 [19], we describe a highly efficient BFS using column-wise partitioning of the adjacency list while carefully considering the non-uniform memory access (NUMA) architecture. We explicitly manage the way in which each working thread accesses a partial adjacency list in local memory during BFS traversal. Our project team have challenged the Graph500 and Green Graph500 benchmarks, which are designed to measure the performance of a computer system for applications that require irregular memory and network access [12–16, 19–22, 24, 25].

In 2013, our project team gained first place in both the big and small data categories in the second Green Graph 500 benchmarks. The Green Graph 500 list collects TEPS-per-watt metrics. Our other implementation, which uses both DRAM and NVM devices and whose objective is to analyze extremely large-scale graphs that exceed the DRAM capacity of the nodes, which gained fourth place in the big data category in the second Green Graph500 list. Figure 1 shows our major achievements in Graph500 benchmark, which are mentioned in this Section. In 2014 and 2015, our project team was a winner of the 8th, 10th, and 11th Graph500 (Fig. 2) and the 3rd to 6th Green Graph500 benchmarks, respectively.

Fig. 2. Our project team were awarded the first place in the 10th and 11th Graph500 benchmarks.

3   High-Performance Computing for Semidefinite Programming Problems

We also present our parallel implementation for large-scale mathematical optimization problems [9–11,17,18,23]. In the last decade, mathematical optimization programming (MOP) problems have been intensively studied in both their theoretical and practical aspect in a wide range of fields, such as combinatorial optimization, structural optimization, control theory, economics, quantum chemistry, sensor network location, data mining, and machine learning. The semidefinite programming (SDP) problem is a predominant problem in mathematical optimization. The primal-dual interior-point method (PDIPM) is one of the most powerful algorithms for solving SDP problems, and many research groups have employed it for developing software packages. However, two well-known major bottleneck parts (the generation of the Schur complement matrix (SCM) and its Cholesky factorization) exist in the algorithmic framework of PDIPM. These two parts where bottlenecks occur are called ELEMENTS and CHOLESKY, respectively. The standard-form SDP has the following primal-dual form.

$$\begin{aligned}
\mathcal{P} : \text{minimize} \quad & \textstyle\sum_{k=1}^{m} c_k x_k \\
\text{subject to} \quad & \boldsymbol{X} = \textstyle\sum_{k=1}^{m} \boldsymbol{F}_k x_k - \boldsymbol{F}_0, \quad \boldsymbol{X} \succeq \boldsymbol{O}. \\
\mathcal{D} : \text{maximize} \quad & \boldsymbol{F}_0 \bullet \boldsymbol{Y} \\
\text{subject to} \quad & \boldsymbol{F}_k \bullet \boldsymbol{Y} = c_k \ (k = 1, \ldots, m), \quad \boldsymbol{Y} \succeq \boldsymbol{O}.
\end{aligned}$$

We denote by $\mathbb{S}^n$ the space of $n \times n$ symmetric matrices. The notation $\boldsymbol{X} \succeq \boldsymbol{O}$ ($\boldsymbol{X} \succ \boldsymbol{O}$) indicates that $\boldsymbol{X} \in \mathbb{S}^n$ is a positive semidefinite (positive definite) matrix. The inner-product between $\boldsymbol{U} \in \mathbb{S}^n$ and $\boldsymbol{V} \in \mathbb{S}^n$ is defined by $\boldsymbol{U} \bullet \boldsymbol{V} = \sum_{i=1}^{n} \sum_{j=1}^{n} U_{ij} V_{ij}$.

In most SDP applications, it is common for the input data matrices $F_0, \ldots, F_m$ to share the same diagonal block structure $(n_1, \ldots, n_h)$. Each input data matrix F_k $(k = 1, \ldots, m)$ consists of sub-matrices in the diagonal positions as follows:

$$F_k = \begin{pmatrix} F_k^1 & O & O & O \\ O & F_k^2 & O & O \\ O & O & \ddots & O \\ O & O & O & F_k^h \end{pmatrix}$$

where $F_k^1 \in \mathbb{S}^{n_1}, F_k^2 \in \mathbb{S}^{n_2}, \ldots, F_k^h \in \mathbb{S}^{n_h}$.

Note that $\sum_{\ell=1}^{h} n_\ell = n$ and the variable matrices X and Y share the same block structure. We define $n_{\max}$ as $\max\{n_1, \ldots, n_h\}$. For the blocks where $n_\ell = 1$, the constraints of positive semidefiniteness are equivalent to the constraints of the non-negative orthant. Such blocks are sometimes called linear programming (LP) blocks.

The size of a given SDP problem can be approximately measured in terms of four metrics.

1. m: the number of equality constraints in the dual form $\mathcal{D}$ (which equals the size of the SCM)
2. n: the size of the variable matrices X and Y
3. $n_{\max}$: the size of the largest block of input data matrices
4. nnz: the total number of nonzero elements in all data matrices.

We denote the time complexities of ELEMENTS and CHOLESKY by $O(mn^3 + m^2n^2)$ and $O(m^3)$, respectively.

We have developed a new version of the semidefinite programming algorithm parallel version (SDPARA), which is a parallel implementation on multiple CPUs and GPUs for solving extremely large-scale SDP problems that have over a million constraints [9,11]. SDPARA can automatically extract the unique characteristics from an SDP problem and identify the bottleneck. The key to the high performance of SDPARA is the acceleration of ELEMENTS and CHOLESKY by using thousands of CPUs and GPUs, respectively. SDPARA could attain high scalability using 16,320 CPU cores on the TSUBAME 2.0 supercomputer and some processor affinity and memory interleaving techniques when the generation of the SCM constituted a bottleneck [11]. SDPARA can also perform parallel Cholesky factorization using thousands of GPUs and techniques to overlap computation and communication if an SDP problem has over two million constraints and Cholesky factorization constitutes a bottleneck. We demonstrated that SDPARA is a high-performance general solver for SDPs in various application fields through numerical experiments at the TSUBAME 2.5 supercomputer, and we solved the largest SDP problem (which has over 2.33 million constraints), thereby creating a new world record. Our implementation also achieved 1.713 PFlops and 1.774 PFlops in double precision for large-scale Cholesky factorization using 2,720 CPUs and 4,080 GPUs [11,23], respectively.

SDP has many applications that involve SDP problems with special structures. We initiated the SDPA project[3], which aims to develop high-performance software packages for SDP, and we have solved a large number of SDP problems since 1995; therefore, we can classify the various types of SDP problems into the following three cases:

1. Case 1: SDP problems are sparse and satisfy the property of correlative sparsity; therefore, SCM tends to become sparse (e.g., the sensor network location problem and the polynomial optimization problem). In this case, CHOLESKY is the bottleneck part of PDIPM.
2. Case 2: m is less or not considerably greater than n and SCM is fully dense (e.g., the quantum chemistry problem and the truss topology problem). In this case, ELEMENTS is the bottleneck part of PDIPM, so we can decrease the time complexity of ELEMENTS $O(mn^3 + m^2n^2)$ to $O(m^2)$ by exploiting the sparsity of the data matrix. ELMENTS for large-scale SDP problems generally requires significant computational resources in terms of CPU cores and memory bandwidth.
3. Case 3: m is considerably greater than n and SCM is fully dense (e.g., the combinatorial optimization problem and quadratic assignment problem(QAP) [9]). In this case, CHOLESKY is the bottleneck part of PDIPM. We accelerated CHOLESKY by using massively parallel GPUs with computational performance much higher than that of CPUs. In order to achieve scalable performance with thousands of GPUs, we utilized a high-performance BLAS kernel along with optimization techniques to overlap computation, PCI-Express communication, and MPI communication [9].

Table 1 shows the performance record of CHOLESKY of SDPARA. In 2003 [7], we have released the SDPARA 1.0.1 and achieved 78.58 GFlops. Our implementation with GPUs achieved 1.713 and 1.774 PFlops for large-scale Cholesky factorization using 4,080 GPUs in 2014 [11] and 2015 [23], respectively.

We previously reported that SDPARA can certainly determine whether the SCM of an input SDP problem becomes sparse (Case 1) or not (Cases 2 and 3). In the present study, we mainly focused on parallel computation of ELEMENTS and CHOLESKY in Cases 2 and 3, respectively. We also demonstrated that SDPARA is a high-performance general solver for SDPs in various application fields through numerical experiments on the TSUBAME 2.5 supercomputer and solved the largest SDP problem (QAP10), which has over 2.33 million constraints [11]; and we created a new world record. Figure 3 and Table 2 show the speed of the CHOLESKY component in teraflops. "New" corresponds to the latest algorithm [11], while "org" denotes the original algorithm in our previous paper [9]. Our implementation also achieved 1.713 PFlops in double precision for large-scale Cholesky factorization using 2,720 CPUs and 4,080 GPUs. Table 3 shows that the speed of the CHOLESKY component in teraflops on the CX400 supercomputer at Kyushu University. We have achieved 294.2 TFlops when using 384 GPUs.

---

[3] http://sdpa.sourceforge.net/.

Table 1. Performance record of CHOLESKY of SDPARA

Year	Paper	n	m	CHOLESY (Flops)
2003	[7]	630	24,503	78.58 Giga
2010	[8]	10,462	76,554	2.414 Tera
2012	[9]	1,779,204	1,484,406	0.533 Peta
2014	[11]	2,752,649	2,339,331	1.713 Peta
2015	[23]	2,322,988	1,962,225	1.774 Peta

Table 2. Performance (teraflops) of GPU CHOLESKY obtained by using up to 1360 nodes (4080 GPUs) on TSUBAME 2.0 [9] and 2.5 [11].

(a) 400 nodes (1200 GPUs)

Name	m	org(2.0)	new(2.0)	new(2.5)
QAP6	709,275	223.0	233.0	314.5
QAP7	1,218,400	248.8	306.2	505.8

(b) 700 nodes (2100 GPUs)

Name	m	org(2.0)	new(2.0)	new(2.5)
QAP6	709,275	309.5	329.0	387.5
QAP7	1,218,400	440.0	470.0	707.1
QAP8	1,484,406	463.8	512.9	825.1

(c) 1360 nodes (4080 GPUs)

Name	m	org(2.0)	new(2.0)	new(2.5)
QAP6	709,275	439.6	437.8	508.7
QAP7	1,218,400	695.2	718.8	952.0
QAP8	1,484,406	779.3	825.6	1186.4
QAP9	1,962,225	–	964.4	1526.5
QAP10	2,339,331	–	1018.5	1713.0

Table 3. Performance (teraflops) of GPU CHOLESKY obtained by using up to 384 nodes (384 GPUs) on CX400

(a) 128 nodes (128 GPUs)

Name	m	org(2.0)	new(2.0)	new(2.5)
QAP6	709,275	–	–	90.1
QAP7	1,218,400	–	–	98.7

(b) 384 nodes (384 GPUs)

Name	m	org(2.0)	new(2.0)	new(2.5)
QAP8	1,484,406	–	–	288.0
QAP8-1	1,495,602	–	–	294.2

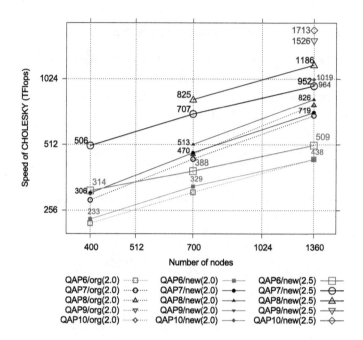

Fig. 3. Performance of GPU CHOLESKY obtained by using up to 1360 nodes (4080 GPUs) on TSUBAME 2.0 and 2.5.

Acknowledgment. This research project was supported by the Japan Science and Technology Agency (JST), the Core Research of Evolutionary Science and Technology (CREST), the Center of Innovation Science and Technology based Radical Innovation and Entrepreneurship Program (COI Program), the TSUBAME 2.0 & 2.5 Supercomputer Grand Challenge Program at the Tokyo Institute of Technology, and "Advanced Computational Scientific Program" of Research Institute for Information Technology, Kyushu University.

References

1. Beamer, S., Asanović, K., Patterson, D.A.:Searching for a parent instead of fighting over children: a fast breadth-first search implementation for Graph500. EECS Department, University of California, Berkeley, CA, UCB/EECS-2011-117 (2011)
2. Beamer, S., Asanović, K., Patterson, D.A.: Direction-optimizing breadth-first search. In: Proceedings of ACM/IEEE International Conference High Performance Computing, Networking, Storage and Analysis (SC12). IEEE Computer Society (2012)
3. Yoo, A., Chow, E., Henderson, K., McLendon, W., Hendrickson, B., Catalyurek, U.: A scalable distributed parallel breadth-first search algorithm on BlueGene/L. In: Proceedings of the 2005 ACM/IEEE Conference on Supercomputing, ser. SC 2005, pp. 25–43. IEEE Computer Society (2005)

4. Checconi, F., Petrini, F., Willcock, J., Lumsdaine, A., Choudhury, A.R., Sabharwal, Y.: Breaking the speed and scalability barriers for graph exploration on distributed-memory machines. In: Proceedings of the International Conference on High Performance Computing, Networking, Storage and Analysis, ser. SC 2012, pp. 13:1–13:12. IEEE Computer Society Press (2012)
5. Satish, N., Kim, C., Chhugani, J., Dubey, P.: Large-scale energy efficient graph traversal: a path to efficient data-intensive supercomputing. In: Proceedings of the International Conference on High Performance Computing, Networking, Storage and Analysis, ser. SC 2012, pp. 14:1–14:11. IEEE Computer Society Press (2012)
6. Checconi, F., Petrini, F.: Traversing trillions of edges in real time: graph exploration on large-scale parallel machines. In: Proceeding of IEEE 28th International Parallel and Distributed Processing Symposium, ser. IPDPS 2014, pp. 425–434. IEEE (2014)
7. Yamashita, M., Fujisawa, K., Kojima, M.: SDPARA: semidefinite programming algorithm parallel version. J. Parallel Comput. 29(8), 1053–1067 (2003)
8. Nakata, M., Fukuda, M., Fujisawa, K.: Variational approach to electronic structure calculations on second-order reduced density matrices and the N-representability problem. In: Siedentop, H. (ed.), Complex Quantum Systems - Analysis of Large Coulomb Systems, Institute of Mathematical Sciences, National University of Singapore, pp. 163–194 (2013)
9. Fujisawa, K., Endo, T., Sato, H., Yamashita, M., Matsuoka, S., Nakata, M.: High-performance general solver for extremely large-scale semidefinite programming problems. In: Proceedings of the 2012 ACM/IEEE Conference on Supercomputing, SC 2012 (2012)
10. Fujisawa, K., Endo, T., Sato, H., Yasui, Y., Matsuzawa, N., Waki, H.: Peta-scale general solver for semidefinite programming problems with over two million constraints. In: International Conference for High Performance Computing, Networking, Storage and Analysis 2013, SC 2013 Regular, Electronic, and Educational Poster (SC 2013) (2013)
11. Fujisawa, K., Endo, T., Yasui, Y., Sato, H., Matsuzawa, N., Matsuoka, S., Waki, H.: Peta-scale general solver for semidefinite programming problems with over two million constraints. In: The 28th IEEE International Parallel & Distributed Processing Symposium (IPDPS 2014) (2014)
12. Iwabuchi, K., Sato, H., Mizote, R., Yasui, Y., Fujisawa, K., Matsuoka, S.: Hybrid BFS approach using semi-external memory. In: International Workshop on High Performance Data Intensive Computing (HPDIC2014) in Conjunction with IEEE IPDPS 2014 (2014)
13. Iwabuchi, K., Sato, H., Yasui, Y., Fujisawa, K., Matsuoka, S.: NVM-based Hybrid BFS with memory efficient data structure. In: The Proceedings of the IEEE Big-Data2014 (2014)
14. Iwabuchi, K., Sato, H., Yasui, Y., Fujisawa, K.: Performance analysis of hybrid BFS approach using semi-external memory. In: International Conference for High Performance Computing, Networking, Storage and Analysis, SC 2013 Regular, Electronic, and Educational Poster (SC 2013) (2013)
15. Suzumura, T., Ueno, K., Sato, H., Fujisawa, K., Matsuoka, S.: A performance characteristics of Graph500 on large-scale distributed environment. In: The Proceedings of the 2011 IEEE International Symposium on Workload Characterization (2011)
16. Ueno, K., Suzumura, T.: Highly scalable graph search for the Graph500 benchmark. In: The 21st International ACM Symposium on High-Performance Parallel and Distributed Computing, HPDC 2012. Delft, Netherlands (2012)

17. Yamashita, M., Fujisawa, K., Fukuda, M., Kobayashi, K., Nakata, K., Nakata, M.: Latest developments in the SDPA family for solving large-scale SDPs. In: Anjos, M.F., Lasserre, J.B. (eds.) Handbook on Semidefinite, Conic and Polynomial Optimization, International Series in Operations Research & Management Science. Springer, Heidelberg (2011)
18. Yamashita, M., Fujisawa, K., Fukuda, M., Nakata, K., Nakata, M.: Parallel solver for semidefinite programming problem having sparse Schur complement matrix. The ACM Trans. Math. Softw. **39**(12), 6 (2012)
19. Yasui, Y., Fujisawa, K., Goto, K.: NUMA-optimized parallel breadth-first search on multicore single-node system. In: The Proceedings of the IEEE BigData2013 (2013)
20. Yasui, Y., Fujisawa, K., Goto, K., Kamiyama, N., Takamatsu, M.: NETAL: high-performance implementation of network analysis library considering computer memory hierarchy. J. Oper. Res. Soc. Jpn. **54**(4), 259–280 (2011)
21. Yasui, Y., Fujisawa, K., Sato, Y.: Fast and energy-efficient breadth-first search on a single NUMA system. In: Kunkel, J.M., Ludwig, T., Meuer, H.W. (eds.) ISC 2014. LNCS, vol. 8488, pp. 365–381. Springer, Heidelberg (2014)
22. Yasui, Y., Fujisawa, K.: Fast and scalable NUMA-based thread parallel breadth-first search. In: The 2015 International Conference on High Performance Computing & Simulation (HPCS 2015) (2015). doi:10.1109/HPCSim.2015.7237065
23. Tsujita, Y., Endo, T., Fujisawa, K.: The scalable petascale data-driven approach for the cholesky factorization with multiple GPUs. In: First International Workshop on Extreme Scale Programming Models and Middleware. In Conjunction with International Conference for High Performance Computing, Networking, Storage and Analysis (SC 2015) (2015). doi:10.1145/2832241.2832245
24. Fujisawa, K., et al.: Advanced computing & optimization infrastructure for extremely large-scale graphs on post peta-scale supercomputers. In: Fujisawa, K., Shinano, Y., Waki, H. (eds.) Optimization in the Real World: Toward Solving Real-World Optimization Problems. Mathematics for Industry. Springer, Heidelberg (2015). doi:10.1007/978-4-431-55420-2_1
25. Yasui, Y., Fujisawa, K.: NUMA-aware scalable graph traversal on SGI UV systems. In: The Proceedings of 1st High Performance Graph Processing Workshop. In Conjunction with The International ACM Symposium on High-Performance Parallel and Distributed Computing (HPDC 2016) (2016)

DSJM: A Software Toolkit for Direct Determination of Sparse Jacobian Matrices

Mahmudul Hasan, Shahadat Hossain$^{(\boxtimes)}$, Ahamad Imtiaz Khan,
Nasrin Hakim Mithila, and Ashraful Huq Suny

University of Lethbridge, Lethbridge, AB, Canada
{hasan,shahadat.hossain,ai.khan,mithila,a.suny}@uleth.ca

Abstract. We describe the main design features of **DSJM** (Determine Sparse Jacobian Matrices), a software toolkit written in standard C++ that enables direct determination of sparse Jacobian matrices. Our design exploits the recently proposed unifying framework "pattern graph" and employs cache-friendly array-based sparse data structures. The **DSJM** implements a greedy grouping (coloring) algorithm and several ordering heuristics. In our numerical testing on a suite of large-scale test instances **DSJM** consistently produced better timing and partitions compared with a similar software.

Keywords: Sparse Jacobian matrix · Compression-reconstruction · Direct determination

1 Introduction

Problems in optimization, simulation, and differential equations often require the computation or estimation of mathematical derivatives of the underlying model components. In many systems analysis tasks system parameters are "optimized" by studying the effect of noise in parameters to the model output. Mathematically, such effects can be captured and quantified by the derivative of the objective with respect to the parameters in question. Optimizing a multivariate scalar function using descent or Newton's method requires the evaluation of gradient and Hessian matrix of the function. Solving a system of nonlinear equations or nonlinear least-squares problem by Newton's method or one of its many variants require the evaluation of its Jacobian matrix and solving the associated linear system at each iterative step. With the advent of faster computers and sophisticated software there is an ever increasing demand for solving larger and more complex problems. Fortunately, many real-life problems are sparse or otherwise "structured". In some optimization models sparsity occur naturally because only a small subset of variables interact nonlinearly. An example of special structure is a partially separable function; the gradient of a partially separable function can be evaluated at a cost that is a small constant multiple of the cost of evaluating the function [9]. If the source code for the function evaluation program is available one may write code for analytic derivatives by hand. For large and/or

© Springer International Publishing Switzerland 2016
G.-M. Greuel et al. (Eds.): ICMS 2016, LNCS 9725, pp. 275–283, 2016.
DOI: 10.1007/978-3-319-42432-3_34

complicated functions this process is time-consuming and highly error-prone. An alternative to hand-coded derivatives is to approximate the Jacobian by a finite-difference (FD) scheme. In the recent years the techniques of automatic or algorithmic differentiation (AD) (see [9] for a comprehensive treatment of AD techniques and their software implementation) have emerged as the method of choice where the derivative quantities can be evaluated with an accuracy up to the machine precision; AD methods do not incur truncation errors of FD schemes. In either method, great savings in computation can be achieved if the sparsity structure of the Jacobian is known a priori or can be computed easily [10], and does not change from iteration to iteration.

There are two main ways in which sparsity can be exploited in evaluating derivatives. In the first, the entries of the Jacobian or Hessian matrix that are known to vanish identically need not be stored explicitly in a data structure. Secondly, operations involving known zeros must be avoided so as to speedup the calculation on sparse data. On the other hand, sparse matrix algorithms often require special techniques that are quite distinct and are usually more complex than their dense counterpart. The computational complexity of sparse matrix operations is affected by factors such as memory traffic and the size and organization of fast cache memory, in addition to the number of floating-point operations. For large-scale problems it is usually the case that the faster cache memory is not large enough to hold the input data in its entirety. The term *locality of reference* is used to signify the access pattern of data during the execution of an algorithm in a hierarchical memory computing system. The principle of data locality postulates that "recently accessed data (temporal) and nearby data (spatial) are likely to be accessed in the near future". Generally, better the reference locality for data fewer the cache-misses.

The purpose of this short paper is to present the software tool **DSJM** that can be used to compress and determine a sparse Jacobian matrix from its sparsity pattern using FD or AD. We exploit the relationship between graph and sparse matrix operations and use a standard array-based computer representation for sparse matrices. Graphs that have been used to model and solve combinatorial problems in Jacobian and Hessian matrix determination are rather dependent on specific properties e.g., structural orthogonality, symmetry etc. the methods try to exploit. For example an adjacency graph representation is a natural choice for a symmetric Hessian matrix while a Jacobian matrix with m rows and n columns can be conveniently represented with a bipartite graph. The pattern graph associated with a sparse matrix is structurally close to its actual computer representation and remains invariant for graph operations required in one-sided [12,14] and two-sided determination [2,11,13]. Further, the Jacobian matrix determination being a sparse matrix problem, we find it advantageous to accurately express the computational cost of graph operations in terms of the nonzero unknowns to be determined. Thus, for the basic sparse matrix operations the chosen representation of the input sparse matrix enables the same asymptotic computational cost as indicated by its graph abstraction. The paper is structured in the following way. In Sect. 2 we briefly review the compression-reconstruction method for the determination of sparse Jacobian matrices. An example is used to illustrate the central ideas in our approach to exploit known sparsity. The user interface to **DSJM** is discussed in Sect. 3. Results from numerical experiments on a suite of large-scale test problems together with some remarks on the comparative performance of **DSJM** and software **ColPack** [7] is also reported in Sect. 3. The paper is concluded in Sect. 4 with a discussion on extension of **DSJM** that are currently being implemented.

2   Direct Determination

We consider the problem of determining the Jacobian matrix $F'(x)$ of a mapping $F : \Re^n \rightarrow \Re^m$. Using differences, the product of the Jacobian matrix with a vector s may be approximated as

$$\left.\frac{\partial F(x + ts)}{\partial t}\right|_{t=0} = F'(x)s \equiv As \approx \frac{1}{\epsilon}[F(x + \epsilon s) - F(x)] \equiv b, \qquad (1)$$

with one extra evaluation of F at $(x + \epsilon s)$ assuming $F(x)$ has already been computed, where $\epsilon > 0$ is a small increment. Also, Algorithmic (or Automatic) Differentiation (AD) [9] forward mode gives $b = F'(x)s$ accurate up to the machine round-off, at a cost which is a small multiple of the cost of one function evaluation. The Jacobian matrix determination problem (JMDP) based on matrix-vector or vector-matrix products can be stated as below.

Obtain vectors $s_j \in \Re^n, j = 1, \ldots, p$ and $w_k \in \Re^m, k = 1, \ldots, q$ with $p + q$ minimized such that the products $b_j = As_j, j = 1, \ldots, p$ or $B = AS$ and $c_k^\top = w_k^\top A, k = 1, \ldots, q$ or $C^\top = W^\top A$ determine the matrix A uniquely.

In absence of any sparsity information, one may use the Cartesian basis vectors $e_i, i = 1, \ldots, n$ in (1) using n extra function evaluations, or in the case of AD [9], m reverse mode calculations or n forward mode calculations, whichever is smaller. On the other hand, if a group of columns of matrix A, say columns j and l, are *structurally orthogonal*, i.e., no two columns have nonzero entries in the same row position, only one extra function evaluation,

$$F_j' + F_l' = A(:, j) + A(:, l) \approx \frac{1}{\epsilon}[F(x + \epsilon(e_j + e_l)) - F(x)], \qquad (2)$$

is sufficient to read-off the nonzero entries from the product $b = As, s = e_j + e_l$. A group of columns that are structurally orthogonal will henceforth be called a *(structurally orthogonal) group*, where the word "structurally orthogonal" is suppressed for brevity when the context is clear. A group of nonzero unknowns are said to be *directly determined* or read-off from the product $b = As$ if no floating point arithmetic operation is needed to compute them. Thus, if the columns (rows) can be partitioned into p (q) structurally orthogonal groups then the Jacobian matrix is directly determined from the *row-compressed (column-compressed)* matrix $B = AS$ ($C^\top = W^\top A$). In this case we have a *one-sided*

$$
\begin{pmatrix}
0 & a_{12} & 0 & 0 & 0 \\
a_{21} & 0 & 0 & a_{24} & 0 \\
0 & 0 & a_{33} & 0 & a_{35} \\
0 & a_{42} & 0 & 0 & a_{45} \\
a_{51} & 0 & 0 & a_{54} & a_{55}
\end{pmatrix}
\begin{pmatrix}
1 & 0 & 0 \\
1 & 0 & 0 \\
1 & 0 & 0 \\
0 & 1 & 0 \\
0 & 0 & 1
\end{pmatrix}
\approx
\begin{pmatrix}
\tilde{a}_{12} & 0 & 0 \\
\tilde{a}_{21} & \tilde{a}_{24} & 0 \\
\tilde{a}_{33} & 0 & \tilde{a}_{35} \\
\tilde{a}_{42} & 0 & \tilde{a}_{45} \\
\tilde{a}_{51} & \tilde{a}_{54} & \tilde{a}_{55}
\end{pmatrix}
$$

$$A \qquad\qquad S \qquad\qquad B$$

Fig. 1. A Jacobian matrix.

compression of matrix A. Determining a sparse Jacobian matrix by exploiting sparsity as in (2) is due to Curtis, Powell, and Reid [3] (henceforth the CPR method). Coleman and Moré [1] further analyze the column partitioning problem and formally show that the problem of finding a minimum cardinality column partitioning *consistent with direct determination* is equivalent to a vertex coloring problem of an associated graph and that the problem is NP-Hard. **DSJM** accepts, as input, a specification of the sparsity pattern of the Jacobian to be determined. As output, it yields matrices S or W.

Let matrix A in Fig. 1 be the Jacobian matrix of some function F at x. Using FD the nonzero unknowns in columns $1, 2$ and 3 can be approximated as,

$$
\frac{1}{\epsilon}\left(F(x + \epsilon(e_1 + e_2 + e_3)) - F(x)\right)^\top \approx \begin{pmatrix} \tilde{a}_{12} & \tilde{a}_{21} & \tilde{a}_{33} & \tilde{a}_{42} & \tilde{a}_{51} \end{pmatrix}
$$

where $\epsilon > 0$ is a small increment (step-size), and $\tilde{(.)}$ indicates that the entry is an approximation to the true value. Thus, matrix A can be approximated with only three extra function evaluations of the form $F(x + \epsilon s)$ for direction s set to $e_1 + e_2 + e_3$, e_4, and e_5 in succession, in addition to evaluating F at x. With AD technology, the entries of A are obtained as three forward evaluations in the form of products As where each product costs approximately 3 function evaluations. The central point here is that if the sparsity pattern of a group of columns is such that for every pair of column indices $j \neq l$ in the group $\mathcal{S}(A(:, j)) \cap \mathcal{S}(A(:, l))$ where $\mathcal{S}(A) = \{(i, j) | a_{ij} \neq 0\}$ denotes the sparsity pattern for matrix A, is empty, then only one AD forward accumulation or one extra function evaluation will suffice to determine the nonzero entries in those columns. In other words, columns $A(: j)$ and $A(:, k)$ are *structurally orthogonal* if there is no index i for which $a_{ij} \neq 0$ and $a_{il} \neq 0$. Columns $1, 2, 3$ as shown above is a structurally orthogonal group. The CPR method groups or colors the columns such that columns in each group are structurally orthogonal. It has been observed frequently that scanning columns in specific order during grouping operation may lead to fewer groups. **DSJM** implements grouping algorithms CPR Grouping (CPR), Recursive Largest-First grouping (RLF), Saturation-Degree Grouping (SD) and ordering algorithms Largest-First Order (LFO), Smallest-Last Order (SLO), Incidence-Degree Order (IDO). In the next section we elaborate on data structure and implementation of main algorithmic tasks.

3   Interface and Implementation

3.1   Data Structure

The sparsity pattern of sparse matrix A can be specified to **DSJM** as index pairs (i, j) where i and j denote, respectively, row index and column index of a matrix nonzero. Internally, the pattern is represented by compressed sparse vectors corresponding to rows and columns. A sparse vector can be stored in compressed form using two arrays: one array to hold the indices of the nonzero entries and the other to hold the corresponding nonzero values. Compressed Sparse Row (CSR) storage scheme is a popular data structure where the sparse row vectors are stored contiguously. A simple implementation of CSR can be provided using three arrays: *rowptr* array indexes into *colind* and *value*, with *rowptr*(i) indicating the location (index) of the first element of row i in the arrays *colind* and *value*. Thus, elements in row i are accessed as

$$value(k) \text{ and are located in } (i, colind(k)), k = rowptr(i) \text{ to } rowptr(i+1) - 1.$$

Compressed Sparse Column (CSC) is simply the transposed matrix stored using the CSR. Sparse matrices in Harwell-Boeing collection [5] are given in CSC. The MATLAB$^{\circledR}$ computing environment [8] and CSparse [4] software use CSC representation for sparse matrices and the associated operations. **DSJM** enables access to columns and rows by maintaining both CRS and CCS structures for a sparse matrix.

3.2   Algorithms

Algorithm CPR takes as input the sparsity pattern of matrix A and an array, *group*, initialized to zero. The index set $\mathcal{L}$ in line 4 represents columns that cannot be grouped with column j; c_m represents the least-numbered structurally orthogonal column group that can include column j. The tasks in lines 4 and 5 are representative of kernel computational operations that are carefully implemented to enable efficiency. For example, the index set $\mathcal{L}$ is computed as l = colind(indj) where indj = rowptr(i) : rowptr(i+1) - 1, i = rowind(indi), indi = colptr(j) : colptr(j+1) - 1. On return from the algorithm column j belongs to the group *group*(j), and *ngroup* holds the number of structurally orthogonal groups. The matrix elements stored in data structures CSR and CSC in the implementation of line 4 are placed in contiguous locations in the computer memory thus ensuring maximum spatial locality [15] and resulting in better cache memory utilization. Also, the computational cost of the above kernel operation is proportional to the size of the data accessed and the number of nonzero arithmetic operations. Indeed, except for SD and RLF algorithms all grouping and ordering heuristics run in time proportional to $\sum_{i=1}^{m} \rho_i^2$ where ρ_i denotes the number of nonzero entries in row i of A. Borrowing ideas from [8] **DSJM** utilizes a dense "work vector" and an efficient "tagging scheme" in implementing many of the sparse linear algebra primitives.

CPR($\mathcal{S}(A)$, *group*, *ngroup*)

```
1   ngroup ← 1
2   for j ← 1 to n
3       do
4           let L = {l | A(:,j) is not structurally orthogonal to A(:,l)} and
5               c_m = min{c | c ∈ {1,...,ngroup +1} ≠ group(l), l ∈ L}
6           group(j) ← c_m
7           if c_m > ngroup
8               then
9                   ngroup ← ngroup +1
```

Algorithm FD-SPJMS can be used to obtain an approximation to the nonzero entries of the directional derivative $F'(x)S(:,k) \equiv B(:,k)$ corresponding to structurally orthogonal group k.

FD-SPJMS(*gptr*, *gcolind*, k, η, B)

```
1   w ← FD-SPJD(F, x, η, gptr, gcolind, k)
2   for ind ← gptr(k) to gptr(k + 1) − 1
3       do
4           j ← gcolind(ind)
5           for each i for which F'(i, j) ≠ 0
6               do
7                   B(i, k) ← w(i)/η(j)
```

FD-SPJD is a user-defined function to compute the difference $F(x + \eta) - F(x)$ where, array η initially containing zeros, can be defined to contain the finite-difference increments $\epsilon(j)$ corresponding to columns j in column group k:

```
1   for ind ← gptr(k) to gptr(k + 1) − 1
2       do
3           j ← gcolind(ind)
4           η(j) ← ε(j)
```

3.3   Numerical Testing

In Table 1 we provide partitioning results for a selection of problems [6] where there is a large gap between the maximum number of nonzero entries in any row (ρ is a lower bound on the number of groups) and the number of groups in the partition (p). In the table m, n, and nnz denote respectively, the number of rows, columns, and nonzero entries in the test matrix. The experiments were performed using an IBM PC with 2.8 GHz Intel Pentium CPU, 1 GB RAM, and 512 KB L2 cache running Linux. Of 8 test problems **DSJM** yields better partition (indicated in **bold face**. A comparison of the running time for the implementations **DSJM** and **ColPack** we find that **DSJM** is about $1.6 - 3.0$ times faster than the **ColPack** (see [13]).

Table 1. Partitioning Results.

Matrix	m	n	nnz	ρ	ColPack [7]	DSJM
af23560	23560	23560	484256	21	41	**37**
cage11	39082	39082	559722	31	62	**54**
cage12	130228	130228	2032536	33	68	**56**
e30r2000	9661	9661	306356	62	68	**65**
e40r0100	17281	17281	553956	62	**66**	67
lpken11	14694	21349	49058	122	123	**122**
lpken13	28632	42659	97246	170	171	**170**
lpmarosr7	3136	9408	144848	48	**70**	76

3.4   DSJM Interface

The functionalities of the software are made available to the client through objects of C++ class `Matrix` implemented in **DSJM**. Client code defines a `Matrix` object to read in sparse matrix data: pairs of indices. Duplicate nonzero entries can be removed by calling the member function `compress`. To set up the CSR and CSC data structures users need to call member functions `computeCRS` and `computeCCS`, respectively. Once the data structure is set up, the columns of the matrix object can be ordered and partitioned into groups consistent with direct determination.

3.5   Example Usage of Matrix Object

Displayed below is a code snippet for structurally orthogonal grouping of the matrix columns using smallest-last order and CPR greedy partitioning algorithm.

```
Matrix matrix(M,N,nnz, false); // Boolean argument false indicates that nonzero
                               // values are not to be stored

// ! C++ code to read-in the sparsity pattern omitted

int nnz = matrix.compress();   // remove duplicate entries
matrix.computeCCS();           // set-up CSC data structure
matrix.computeCRS();           // set-up CSR  data structure

int *order = new int[N+1];     // array to store order information
matrix.slo(order);             // columns in smallest-last order

int *color = new int[N+1];     // array to store the partitioning
int maxgrp = matrix.greedycolor(order,color);
for (int j = 1; j <= N; j++)   // print the partition information
    printf("Column j belongs to %d partition\n",color[i]);
```

DSJM grouping algorithms can be accessed from within MATLAB environment using MATLAB's MEX interface as below:

```
B = dsjmcolor(A,'slo');
```

In the call to *dsjmcolor*(), A is a MATLAB sparse matrix. The second parameter "slo" denotes that the the matrix columns be CPR grouped using smallest-last ordering.

4   Concluding Remarks

DSJM is being developed to address the need for efficient software toolkits that are simple and intuitive to use. The current implementation provides a collection of stand-alone column ordering and grouping algorithms and driver routines for efficient direct determination of sparse Jacobian matrices through a sparse matrix class interface. The tool is currently being extended with two-sided compression algorithms [11].

Acknowledgements. This research was supported in part by Natural Sciences and Engineering Research Council of Canada (NSERC) Discovery Grant (Individual).

References

1. Coleman, T.F., Moré, J.J.: Estimation of sparse Jacobian matrices and graph coloring problems. SIAM J. Numer. Anal. **20**(1), 187–209 (1983)
2. Coleman, T.F., Verma, A.: The efficient computation of sparse Jacobian matrices using automatic differentiation. SIAM J. Sci. Comput. **19**(4), 1210–1233 (1998)
3. Curtis, A.R., Powell, M.J.D., Reid, J.K.: On the estimation of sparse Jacobian matrices. J. Inst. Math. Appl. **13**, 117–119 (1974)
4. Davis, T.A.: Direct Methods for Sparse Linear Systems (Fundamentals of Algorithms 2). Society for Industrial and Applied Mathematics, Philadelphia, PA, USA (2006)
5. Duff, I.S., Grimes, R.G., Lewis, J.G.: Sparse matrix test problems. ACM Trans. Math. Softw. **15**(1), 1–14 (1989)
6. Gebremedhin, A.H., Manne, F., Pothen, A.: What color is your jacobian? graph coloring for computing derivatives. SIAM Rev. **47**(4), 629–705 (2005)
7. Gebremedhin, A.H., Nguyen, D., Patwary, M.M.A., Pothen, A.: ColPack: Software for graph coloring and related problems in scientific computing. ACM Trans. Math. Softw. **40**(1), 1–31 (2013)
8. Gilbert, J.R., Moler, C., Schreiber, R.: Sparse matrices in matlab: design and implementation. SIAM J. Matrix Anal. Appl. **13**(1), 333–356 (1992)
9. Griewank, A., Walther, A.: Evaluating Derivatives: Principles and Techniques of AlgorithmicDifferentiation, 2nd edn. Society for Industrial and Applied Mathematics, Philadelphia, PA, USA (2008)
10. Griewank, A., Mitev, C.: Detecting Jacobian sparsity patterns by Bayesian probing. Math. Prog. **93**(1), 1–25 (2002)
11. Hossain, A.S., Steihaug, T.: Computing a sparse Jacobian matrix by rows and columns. Optim. Methods Softw. **10**, 33–48 (1998)
12. Hossain, S., Steihaug, T.: Graph coloring in the estimation of sparse derivative matrices: Instances and applications. Discrete Appl. Math. **156**(2), 280–288 (2008)
13. Hossain, S., Steihaug, T.: Graph models and their efficient implementation for sparse jacobian matrix determination. Discrete Appl. Math. **161**(12), 1747–1754 (2013)

14. Newsam, G.N., Ramsdell, J.D.: Estimation of sparse Jacobian matrices. SIAM J. Alg. Disc. Meth. **4**(3), 404–417 (1983)
15. Park, J.-S., Penner, M., Prasanna, V.K.: Optimizing graph algorithms for improved cache performance. IEEE Trans. Parallel Distrib. Syst. **15**(9), 769–782 (2004)

Software for Cut-Generating Functions in the Gomory–Johnson Model and Beyond

Chun Yu Hong[1], Matthias Köppe[2(✉)], and Yuan Zhou[2]

[1] Department of Statistics, University of California, Berkeley, Berkeley, CA, USA
jcyhong@berkeley.edu
[2] Department of Mathematics, University of California, Davis, Davis, CA, USA
{mkoeppe,yzh}@math.ucdavis.edu
http://statistics.berkeley.edu/people/chun-yu-hong
http://www.math.ucdavis.edu/~mkoeppe
http://www.math.ucdavis.edu/~yzh

Abstract. We present software for investigations with cut-generating functions in the Gomory–Johnson model and extensions, implemented in the computer algebra system SageMath.

Keywords: Integer programming · Cutting planes · Group relaxations

1 Introduction

Consider the following question from the theory of linear inequalities over the reals: Given a (finite) system $Ax \leq b$, exactly which linear inequalities $\langle a, x \rangle \leq \beta$ are *valid*, i.e., satisfied for every x that satisfies the given system? The answer is given, of course, by the Farkas Lemma, or, equivalently, by the strong duality theory of linear optimization. As is well-known, this duality theory is symmetric: The dual of a linear optimization problem is again a linear optimization problem, and the dual of the dual is the original (primal) optimization problem.

The question becomes much harder when all or some of the variables are constrained to be integers. The theory of valid linear inequalities here is called *cutting plane theory*. Over the past 60 years, a vast body of research has been carried out on this topic, the largest part of it regarding the polyhedral combinatorics of integer hulls of particular families of problems. The general theory again is equivalent to the duality theory of integer linear optimization problems. Here the dual objects are not linear, but *superadditive* (or subadditive) functionals, making the general form of this theory infinite-dimensional even though the original problem started out with only finitely many variables.

The authors gratefully acknowledge partial support from the National Science Foundation through grant DMS-1320051 awarded to M. Köppe.

C.Y. Hong—The first author's contribution was done during a Research Experiences for Undergraduates at the University of California, Davis. He was partially supported by the National Science Foundation through grant DMS-0636297 (VIGRE).

© Springer International Publishing Switzerland 2016
G.-M. Greuel et al. (Eds.): ICMS 2016, LNCS 9725, pp. 284–291, 2016.
DOI: 10.1007/978-3-319-42432-3_35

These superadditive (or subadditive) functionals appear in integer linear optimization in various concrete forms, for example in the form of *dual-feasible functions* [1], *superadditive lifting functions* [12], and *cut-generating functions* [6].

In the present paper, we describe some aspects of our software [10] for cut-generating functions in the classic 1-row Gomory–Johnson [7,8] model. In this theory, the main objects are the so-called *minimal valid functions*, which are the $\mathbb{Z}$-periodic, subadditive functions $\pi \colon \mathbb{R} \to \mathbb{R}_+$ with $\pi(0) = 0$, $\pi(f) = 1$, that satisfy the *symmetry condition* $\pi(x) + \pi(f - x) = 1$ for all $x \in \mathbb{R}$. (Here f is a fixed number.) We refer the reader to the recent survey [4,5].

Our software is a tool that enables mathematical exploration and research in this domain. It can also be used in an educational setting, where it enables hands-on teaching about modern cutting plane theory based on cut-generating functions. It removes the limitations of hand-written proofs, which would be dominated by tedious case analysis.

The first version of our software [10] was written by the first author, C. Y. Hong, during a Research Experience for Undergraduates in summer 2013. It was later revised and extended by M. Köppe and again by Y. Zhou. The latter added an electronic compendium [11] of extreme functions found in the literature, and added code that handles the case of discontinuous functions. Version 0.9 of our software was released in 2014 to accompany the survey [4,5]; the software has received continuous updates by the second and third authors since.[1]

Our software is written in Python, making use of the convenient framework of the open-source computer algebra system SageMath [14]. It can be run on a local installation of SageMath, or online via *SageMathCloud*.

2   Continuous and Discontinuous Piecewise Linear $\mathbb{Z}$-periodic Functions

The main objects of our code are the $\mathbb{Z}$-periodic functions $\pi \colon \mathbb{R} \to \mathbb{R}$. Our code is limited to the case of piecewise linear functions, which are allowed to be discontinuous; see the definition below. In the following, we connect to the systematic notation introduced in [3, Section 2.1]; see also [4,5]. In our code, the periodicity of the functions is implicit; the functions are represented by their restriction to the interval $[0, 1]$.[2] They can be constructed in various ways using Python functions named `piecewise_function_from_breakpoints_and_values` etc.; see the source code of the electronic compendium for examples. We also suppress the details of the internal representation; instead we explain the main ways in which the data of the function are accessed.

π.`end_points()` is a list $0 = x_0 < x_1 < \cdots < x_{n-1} < x_n = 1$ of possible breakpoints of the function in $[0, 1]$. In the notation from [3–5], these endpoints

---

[1] Two further undergraduate students contributed to our software. P. Xiao contributed some documentation and tests. M. Sugiyama contributed additional functions to the compendium, and added code for superadditive lifting functions.

[2] The functions are instances of the class `FastPiecewise`, which extends an existing SageMath class for piecewise linear functions.

are extended periodically as $B = \{ x_0 + t, x_1 + t, \ldots, x_{n-1} + t : t \in \mathbb{Z} \}$. Then the set of 0-dimensional faces is defined to be the collection of singletons, $\{ \{x\} : x \in B \}$, and the set of one-dimensional faces to be the collection of closed intervals, $\{ [x_i + t, x_{i+1} + t] : i = 0, \ldots, n - 1 \text{ and } t \in \mathbb{Z} \}$. Together, we obtain $\mathcal{P} = \mathcal{P}_B$, a locally finite polyhedral complex, periodic modulo $\mathbb{Z}$.

π.values_at_end_points() is a list of the function values $\pi(x_i)$, $i = 0, \ldots, n$. This list is most useful for continuous piecewise linear functions, as indicated by π.is_continuous(), in which case the function is defined on the intervals $[x_i, x_{i+1}]$ by linear interpolation.

π.limits_at_end_points() provides data for the general, possibly discontinuous case in the form of a list limits of 3-tuples, with

$$\text{limits}[i][0] = \pi(x_i)$$
$$\text{limits}[i][1] = \pi(x_i^+) = \lim_{x \to x_i, x > x_i} \pi(x)$$
$$\text{limits}[i][-1] = \pi(x_i^-) = \lim_{x \to x_i, x < x_i} \pi(x).$$

The function is defined on the open intervals (x_i, x_{i+1}) by linear interpolation of the limit values $\pi(x_i^+)$, $\pi(x_{i+1}^-)$.

$\pi(x)$ and π.limits(x) evaluate the function at x and provide the 3-tuple of its limits at x, respectively.

π.which_function(x) returns a linear function, denoted $\pi_I : \mathbb{R} \to \mathbb{R}$ in [3–5], where I is the smallest face of $\mathcal{P}$ containing x, so $\pi(x) = \pi_I(x)$ for $x \in \text{rel int}(I)$.

Functions can be plotted using the standard SageMath function plot(π), or using our function plot_with_colored_slopes(π), which assigns a different color to each different slope value that a linear piece takes.[3] Examples of such functions are shown in Figs. 2 and 3.

3   The Diagrams of the Decorated 2-Dimensional Polyhedral Complex $\Delta\mathcal{P}$

We now describe certain 2-dimensional diagrams which record the subadditivity and additivity properties of a given function. These diagrams, in the continuous case, have appeared extensively in [4,5,11]. An example for the discontinuous case appeared in [11]. We have engineered these diagrams from earlier forms that can be found in [9] (for the discussion of the merit_index) and in [3], to become power tools for the modern cutgeneratingfunctionologist. Not only is the minimality of a given function immediately apparent on the diagram, but also the extremality proof for a given class of piecewise minimal valid functions follows a standard pattern that draws from these diagrams. See [5, prelude] and [11, Sections 2 and 4] for examples of such proofs.

---

[3] See also our function number_of_slopes. We refer the reader to [4, Section 2.4] for a discussion of the number of slopes of extreme functions, and [2] and bcdsp_arbitrary_slope for the latest developments in this direction.

3.1   The Polyhedral Complex and Its Faces

Following [3–5], we introduce the function

$$\Delta\pi \colon \mathbb{R} \times \mathbb{R} \to \mathbb{R}, \quad \Delta\pi(x, y) = \pi(x) + \pi(y) - \pi(x + y),$$

which measures the slack in the subadditivity condition.[4] Thus, if $\Delta\pi(x, y) < 0$, subadditivity is violated at (x, y); if $\Delta\pi(x, y) = 0$, additivity holds at (x, y); and if $\Delta\pi(x, y) > 0$, we have strict subadditivity at (x, y). The piecewise linearity of $\pi(x)$ induces piecewise linearity of $\Delta\pi(x, y)$. To express the domains of linearity of $\Delta\pi(x, y)$, and thus domains of additivity and strict subadditivity, we introduce the two-dimensional polyhedral complex $\Delta\mathcal{P}$. The faces F of the complex are defined as follows. Let $I, J, K \in \mathcal{P}$, so each of I, J, K is either a breakpoint of π or a closed interval delimited by two consecutive breakpoints. Then

$$F = F(I, J, K) = \{ (x, y) \in \mathbb{R} \times \mathbb{R} : x \in I, \, y \in J, \, x + y \in K \}.$$

In our code, a face is represented by an instance of the class **Face**. It is constructed from I, J, K and is represented by the list of vertices of F and its projections $I' = p_1(F)$, $J' = p_2(F)$, $K' = p_3(F)$, where $p_1, p_2, p_3 \colon \mathbb{R} \times \mathbb{R} \to \mathbb{R}$ are defined as $p_1(x, y) = x$, $p_2(x, y) = y$, $p_3(x, y) = x + y$. The vertices $\mathrm{vert}(F)$ are obtained by first listing the basic solutions (x, y) where x, y, and $x + y$ are fixed to endpoints of I, J, and K, respectively, and then filtering the feasible solutions. The three projections are then computed from the list of vertices. Due to the $\mathbb{Z}$-periodicity of π, we can represent a face as a subset of $[0, 1] \times [0, 1]$. See Fig. 1 for an example. Because of the importance of the projection $p_3(x, y) = x + y$, it is convenient to imagine a third, $(x + y)$-axis in addition to the x-axis and the y-axis, which traces the bottom border for $0 \le x + y \le 1$ and then the right border for $1 \le x + y \le 2$. To make room for this new axis, the x-axis should be drawn on the top border of the diagram.

3.2   plot_2d_diagram_with_cones

We now explain the first version of the 2-dimensional diagrams, plotted by the function plot_2d_diagram_with_cones(π); see Fig. 2. At the border of these diagrams, the function π is shown twice (*blue*), along the x-axis (*top border*) and along the y-axis (*left border*). The solid grid lines in the diagrams are determined by the breakpoints of π: vertical, horizontal and diagonal grid lines correspond to values where x, y and $x + y$ are breakpoints of π, respectively. The vertices of the complex $\Delta\mathcal{P}$ are the intersections of these grid lines.

   In the continuous case, we indicate the sign of $\Delta\pi(x, y)$ for all vertices by colored dots on the diagram: *red* indicates $\Delta\pi(x, y) < 0$ (subadditivity is violated); *green* indicates $\Delta\pi(x, y) = 0$ (additivity holds).

Example 1. In Fig. 2 (left), the vertex $(x, y) = (\frac{1}{5}, \frac{3}{5})$ is marked green, since

$$\Delta\pi(\tfrac{1}{5}, \tfrac{3}{5}) = \pi(\tfrac{1}{5}) + \pi(\tfrac{3}{5}) - \pi(\tfrac{4}{5}) = \tfrac{1}{5} + \tfrac{4}{5} - 1 = 0.$$

---

[4] It is available in the code as delta_pi(π, x, y); in [7], it was called $\nabla(x, y)$.

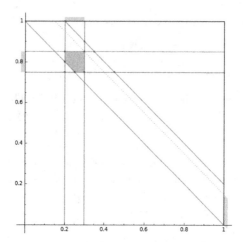

Fig. 1. An example of a face $F = F(I, J, K)$ of the 2-dimensional polyhedral complex $\Delta\mathcal{P}$, set up by `F = Face([[0.2, 0.3], [0.75, 0.85], [1, 1.2]])`. It has vertices (*blue*) $(0.2, 0.85), (0.3, 0.75), (0.3, 0.85), (0.2, 0.8), (0.25, 0.75)$, whereas the other basic solutions (*red*) $(0.2, 0.75), (0.2, 1), (0.3, 0.9), (0.35, 0.85), (0.45, 0.75)$ are filtered out because they are infeasible. The face F has projections (*gray shadows*) $I' = p_1(F) = [0.2, 0.3]$ (*top border*), $J' = p_2(F) = [0.75, 0.85]$ (*left border*), and $K' = p_3(F) = [1, 1.15]$ (*right border*). Note that $K' \subsetneq K$.

In the discontinuous case, beside the subadditivity slack $\Delta\pi(x, y)$ at a vertex (x, y), one also needs to study the limit value of $\Delta\pi$ at the vertex (x, y) approaching from the interior of a face $F \in \Delta\mathcal{P}$ containing the vertex (x, y). This limit value is defined by

$$\Delta\pi_F(x, y) = \lim_{\substack{(u,v) \to (x,y) \\ (u,v) \in \mathrm{rel\,int}(F)}} \Delta\pi(u, v), \quad \text{where } F \in \Delta\mathcal{P} \text{ such that } (x, y) \in F.$$

We indicate the sign of $\Delta\pi_F(x, y)$ by a colored cone inside F pointed at the vertex (x, y) on the diagram. There could be up to 12 such cones (including rays for one-dimensional F) around a vertex (x, y).

Example 2. In Fig. 2 (right), the lower right corner $(x, y) = (\frac{2}{5}, \frac{4}{5})$ of the face $F = F(I, J, K)$ with $I = [\frac{1}{5}, \frac{2}{5}]$, $J = [\frac{4}{5}, 1]$, $K = [1, \frac{6}{5}]$ is green, since

$$\Delta\pi_F(x, y) = \lim_{\substack{(u,v) \to (\frac{2}{5}, \frac{4}{5}) \\ (u,v) \in \mathrm{rel\,int}(F)}} \Delta\pi(u, v)$$

$$= \lim_{u \to \frac{2}{5}, \, u < \frac{2}{5}} \pi(u) + \lim_{v \to \frac{4}{5}, \, v > \frac{4}{5}} \pi(v) - \lim_{w \to \frac{6}{5}, \, w < \frac{6}{5}} \pi(w)$$

$$= \pi(\tfrac{2}{5}^-) + \pi(\tfrac{4}{5}^+) - \pi(\tfrac{1}{5}^-) \quad (\text{as } \pi(\tfrac{6}{5}^-) = \pi(\tfrac{1}{5}^-) \text{ by periodicity})$$

$$= 0 + 1 - 1 = 0.$$

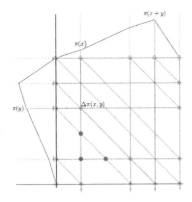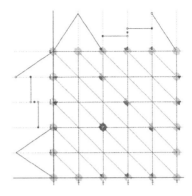

Fig. 2. Two diagrams of functions and their polyhedral complexes $\Delta\mathcal{P}$ with colored cones at vert($\Delta\mathcal{P}$), as plotted by the command `plot_2d_diagram_with_cones(h)`. *Left*, continuous function `h = not_minimal_2()`. *Right*, a random discontinuous function generated by `h = random_piecewise_function(xgrid=5, ygrid=5, continuous_proba=1/3, symmetry=True)`.

The horizontal ray to the left of the same vertex $(x, y) = (\frac{2}{5}, \frac{4}{5})$ is red, because approaching from the one-dimensional face $F' = F(I', J', K')$ that contains (x, y), with $I' = [\frac{1}{5}, \frac{2}{5}]$, $J' = \{\frac{4}{5}\}$, $K' = [1, \frac{6}{5}]$, we have the limit value

$$\Delta\pi_{F'}(x, y) = \lim_{\substack{(u,v) \to (\frac{2}{5}, \frac{4}{5}) \\ (u,v) \in \mathrm{rel\,int}(F')}} \Delta\pi(u, v) = \lim_{\substack{u \to \frac{2}{5} \\ u < \frac{2}{5}}} \pi(u) + \pi(\tfrac{4}{5}) - \lim_{\substack{w \to \frac{6}{5} \\ w < \frac{6}{5}}} \pi(w) = 0 + \tfrac{3}{5} - 1 < 0.$$

3.3 `plot_2d_diagram` and additive faces

Now assume that π is a subadditive function. Then there are no red dots or cones on the above diagram of the complex $\Delta\mathcal{P}$. See Fig. 3.

For a **continuous subadditive function** π, we say that a face $F \in \Delta\mathcal{P}$ is *additive* if $\Delta\pi = 0$ over all F. Note that $\Delta\pi$ is affine linear over F, and so the face F is additive if and only if $\Delta\pi(x, y) = 0$ for all $(x, y) \in \mathrm{vert}(F)$. It is clear that any subface E of an additive face F ($E \subseteq F$, $E \in \Delta\mathcal{P}$) is still additive. Thus the additivity domain of π can be represented by the list of inclusion-maximal additive faces of $\Delta\mathcal{P}$; see [4, Lemma 3.12].[5]

For a **discontinuous subadditive function** π, we say that a face $F \in \Delta\mathcal{P}$ is *additive* if F is contained in a face $F' \in \Delta\mathcal{P}$ such that $\Delta\pi_{F'}(x, y) = 0$ for any $(x, y) \in F$.[6] Since $\Delta\pi$ is affine linear in the relative interiors of each face of $\Delta\mathcal{P}$, the last condition is equivalent to $\Delta\pi_{F'}(x, y) = 0$ for any $(x, y) \in \mathrm{vert}(F)$. Depending on the dimension of F, we do the following.

---

[5] This list is computed by `generate_maximal_additive_faces(π)`.

[6] Summarizing the detailed additivity and additivity-in-the-limit situation of the function using the notion of additive faces is justified by [3, Lemmas 2.7 and 4.5] and their generalizations.

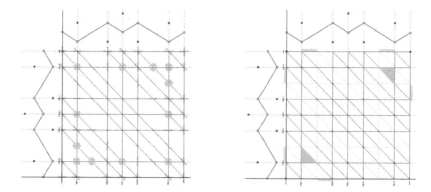

Fig. 3. Diagrams of $\Delta\mathcal{P}$ of a discontinuous function `h = hildebrand_2_sided_discont_2_slope_1()`, with (*left*) additive limiting cones as plotted by the command `plot_2d_diagram_with_cones(h)`; (*right*) additive faces as plotted by the command `plot_2d_diagram(h)`.

1. Let F be a two-dimensional face of $\Delta\mathcal{P}$. If $\Delta\pi_F(x,y) = 0$ for any $(x,y) \in \text{vert}(F)$, then F is additive. Visually on the 2d-diagram with cones, each vertex of F has a green cone sitting inside F.
2. Let F be a one-dimensional face, i.e., an edge of $\Delta\mathcal{P}$. Let $(x_1,y_1),(x_2,y_2)$ be its vertices. Besides F itself, there are two other faces $F_1, F_2 \in \Delta\mathcal{P}$ that contain F. If $\Delta\pi_{F'}(x_1,y_1) = \Delta\pi_{F'}(x_2,y_2) = 0$ for $F' = F$, F_1, or F_2, then the edge F is additive.
3. Let F be a zero-dimensional face of $\Delta\mathcal{P}$, $F = \{(x,y)\}$. If there is a face $F' \in \Delta\mathcal{P}$ such that $(x,y) \in F'$ and $\Delta\pi_{F'}(x,y) = 0$, then F is additive. Visually on the 2d-diagram with cones, the vertex (x,y) is green or there is a green cone pointing at (x,y).

On the diagram Fig. 3 (right), the additive faces are shaded in green. The projections $p_1(F)$, $p_2(F)$, and $p_3(F)$ of a two-dimensional additive face F are shown as gray shadows on the x-, y- and $(x+y)$-axes of the diagram, respectively.

4   Additional Functionality

`minimality_test(`π`)` implements a fully automatic test whether a given function is a minimal valid function, using the information that the described 2-dimensional diagrams visualize. The algorithm is equivalent to the one described, in the setting of discontinuous pseudo-periodic superadditive functions, in Richard, Li, and Miller [13, Theorem 22].

`extremality_test(`π`)` implements a grid-free generalization of the automatic extremality test from [3], which is suitable also for piecewise linear functions with rational breakpoints that have huge denominators. Its support for functions with algebraic irrational breakpoints such as `bhk_irrational` [3, Section 5] is experimental and will be described in a forthcoming paper.

generate_covered_intervals(π) computes connected components of covered (affine imposing [3]) intervals. This is an ingredient in the extremality test.

extreme_functions is the name of a Python module that gives access to the electronic compendium of extreme functions; see [11] and [4, Tables 1–4].

procedures provides transformations of extreme functions; see [4, Table 5].

random_piecewise_function() generates a random piecewise linear function with prescribed properties, to enable experimentation and exploration.

demo.sage demonstrates further functionality and the use of the help system.

References

1. Alves, C., Clautiaux, F., de Carvalho, J.V., Rietz, J.: Dual-feasible functions for integer programming and combinatorial optimization: Basics, extensions and applications. EURO Advanced Tutorials on Operational Research. Springer, Heidelberg (2016). doi:10.1007/978-3-319-27604-5. ISBN 978-3-319-27602-1
2. Basu, A., Conforti, M., Di Summa, M., Paat, J.: Extreme functions with an arbitrary number of slopes. In: Louveaux, Q., Skutella, M. (eds.) IPCO 2016. LNCS, vol. 9682, pp. 190–201. Springer, Heidelberg (2016). doi:10.1007/978-3-319-33461-5_16
3. Basu, A., Hildebrand, R., Köppe, M.: Equivariant perturbation in Gomory and Johnson's infinite group problem. I. The one-dimensional case. Math. Oper. Res. **40**(1), 105–129 (2014). doi:10.1287/moor.2014.0660
4. Basu, A., Hildebrand, R., Köppe, M.: Light on the infinite group relaxation I: foundations and taxonomy. 4OR **14**(1), 1–40 (2016). doi:10.1007/s10288-015-0292-9
5. Basu, A., Hildebrand, R., Köppe, M.: Light on the infinite group relaxation II: sufficient conditions for extremality, sequences, and algorithms. 4OR **14**(2), 107–131 (2016). doi:10.1007/s10288-015-0293-8
6. Conforti, M., Cornuéjols, G., Daniilidis, A., Lemaréchal, C., Malick, J.: Cut-generating functions and S-free sets. Math. Oper. Res. **40**(2), 253–275 (2013). http://dx.doi.org/10.1287/moor.2014.0670
7. Gomory, R.E., Johnson, E.L.: Some continuous functions related to corner polyhedra I. Math. Program. **3**, 23–85 (1972). doi:10.1007/BF01584976
8. Gomory, R.E., Johnson, E.L.: Some continuous functions related to corner polyhedra, II. Math. Programm. **3**, 359–389 (1972). doi:10.1007/BF01585008
9. Gomory, R.E., Johnson, E.L.: T-space and cutting planes. Math. Program. **96**, 341–375 (2003). doi:10.1007/s10107-003-0389-3
10. Hong, C.Y., Köppe, M., Zhou, Y.: Sage program for computation and experimentation with the 1-dimensional Gomory-Johnson infinite group problem (2014). https://github.com/mkoeppe/infinite-group-relaxation-code
11. Köppe, M., Zhou, Y.: An electronic compendium of extreme functions for the Gomory-Johnson infinite group problem. Oper. Res. Lett. **43**(4), 438–444 (2015). doi:10.1016/j.orl.2015.06.004
12. Louveaux, Q., Wolsey, L.A.: Lifting, superadditivity, mixed integer rounding and single node flow sets revisited. Q. J. Belg. Fr. Ital. Oper. Res. Soc. **1**(3), 173–207 (2003). doi:10.1007/s10288-003-0016-4
13. Richard, J.-P.P., Li, Y., Miller, L.A.: Valid inequalities for MIPs and group polyhedra from approximate liftings. Math. Program. **118**(2), 253–277 (2009). doi:10.1007/s10107-007-0190-9
14. Stein, W.A., et al.: Sage Mathematics Software (Version 7.1), The Sage Development Team (2016). http://www.sagemath.org

Mixed Integer Nonlinear Program for Minimization of Akaike's Information Criterion

Keiji Kimura[1]([✉]) and Hayato Waki[2]

[1] Faculty of Mathematics, Kyushu University, Fukuoka, Japan
k-kimura@math.kyushu-u.ac.jp
[2] Institute of Mathematics for Industry, Kyushu University, Fukuoka, Japan
waki@imi.kyushu-u.ac.jp

Abstract. Akaike's information criterion (AIC) is a measure of the quality of a statistical model for a given set of data. We can determine the best statistical model for a particular data set by the minimization based on the AIC. Since it is difficult to find the best statistical model from a set of candidates by this minimization in practice, stepwise methods, which are local search algorithms, are commonly used to find a better statistical model though it may not be the best.

We formulate this AIC minimization as a mixed integer nonlinear programming problem and propose a method to find the best statistical model. In particular, we propose ways to find lower and upper bounds and a branching rule for this minimization. We then combine them with SCIP, which is a mathematical optimization software and a branch-and-bound framework. We show that the proposed method can provide the best statistical model based on AIC for small-sized or medium-sized benchmark data sets in UCI Machine Learning Repository. Furthermore, we show that this method can find good quality solutions for large-sized benchmark data sets.

Keywords: Mixed integer nonlinear program · SCIP · Akaike's information criterion

1   Introduction

Selecting the best statistical model from a number of candidate statistical models for a given set of data is one of the most important problems solved in statistical applications, e.g. regression analysis. This is called *variable selection*. Variable selection provides a simplest statistical model for a given data set and improves the prediction performance while keeping the goodness-of-fit for a given data set. See [5] for more details on variable selection.

In variable selection based on *an information criterion*, all the candidates are evaluated by the information criterion and select a statistical model by using evaluations. Akaike's information criterion (AIC) is one of the information criteria and proposed in [2]. An AIC value is computed for each candidate, and the model whose AIC value is the smallest is selected as the best statistical model.

© Springer International Publishing Switzerland 2016
G.-M. Greuel et al. (Eds.): ICMS 2016, LNCS 9725, pp. 292–300, 2016.
DOI: 10.1007/978-3-319-42432-3_36

Since we often need to handle too many candidates of statistical models in practical applications, the global minimization based on AIC is not practical. Instead of the global minimization, stepwise methods, which are local search algorithms, are commonly used to find a statistical model which has as small AIC as possible though it may not be the smallest.

The contribution of our study is to provide a mixed integer nonlinear programming (MINLP) formulation for the minimization based on AIC in linear regression and a method to solve it efficiently via SCIP. SCIP is a mathematical optimization software and a branch-and-bound framework. SCIP has high flexibility of user plugin and control on various parameters in the branch-and-bound framework for efficient computation. We implement a *relaxator*, which is a procedure to find lower bounds of the MINLP problems, and define a branching rule. By applying our proposed method to benchmark data sets in [10], we can obtain the best statistical models for some of them.

We introduce related work. Miyashiro and Takano [7] propose a mixed integer second-order cone programming (MISOCP) formulation for variable selection based on some information criteria in linear regression. In our numerical comparison with them, our proposed approach outperforms MISOCP formulation.

The organization is as follows: We give a brief introduction of AIC-based linear regression in Sect. 2. We introduce the MINLP formulation of the AIC minimization and propose a procedure to compute lower and upper bounds used in the branch-and-bound framework in Sect. 3. Section 4 introduces techniques for more efficient computation. We present a numerical result in Sect. 5 and discuss future work of our proposed approach in Sect. 6.

2   Preliminary on Akaike's Information Criterion in Linear Regression

We explain how to select the best statistical model via AIC in linear regression analysis. Linear regression is a fundamental statistical tool which determines coefficients $\beta_0, \ldots, \beta_p \in \mathbb{R}$ for the following equation from a given set of data:

$$y = \beta_0 + \sum_{j=1}^{p} \beta_j x_j. \tag{1}$$

Here $x_1, \ldots, x_p$ and y are called *the explanatory variables* and *the response variable* respectively. In fact, we adopt coefficients $\beta_0, \ldots, \beta_p$ which minimize $\sum_{i=1}^{n} \epsilon_i^2$ for a set of data $(x_{i1}, \ldots, x_{ip}, y_i) \in \mathbb{R}^p \times \mathbb{R}$ $(i = 1, \ldots, n)$, where ϵ_i is the ith residual and defined by $\epsilon_i = y_i - \beta_0 - \sum_{j=1}^{p} \beta_j x_{ij}$.

In linear regression analysis, variable selection based on AIC corresponds to the selection of a subset of the set of explanatory variables in (1) via AIC. More precisely, for a set $S \subseteq \{1, \ldots, p\}$ of candidates of explanatory variables in the statistical model (1), AIC is defined in [2] as follows:

$$\text{AIC}(S) = -2 \max_{\beta, \sigma^2} \{\ell(\beta, \sigma^2) : \beta_j = 0 \ (j \in \{1, \ldots, p\} \setminus S)\} + 2(\#(S) + 2) \tag{2}$$

where $\beta = (\beta_0, \ldots, \beta_p) \in \mathbb{R}^{p+1}$, $\#(S)$ stands for the number of elements in the set S and $\ell(\beta, \sigma^2)$ is the loglikelihood function defined by

$$\ell(\beta, \sigma^2) = -\frac{n}{2} \log(2\pi\sigma^2) - \frac{1}{2\sigma^2} \sum_{i=1}^{n} \epsilon_i^2.$$

Here we assume that all the residual ϵ_i are independent and normally distributed with zero means and variance σ^2. Computing AIC values for all subsets S of the explanatory variables in (1), we can obtain the best AIC-based subset. However since the number of subsets is 2^p, the computation of all subsets is not practical.

3   MINLP Formulation for the Minimization of AIC

We provide a mathematical optimization formulation via mixed integer nonlinear programming (MINLP) in this section. For this we focus on the first term of (2). Let S be a set of candidates of explanatory variables in (1). By substituting $\beta_j = 0$ $(j \in \{1, \ldots, p\} \setminus S)$ to the objective function, the first term can be regarded as the unconstrained minimization. The minimum solution satisfies $\sigma^2 = \frac{1}{n} \sum_{i=1}^{n} \epsilon_i^2$. This is obtained by calculating the partial derivative of the objective function with σ^2. Substituting this equation to (2), we simplify (2) as follows:

$$\text{AIC}(S) = \min_{\beta_j} \left\{ n \log \left(\sum_{i=1}^{n} \epsilon_i^2 \right) : \beta_j = 0 \ (j \in \{1, \ldots, p\} \setminus S) \right\} \tag{3}$$
$$+ 2(\#(S) + 2) + n \left(\log(2\pi/n) + 1 \right),$$

where $\epsilon_i = y_i - \beta_0 - \sum_{j \in S} \beta_j x_{ij}$ for all $i = 1, \ldots, n$.

We formulate the minimization of AIC(S) over $S \subseteq \{1, \ldots, p\}$ by the following MINLP formulation:

$$\min_{\substack{\beta_j, z_j, \\ \epsilon_i, k}} \left\{ n \log \left(\sum_{i=1}^{n} \epsilon_i^2 \right) + 2k : \begin{array}{l} \epsilon_i = y_i - \beta_0 - \sum_{j=1}^{p} \beta_j x_{ij} \ (i = 1, \ldots, n), \\ \sum_{j=1}^{p} z_j = k, \beta_0 \in \mathbb{R}, \\ z_j \in \{0, 1\}, |\beta_j| \le M z_j \ (j = 1, \ldots, p), \end{array} \right\}, \tag{4}$$

where M is a sufficiently large positive number. We remove the constant in (3) in the objective function of (4). If $z_j = 0$, then β_j must be zero, which implies that the jth explanatory variable is not used in the statistical model (1). Note that the minimization of the first term in (3) has an optimal solution with a finite value for any subset $S \subseteq \{1, \ldots, p\}$. This can be proved by using the monotonicity of the logarithm function and the positive semidefiniteness of the quadratic function $\sum_{i=1}^{n} \epsilon_i^2$. Hence we can ensure that (4) has the same optimal solution as the minimization in (3) by taking a sufficiently large positive number

M in (4). In addition, we use M only for the construction of the relaxation problem (6), but does not use for the practical computation.

Next we provide a procedure to find lower bounds of the subproblem of (4) at each node in the branch-and-bound tree. Some variables z_j in (4) are fixed to zero or one at each node of the tree. We define the sets Z_0, Z_1 and Z for a given node as follows:

$$Z_1 = \{j \in \{1,\ldots,p\} : z_j \text{ is fixed to } 1\}, Z_0 = \{j \in \{1,\ldots,p\} : z_j \text{ is fixed to } 0\},$$
$$Z = \{j \in \{1,\ldots,p\} : z_j \text{ is not fixed}\}.$$

We remark that $Z_1 \cup Z_0 \cup Z = \{1,\ldots,p\}$ and that each set is disjoint with one another. Then the subproblem at the node is formulated as follows:

$$\begin{cases} \min_{\beta_j, z_j} \quad n \log \left(\sum_{i=1}^{n} \left(y_i - \beta_0 - \sum_{j=1}^{p} \beta_j x_{ij} \right)^2 \right) + 2 \sum_{j=1}^{p} z_j \\ \text{subject to } z_j = 1 \ (j \in Z_1), z_j = 0 \ (j \in Z_0), z_j \in \{0,1\} \ (j \in Z), \\ \qquad \qquad \beta_0 \in \mathbb{R}, |\beta_j| \leq M z_j \end{cases} \tag{5}$$

We deal with the following problem to obtain a lower bound of the optimal value of (5):

$$\begin{cases} \min_{\beta_j} \quad n \log \left(\sum_{i=1}^{n} \left(y_i - \beta_0 - \sum_{j=1}^{p} \beta_j x_{ij} \right)^2 \right) + 2\#(Z_1) \\ \text{subject to } \beta_0, \beta_j \in \mathbb{R} \ (j \in Z \cup Z_1), \beta_j = 0 \ (j \in Z_0) \end{cases} \tag{6}$$

Recall that $\#(Z_1)$ stands for the number of elements in the set Z_1.

We can obtain the lower bound of the optimal value of of (5) after solving (6). In fact, it is clear that any feasible solution of (5) is also feasible for (6). Furthermore, we have $\sum_{j=1}^{p} z_j = \sum_{j \in Z_1 \cup Z} z_j \geq \sum_{j \in Z} z_j$ for any feasible solution (β, z) of (5). This implies that the optimal value of (6) is less than or equal to (5), and thus we can obtain the lower bound of (5) after solving (6). We deal with (6) as *the relaxation problem of (5)* in this sense.

We can freely remove the constant $2\#(Z_1)$ and the logarithm by the monotonicity of the logarithm function in (6), and obtain the following problem from (6):

$$\min_{\beta_j} \left\{ \sum_{i=1}^{n} \left(y_i - \beta_0 - \sum_{j=1}^{p} \beta_j x_{ij} \right)^2 : \begin{array}{l} \beta_0, \beta_j \in \mathbb{R} \ (j \in Z \cup Z_1) \\ \beta_j = 0 \ (j \in Z_0) \end{array} \right\}. \tag{7}$$

Since (7) is the unconstrained minimization of a quadratic function, we can obtain an optimal solution of (7) by solving a linear system. We denote the optimal value of (7) by ξ^*. The optimal value of (6) is $n \log(\xi^*) + 2\#(Z_1)$, which is used as a lower bound of (5).

We provide a procedure that constructs a feasible solution of (5) and computes an upper bound of θ^*. For this we use a solution obtained after solving (7). Let $\tilde{\beta} \in \mathbb{R}^{p+1}$ be an optimal solution of (7). We define $\tilde{z}_j$ by $\tilde{z}_j = 1$ if $\tilde{\beta}_j \neq 0$, otherwise $\tilde{z}_j = 0$ for all $j = 1, \ldots, p$. It is easy to see that $(\tilde{\beta}_j, \tilde{z}_j)$ is feasible for (5) and the objective value is $n \log(\xi^*) + 2\#(Z \cup Z_1)$. If the objective value is smaller than the current best upper bound, then we update the current best upper bound and the current best solution.

4   Some Techniques to Improve the Numerical Performance

4.1   SCIP

In order to implement our proposed approach, we use SCIP [1,8,11], which is a mathematical optimization software and a branch-and-bound framework. In fact, it has high user plug-in flexibility which helps to solve (4) efficiently. We implement a procedure, which is called *relaxator* or *relaxation handler*, to obtain lower bounds, upper bounds and feasible solutions of (4) at each node and to define a branching rule described in Sect. 4.3.

4.2   Handling the Linear Dependency in Data

We illustrate that we can efficiently compute the optimal value of (4) by using the linear dependency in data. Although linearly independent data is often the assumption in standard statistical textbooks, practical data has often linear dependency, e.g. `servo` and `auto-mpg` in UCI Machine Learning Repository [10]. For given data $(x_{i1}, x_{i2}, \ldots, x_{ip}, y_i) \in \mathbb{R}^p \times \mathbb{R}$ $(i = 1, \ldots, n)$, we denote

$$x^0 = \begin{pmatrix} 1 \\ \vdots \\ 1 \end{pmatrix}, x^j = \begin{pmatrix} x_{1j} \\ \vdots \\ x_{nj} \end{pmatrix} \quad (j = 1, \ldots, p).$$

We say that data has linear dependency if the vectors $x^0, x^1, \ldots, x^p \in \mathbb{R}^n$ are linearly dependent.

The following lemma ensures that we do not need to branch $z_j = 1$ for some $j \in Z$ if the data has the linear dependency. Thus we need to handle only $z_j = 0$ in this case.

Lemma 1. *Assume that in (5), there exists $q \in Z$ such that the vector x^q and vectors $\{x^j : j \in Z_1 \cup \{0\}\}$ are linearly dependent. Then an optimal solution of (5) satisfies $z_q = 0$.*

If a set of data has linear dependency, there exist $q \in Z$ and nonempty set $Z_1 \subseteq \{1, \ldots, p\}$ which satisfy the assumption of Lemma 1. To implement Lemma 1, we collect sets of linearly dependent vectors from a given set of data before solving (4). We then add the constraints $\sum_{j \in S_\ell} z_j \leq \#(S_\ell) - 1$ $(\ell = 1, \ldots, m)$ in (4), where $S_1, \ldots, S_m$ denote the sets of linearly dependent vectors. By this addition, we do not generate the node in which $S_\ell \subseteq Z_1$ holds.

4.3 Most Frequent Branching

We define a branching rule for variables z_j to improve the performance of our implementation. We call it the *most frequent branching*. In our preliminary numerical experiment, we observe that the explanatory variables used in the best statistical model are also used in the statistical model whose AIC value is close to the smallest AIC value. By branching variables z_j in (5) which correspond to such explanatory variables, we can expect that (5) at the node generated by $z_j = 0$ is pruned as early as possible. To find such explanatory variables, we use stored feasible solution obtained in our procedure to compute upper bounds. We describe most frequent branching rule at the current node in Algorithm 1.

Algorithm 1. Most frequent branching rule

Input: a positive integer N, a set Z of unfixed variables in the node and all feasible solutions of (4) found from the root node through the current node

Output: $\tilde{j} \in Z$

1. Choose N feasible solutions $(\beta^1, z^1), \ldots, (\beta^N, z^N)$ out of all stored feasible solutions. Here (β^i, z^i) is a feasible solution of (4) whose objective value is the ith smallest in all the stored solutions.
2. Compute score value s_j for all $j \in Z$ defined by $s_j = \#(T_j)$, where $T_j = \{\ell \in \{1, \ldots, N\} : z_j^\ell = 1\}$.
3. Return $J \in Z$ with $s_J = \max\limits_{j \in Z}\{s_j\}$

We observe in our preliminary numerical experiment that the obtained lower bound at the child node generated by $z_J = 0$ tends to be relatively bigger and that the pruning process tends to work earlier in our branching rule than in branching rules implemented in SCIP. As a result, the number of nodes visited in the branch-and-bound tree often decreases.

5 Numerical Experiment

We apply our proposed method to benchmark data sets in [10]. We apply our implementation to standardized data sets, *i.e.* the data is transformed to have the zero mean and unit variance. Note that the standardized data has also linear dependency even if we apply the standardization to the original data which has linear dependency. The specification of the computer is CPU: 3.5 GHz Intel Core i7, Memory: 16GB and OS: OS X 10.9.5.

We compare our proposed method with the MISOCP approach proposed in [7] via CPLEX [6]. This approach is also obtained from (4). Although the objective function of (4) is non-convex, the difficulty due to the non-convexity is overcome by using the identity $\exp(\log(x)) = x$ and the monotonicity of the exponential function $\exp(x)$. See [7, Section 3.2] for the detail. The resulting problem is formulated as MISOCP and is tractable by CPLEX.

Table 1. Summary of numerical results by MINLP, MISOCP and SW

Name	n	p	Methods	AIC	k	time(sec)	gap(%)
housing	506	13	MINLP	**776.21**	11	0.04	0.0
			MISOCP	**776.21**	11	7.96	0.0
			SW	**776.21**	11	0.10	–
•servo	167	19	MINLP	**258.35**	9	0.78	0.0
			MISOCP	**258.35**	9	7.99	0.0
			SW	260.16	10	0.18	–
•auto-mpg	392	25	MINLP	**332.88**	15	1.64	0.0
			MISOCP	**332.88**	15	303.83	0.0
			SW	337.96	18	0.32	–
•solarflareC	1066	26	MINLP	**2816.29**	9	9.81	0.0
			MISOCP	**2816.29**	9	304.51	0.0
			SW	2821.61	12	1.08	–
•solarflareM	1066	26	MINLP	**2926.90**	7	7.76	0.0
			MISOCP	**2926.90**	7	255.02	0.0
			SW	2930.91	9	1.16	–
•solarflareX	1066	26	MINLP	**2882.80**	3	1.27	0.0
			MISOCP	**2882.80**	3	19.39	0.0
			SW	2891.56	9	1.20	–
breastcancer	194	32	MINLP	**508.40**	10	450.05	0.0
			MISOCP	508.62	10	>5000	3.72
			SW	509.96	14	0.60	–
•forestfires	517	63	MINLP	**1429.64**	12	>5000	0.71
			MISOCP	1431.32	12	>5000	6.44
			SW	1447.36	21	7.43	–
•automobile	159	65	MINLP	**-60.34**	34	>5000	16.45
			MISOCP	-55.83	34	>5000	27.22
			SW	-47.61	40	2.64	–
crime	1993	100	MINLP	**3410.25**	50	>5000	0.53
			MISOCP	3469.34	74	>5000	8.51
			SW	**3410.25**	50	105.40	–

In addition, we also compare our proposed method with the stepwise method with backward elimination. This method starts with all explanatory variables and removes one explanatory variable at a time until the AIC value does not decrease. More precisely, for the current set S of explanatory variables, we choose an explanatory variable so that AIC($S \setminus \{j\}$) is minimum over all $j \in S$.

Table 1 shows the summary of numerical comparisons. The mark • in the first column indicates that the data has linear dependency. The second, third, and sixth columns indicate the numbers of data, the explanatory variables in the statistical model (1), and the ones in the models found by using each method. The fifth column indicates the obtained AIC values by each method. The values with the bold font are the best among three values. The seventh column indicates the CPU time in seconds to compute the optimal value. ">5000" means that the method cannot find the optimal value within 5000 s. The last column indicates the gap in the percent as follows:

$$\text{gap} = \frac{\text{upper bound} - \text{lower bound}}{\max\{1, |\text{upper bound}|\}} \times 100.$$

It should be noted that if the gap is sufficiently close to zero, then the obtained value is optimal. MINLP, MISOCP and SW indicate the results obtained by our proposed method, MISOCP approach and the stepwise method, respectively.

We observe from Table 1 that (i) MINLP computes the optimal value much faster than MISOCP. MINLP finds smaller AIC values than MISOCP even when MINLP cannot find them within 5000 s, and that (ii) the values found by using MINLP is smaller than or equal to SW in some data sets. In fact, since SW is a local search algorithm, it does not always find the global optimal values. In contrast, MINLP finds the global optimal value. In particular, MINLP can obtain the best statistical models (1) for $p \leq 32$.

6   Future Work

Future work involves to applying our implementation to data sets with more large p and/or n. A possible choice to accomplish this involves the use of parallel computation via ParaSCIP and FiberSCIP [9].

Another future work is to compare our proposed method with a mixed integer quadratic programming (MIQP) formulation for linear regression with a cardinality constraint proposed by Bertsimas and Shioda [4] and Bertsimas et al. [3]. Their formulation is available to our problems by fixing the number of explanatory variables from 0 to p.

Acknowledgements. The second author was supported by JSPS KAKENHI Grant Numbers 26400203. We would like to thank the anonymous referees for providing significant comments on the presentation of the manuscript.

References

1. Achterberg, T.: SCIP solving constraint integer programs. Math. Prog. Comp. **1**(1), 1–41 (2009)
2. Akaike, H.: A new look at the statistical model identification. IEEE Trans. Autom. Control **19**(6), 716–723 (1974)
3. Bertsimas, D., King, A., Mazumder, R.: Best subset selection via a modern optimization lens. Ann. Stat. **44**(2), 813–852 (2016)

4. Bertsimas, D., Shioda, R.: Algorithm for cardinality-constrained quadratic optimization. Comput. Optim. Appl. **43**, 1–22 (2009)
5. Guyon, I., Elisseeff, A.: An introduction to variable and feature selection. J. Mach. Learn. Res. **3**, 1157–1182 (2003)
6. IBM ILOG CPLEX Optimizer 12.6.2, IBM ILOG (2015)
7. Miyashiro, R., Takano, Y.: Mixed integer second-order cone programming formulations for variable selection. Eur. J. Oper. Res. **247**, 721–731 (2015)
8. SCIP: Solving Constraint Integer Programs. http://scip.zib.de/
9. Shinano, Y., Achterberg, T., Berthold, T., Heinz, S., Koch, T.: ParaSCIP: A parallel extension of SCIP. In: Bischof, C., Hegering, H.-G., Nagel, W.E., Wittum, G. (eds.) Competence in High Performance Computing 2010, pp. 135–148. Springer, Heidelberg (2012)
10. UCI Machine Learning Repository. http://archive.ics.uci.edu/ml/
11. Vigerske, S., Gleixner, A.: SCIP: Global Optimization of Mixed-Integer Nonlinear Programs in a Branch-and-Cut Framework. ZIB-Report 16–24, Zuse Institute Berlin, May 2016

PySCIPOpt: Mathematical Programming in Python with the SCIP Optimization Suite

Stephen Maher[1], Matthias Miltenberger[1]([✉]), João Pedro Pedroso[2],
Daniel Rehfeldt[1], Robert Schwarz[1], and Felipe Serrano[1]

[1] Zuse Institute Berlin, Berlin, Germany
{maher,miltenberger,rehfeldt,schwarz,serrano}@zib.de
[2] Faculdade de Ciências da Universidade do Porto, Porto, Portugal
jpp@fc.up.pt
http://www.zib.de
http://www.dcc.fc.up.pt/~jpp/

Abstract. SCIP is a solver for a wide variety of mathematical optimization problems. It is written in C and extendable due to its plug-in based design. However, dealing with all C specifics when extending SCIP can be detrimental to development and testing of new ideas. This paper attempts to provide a remedy by introducing PySCIPOpt, a Python interface to SCIP that enables users to write new SCIP code entirely in Python. We demonstrate how to intuitively model mixed-integer linear and quadratic optimization problems and moreover provide examples on how new Python plug-ins can be added to SCIP.

Keywords: SCIP · Mathematical optimization · Python · Modeling

1 Introduction

Since its initial release in 2005, SCIP has matured into a powerful solver for various classes of optimization problems and has achieved considerable acclaim in academia and industry. It is distributed as part of the SCIP Optimization Suite [13], along with the LP solver SoPlex [6,15], the modeling language ZIMPL [17], the generic column generation solver GCG [11], and the parallelization framework UG [16]. For an in-depth description of SCIP and the Optimization Suite in general we refer to the original publication [1] and the report about the latest release 3.2 [3]. SCIP is available in source code and provides tutorials and comprehensive documentation for researchers and practitioners on its web page [13], thereby allowing users to extend its functionality and write custom plug-ins. As yet, however, such extensions required not only knowledge of the C programming language, but furthermore impeded fast prototyping, an obstacle keeping some (potential) users from implementing their ideas within a reasonable amount of time. To overcome these impediments, we have developed PySCIPOpt, a Python interface to SCIP that allows for fast prototyping of new algorithmic concepts and concurrently benefits from the underlying high

© Springer International Publishing Switzerland 2016
G.-M. Greuel et al. (Eds.): ICMS 2016, LNCS 9725, pp. 301–307, 2016.
DOI: 10.1007/978-3-319-42432-3_37

performance C code. The interface is implemented by means of the programming language Cython [8] and documented with the Python standard Docstring [10].

Finally, it is certainly worth mentioning that with JuMP [4] there is already an optimization software package available for the Julia language that provides both fast implementation possibilities and high performance. However, this software does not yet fully support SCIP.[1]

2   Modeling

The Python interface supports an intuitive modeling syntax using linear and quadratic expressions. The following example shows how to build up and subsequently solve a small mixed-integer quadratic programming problem:

<table>
<tr><td>Mathematical formulation:</td><td>Python code:</td></tr>
</table>

minimize $x + 3y$

subject to $2x - y^2 \geq 10$

$$x, y \quad \geq 0$$

$$x \in \mathbb{R}$$

$$y \in \mathbb{Z}$$

```python
from pyscipopt import Model
scip = Model()
x = scip.addVar('x', vtype='C')
y = scip.addVar('y', vtype='I')
scip.setObjective(x + 3y)
scip.addCons(2x - y*y >= 10)
scip.optimize()
```

Most methods can be called with a small number of parameters, leaving it to the interface to fill in the remaining parameters with default values. If necessary, these parameters can be set in any order to provide a more flexible interface. This feature is exemplified by the **addVar()** method, which is used in the example above with only two specified parameters, while in fact six may be provided:

```python
addVar(self,              # the SCIP model
       name = '',         # name of the variable
       vtype = 'C',       # variable type ('C', 'I', or 'B')
       lb = 0.0,          # lower bound
       ub = None,         # upper bound
       obj = 0.0,         # objective coefficient
       pricedVar = False  # is it a pricing candidate?
       )
```

In this way, a minimalistic and intuitive code can be designed without surrendering any functionality of the wrapped SCIP method.

3   Extending SCIP: Writing Plug-Ins in Python

Every plug-in supported by PYSCIPOPT is encapsulated in a separate file that declares its interface methods to SCIP. The file defines the Python base class

---

[1] SCIP can already be used to solve models formulated in JuMP via AMPL's nl format [7]. Furthermore, there is an ongoing development effort to develop an interface that supports callbacks [14].

for the respective plug-in, including its callbacks. These callback definitions can be left empty for optional callbacks, but need to be implemented by the user in case of fundamental ones. In line with the overall approach of PYSCIPOPT, the user can thereby implement customized plug-ins with minimal effort, but access to more intricate functionalities of SCIP is still provided.

3.1   Constraint Handler Example: TSP

In this section we show how to design a simple constraint handler that can solve the traveling salesman problem (TSP).

The following model for solving the TSP has been proposed by Dantzig–Fulkerson–Johnson (DFJ) [2]. Let $G = (V, E)$ be a complete, undirected graph with $V = \{1, \cdots, n\}$ being the vertex set and E the edge set. Furthermore, let c_{ij} be the weight of the edge (i, j). We associate with each edge (i, j) a variable x_{ij}, with $x_{ij} = 1$ if edge (i, j) is used in the solution and $x_{ij} = 0$ otherwise. Thereupon, the DFJ integer programming formulation can be stated as follows:

$$\text{minimize} \sum_{i,j} c_{ij} x_{ij} \tag{1}$$

$$\text{subject to} \sum_{j} x_{ij} = 2 \qquad \forall i \in \{1, \cdots, n\} \tag{2}$$

$$\sum_{i,j \in S} x_{ij} \leq |S| - 1 \quad \forall S \subsetneq \{1, \cdots, n\}, \ |S| \geq 2 \tag{3}$$

$$x_{ij} \in \{0, 1\} \qquad \forall i < j \tag{4}$$

The constraints (3) exclude subtours by imposing that for any proper subset S of the vertex set V such that $|S| \geq 2$ a solution cannot encompass a cycle within S. However, as there is an exponential number of subsets of V, it is impractical to specify all of these constraints. A possible approach is to iteratively solve the problem, starting without these constraints and after each solving round add constraints (3) violated by the current solution.

The constraint handler in our PYSCIPOPT TSP implementation does not generate its own constraints, but instead SCIP is querying whether the current solution is feasible and in case it is not, how feasibility can be achieved. This is accomplished by setting the **needscons** flag to **False** when including the constraint handler into SCIP.

Our example[2] uses the external Python library `networkx` to compute the connected components of a graph. The constraint handler then adds the corresponding subtour elimination constraints as described above when called in the `consenfolp()` callback. An integer feasible solution that satisfies the first set of constraints will be checked for subtours in the `conscheck()` callback of the `TSPconshdlr` class. The `conslock()` callback sets up locks on the variables of the constraint and is required for a correct implementation. In a nutshell, it tells SCIP how the variables can be rounded without violating the constraint.

For more information about the different callbacks we refer to the SCIP documentation [13].

When adding constraints via linear or quadratic expressions we recommend to use the `quicksum()` function. This eliminates the overhead of intermediate creation/destruction of multiple expressions and instead sets up one instance that will be iteratively extended. A similar implementation is also provided in the Python interface of the commercial optimization software Gurobi [12].

4   Conclusion and Outlook

With PYSCIPOPT we provide a SCIP based optimization tool that allows for fast, minimalistic and intuitive programing, while still having the more intricate functionalities of SCIP up its sleeve. We hope that the availability of such a device will help mathematical programming experts set up prototypes more efficiently and moreover allow less experienced users to more easily make use of the wide range of capabilities provided by SCIP. In this way, PYSCIPOPT may also serve as coherent tool to make (undergraduate) mathematical optimization students familiar with the subject, as has already successfully been the case for instance at the University of Porto [9].

Future developments naturally encompass the extension of PYSCIPOPT to cover even more functionalities provided by SCIP. Missing functions can be easily added by specifying their header declaration in the `scip.pxd` file and defining a corresponding wrapper function in the class `Model` in the `scip.pyx` file. Of special interest in this context are additional general nonlinear programming methods [5]: Our objective is to make them blend in with the already existing features, while maintaining the underlying minimalistic design of PYSCIPOPT.

---

[2]   A proper implementation of this constraint handler would implement the callback `conssepalp` to separate the LP solution. The callback `consenfolp` is called at the end of the node processing loop, where possibly several calls to the `conssepalp` callback of the different constraint handlers have already been made.

For reasons of space we provide a concise, although rather inefficient, implementation.

tsp_example.py

```python
import networkx
from pyscipopt import Model, Conshdlr, quicksum, SCIP_RESULT


EPS = 1.e-6


class TSPconshdlr(Conshdlr):

  def __init__(self, variables):
    self.variables = variables

  def find_subtours(self, solution=None):
    edges = []
    x = self.variables
    for (i,j) in x:
      if self.model.getSolVal(solution, x[i,j]) > EPS:
        edges.append((i,j))
    G = networkx.Graph()
    G.add_edges_from(edges)
    components = list(networkx.connected_components(G))
    return [] if len(components) == 1 else components

  def conscheck(self, constraints, solution, check_integrality,
                check_lp_rows, print_reason):
    if self.find_subtours(solution):
      return {"result": SCIP_RESULT.INFEASIBLE}
    else:
      return {"result": SCIP_RESULT.FEASIBLE}

  def consenfolp(self, constraints, n_useful_conss, sol_infeasible):
    subtours = self.find_subtours()
    if subtours:
      x = self.variables
      for subset in subtours:
        self.model.addCons(quicksum(x[i,j] for(i,j) in pairs(subset))
                    <= len(subset) - 1)
        print("cut: len(%s) <= %s" % (subset, len(subset) - 1))
      return {"result": SCIP_RESULT.CONSADDED}
    else:
      return {"result": SCIP_RESULT.FEASIBLE}

  def conslock(self, constraint, nlockspos, nlocksneg):
    x = self.variables
    for (i,j) in x:
      self.model.addVarLocks(x[i,j], nlocksneg, nlockspos)
```

```
def create_tsp(vertices, distance):
  model = Model("TSP")
  x = {}
  for (i,j) in pairs(vertices):
    x[i,j] = model.addVar(vtype = "B",name = "x(%s,%s)" % (i,j))
  for i in vertices:
    model.addCons(
      quicksum(x[j,i] for j in vertices if j < i) +
      quicksum(x[i,j] for j in vertices if j > i) == 2)
  conshdlr = TSPconshdlr(x) # set up conshdlr with all variables
  model.includeConshdlr(conshdlr, "TSP", "TSP subtour eliminator",
                        needscons=False)
  model.setObjective(quicksum(distance[i,j] * x[i,j]
                      for (i,j) in pairs(vertices)), "minimize")
  return model, x

def solve_tsp(vertices, distance):
  model, x = create_tsp(vertices, distance)
  model.optimize()
  edges = []
  for (i,j) in x:
    if model.getVal(x[i,j]) > EPS:
      edges.append((i,j))
  return model.getObjVal(), edges

def pairs(vertices):
  for i in vertices:
    for j in vertices:
      if i < j:
        yield (i,j)

def test_main():
  vertices = [1, 2, 3, 4, 5, 6]
  distance = {(u,v):1 for (u,v) in pairs(vertices)}
  for u in vertices[:3]:
    for v in vertices[3:]:
      distance[u,v] = 10
  objective_value, edges = solve_tsp(vertices, distance)
  print("Optimal tour:", edges)
  print("Optimal cost:", objective_value)

if __name__ == "__main__":
  test_main()
```

References

1. Achterberg, T.: Constraint integer programming. Ph.D. thesis, Technische Universität Berlin (2007)
2. Dantzig, G.B., Fulkerson, D.R., Johnson, S.M.: Solution of a large-scale traveling-salesman problem. Oper. Res. **3**, 393–410 (1954)
3. Gamrath, G., Fischer, T., Gally, T., Gleixner, A.M., Hendel, G., Koch, T., Maher, S.J., Miltenberger, M., Müller, B., Pfetsch, M.E., Puchert, C., Rehfeldt, D., Schenker, S., Schwarz, R., Serrano, F., Shinano, Y., Vigerske, S., Weninger, D., Winkler, M., Witt, J.T., Witzig, J.: The SCIP optimization suite 3.2. Technical report 15–60, ZIB, Takustr. 7, 14195 Berlin (2016)
4. Lubin, M., Dunning, I.: Computing in operations research using Julia. INFORMS J. Comput. **27**(2), 238–248 (2015)
5. Vigerske, S., Gleixner, A., SCIP: global optimization of mixed-integer nonlinear programs in a branch-and-cut framework. Technical report 16–24, ZIB, Takustr. 7, 14195 Berlin (2016)
6. Wunderling, R.: Paralleler und objektorientierter Simplex-Algorithmus. Ph.D. thesis, Technische Universität Berlin (1996)
7. AmplNLWriter.jl. https://github.com/JuliaOpt/AmplNLWriter.jl
8. Cython. http://www.cython.org/
9. Decision support methods, Universidade do Porto. http://www.dcc.fc.up.pt/~jpp/mad/
10. PEP 0257 - Docstring Conventions. https://www.python.org/dev/peps/pep-0257/
11. GCG: Generic Column Generation. http://www.or.rwth-aachen.de/gcg/
12. Gurobi optimizer reference manual. http://www.gurobi.com/
13. SCIP: Solving Constraint Integer Programs. http://scip.zib.de/
14. SCIP.jl. https://github.com/ryanjoneil/SCIP.jl
15. SoPlex: Sequential object-oriented simPlex. http://soplex.zib.de/
16. UG: Ubiquity Generator framework. http://ug.zib.de/
17. ZIMPL: Zuse Institute Mathematical Programming Language. http://zimpl.zib.de/

A First Implementation of ParaXpress: Combining Internal and External Parallelization to Solve MIPs on Supercomputers

Yuji Shinano[1], Timo Berthold[2(✉)], and Stefan Heinz[2]

[1] Zuse Institute Berlin, Berlin, Germany
shinano@zib.de
[2] Fair Isaac Germany GmbH, Berlin, Germany
{timoberthold,stefanheinz}@fico.com

Abstract. The Ubiquity Generator (UG) is a general framework for the external parallelization of mixed integer programming (MIP) solvers. It has been used to develop ParaSCIP, a distributed memory, massively parallel version of the open source solver SCIP, running on up to 80,000 cores. In this paper, we present a first implementation of ParaXpress, a distributed memory parallelization of the powerful commercial MIP solver FICO Xpress. Besides sheer performance, an important difference between SCIP and Xpress is that Xpress provides an internal parallelization for shared memory systems. When aiming for a best possible performance of ParaXpress on a supercomputer, the question arises how to balance the internal Xpress parallelization and the external parallelization by UG against each other. We provide computational experiments to address this question and we show preliminary computational results for running a first version of ParaXpress on 6,144 cores in parallel.

Keywords: Mixed integer programming · Distributed memory parallelization

1 Introduction

This paper deals with solving mixed integer programming (MIP) problems in parallel. Many optimization problems arising in practice can be modeled as MIP, see, e.g., [7]. A MIP is given in the following general form:

$$\min\{c^\top x : Ax \le b, l \le x \le u, x_j \in \mathbb{Z}, \text{ for all } j \in I\}, \tag{1}$$

with matrix $A \in \mathbb{R}^{m \times n}$, vectors $b \in \mathbb{R}^m$ and $c, l, u \in \mathbb{R}^n$, and a subset $I \subseteq \{1, \ldots, n\}$. The standard algorithm used to solve MIP is an LP-based branch-and-bound, which implicitly enumerates the whole solution space to find an optimal solution x^* that gives the minimum value of (1).

Over the past twenty years, substantial progress has been made in the solvability of large-scale mixed integer programs [2]. As a result, state-of-the-art

© Springer International Publishing Switzerland 2016
G.-M. Greuel et al. (Eds.): ICMS 2016, LNCS 9725, pp. 308–316, 2016.
DOI: 10.1007/978-3-319-42432-3_38

MIP solvers are nowadays capable of solving a variety of different types of MIP instances which arise from real-world applications within reasonable time and mixed integer programming has become a standard technique in operations and in logistics industry. At the same time, we witnessed big progress in the development of scalable, massively parallel branch-and-bound system, such as, PEBBL [4], ParSSSE [13] and ALPS [14]. Surprisingly, only a handful of publications [3,9,11] address the combination of both lines of research, i.e., an effective parallelization of state-of-the-art MIP solvers, capable of running on modern supercomputers. Two of these publications are based on using Cplex as underlying MIP solver [3,11].

ParaSCIP is an external parallelization of the open source solver SCIP and has been developed using the *Ubiquity Generator (UG) framework* [12]. UG is a general software framework to parallelize MIP solvers externally; it is the framework that we used to implement the FiberXpress and ParaXpress solvers which we present in this paper. FiberXpress is the shared memory pendant to ParaXpress, see Sect. 2.

The FICO Xpress Optimization Suite features software tools used to model and solve linear, integer, quadratic, nonlinear, and robust optimization problems [5,6]. The core solver of this suite is the FICO Xpress-Optimizer, which comes with a callable library that is accessible from all the major programming platforms. It combines flexible data access functionality and optimization algorithms, using state-of-the-art methods, which enable the user to handle the increasingly complex problems arising in industry and academia.

The Xpress C API provides all necessary functionality for an external parallelization by UG. In this study, we examine the feasibility of an external parallelization of Xpress by making a preliminary implementation of it by using UG. The key point addressed in this paper is how to combine UG's external parallelization with Xpress's internal parallelization. We would like to answer two questions: (i) what is the best combination of those two parallelization approaches and (ii) what is the "price of externalization", i.e., what potential is lost by having a *pure* external parallelization like, e.g., ParaSCIP? We show computational experiments for a comparison between internal and external parallelization of Xpress. We further show preliminary computational results for a massively parallel ParaXpress using up to 6,144 cores on a supercomputer.

2   Features of FiberXpress and ParaXpress

In this section, we introduce a preliminary implementation of FiberXpress and ParaXpress by using UG. Figure 1 shows the design structure of UG. UG is written in C++. It consists of a set of base classes to instantiate parallel branch-and-bound based solvers. Both, the MIP solver and the parallelization library used for communications are abstract classes. FiberXpress is the instantiated parallel solver where Xpress is used as MIP solver and Pthreads is used as the parallelization library, and ParaXpress is the instantiated parallel solver where Xpress is used as MIP solver and MPI is used as the parallelization library. FiberXpress addresses

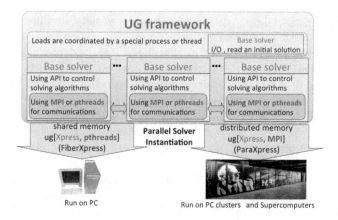

Fig. 1. Design structure of ubiquity generator framework.

shared memory systems, ParaXpress addresses distributed memory systems. There is a single LOADCOORDINATOR (abbreviated to LC throughout this paper), which makes all decisions concerning the dynamic load balancing and distributes subproblems of MIP instances to be solved (so-called sub-MIPs). Further, there is a set of parallel SOLVERs that solve the distributed sub-MIPs. Both, the LC and an abstract implementation of a SOLVER, are part of the UG framework.

The following three features have been implemented in UG and can therefore be directly used for the external parallelization of Xpress.

Ramp-up The phase that lasts until all solvers have become busy.

Dynamic Load Balancing A work load coordination mechanism to keep all available Xpress SOLVERs busy.

Checkpointing and restarting A mechanism to restart a solving process of an interrupted job from the last previous checkpoint.

We will briefly explain these features and special adaptations for ParaXpress in the following.

In the initialization, the LC and all the SOLVERs read the instance data of the MIP to solve. We refer to the resulting instance as the *original instance*, that is (P). The original instance is embedded into the (local) Xpress environment of each SOLVER. Later, only differences between a sub-MIP and the original instance will be communicated.

An important technical difference between ParaSCIP and ParaXpress is the following. In ParaSCIP, the original instance is presolved once within the LC and the presolved instance is transferred to the individual SOLVERs. In ParaXpress, the original instance is transferred to each SOLVER and presolved locally. The reason for this design decision is that is currently not possible to extract the presolved model from Xpress. Further, in the current preliminary implementation of Para-Xpress, possible pruning of the transferred problem is checked for the first time after complete root node processing, including solving the root LP and cutting.

This is a potential performance bottleneck, in particular during ramp-down, and will be improved in future implementations.

After the initialization step, the LC creates the root node of the branch-and-bound tree. Each node transferred through the system—called a PARANODE—acts as the root of a subtree. The information sent to a PARANODE only consists of variable bound changes. The SOLVER that receives a new branch-and-bound node instantiates the corresponding sub-MIP using the original instance, which was distributed in the initialization step, and the received bound changes. Therefore, PARANODE is considered as a representation of a sub-MIP in ParaXpress.

UG provides two ramp-up mechanisms [12]. However, in this preliminary implementation, we only realized the *normal ramp-up*. Normal ramp-up works as follows. SOLVERs that are already solving a sub-MIP transfer every second child node back to the LC. The LC maintains a *node pool* from which it assigns nodes to idle SOLVERs. If no idle SOLVER exists, the LC keeps collecting nodes from SOLVERs until it has p "heavy" (promising to have a large subtree underneath) unassigned nodes in its node pool. As soon as the LC's node pool has accumulated p "heavy" nodes, it sends a message to all SOLVERs to stop sending nodes.

After the ramp-up, the dynamic load balancing mechanism works as follows. Periodically, each SOLVER notifies the LC about the number of unexplored nodes in its Xpress environment and the lower bound of its subtree; we call this information the *solver status*. If a SOLVER becomes idle, the LC sends one of the nodes from the pool to the idle SOLVER. In order to keep all SOLVERs busy, the LC aims to always have a sufficient number of unprocessed nodes left in its node pool. Further, the LC aims to keep at least p "heavy" nodes in the node pool by employing a *collecting mode*. We call a node *heavy*, if the lower bound value of its subtree (NODEBOUND) is sufficiently close to the lower bound value of the complete search tree (GLOBALBOUND). This is evaluated by the comparison

$$\frac{\text{NODEBOUND} - \text{GLOBALBOUND}}{\max\{|\text{GLOBALBOUND}|, 1.0\}} < \text{THRESHOLD}. \tag{2}$$

When a SOLVER receives the message to switch into collecting mode, it changes the search strategy to "best bound order" (see [1]). Similar to normal ramp-up, the SOLVER alternates between solving nodes and transferring them to the LC.

SOLVERs switch to collecting mode in ascending order of the minimum lower bound of their open nodes. The collecting mode is stopped as soon as the number of heavy nodes in the pool is larger than $1.5 \cdot p$.

The termination phase starts when the node pool is empty and all SOLVERs are idle. In this phase, the LC collects statistical information from all SOLVERs and outputs the optimal solution and statistics.

The checkpointing and restarting mechanism is identical to that of ParaSCIP [8]. ParaXpress saves only *primitive* nodes; these are nodes that have no ancestor nodes in the LC. The restart involves ParaXpress reading the nodes saved in the checkpoint file and restoring them into the node pool of the LC. The LC subsequently distributes these nodes to the SOLVERs in an order determined by their lower bounds. The discussion of this mechanism can be found in [8,10].

3 Computational Experiments

Our computational experiments consist of two parts. One is running FiberXpress on a shared memory system in order to analyze the interplay of the Xpress internal parallelization with UG's external parallelization. The other is running ParaXpress on a supercomputer as a proof-of-concept.

For both experiments, we used Xpress release 7.9 and UG 0.8.2. The experiments in Sect. 3.1 were conducted on a cluster of 20 core Intel Xeon CPU E5-2670 v2 CPUs at 2.50 GHz with 24 MB cache and 128 GB main memory, running an Ubuntu 14.04 with a gcc 4.8.4 compiler. The experiments in Sect. 3.2 were conducted on the HLRN III supercomputer that is a Cray XC30 with 24 core Intel Xeon E5-2695 v2 CPUs at 2.400 GHz, running SLSE v.11 with an Intel Parallel Studio Composer Edition 15.0.1.133.

Table 1. Comparison of internal and external parallelization

Instance	Xpress 16		UG 2 Xpress 8		UG 4 Xpress 4		UG 8 Xpress 2		UG 16 Xpress 1	
	Nodes	Time	Nodes	Time	Nodes	Time	Nodes	Time	Nodes	Time
csched007	571993	553	674247	537	* 412393	* 409	686640	913	435195	809
gmu-35-40	* 304215	76	349087	* 58	522770	112	483199	110	1147446	387
k16x240	903579	112	894575	* 102	767408	143	731922	158	* 669815	267
momentum1	966771	* 2217	968874	2584	1035816	4051	818913	5046	* 266026	2280
neos15	1362829	* 473	2006103	553	* 1011992	508	1426355	586	1617498	962
neos-1616732	1597293	555	1275256	* 499	1616314	910	880811	838	* 694200	676
noswot	36563	14	* 24623	* 5	43007	11	30818	8	47413	18
ns1766074	526817	42	523108	* 32	521909	87	* 510014	189	513897	291
ns894788	25124	83	16356	79	5596	* 32	28150	139	* 4378	33
pg	* 10465	* 10	11679	* 10	14431	13	19346	19	48407	43
pigeon-10	2090347	* 78	* 1810887	159	2053671	305	2081876	567	2609692	969
pigeon-11	26725387	1320	* 14703728	* 1251	16767144	2464	19286131	3897	20079415	3610
ran14x18-disj-8	* 739241	235	854131	* 193	1884385	429	1606310	712	3983341	1312
ran14x18	* 1069177	* 214	1416210	250	1141378	238	1450989	326	1229049	432
reblock166	* 60565	* 60	96409	96	313183	431	118985	293	278644	896
timtab1	* 106737	42	158661	35	148651	* 33	296132	66	337464	165
umts	188753	* 79	201974	87	185982	94	287795	220	* 120737	127
shifted geom mean	385513	145.1	393505	145.4	415971	206.3	462828	299.0	455117	381.4

3.1 Comparison Between FiberXpress and Xpress Internal Parallelization

In contrast to FiberSCIP and ParaSCIP, each SOLVER of FiberXpress and ParaXpress can run in multi-threaded mode. Since on modern supercomputers, each single computing node itself is a multi-core shared memory system, one could either run a single, multi-threaded Xpress instance per computing node, or as many UG SOLVERs as cores, each with a single-threaded Xpress— or any combination in between. In this section, we analyze which relationship between UG SOLVER threads and Xpress threads is most beneficial.

Therefore, we chose a test set that it is suitable for studying MIP solver parallelization: the Tree test set of the MIPLIB 2010. We removed all instances that Xpress 7.9 can solve at the root node and all for which at least one of the tested solvers hit a memory limit. This left us with 17 MIP instances on which we conducted the following experiment.

We compare a run of standalone Xpress with 16 threads against four different FiberXpress versions: two UG SOLVER threads, each running Xpress with eight threads (UG 2 Xpress 8), four UG SOLVER threads, each running Xpress with four threads (UG 4 Xpress 4), eight UG SOLVER threads, each running Xpress with two threads (UG 8 Xpress 2), and 16 UG SOLVER threads, each running Xpress with one thread (UG 16 Xpress 1). Note that for this experiment, we ran both, FiberXpress and Xpress, in opportunistic mode. A summary of the results is given in Table 1. The Columns Nodes and Time depict the shifted geometric means of the number of branch-and-bound nodes and the running time for each of the five solvers. The shifted geometric mean of values $t_1, \ldots, t_n$ with shift s is defined as $\sqrt[n]{\prod(t_i + s)} - s$. We use a shift of $s = 10$ for time and $s = 100$ for nodes. For each instance, we mark the minimal time and the minimal number of branch-and-bound nodes by an asterisk "*".

From Table 1, we see a clear tendency that an internal parallelization like the one of Xpress outperforms an external parallelization like the one by UG. This is not surprising, since the internal parallelization directly benefits from sharing different kind of statistics and the possibility to use parallelization for different aspects than the tree search, e.g., LP solving. Notably, the (UG 2 Xpress 8) setting almost showed the same performance as standalone Xpress which emphasizes that with a growing number of CPU cores per computing, it might indeed make sense to split the resources between UG and Xpress.

Looking at individual instances, the minimal running time was almost always achieved by standalone Xpress or using two UG threads with an eight-threaded Xpress each. However, when considering the number of branch-and-bound nodes, we see that the minimal value is most often achieved by either standalone Xpress or the "other extreme", 16 UG threads with single-threaded Xpress. A major difference between these two solvers is that for (UG 16 Xpress 1), much more nodes will undergo a complete presolving and "root-node-like" cutting plane separation, which for some instances can help to reduce the size of the branch-and-bound tree significantly. As mentioned in Sect. 2, this is also a potential bottleneck w.r.t. running time in our current implementation. This effect might come even more into play when running on a supercomputer with thousands of cores. Notably, all instances were solved within the time limit of two hours by all solvers.

Concerning the mean running time, the mean values in Table 1 sketch a clear picture: (UG 4 Xpress 4) is about 30 % slower than (UG 2 Xpress 8); (UG 8 Xpress 2) is about 30 % slower than (UG 4 Xpress 4), and (UG 16 Xpress 1) is about 20 % slower than (UG 8 Xpress 2). Altogether, this gives a factor of 2.6 between 16-threaded UG and 16-threaded Xpress. One could call this the *price of externalization*: for this particular test set and solver combination, using an external parallelization is about 2.6 times slower than using an internal parallelization. Note that (UG 16 Xpress 1) suffers the most from missing intermediate pruning checks at the root node, see Sect. 2. Therefore, a factor of 2.6 is probably an over-estimation of the value that a more mature implementation of FiberXpress will give.

Finally, the mean number of branch-and-bound nodes required to prove optimality reaches its minimum value for standalone Xpress. Not surprisingly, nodes get pruned faster when the solution process does not have to go through a second level of synchronization. There are a couple of exceptions, as mentioned above, where the reductions by local presolving and cutting plane separation outweigh this effect. All of the instances and solvers produced rather large branch-and-bound nodes, there were only two cases where the number of nodes was smaller than 10,000. The largest tree in this experiment consisted of more than 26 million nodes.

3.2   First Impression of Large Scale ParaXpress Computations

We conducted computational experiments with ParaXpress on the HLRN III supercomputer, using 256 computing nodes with 24 cores each, giving a total of 6144 cores.

With a version of ParaXpress that used 12 UG processes per node, each running Xpress with two threads, could solve the notoriously hard MIPLIB 2010 instance timtab2 within two hours, processing 1,538,064,260 branch-and-bound nodes. It is to be said that for this particular instance, running more UG processes and few Xpress threads performed better than running few UG threads with many Xpress threads. For more details, see Table 2. From Table 2, we can observe that more UG threads further helps to find the optimal solution fast.

During ramp-own, the vast majority of sub-MIPs that are processed by the Solvers can often be solved within a single branch-and-bound node, i.e., at the root. Root node processing in multi-threaded Xpress does not benefit as much as from parallelization as the tree search. Thus, having more UG processes with less Xpress threads each gives us the possibility to process more such trivial sub-MIPs in parallel. However, when sub-MIPs are hard enough to generate a reasonable size of search tree, the results in Table 1 imply that using multi-threaded Xpress is more beneficial.

Table 2. Comparison of internal and external parallelization for solving `timtab2` in 2 hours on HLRN III (256 compute nodes with 24 cores each)

Settings	UG 256 Xpress 24	UG 512 Xpress 12	UG 1024 Xpress 6	UG 2048 Xpress 3	UG 3072 Xpress 2	UG 6144 Xpress 1
Comp. time (sec.)	7200	7200	7200	7200	6432	7200
Opt. found time (sec.)	-	-	6589	1573	374	308
# of nodes solved	737466918	1364575076	1587257539	1419599385	1538064260	1821264002
# of nodes remained	47010556	95829812	57527687	21321739	0	18198038
Primal bound	1106555.0000	1111174.0000	1096557.0000	1096557.0000	1096557.0000	1096557.0000
Dual bound	893103.8671	941191.1516	1022324.7668	1068745.3351	1096557.0000	1025787.6759
Gap (%)	23.90	18.06	7.26	2.60	0.00	6.90

4   Concluding Remarks

There are two main observations that we made from our computational experiments with a preliminary implementation of ParaXpress. First, ParaXpress can be used to solve hard MIP instances that need more than a billion branch-and-bound node to prove optimality within reasonable time. Second, finding a good balance between external and internal parallelization can be difficult. The made observations give rise to the idea of having an adaptive scheme that intensifies the usage of external parallelization towards the end of the search. Finally, the renewed parallelization framework in Xpress 8.0 might help to further improve the performance of ParaXpress. Within ParaXpress itself, there is potential for improvement by a more effective pruning of trivial subproblems and by implementing features such as racing ramp-up.

References

1. Achterberg, T.: Constraint Integer Programming. Ph.D. thesis, Technische Universität Berlin (2007)
2. Achterberg, T., Wunderling, R.: Mixed integer programming: Analyzing 12 years of progress. In: Jünger, M., Reinelt, G. (eds.) Facets of Combinatorial Optimization - Festschrift for Martin Grötschel, pp. 449–481. Springer, Heidelberg (2013)
3. Bussieck, M.R., Ferris, M.C., Meeraus, A.: Grid-enabled optimization with GAMS. IJoC **21**(3), 349–362 (2009)
4. Eckstein, J., Hart, W.E., Phillips, C.A.: Pebbl: an object-oriented framework for scalable parallel branch and bound. Math. Program. Comput. **7**(4), 429–469 (2015). http://dx.doi.org/10.1007/s12532-015-0087-1
5. FICO Xpress-Optimizer. http://www.fico.com/en/Products/DMTools/xpress-over/viewPages/Xpress-Optimizer.aspx
6. Laundy, R., Perregaard, M., Tavares, G., Tipi, H., Vazacopoulos, A.: Solving hard mixed-integer programming problems with Xpress-MP: a MIPLIB 2003 case study. INFORMS J. Comput. **21**(2), 304–313 (2009)
7. Nemhauser, G.L., Wolsey, L.A.: Integer and combinatorial optimization. Wiley, New York (1988)
8. Shinano, Y., Achterberg, T., Berthold, T., Heinz, S., Koch, T., Winkler, M.: Solving hard MIPLIB2003 problems with ParaSCIP on supercomputers: An update. In: 2014 IEEE International Parallel Distributed Processing Symposium Workshops (IPDPSW), pp. 1552–1561, May 2014
9. Shinano, Y., Achterberg, T., Berthold, T., Heinz, S., Koch, T.: ParaSCIP - a parallel extension of SCIP. In: Bischof, C., Hegering, H.G., Nagel, W.E., Wittum, G. (eds.) Competence in High Performance Computing 2010, pp. 135–148. Springer, Heidelberg (2012)
10. Shinano, Y., Achterberg, T., Berthold, T., Heinz, S., Koch, T., Winkler, M.: Solving open MIP instances with ParaSCIP on supercomputers using up to 80,000 cores. In: Proceedings of 30th IEEE International Parallel & Distributed Processing Symposium, to appear (2016)
11. Shinano, Y., Achterberg, T., Fujie, T.: A dynamic load balancing mechanism for new ParaLEX. Proc. ICPADS **2008**, 455–462 (2008)

12. Shinano, Y., Heinz, S., Vigerske, S., Winkler, M.: FiberSCIP - a shared memory parallelization of SCIP. Technical Report ZR 13–55, Zuse Institute Berlin (2013)
13. Sun, Y., Zheng, G., Jetley, P., Kalé, L.V.: ParSSSE: An adaptive parallel state space search engine. Parallel Process. Lett. **21**(3), 319–338 (2011)
14. Xu, Y., Ralphs, T.K., Ladányi, L., Saltzman, M.: Alps version 1.5.2 (2015)

Interactive Operation to Scientific Artwork and Mathematical Reasoning

CindyJS

Mathematical Visualization on Modern Devices

Martin von Gagern[1], Ulrich Kortenkamp[1], Jürgen Richter-Gebert[2],
and Michael Strobel[2(✉)]

[1] University of Potsdam, Potsdam, Germany
{gagern,ulrich.kortenkamp}@uni-potsdam.de
[2] Technical University of Munich, Munich, Germany
{richter,strobel}@ma.tum.de
http://www.math.uni-potsdam.de/professuren/didaktik-der-mathematik,
http://www-m10.ma.tum.de

Abstract. The *CindyJS* Project brings interactive mathematical visualization to a broad variety of devices. Using projective geometry, homotopy methods and well tuned algorithms the CindyJS project is one of the first real time capable software projects in this field that at the same time approaches high-level mathematical descriptions and performance.

Keywords: Dynamic geometry · Interactive visualization · Projective geometry · Homotopy methods · Web technology

1 Introduction

Visualization and real-time interactive simulation play an important role both in mathematical research and in mathematical communication. The *CindyJS* Project aims at the development of a software platform and its mathematical foundation that allows a versatile and fast prototyping of mathematical experiments and visualizations which can be used for research and demonstration. The project attacks both the mathematical and the software related aspects of such a platform. In particular, the system should be usable as a flexible authoring system for providing mathematical content that can run in contemporary web browsers, taking advantage of modern hardware and software technologies.

To understand the relevance and challenges of the creation of such a visualization system in the current decade one has to take recent developments in the landscape of browser based interaction possibilities into account. The recent past showed a dramatic change of possible environments for such a general math visualization system. One of the major achievements of the *Cinderella* platform as

M. von Gagern, J. Richter-Gebert and M. Strobel were supported by the DFG Collaborative Research Center TRR 109, "Discretization in Geometry and Dynamics". M. von Gagern and U. Kortenkamp were supported by the project "M C Squared" which has received funding from the EU 7th Framework Programme (FP7/2007-2013) under grant agreement no. 610467.

G.-M. Greuel et al. (Eds.): ICMS 2016, LNCS 9725, pp. 319–326, 2016.
DOI: 10.1007/978-3-319-42432-3_39

developed by Kortenkamp and Richter-Gebert was the ability to provide interactive math-related content via webpages [8]. With this philosophy, in recent
years extensive collections of interactive applets have been created and used as
tools for teaching and research [6,11] as well as for communication to the general
public. However, the underlying technology of *Cinderella* on the web was Java
in browsers, which became increasingly difficult to use due to security concerns.
Recently, Oracle even announced the discontinuation of the Java Plugin.

As a consequence to the gradual demise of Java Applet usability, a new
project called *CindyJS* was started. Its aim is allowing the creation of such
mathematical visualizations and embedding these into web pages using modern
plugin-less web technology like JavaScript, HTML5 and WebGL. The implementation of such a system poses several challenges, and while some of them have
been resolved already, others still remain open and challenging. By now, the
present prototype is already being used in production for the next generation of
interactive content.

CindyJS aims to be a viewer for interactive mathematical content (generated
by *Cinderella* or by explicit coding) in modern web browsers. In particular,
mobile devices like cell phones and tablet computers were included as target
platforms. The *CindyJS* project started in late 2013 and has already achieved
about 80 % compatibility to *Cinderella*: an implementation of an interpreter for
the scripting language, the implementation of the physics simulation engine, and
the geometry kernel are already usable for every day work. All of these have been
implemented with extensive optimizations for fast real-time interaction. Special
care was paid to software-ergonomic aspects in relation to usability on touch
devices. In particular, the experiences collected in the project *iOrnament*, as
described in [10], influenced the design of *CindyJS*. Like *Cinderella*, *CindyJS* is
intended to be an open-ended project (Fig. 1).

2   Project Guidelines

Our system has intentionally been written and designed from scratch for several reasons. Based on the experience of prior projects, like the Java version of
Cinderella, the major development guidelines are:

– Performance: A detailed performance analysis of implementation patterns was
  preceding the core development to avoid performance bottlenecks. Cross compiling techniques, as used in other visualization oriented projects, lead to poor
  execution speed which makes these less suitable for advanced examples.
– Extendability: It is very easy to add additional functionality to the system,
  while still maintaining global consistency (also under mathematical aspects).
– Interoperability: The system has thin and lightweight interfaces to communicate with other software components in the browser. By this it is relatively
  easy to include plugins and third party frameworks.
– Mathematical expressiveness: Many commonly used mathematical functionality is already included in the system as primitive operations. The scripting
  language allows for a high level implementation of mathematical concepts.

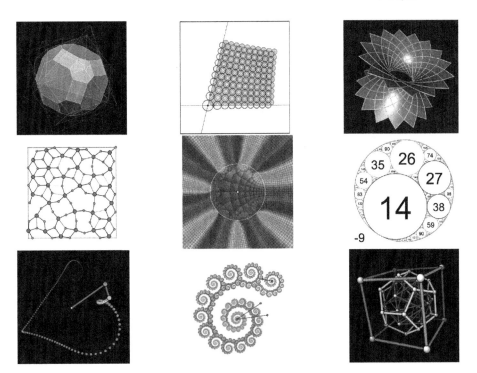

Fig. 1. Snapshots of interactive animations with *CindyJS*: Different applications of the *CindyJS* viewer, demonstrating its versatility. All these images, ordered left to right, are snapshots of dynamic interactive visualizations. (a) Intersection of polytopes (Deutsches Museum), (b) z^α grid (DGD GALLERY), (c) Discrete CMC Surface (DGD GALLERY), (d) Crystallization processes (MiMa, Oberwolfach), (e) Complex function plot, (f) Apollonean gasket (Mathe-Vital), (g) Double pendulum (Deutsches Museum), (h) Fractal (with WebGL support), (i) Nested polyhedra (ix-quadrat). More examples can be found on [3].

– Mathematical consistency: Implementing a visualization system for mathematical contexts is often preceded by a (rather subtle) modeling step in which a mathematical realm is mapped to algorithmic representations. One of the biggest achievements of the development of *Cinderella* and *CindyJS* was the capability to resolve ambiguities in geometric constructions by expanding the domain of $\mathbb{RP}^2$ using complex detours.

3 Architecture

Over the last two years a first prototypical implementation of the system has been created that contained many of the necessary key components and adheres to the above design principles. It inherits some features of the overall architecture of *Cinderella*, however it is designed to be even more modular and open for

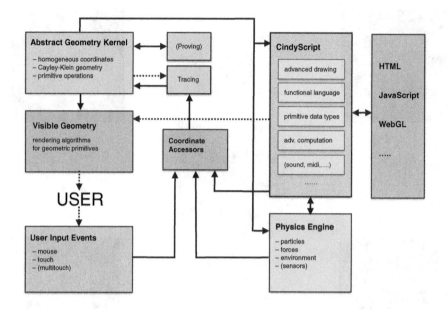

Fig. 2. Current top-level architecture of *CindyJS*

interoperation with other programs. Among the main components implemented so far there are a version of the scripting language *CindyScript*, a geometry kernel based on projective geometry, a first implementation of tracing strategies for homotopy continuation of geometric configurations, a basic physics engine, a 3D viewer and many other components. For an overview see Fig. 2.

4   Geometric Primitive Operations and Tracing

All geometric operations in *CindyJS* are strictly based on principles of projective geometry and Caley-Klein geometries. This ensures a consistent treatment that has to take care only of a minimum number of special cases. It also allows for an easy generalization to other geometries like for instance hyperbolic geometry. Most of the concepts are explained in [7], for examples see Fig. 3. *Cinderella* was the first software to resolve the problem of continuity in presence of ambiguous operations in geometric constructions like the intersection of a circle and a line. Singularities which naturally arise are resolved via complex detours. We won't go into details here, but a short example may illustrate the core idea. Assume we are moving a free element of the construction in such a way that two points of intersection merge into a single one before becoming distinct again (and perhaps complex-valued). At the point of the singularity, where the two points became one, we can no longer tell them apart. But by choosing a complex parametrization of the movement, we can move from the situation before the singularity to that after the singularity in such a way that the points remain distinguishable

Fig. 3. Geometric operations: intersection of conics, traced locus and angle bisector (Color figure online)

at all times. An instance of locus generation using this tracing technique can be found in Fig. 3, and we refer to [4] for further details.

5   *CindyScript* – Programming on a Napkin

Have you ever explained something to another mathematician, while sitting in a loud pub and having nothing but a pen and a napkin? *CindyScript* is the programming equivalent of the napkin. In other words, in *CindyScript* concepts should be expressible in sketchy, sometimes non-formal, nevertheless complete and most of all understandable way. It should be easy to write down simple concepts. Close to what you have in mind, also with respect to notation. It should be flexible and forgiving with respect to small glitches. Code should be easily explainable to non-experts. The size of a napkin is limited. *CindyScript* has (originally) never been intended to create huge software systems. It is not really a programming language for programmers, instead it is more a programming language for people who have little to no programming experience, but perhaps have some mathematical background. At the same time it is not an educational tool for "learning how to program" (at least it was not intended to be). It is a language in which small and easy things should be expressible in short an easy terms. In a sense, it is a "special purpose language" that on purpose avoids concepts that are complicated to explain to a non-expert. The core of the sunflower example in Fig. 5 is essentially the following:

```
repeat(500,i,
    w=i*pi/180*(137.508+B.x*0.5);
    p=A+(cos(w),sin(w))*0.3*sqrt(i);
    draw(p,size->sqrt(i)*.4,color->hue(i/34));
);
```

These lines should show exemplary the easiness and the power of expression that comes with *CindyScript*. In particular *CindyScript* has dynamic typing and internal objects (like numbers, vectors, matrices, etc.) that are close to the mathematical realm. For further details we refer to the *Cinderella 2* handbook [9].

6 Plugins

CindyJS was designed with extendability and interoperability in mind. This enables us to attach plugins easily which extend the features of *CindyJS* without bloating the core functionality. This keeps the essence of our project swift and lightweight. Details of some plugins will be discussed in separate articles [5,12], we just highlight some of them here.

Cindy3D is our OpenGL binding to *CindyScript*. Employing WebGL, Cindy3D is used to display 3D content using the GPU.

CindyGL gives the user access to the GPU through WebGL in *CindyScript*. While Cindy3D serves for rather classic displaying purposes, CindyGL provides access to the GPU fragment shader in *CindyScript*, which can be used for color-plots and advanced custom image manipulation. It aims to overcome technical obstacles by integrating a pipeline which translates easy-to-write *CindyScript* code into highly parallelized shader code driving GPU calculations.

A version of TEX was also integrated in our system. After some performance tests we decided to drop MathJax and switched to KaTeX [1]. With some modifications to KaTeX, the most common tasks using TEX are computed in real time.

We provide some selected examples in Fig. 4.

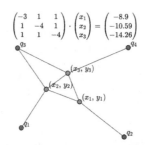
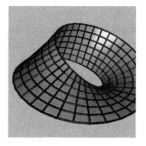

Fig. 4. *CindyJS* plugins: CindyGL raycaster, embedded TEX, Cindy3D Möbius strip

7 Technical Aspects

Modern browsers are a blessing and a curse at the same time. They are very well attuned for displaying content and since HTML5 broadened their capabilities introducing features like touch input or drag and drop. Also the execution speed of JavaScript increased dramatically over the years which is crucial for numerics. We implemented a feature-rich linear algebra package from scratch, which was pioneering work in complex arithmetics (for browsers). When possible, we used typed arrays in order to improve performance. Another crucial point was the embedding of *CindyScript* into HTML and the browser environment. We employ `<script>` tags which are directly written into the HTML code and will be interpreted by our libraries. Figure 5 shows a full example of the integration.

```
<!DOCTYPE html>
<html>
    <head>
        <title>Sunflower</title>
        <script type="text/javascript" src="Cindy.js"></script>
    </head>
<body>
<script id='csmove' type='text/x-cindyscript'>
    repeat(500,i,
            w=i*pi/180*(137.508+B.x*0.5);
            p=A+(cos(w),sin(w))*0.3*sqrt(i);
            draw(p,size->sqrt(i)*.4,color->hue(i/34));
    );
</script>

<script type="text/javascript">
    var gslp=[ {name:"A", kind:"P", type:"Free", pos:[0,0]},
               {name:"B", kind:"P", type:"Free", pos:[0,9]}];

    createCindy({canvasname:"CSCanvas",
                 movescript:"csmove",
                 geometry:gslp});
</script>
<canvas id="CSCanvas" width=500 height=500></canvas>
</body>
</html>
```

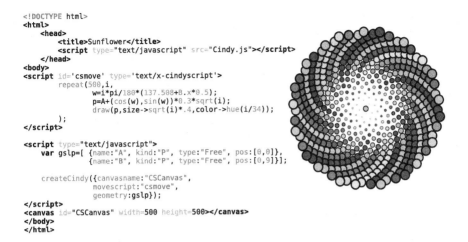

Fig. 5. Integration of *CindyJS* in HTML5

8   *CindyJS* in the Wild, a Selection

CindyJS is already suitable for every day work and is used in a large variety of places. We mention just a few of them:

Mathe-Vital. The teaching platform Mathe-Vital focuses on interactive mathematical content at university level. Currently it consists of roughly 500 applets for scientific education. Within the next few years the platform will be entirely migrated to the *CindyJS* viewer.

Bruchrechnen! Funded by the Heinz Nixdorf Stiftung and in collaboration with the TUM School of Education, we are currently developing an interactive schoolbook about fractions. The schoolbook contains many interactive exercises almost all of which are realized within the *CindyJS* framework.

Teach@TUM. The BMBF funded "Qualitätsoffensive Lehrerbildung" supports a project for the creation of structures and material to increase the quality of teacher education. In this context the *CindyJS* viewer is a major component for interactive teaching materials and micro laboratories.

ix-quadrat. All interactive exhibits in our campus maths exhibition ix-quadrat are nowadays realized based with *CindyJS*. They are of particularly high complexity and form benchmark cases for much of the functionality of *CindyJS*.

M C Squared. This EU-funded project for the creation of creative interactive content allows the use of *Cinderella* as a widget factory. It automatically converts the resulting widgets to *CindyJS* if the user is using the HTML5 player, e.g. on a tablet.

Mathe-Werkstatt. The project Mathe-Werkstatt (Leuders, Prediger et al.) provides interactive material for all grades of mathematical school education. Over the last few months the team of authors there started to also support a *Cinderella/CindyJS* based environment. *CindyJS* formed an essential component in the design of this course.

DGD Gallery. *CindyJS* is also used as a visualization component in the context of the SFB TR109 gallery. Besides 2D interaction *CindyJS* is also used as a 3D viewing engine [2].

9   Conclusion

Overall the *CindyJS* project aims to be one of the first medium to large scale mathematical visualization projects based on an HTML5 framework. Our project is licensed under the Apache 2 license and can be obtained from https://github.com/cindyjs.

References

1. Alpert, B., Eisenberg, E.: Katex (2013). https://khan.github.io/KaTeX/
2. Joswig, M., Mehner, M., Sechelmann, S., Techter, J., Bobenko, A.I.: DGD gallery: Storage, sharing, and publication of digital research data. CoRR, abs/1512.04364 (2015). http://gallery.discretization.de/
3. Kortenkamp, U., Kranich, S., Richter-Gebert, J., Strobel, M., von Gagern, M., Wurster, J.: CindyJS: A growing collection of interactive demonstrations (2015). http://science-to-touch.com/DGD
4. Kortenkamp, U.: Foundations of Dynamic Geometry. Ph.D. thesis, ETH Zürich, Zurich (1999)
5. Montag, A., Richter-Gebert, J.: CindyGL: authoring GPU-based interactive mathematical content. In: Greuel, G.-M., Koch, T., Paule, P., Sommese, A. (eds.) ICMS 2016. LNCS, vol. 9725, pp. 359–365. Springer, Heidelberg (2016)
6. Richter-Gebert, J.: Mathe-vital (2007–2015). http://mathe-vital.de
7. Richter-Gebert, J.: Perspectives on Projective Geometry: A Guided Tour Through Real and Complex Geometry. Springer, Heidelberg (2011)
8. Richter-Gebert, J., Kortenkamp, U.: The power of scripting: DGS meets programming. Acta Didactica Napocensia 3(2), 67–78 (2010)
9. Richter-Gebert, J., Kortenkamp, U.: The Cinderella.2 Manual: Working with the Interactive Geometry Software. Springer, Heidelberg (2012)
10. Richter-Gebert, J.: Touch und Tablet.: Stationen einer Designstudie. Preprint, 11 Seiten, erscheint 2015 In: Heintz, G., Pinkernell, G., Schacht, F. (Hrsg.): Digitale Werkzeuge für den Mathematikunterricht. Festschrift für Hans-Jürgen Elschenbroich. Verlag Seeberger, Neuss ISBN 978-3-940516-20-6, zugreifbar unter. http://science-to-touch.com/Articles/pdf/HJE_Artikel.pdf
11. Richter-Gebert, J.: Mikrolaboratorien und virtuelle Modelle im universitären Mathematikunterricht. In: Ableitinger, C., Kramer, J., Prediger, S. (eds.) Zur doppelten Diskontinuität in der Gymnasiallehrerbildung, Konzepte und Studien zur Hochschuldidaktik und Lehrerbildung Mathematik, pp. 169–186. Springer Fachmedien, Wiesbaden (2013)
12. von Gagern, M., Richter-Gebert, J.: CindyJS Plugins - Extending the mathematical visualization framework. In: Greuel, G.-M., Koch, T., Paule, P., Sommese, A. (eds.) ICMS 2016. LNCS, vol. 9725, pp. 327–334. Springer, Heidelberg (2016)

CindyJS Plugins

Extending the Mathematical Visualization Framework

Martin von Gagern[1(✉)] and Jürgen Richter-Gebert[2]

[1] University of Potsdam, Potsdam, Germany
gagern@uni-potsdam.de
[2] Technical University of Munich, Munich, Germany
richter@ma.tum.de
http://www.math.uni-potsdam.de/professuren/didaktik-der-mathematik
http://www-m10.ma.tum.de

Abstract. *CindyJS* is a framework for creating interactive (mathematical) content for the web. It can be extended using plugins, two of which are presented here.
– *Cindy3D* enables displaying 3D content via WebGL.
– The *KaTeX* plugin typesets formulas within *CindyJS*.
We also discuss the general structure of plugins in CindyJS.

Keywords: Interactive visualization · Web technologies · 3D · Geometry · WebGL · OpenGL · Typesetting · TeX · KaTeX

1 Introduction

The *CindyJS* project is a system for the presentation of visual and interactive mathematical web content (see [8]). It aims to be feature compatible with *Cinderella* [7], a Java-based authoring system. It is possible to extend *Cinderella* using custom plugins. *CindyJS* provides a similar mechanism to allow extension by plugins. Compared to the original Cinderella plugin interface, the plugin api of *CindyJS* offers more possibilities: while a plugin to *Cinderella* is essentially restricted to providing new functions for the built-in scripting language *CindyScript*, plugins in *CindyJS* can perform additional tasks, by accessing selected portions of the internal data of a visualization, like the canvas of the construction or unevaluated expressions passed to a plugin-provided function.

2 Plugin Interface

At the JavaScript level, a plugin for *CindyJS* is simply a callback method registered with either the *CindyJS* framework as a whole, or one specific widget.

M. von Gagern was supported by the project "M C Squared" which has received funding from the EU 7th Framework Programme (FP7/2007-2013) under grant agreement no. 610467.
M. von Gagern and J. Richter-Gebert were supported by the DFG Collaborative Research Center TRR 109, "Discretization in Geometry and Dynamics".

G.-M. Greuel et al. (Eds.): ICMS 2016, LNCS 9725, pp. 327–334, 2016.
DOI: 10.1007/978-3-319-42432-3_40

That function can interact with *CindyJS* using a set of API functions, the most important of which are probably a function to define new *CindyScript* functions and a function to evaluate a given expression. In order to help maintain backwards compatibility, plugins must declare which version of the API they are using, so that the framework can apply a compatibility layer if its internal representation were to change. Here is a simple example:

```
// Register a plugin called "hello", using plugin API version 1
CindyJS.registerPlugin(1, "hello", function(api) {

  // Define a CindyScript function called "greet"
  // that takes a single argument
  api.defineFunction("greet", 1, function(args, modifs) {

    // Evaluate the argument expression
    // (as opposed to inspecting the unevaluated formula)
    var arg0 = api.evaluate(args[0]);

    // Return string as a CindyScript value object,
    // we might want to offer some API for this one day
    return {
      ctype: "string",
      value: "Hello, " + api.instance.niceprint(arg0)
    };

  });
});
```

3   Viewing Spatial Objects Using *Cindy3D* and WebGL

The goal of the *Cindy3D* plugin is visualizing spatial mathematical objects using WebGL.

Cinderella (the Java ancestor of *CindyJS*) has a plugin of that same name [6], based on JOGL to provide OpenGL bindings. The *Cindy3D* plugin for *CindyJS* started out as a port of that Java plugin, but by now most of the code has been rewritten, so the main connection is a common API that is exposed to *CindyScript*. In the long run, it is expected to port large ports of the plugin back to the Java version, for consistent results, easier maintainance and common sets of features. In the subsequent text, the term *Cindy3D* will refer to the *CindyJS* plugin only.

Cindy3D features four kinds of geometric primitives: spheres, rods, polygons and meshes. A sphere is often used to denote a point in a 3D setup, just as disks denote points in the planar view of *CindyJS*. A rod is a cylinder with two spherical endpoints, and therefore the 3D counterpart to a line segment. A polygon is a planar surface (although it's the user's task to ensure that the vertices actually lie within a single plane). Non-planar surfaces are modeled as meshes, which are triangle meshes internally but quadrilateral meshes in the *CindyScript* API (Fig. 1).

```
// Set up 3D environment
begin3d();
background3d([0,0,0]);
size3d(2.4); // Default size of points and segments

// Declare some constants which may be used to tune appearance
n = 300; r1 = 1; r2 = 0.3; k = 5; l = 3;

// Compute point for given parameter value w
f(w):=(sin(l*w), cos(l*w),          0)*(r1 + r2*cos(k*w))
    + (        0,        0, r2*sin(k*w));

// Connect consecutive points with a colored rod
repeat(n, i,
    w1 = i/n*360°;
    w2 = (i+1)/n*360°;
    color3d(hue(i/n));
    draw3d(f(w1), f(w2));
);

end3d();
```

Fig. 1. Code and result for a simple 3D object

Contrary to many other 3D rendering environments, spheres and rods are not subdivided into triangle meshes. Instead, an object which covers the sphere or rod is drawn with a custom fragment shader to render a high-fidelity representation of the actual geometric object using a raycasting approach. In the case of a sphere, the covering object is a square facing the camera and just in front of the sphere. For the rod, the covering object is a box containing the actual rod. One consequence of this rendering approach is that the position of the geometric primitive being rendered does not correspond to the position of the final point once it's rendered: the depth may differ, so the shader code has to update the fragment depth. This requires the use of a WebGL extension called `EXT_frag_depth` which isn't available on all devices yet, although the percentage of supported devices is ever increasing.

Cindy3D employs some simple raycasting to provide cheap yet realistic lighting of the scene. The details of the appearance can be controlled by placing lights, by controlling object material properties like color or shininess, and of course by placing the camera and configuring its lens.

Cindy3D supports translucent objects, which is important for many mathematical visualizations Fig. 2. This is a challenging task, since for accurate results the scene has to be rendered strictly from back to front. As of this writing, *Cindy3D* doesn't sort its primitives yet. But it already represents all its triangles as distinct objects, which greatly simplifies the task of back-to-front rendering, since the triangles of different meshes can then be sorted as a whole, giving correct order in regions where both meshes twist around one another. Even more demanding would be the task of accurate back-to-front rendering in places where the constituent triangles intersect. That would require computing the corresponding intersections. No such endeavor is planned for the near future. In commonly used examples it is often surprising how close one has to look to spot errors due to the incorrect rendering order of the current naive implementation.

Cindy3D is not designed to allow manipulation of displayed objects. The objects are constructed from a sequence of drawing commands in *CindyScript*, and then viewed in *Cindy3D* as they are. What *can* be controlled interactively is the configuration of the camera. It can be rotated around the object, which for every point and purpose is the same as rotating the object around the look-at point of the camera since light sources can be fixed to the camera or the object at the user's discretion. It can also be rotated around itself, or translated in three spatial directions. The camera can move closer to the object, or farther out, which is colloquially called a zoom. It can also perform an actual photographic zoom, i.e. change the field of view. All of these operations are accessible by mouse movements, in combination with modifier keys.

The source code of *Cindy3D* is clearly structured into code which provides specific bindings to the *CindyJS* API, and code which does the internal data representation and manipulation. It would be fairly easy to replace the former by bindings for some other software package, in order to turn *Cindy3D* into a 3D model viewer for a different web-based (or at least browser-based) application.

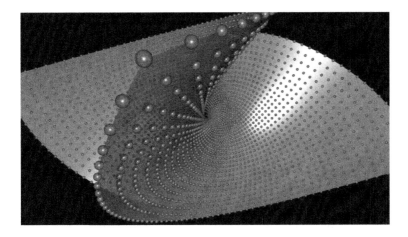

Fig. 2. Translucent enneper surface with spheres indicating grid points

4   Typesetting Formulas Using *KaTeX* bindings

Some mathematical content can be best described using a combination of geometric elements and formulas. Sometimes it is enough to include the formulas in the text surrounding a given widget. But if the text has to be positioned with respect to a given element of the widget, or perhaps contains numbers which change during interaction, then it is important to typeset these formulas within the widget Fig. 3.

The lingua franca for entering formulas is TEX for most mathematical communities. *Cinderella* comes with its own home-grown parser and renderer for TEX-like formulas. For *CindyJS* it was decided not to port that parser, but instead build on one of the existing efforts to bring math typesetting to the web. While MathML has been intended as default representation for formulas, browser support for it is severely lacking, with no change anticipated in the near future.

The most common solution is *MathJax* [2], a JavaScript library to render formulas. But *MathJax* has several problems which make it unsuitable for the application at hand. It operates on the HTML DOM tree, so one would have to somehow position the text above the widget, instead of drawing it to the widget canvas the same way geometric objects are being drawn. It operates asynchronously, so there is a delay between the time when drawing a formula is requested and the time when the typeset version of said formula is actually ready for display. This fits in poorly with the synchronous drawing paradigma employed by *CindyScript*.

Looking for an alternative, we found the *KaTeX* project [1]. It provides synchroneous operations, and usually renders significantly faster than *MathJax*. The main drawback is its lack of features: many things supported by *MathJax* are (or at least were) not available in *KaTeX*. We identified those features whose absence would cause the most trouble for existing or envisioned content, and

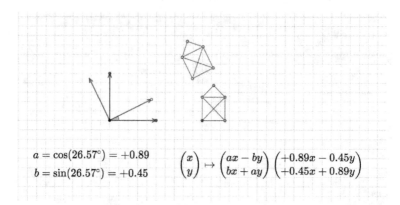

$$a = \cos(26.57°) = +0.89$$
$$b = \sin(26.57°) = +0.45$$
$$\begin{pmatrix} x \\ y \end{pmatrix} \mapsto \begin{pmatrix} ax - by \\ bx + ay \end{pmatrix} \begin{pmatrix} +0.89x - 0.45y \\ +0.45x + 0.89y \end{pmatrix}$$

Fig. 3. Educational widget using typeset formula

had those features implemented for *KaTeX*. Foremost on that list is support for matrices, which was developed for *CindyJS* but has been merged into the official *KaTeX* code base as well.

Just like *MathJax*, *KaTeX* is designed to modify the HTML DOM tree. But it has an internal intermediate representation of how to arrange and nest its boxes, from which the actual HTML elements are being generated. Using this representation it was possible to modify the code in such a way that instead of creating HTML elements it creates canvas drawing commands. Since the internal representation only provides vertical placement information, horizontal positioning has to measure text dimensions. For this reason, the render-to-canvas process has two phases. In the first phase, the individual blocks are measured and positioned relative to one another. The result is an object which contains enough data to render all required glyphs, but which also has overall measurements of the whole formula. These can be used to position the box, e.g. for horizontal alignment, before the glyphs get actually drawn to canvas. The draw-to-canvas feature hasn't been merged into the official *KaTeX* (yet?), mainly since it caters for some very specific use cases only. Nevertheless, this feature can be of use to other projects facing the same problem of how to place high-quality math typesetting in a web application based on the HTML canvas element.

The *KaTeX* plugin for *CindyJS* is a fairly thin layer binding *CindyJS* to a version of the *KaTeX* library which includes our customizations like render-to-canvas. It is different from other plugins in that it doesn't provide any new *CindyScript* commands. Instead it modifies the behavior of existing commands using a hook in the *CindyJS* rendering pipeline. Much of the complexity of the plugin, however, is spent on ensuring the automated loading of required resources. Fonts in particular are difficult to handle, since on some browsers loading of these only starts once they are actually being used, so they won't be available upon that first use. In this situation, the *KaTeX* plugin will not render the text but instead wait for the required fonts and trigger a repaint once they become ready.

5   Other Plugins

There have been other successfull applications of the plugin infrastructure as well. The following list demonstrates the high flexibility of the plugin infrastructure.

CindyGL is a tool which allows running a subset of the *CindyScript* language on the GPU. It is described in a separate article, [5].

QuickHull3D is an algorithm [2] and a Java library [4] used by *Cinderella* to implement its `convexhull3d` operation. That Java code was compiled to JavaScript using GWT, and made available as a plugin, in order to provide compatibility. This is a temporary solution since it would be preferable to have a native JavaScript implementation inside the core of *CindyJS*, which is something currently being worked at. However, this setup demonstrates that plugins can be used to build connections even to some Java libraries.

MC Squared (or Mathematical Creativity Squared) is a project which allows authors to combine widgets from various sources into so-called C-Books. When viewing them in a HTML5 environment, *CindyJS* is used to display *Cinderella* widgets, and a plugin is used to realize the integration into that specific environment, in particular the communication with other widgets.

Metadata extraction from images generated by the ornament drawing app *iOrnament* was demonstrated in a proof-of-concept plugin implementation. This made it possible to post-process the ornaments in a way compatible with the symmetry group used for their creation.

Tests in the *CindyJS* test suite sometimes use plugins to allow a *CindyJS* instance to report results back to the testing framework.

Development of the complex tracing implementation within *CindyJS* itself was helped by visualizations of the tracing process. One instance of *CindyJS* was showing some construction to be interacted with, while a second instance turned debug logs of the operations behind the scenes into helpful visualizations, providing a real-time view of the corresponding internal computations. Those logs were generated by the first instance and made available to the second via a custom plugin.

6   Conclusion

Plugins are a useful way to extend a software framework in order to adapt it to new requirements. Often the people writing the plugins are distinct from those writing the core framework. So far, *CindyJS* has not attracted any third party plugins that we know of. But the fact that plugins are being used by the developers themselves ensures that the plugin infrastructure is powerful and flexible enough to accomodate various requirements. They are particularly useful for connecting optional components that provide their own interfaces, translating between different conventions and representations. Plugins are the perfect tool for extending functionality for custom applications without bloating the core implementation for everyone.

References

1. Alpert, B., Eisenberg, E.: Katex (2013). https://khan.github.io/KaTeX/
2. Barber, C.B., Dobkin, D.P., Huhdanpaa, H.: The quickhull algorithm for convex hulls. ACM Trans. Math. Softw. **22**(4), 469–483 (1996)
3. Davide, C., Sorge, V., Perfect, C., Krautzberger, P.: Mathjax (2009). https://www.mathjax.org/
4. Lloyd, J.: Quickhull3d – a robust 3d convex hull algorithm in java (2004). http://www.cs.ubc.ca/~lloyd/java/quickhull3d.html
5. Montag, A., Richter-Gebert, J.: CindyGL: authoring GPU-based interactive mathematicalcontent. In: Greuel, G.-M., et al. (eds.) ICMS 2016. LNCS, vol. 9725, pp. 351–358. Springer, Heildelberg (2016)
6. Reitinger, M., Sommer, J.: Cindy3d (2012). http://gagern.github.io/Cindy3D/
7. Richter-Gebert, J., Kortenkamp, J.: The Cinderella.2 Manual: Working with the Interactive Geometry Software. Springer, Heidelberg (2012)
8. von Gagern, M., Kortenkamp, U., Richter-Gebert, J., Strobel, M.: CindyJS – mathematical visualization on modern devices. In: Greuel, G.-M., et al. (eds.) ICMS 2016. LNCS, vol. 9725, pp. 319–326. Springer, Heildelberg (2016)

Generating Data for 3D Models

Naoki Hamaguchi[1]([⊠]) and Setsuo Takato[2]

[1] National Institute of Technology, Nagano College, Nagano, Japan
hama@nagano-nct.ac.jp
[2] Toho University, Funabashi, Japan
takato@phar.toho-u.ac.jp
http://ketpic.com

Abstract. KeTpic is a macro package which generates the graphical code that can be used in LaTeX. In 2014, commands for generating data in obj format were added to KeTpic. The data can also be converted to stl format, with which 3D printers can make 3D models. KeTCindy is a Cinderella plug-in that generates data for a variety of teaching materials along with 3D models without much additional effort. In this paper, we show a variety of KeTCindy commands for doing that, and present the actual teaching materials that result.

Keywords: KeTCindy · KeTpic · 3D model

1 Introduction

K$_E$Tpic is a macro package which generates graphical code that can be used in LaTeX. The first version was released in 2006, and since then some mathematics teachers in Japan have been using it to make materials for distribution in their classes. In collegiate-level math classes, these materials require not only mathematical formulae but figures such as geometric shapes, graphs, charts, and tables. Teachers sometimes need 3D figures as well, and so we have added commands for polyhedra figures and surfaces. For these figures, with K$_E$Tpic, hidden lines are removed for solid models, and the rear parts of segments are cut for skeleton models. Figures are composed of a small number of lines, and are appropriate for printed materials. However, it still might be more effective on some occasions to show physical models in class. In 2014, commands for generating data in obj format were added to K$_E$Tpic [1]. This data can also be converted to stl format, with which 3D printers can make 3D physical models. These new commands in K$_E$Tpic in conjunction with existing commands enable teachers to make teaching materials in various ways:

1. Handouts to be distributed.
2. Slides to be presented on the screen.
3. Figures to be manipulated by the students on their tablets.
4. Physical models to be displayed or passed around.

© Springer International Publishing Switzerland 2016
G.-M. Greuel et al. (Eds.): ICMS 2016, LNCS 9725, pp. 335–341, 2016.
DOI: 10.1007/978-3-319-42432-3_41

Physical models have the most information about real objects. However, there are cases in which they can hinder students from grasping the point under discussion. Therefore, the above-mentioned materials should be evaluated and combined to suit the contents.

The current version of KₑTpic mainly uses Scilab, which does not support viewing of 3D figures in obj format, so one has to use other software such as Meshlab [3] to do this. As a result, the following procedure, which may be slightly inefficient, is needed.

1. Start Scilab and enter the scripts in the editor.
2. Execute the scripts, then a file in obj format will be generated.
3. Start Meshlab and open the file. The figure will appear on the screen.
4. To modify the figure, close Meshlab and go back to Scilab to revise the scripts.

Materials are made through trial and error, and to repeat the above steps over and again results in the software starting up over and again, and is a drain on both time and processing power. This is common when generating ordinary figures with KₑTpic, as can be seen by replacing LATEX for Meshlab in the above procedure. The graphical user interface (GUI) of KₑTpic had been required for a long while.

Cinderella [2] is dynamic geometry software developed by Gebert and Kortenkamp. We have held several research meetings with Kortenkamp, and the first version of KₑTCindy was released in September, 2014. Cinderella works as a KₑTpic GUI in KₑTCindy. The figure-generating flow in KₑTCindy is as follows:

1. Start Cinderella. It has two screens, the display and the script editor.
2. Add geometric components on the display if necessary.
3. Write scripts in the editor, execute the scripts, and the figures appear on the display.
4. Press two buttons, "Texview" and "Exekc", on the screen for batch processing to execute Scilab, LATEX and Pdf viewer sequentially.
5. One can get the figure in pdf format generated by the LATEX compiler.

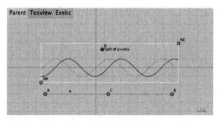

The above screenshots show that

– A slider is displayed. One can move point C to determine the x coordinate, which is used in the editor.
– The last argument of "Plotdata" is optional. Here "do" changes the line style to dotted.

2   K$_E$TCindy commands for 3D models

When explaining spacial figures in class, the addition of materials generated by LATEX, obj format files for tablets or screens, improves students' ability to grasp the concepts. The current version of K$_E$TCindy functions to generate a file in obj format and call up the Meshlab directory from within the program, which allows teachers to make materials easily, outwardly using only Cinderella.

Command for 3D model are as follows.

1. Mkviewobj
2. Mkobjcmd
3. Mkobjthickcmd
4. Mkobjcrvcmd
5. Mkobjsymbcmd
6. Mkobjpolycmd
7. Concatcmd

The point is that the usual commands for generating spacial figures for LATEX documents are also used to generate files in obj format.

Here we explain some of these commands.

2.1   Making Surfaces

At first, set a list to define a surface as follows:

```
fd=["z=x^2-y^2","x=[-2,2]","y=[-2,2]","nesw"];
```

The list fd is available both to make LATEX figures and to make and view files in obj format. For the former, add the following script:

```
Sfbdparadata(fd);
```

the result is the below-left figure.

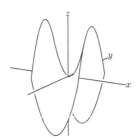 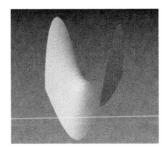

For the latter, add the following command:

```
tmp1=Mkobjcmd("1",fd);
Mkviewobj("ex",tmp1,["make","view"]);
```

then K$_E$TCindy calls up Meshlab, and the figure appears on the Meshlab screen as shown in the upper-right figure.

Surface thickness must be specified to make physical models. In this case, use "Mkobjthickcmd" instead of "Mkobjcmd".

```
tmp1=Mkobjthickcmd("1",fd);
```

Here we insert a comment that a command to call up Maxima, one of a few free CASs, is used inside this command to find the normal vector of the surface. KETCindy can call up several free CASs such as Maxima, Fricas and Risa/Asir.

2.2 Space Curves and Segments

One-dimensional figures, such as space curves and segments, are indispensable for making materials, although it is not easy to make physical models of them. KETCindy uses the following commands to do so:

```
Spacecurve("1","[cos(t),sin(t),0.3*t]","t=[0,6*pi]");
Spaceline("2",[A3d, B3d]);
Xyzax3data("","x=[-5,5]","y=[-5,5]","z=[-5,5]");
Skeletonparadata(1);
```

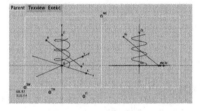

To generate LATEX figures, just press two buttons, "Texview" and "Exekc", on the screen, and use the command "Mkobjcrvcmd" to make a file in obj format.

```
tmp1=Mkobjcrvcmd("1","sc3d1");
tmp2=Mkobjcrvcmd("2","sl3d2");
tmp3=Mkobjcrvcmd("3","ax3d");
tmp=Concatcmd([tmp1,tmp2,tmp3]);
Mkviewobj("ex",tmp,["make","view"]);
```

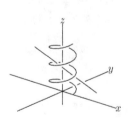

"Concatcmd" concatnates these, making a single file out of them all.

2.3 2D Character and Symbol

Adding character strings to 3D figures is sometimes necessary and facilitates understanding. Use the command "Mkobjsymbcmd" for this purpose.

```
tmp4=Mkobjsymbcmd(path,"x",0.5,0,[0,-1,0],[0,0,5]);
tmp5=Mkobjsymbcmd(path,"P",0.5,0,[1,0,0],[5,0,0]);
Circledata("1",[[0,0],[1,0]],["nodisp"]);
tmp6=Mkobjsymbcmd("cr1",0.5,0,[0,1,0],[0,5,0]);
```

The file read to generate the curves of the characters consists of Bézier curve control points.

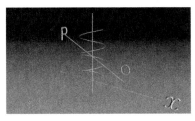

3   Examples of Teaching Materials

In this section, we introduce some 3D models as examples of teaching materials, the first, a PDF file; next, screen shots of obj files displayed in Meshlab; and last, photographs of actual printed out 3D physical models.

Example 1. Function $z = \dfrac{x^2 - y^2}{x^2 + y^2}$

This illustrates the common problem of a two-variable function which is not continuous at the origin.

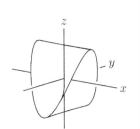
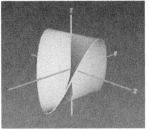

Example 2. Functions $z = x^2 + y^2$ and $z = x^2 - y^2$

The following are scripts to make the file in obj format. Here, one can use fd1 and fd2 to make the figure with LaTeX.

```
fd1=["z=x^2+y^2","x=R*cos(T)","y=R*sin(T)",
    "R=[0,2]","T=[0,2*pi]","e"];
fd2=["Z=X^2-Y^2","X=U","Y=-(1-V)*sqrt(U^2+1)+V*sqrt(U^2+1)",
    "U=[-2,2]","V=[0,1]","news"];
tmp1=Mkobjthickcmd("1",fd1,[0.1,"+n+s-e-w+","assume(R>0)"]);
tmp2=Mkobjthickcmd("2",fd2,[0.1,"+n+s-e-w+"]);
tmp3=Mkobjcrvcmd("3","ax3d");
Mkviewobj("hp",Concatcmd([tmp1,tmp2,tmp3]),["make","view"]);
```

The intersection of these graphs is the spacial curve $z = x^2$, $y = 0$.

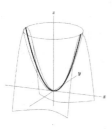  

Example 3. Polyhedron

3D models of polyhedra are sometimes useful for mathematics education. KETCindy can generate two types of them, one a solid type, and the other a skeleton type. In the following scripts, we use the data presented by Mitani [4]. The command "VertexEdgeFace" generates a list consisting of vertices, edges, and faces named "phv3d", "phe3d1", and "phf3d" respectively. We use vertice and face data for solid models, and edge data for skeleton models.

```
polydt=Readobj("r05.obj",["size=-3.5"]);
pd=VertexEdgeFace("1",polydt,["Pt=fix","Edg=nogeo"]);
Nohiddenbyfaces("1","phf3d1",[],["do"]);
tmp=Mkobjpolycmd("1",["phv3d1","phf3d1"],[[0,0,0]]);
Mkviewobj("hp",tmp,["make","view"]);
```

```
polydt=Readobj("r05.obj",["size=-3.5"]);
pd=VertexEdgeFace("1",polydt,["Pt=fix","Edg=nogeo"]);
Skeletonparadata("1",[1.5]);
tmp=Mkobjcrvcmd("2","phe3d1",[0.1]);
Mkviewobj("hp",tmp,["make","view"]);
```

4   Conclusions and Future Work

So, with the advent of K$_E$TCindy, one can streamline the production of teaching materials from print-outs to actual 3D models, getting the job done in a single window, which calls up various software in the background, ultimately saving time and processing power.

More importantly, lesson plans are qualitatively different as one can now display any combination of the above-mentioned 3D teaching materials on a single screen, and manipulate them side by side to determine what is most suitable.

We will follow up to ensure that all the above-mentioned methods are effective, and will make any and all modifications necessary to get them so.

Acknowledgment. This work was supported by JSPS KAKENHI Grant Number 15K00944.

References

1. Takato, S., Hamaguchi, N., Sarafian, H.: Generating data of mathematical figures for 3D printers with KETpic and educational impact of the printed models. In: Hong, H., Yap, C. (eds.) ICMS 2014. LNCS, vol. 8592, pp. 629–634. Springer, Heidelberg (2014)
2. http://cinderella.de
3. http://meshlab.sourceforge.net
4. http://mitani.cs.tsukuba.ac.jp/polyhedron/index.html

The Actual Use of KETCindy in Education

Masataka Kaneko[✉]

Toho University, Funabashi, Japan
masataka.kaneko@phar.toho-u.ac.jp

Abstract. Today, various tools have been developed to visualize mathematical objects dynamically. For example, graphical user interfaces have been implemented on many computer algebra systems like Mathematica in which a dynamic presentation of geometric shapes and function graphs can be generated by using sliders. Among such tools, dynamic geometry software like Cinderella are quite excellent in that they allow us to control those objects more interactively. At the same time, static presentation of those objects on printed matters is also indispensable for mathematical activities since it is through paper and pencil-based activities that we can most easily synchronize computation and observation. Thus, especially for educational purposes, the selection and the usage of these methods at each stage of the learning process is crucial. Since KETCindy, which we have recently developed, serves a direct linkage between interactive presentation of graphics on Cinderella and its exported image into TEX, it can be expected that using KETCindy enables mathematics learners to unify their intuitive reasoning through observation of the interactive presentation on PC and their discursive inference with the use of TEX documents including finely tuned graphics. In this paper, the effect of such unifiability on the learners' reasoning processes is illustrated through time-series detection of learners' activities during some case study in which KETCindy system is used.

Keywords: KETCindy · TEX · Dynamic geometry software · Cinderella · Studiocode system

1 Introduction

There are many obstacles that prevent undergraduate students from grasping the fundamental notions of calculus correctly and using them appropriately in applied areas such as physics and engineering. For instance, the descriptions of a function's derivative in many textbooks can mislead students' understanding [1]. In fact, many textbooks use touch-the-curve expressions to explain a function's derivative which may prevent students from understanding its precise meaning as the limit of secant lines or linearization of function. To address this problem, it is recommended in [1] to "visualize the tangent definition (i.e., the limit of secant lines) as a sequence of magnifications, zooming in on the point of tangency". Also the definition of the definite integral is not easy for undergraduate

© Springer International Publishing Switzerland 2016
G.-M. Greuel et al. (Eds.): ICMS 2016, LNCS 9725, pp. 342–350, 2016.
DOI: 10.1007/978-3-319-42432-3_42

students to grasp. Some previous research [2] illustrates that a high prevalence of area-under-the-curve and anti-derivative ideas and a relatively low occurrence of Riemann sum-based ideas in students' interpreting definite integrals can be observed. This imbalance leads to the students' having difficulty in applying the notion of definite integral in physical situations [3]. To resolve this imbalance, it is suggested in [2] that "increased experience with applied, science-based integral expressions in calculus courses may help strengthen students' activation of the multiplicatively based summation conception in pure mathematics contexts". While paper and pencil-based activities are needed for students' executing their mathematical reasoning and computing, the activities using dynamic geometry software like Cinderella (http://cinderella.de/) and GeoGebra (http://www.geogebra.org/) should be effective especially when students need to grasp the mathematical concepts related to dynamic objects as cited above [4,5]. To smoothly connect these two kinds of activities, we developed the KETCindy plug-in to Cinderella. It enables us to synchronize the interactive presentation of mathematical objects on the Cinderella screen and the generation of the corresponding graphical image in TEX documents [6]. It can be downloaded at the website http://ketpic.com/. Though the author found that using KETCindy often positively influenced students in their learning calculus through his experiences in real classrooms, it has also turned out that insufficient knowledge about which aspects of learners' reasoning processes are worked upon by the dynamic presentation of mathematical objects might cause learners' unexpected confusion. The aim of this research is to detect the qualitative change in learners' reasoning processes as mathematical objects are presented to them dynamically. In this paper, we show some results of our case study in this direction.

2   Generation of TEX graphics via KETCindy

The procedure to generate graphical images both on the Cinderella screen and on the TEX final output through KETCindy can be summarized in Fig. 1.

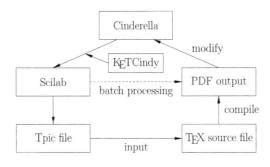

Fig. 1. KETCindy cycle

While graphical images of mathematical objects are generated and handled interactively on the Cinderella screen, KETCindy converts the drawing procedure into the Scilab commands that generate the corresponding graphical data which are subsequently formatted into TeX graphical codes. The compilation of the TeX source file in which these TeX graphical codes are input leads to the generation of high-quality graphical images in TeX final outputs (in PDF form). If there are any points to be corrected, you only need to modify the Cinderella file. The interactive operation on PC can be directly reflected in the generated image on TeX final output through the KeTCindy system. Moreover the generated image can be finely tuned by KeTCindy commands embedded into the scripting language of Cinderella (Cindyscript). Since it is not easy for most teachers to handle a great deal of software, some batch processing, shown by a dashed arrow in Fig. 1, is implemented on KeTCindy system. This process is executed only by clicking a button on the Cinderella screen.

As a simple example, we see the procedure for drawing a circle and a triangle which is inscribed in it via KeTCindy. Figure 2 shows the Cinderella screen and final TeX outputs. As shown in it, we can freely change the position of vertices on the circle by mouse dragging, and the change is directly reflected in the exported image in TeX. This output can be obtained simply by clicking buttons `Texview` and `Exekc` located on the Cinderella screen.

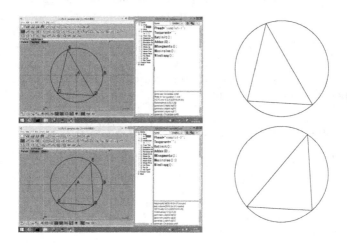

Fig. 2. Drawing circles and segments

Also it is possible to draw function graphs and parametric curves by inputting KeTCindy commands into Cindyscript. Figure 3 is an example which shows the drawing of the graph of a trigonometric function and the parametric circle with moving points on them. Here the point C, located on the segment AB, plays the role of "slider". In fact, the positions of the points D and F are controlled by moving point C.

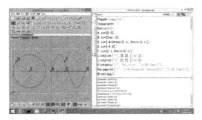

Fig. 3. Drawing graphs and parametric curves

3 Methods

Since paper and pencil-based activities are widely used in collegiate mathematics education, the primary target in this research is to discover how learners' reasoning processes are influenced when presented with mathematical objects dynamically. One possible way to illustrate these influences is to detect the transition of learners' responses and behaviors during some experimental lesson in which they are asked to do similar tasks before and after the above mentioned dynamic presentation. Among other elements, the most probable ones which we can observe should be the followings:

1. The change in the learners' concept which can be identified through their verbal behaviors and gestures.
2. The change in the learners' organizations of the steps to do the tasks which might lead to the improvement of efficiency in their reasoning.

To detect these changes, we videotape all the behaviors of students and classify the students' activities into some characteristic categories. While we watch the recorded video image, we identify which sort of activity the student does at each time interval. We use the Studiocode system (http://www.studiocodegroup. com/) to code the classified behaviors into a time-line and synchronize the coded information with the recorded video image for further analysis.

We choose an elementary topic, drawing the graphs of trigonometric functions, as the theme for this case study. This is because the more complicated the theme, the more varied learners' reasoning processes, which might exceed the capacity of this pilot research. Since the concepts like amplitude, period, and phase difference are needed to draw those graphs and are indispensable for learners to understand many mathematical models in the applied areas, it should be desirable that science majors be proficient in drawing those graphs. However, they often adopt a plan to plot points and to search some curve passing through them which leads to too much time spent and incorrectly drawn shapes.

To search for an effective way to promote those proficiencis, we planned an experimental lesson as a part of a remedial course at a university in Japan. We recruited one female student aged 19 (the first grade in Japanese university). She stated that she was not good at drawing graphs of trigonometric functions though she once studied it in senior high school without any difficulty in calculating the values of trigonometric functions. She also stated that no dynamic presentation was given when she learned this topic. She was asked to draw the graph of a given function onto the worksheet in Fig. 4 together with all calculations.

Fig. 4. Sample of worksheet

In the first stage, the subject was asked to draw the graphs of the following six functions after she reviewed the textbook which she used in high school.

$$(1) \ y = \sin x \quad (2) \ y = \cos x \quad (3) \ y = \sin 2x \quad (4) \ y = \cos \frac{1}{2}x$$

$$(5) \ y = \sin \left(x - \frac{1}{3}\pi \right) \quad (6) \ y = \cos \left(2x + \frac{2}{3}\pi \right)$$

When she made some mistakes in a task, the corresponding advice was given immediately after she completed it. Moreover, the author asked her to clarify her reasoning process after she completed each task or two.

In the second stage, the relationships between the following pairs of graphs

$$(1) \text{ and } (3) \quad (2) \text{ and } (4) \quad (1) \text{ and } (5)$$

were explained regarding the difference of period and phase by using dynamic presentations via KETCindy. Some samples are shown in Fig. 5. In these figures, by dragging the point MF on the slider, we can interactively change the value of x accordingly. Through this presentation, it is expected that the subject can clearly recognize the correspondence between the rotation angle in the circle and the value of x in the graph. Since the recognition might lead to some progress

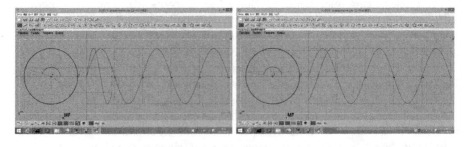

Fig. 5. Cinderella screen used for dynamic presentation

in her understanding the above mentioned relationships between the graphs, she was asked to explain those relationships in her own words.

In the last stage, the subject was asked to draw the graphs of the following two functions which are similar to (6).

$$(7)\ y = \sin\left(2x + \frac{1}{3}\pi\right) \qquad (8)\ y = \cos\left(\frac{1}{2}x - \frac{1}{6}\pi\right)$$

After the experiment, the recorded video data was input into a time-line via the Studiocode system and the video image was studied in order to classify the subject's behavior. The details of the coding procedure are described in the next section.

One week later, we gave a supplementary lesson so the subject could review. In the review lesson, she was asked to draw the graphs of the following functions.

$$(1)\ y = \cos 2x \quad (2)\ y = \sin\frac{1}{2}x \quad (3)\ y = \cos\left(2x + \frac{1}{3}\pi\right) \quad (4)\ y = \sin\left(\frac{1}{2}x - \frac{1}{6}\pi\right)$$

4 Results and Discussions

Though it took much time, the subject could complete the first five tasks without being given any hints. When she did task (6), she at first made mistake in computing the value y for some values of x and drew the wrong graph. However, after the mistake was pointed out to her, she completed the same task correctly. The resulting worksheets for tasks (3) (4) (5) (6) are shown in Fig. 6.

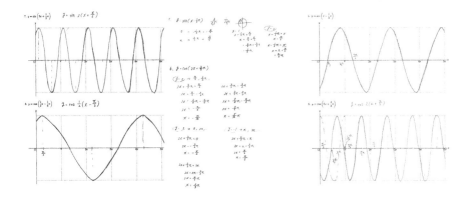

Fig. 6. Resulting worksheets

While the subject completed tasks (1) (2) (3) (4) by substituting several values for x and searching for points where the value of y is equal to 0 or 1, she computed the values of y for some values of x and tried to find the curve fitting the result of those computations in tasks (5) (6), as indicated in the middle sheet in Fig. 6.

This result is compatible with her statements in the interview before the lesson. According to her statements, she remembered that the substitution of x by ax results in the change of period from 2π to $\dfrac{2\pi}{a}$ and that the substitution of x by $x - a$ results in the parallel transport of the graph by a in the x direction, though she was unsure about the reason.

When asked to explain her reasoning during the second stage, she used the word "rotation becomes slow" with her hand rotating to explain (4). Also she used the word "rotation delays" to explain (5). These responses indicate that she could recognize the linkage between rotation and the change of x clearly through dynamic presentation via K$_E$TCindy. She claimed that the graph of (6) should be obtained by parallelly translating the graph of $y = \cos 2x$ by $-\dfrac{2}{3}\pi$ in the x direction which is a misunderstanding induced from the expression $y = \cos\left(2x + \dfrac{2}{3}\pi\right)$. Therefore she was asked to add the graph of $y = \cos 2x$ on the worksheet. She handwrote it with a dashed line as shown in the last image of Fig. 6, and she became aware of her mistake. This result indicates the possibility that dynamic presentation of mathematical object and its exported image in paper play complementary roles in learners' reasoning processes.

Observing the whole recorded video, the behaviors of the subject were classified into the following categories.

1. Calculating the transformed expression of the given function
2. Computing the values y for specific values of x by substitution
3. Computing the value x of the intercept on the x-axis
4. Plotting points through which the graph of the given function passes
5. Trying to find the curve fitting into the plotted points
6. Drawing the curve on the worksheet confidently

We adopted the rule that silent thought between the same kinds of behaviors indicates that behavior continued throughout that silence. Category 5 (fitting the curve) is identified by the subject's moving her hand or pencil without drawing a curve. The typical scene of category 5 together with the resulting classification of her behaviors in the first stage coded on the time-line is shown in Fig. 7. Here category 1 (calculation) was not observed.

The great contrast between the first stage and the third stage (before and after the K$_E$TCindy presentation in the second stage) can be seen in Fig. 8. In fact, categories 2, 3, and 5 disappeared and category 1 appeared. Moreover, the transitions from one category to another largely decreased after the second stage. These results indicate that the subject could master some efficient way of drawing graphs without any trial and error.

The result of the supplementary lesson was quite similar to that of stage three (though the subject made some mistake in task (1)) as shown in Fig. 9. This result indicates that the efficient way of drawing graphs, which she mastered in the third stage, had been transferred to long-term memory.

Since no technical advice was given in the second stage, it can be seen that the resulting progress in drawing graphs was caused not by drilling in the technique

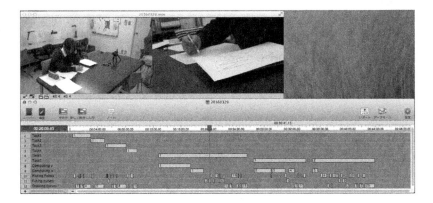

Fig. 7. The subject's behaviors in the first stage

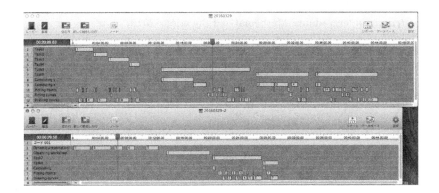

Fig. 8. Comparison between the first stage and the third stage

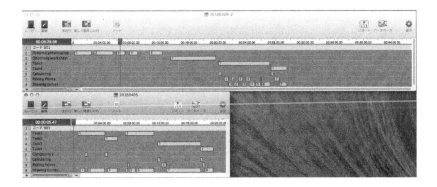

Fig. 9. Comparison with the supplementary lesson

to draw graphs but by promoting conceptual understanding. We can assume that the dynamic presentation via K_ETCindy played a role in triggering it.

Acknowledgements. K$_E$TCindy is developed mainly by Professor Setsuo Takato at Toho University. The author greatly appreciates his cooperation. Also this research is supported by the Japan Society for the Promotion of Science (KAKENHI 15K01037).

References

1. Kajander, A., Lovic, M.: Mathematics textbooks and their potential role in supporting misconceptions. Int. J. Math. Educ. Sci. Technol. **40**(2), 173–181 (2009)
2. Jones, S.: The prevalence of area-under-a-curve and anti-derivative conceptions over Riemann sum-based conceptions in students explanations of definite integrals. Int. J. Math. Educ. Sci. Technol. **46**(5), 721–736 (2015)
3. Christensen, W., Thompson, J.: Investigating student understanding of physics concepts and the underlying calculus concepts in thermodynamics. In: Proceedings of the 13th Special Interest Group of Mathematical Association of America on Research in Undergraduate Mathematics Education (2010)
4. Kortenkamp, U.: Interoperable interactive geometry for Europe. Electron. J. Math. Technol. **5**(1), 1–14 (2011)
5. Lavicza, Z.: Integrating technology into mathematics teaching: a review. Int. J. Math. Educ. **42**(1), 105–119 (2010)
6. Kaneko, M., Yamashita, S., Kitahara, K., Maeda, Y., Nakamura, Y., Kortenkampand, U., Takato, S.: K$_E$TCindy-Collaboration of Cinderella and K$_E$Tpic-. Int. J. Technol. Math. Educ. **22**(4), 179–185 (2015)

Cooperation of KeTCindy and Computer Algebra System

Shigeki Kobayashi[1]([⊠]) and Setsuo Takato[2]

[1] National Institute of Technology, Nagano College, Nagano, Japan
kobayasi@nagano-nct.ac.jp
[2] Toho University, Funabashi, Japan
takato@phar.toho-u.ac.jp

Abstract. In Japan, some wooden plaques presenting geometrical puzzles (SANGAKU) show problems of drawing smaller circles within a larger circle in contact with it or a circle in contact with a quadratic curve, such as an oval. Problems of these kinds can be interesting teaching materials. Dynamic geometry software such Cinderella or GeoGebra is useful to solve these problems. One can use high-quality TeX graphics with KeTCindy (Cinderella plug-in) to draw these figures. One can also draw figures by solving simultaneous equations using a computer algebra system such as Maxima or Risa/Asir. Although simultaneous equations can be solved in Maxima, the solution takes time. Alternatively, it does not work when too many variables are included. In such cases, one can convert it to a system of equations that are easier to solve using a Grobner base in Risa/Asir. Then a user can solve them by giving the result to Maxima. A user can accomplish this on Cinderella through KeTCindy, and can draw circles using the result. This method can also be applied to other difficult circumstances. This paper presents some examples.

Keywords: KeTCindy · Cinderella · Dynamic geometry software · Computer algebra system · Maxima · Risa/Asir

1   Introduction

During the Edo period (1603–1867) Japan was isolated from the western world. However, learned people of all classes, from farmers to samurai, produced theorems in Euclidean geometry. These theorems appeared as beautifully colored drawings on wooden tablets that were hung under one roof in the precincts of a shrine or temple. The tablet was called a mathematics tablet, of SANGAKU in Japanese. Many skilled geometers dedicated a SANGAKU to thank the gods for the discovery of a theorem. A proof of the proposed theorem was rarely given [1]. Problems of these kinds can be interesting teaching materials.

Some SANGAKU present problems of drawing smaller circles within a larger circle in contact with it or a circle in contact with a quadratic curve, such as an oval.

G.-M. Greuel et al. (Eds.): ICMS 2016, LNCS 9725, pp. 351–358, 2016.
DOI: 10.1007/978-3-319-42432-3_43

Dynamic geometry software such as Cinderella or GeoGebra is useful to solve these problems. A user can draw a clean figure. Then we can use a high-quality TeX graphics with KeTCindy (Cinderella plug-in) to draw these figures.

2   SANGAKU of Nara Miminashi Yamaguchi Shrine

Here, we present an example using Cinderella and KeTCindy. This is a SAN-GAKU of Nara Miminashi Yamaguchi Shrine (Fig. 1). We show the living language reason of the original.

Fig. 1. Miminashi yamaguchi shrine

Big circle B and small circle A cross. Circle C touches big circle B internally and is circumscribed to small circle A. There is a center of three circles on a straight line. Circle C internally touches with a big circle B and is circumscribed in small circle A. In contact to big circle and small circle, we write it to be circumscribed sequentially from the former.

We can draw a figure of these circles using a function to picture a quadratic curve of Cinderella in (Fig. 2) [2].

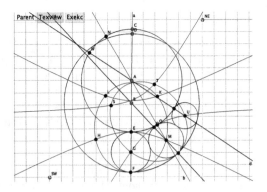

Fig. 2. Cinderella

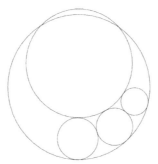

Fig. 3. KeTCindy.

The next figure made Tex file using KetCindy and had pdf (Fig. 3).

We can also draw figures by solving simultaneous equations using Computer Algebra System such as Maxima, Risa/Asir.

Letting a, b, c be the respective radii of circles A, B, and C, and letting r, (x, y) respectively denote a radius and a center coordinate of first circle, then we can describe first circle by deciphering the simultaneous equation below.

$$\begin{cases} x^2 + y^2 = (a + r)^2 \\ (x - (a + 2c - b))^2 + y^2 = (b - r)^2 \\ (x - (a + c))^2 + y^2 = (c + r)^2 \end{cases} \tag{1}$$

The first circle can be written according to the next script. In Cinderella, one can calculate in Maxima when using KeTCindy.

```
if(1==1,
cmdL=[
        "eq1:x^2+y^2-(a+r)^2",[],
        "eq2:(x-(a+2*c-b))^2+y^2-(b-r)^2",[],
        "eq3:(x-(a+c))^2+(y)^2-(c+r)^2",[],
        "ans:solve",["[eq1=0,eq2=0,eq3=0]","[r,x,y]"],
        "ans1:ev",["r","ans[1]"],"ans1:ratsimp",["ans1"],
        "ans2:ev",["x","ans[1]"],"ans2:ratsimp",["ans2"],
        "ans3:ev",["y","ans[2]"],"ans3:ratsimp",["ans3"],
        "ans1::ans2::ans3",[]
];
CalcbyM("ans1",cmdL,["","All=y"]); <-- execute command of Maxima
hankei(a,b,c):=parse(ans1_1); <-- make functions
xzahyo(a,b,c):=parse(ans1_2);
yzahyo(a,b,c):=parse(ans1_3);
);
a=dist(A,C); <-- initial values
b=dist(B,D);
r0=dist(G,E);
```

```
Circledata("1",[[xzahyo(a,b,r0),yzahyo(a,b,r0)],
            [xzahyo(a,b,r0),yzahyo(a,b,r0)]+[hankei(a,b,r0),0]]);
```

Similarly, we do it and produce a simultaneous equation in sequence and can picture a circle.

Circles can be described by deciphering the next simultaneous equation. It is necessary to replace x_0, y_0, z_0 of a provided solution with the center and the radius of the previous circle sequentially.

$$\begin{cases} x^2 + y^2 = (a+z)^2 \\ (x - (a+2c-b))^2 + y^2 = (b-z)^2 \\ (x - x_0)^2 + (y - y_0)^2 = (z_0 + z)^2 \end{cases} \tag{2}$$

Maxima takes time. Alternatively, it does not work when there are too many variables. In such cases, it is possible to convert it to a system of equations easier to solve using a Grobner base in Risa/Asir, and process it correctly by giving the result to Maxima.

```
if(1==1,
cmdL=[
      "load",["'gr'"],
      "Eq1=x^2+y^2-(a+z)^2",[],
      "Eq2=(x-(a+2*c-b))^2+y^2-(b-z)^2",[],
      "Eq3=(x-x0)^2+(y-y0)^2-(z0+z)^2",[],
      "G=nd_gr",["[Eq1,Eq2,Eq3]","[y,x,z]",0,2],  <-- Grobner base
      "G[0]::G[1]::G[2]",[]
];
CalcbyA("ans",cmdL,[""]);  <-- execute command of Risa/Asir
);
if(1==1,
cmdL=[
      "A1:solve",[ans_1+"=0","z"],  <-- hand a result of Risa/Asir
      "ans1:ev",["z","A1[1]"],"ans1:ratsimp",["ans1"],
      "eq2:subst",["ans1","z",ans_2],[];
      "A2:solve",["eq2=0","x"],
      "eq3:subst",["ans1","z",ans_3],[];
      "A3:solve",["eq3=0","y"],
      "ans2:ev",["x","A2[1]"],"ans2:ratsimp",["ans2"],
      "ans3:ev",["y","A3[1]"],"ans3:ratsimp",["ans3"],
      "ans1::ans2::ans3",[]
];
CalcbyM("ans1",cmdL,["","All=y"]);  <-- execute command of Maxima
hankei(a,b,c,x0,y0,z0):=parse(ans1_1);  <-- make functions
xzahyo(a,b,c,x0,y0,z0):=parse(ans1_2);
yzahyo(a,b,c,x0,y0,z0):=parse(ans1_3);
);
a=dist(A,C);  <-- initial values
```

```
b=dist(B,D);
c=dist(G,E);
x0=a+c;
y0=0;
z0=c;
r1=hankei(a,b,c,x0,y0,z0);
x1=xzahyo(a,b,c,x0,y0,z0);
y1=yzahyo(a,b,c,x0,y0,z0);
Circledata("1",[[x1,y1],[x1,y1]+[r1,0]]); <-- first circle
r2=hankei(a,b,c,x1,y1,r1);
x2=xzahyo(a,b,c,x1,y1,r1);
y2=yzahyo(a,b,c,x1,y1,r1);
Circledata("2",[[x2,y2],[x2,y2]+[r2,0]]); <-- second circle
r3=hankei(a,b,c,x2,y2,r2);
x3=xzahyo(a,b,c,x2,y2,r2);
y3=yzahyo(a,b,c,x2,y2,r2);
Circledata("3",[[x3,y3],[x3,y3]+[r3,0]]); <-- third circle
```

We can deal with this on Cinderella through KeTCindy. We can draw circles using the result (Fig. 4).

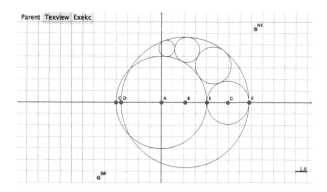

Fig. 4. KeTCindy and CAS(1).

3   SANGAKU of Gunma Kumano Shrine

The next example is a SANGAKU problem from Gunma Kumano Shrine (Fig. 5). It is a problem to draw a figure of circle to touch an oval internally. We can also draw figures by solving simultaneous equations using Maxima, Risa/Asir. When the length of the major axis of the oval is $2a$, the length of the minor axis of the oval is $2b$, the radius of small circle is r, we can describe a circle by deciphering the following simultaneous equation.

$$\begin{cases} 9r^2x(y-r) = a^2(x-3r)y \\ 9r^2(y-r) = y(3rx + ry - 9r^2) \\ 9r^2x^2 + a^2y^2 = 9a^2r^2 \\ (x-3r)^2 + (y-r)^2 = r^2 \end{cases} \tag{3}$$

We can draw circles according to the following script.

Fig. 5. Kumano Shrine

```
if(1==1,
cmdL=[
      "load",["'gr'"],
      "Eq1=9*r^2*x*(y-r)-a^2*y*(x-3*r)",[],
      "Eq2=9*r^2*(y-r)-y*(3*r*x+r*y-9*r^2)",[],
      "Eq3=9*r^2*x^2+a^2*y^2-9*a^2*r^2",[],
      "Eq4=(x-3*r)^2+(y-r)^2-r^2",[],
      "G=nd_gr",["[Eq1,Eq2,Eq3,Eq4],[x,y,a,r]",0,2],
      "G[0]::G[1]::G[2]::G[3]",[]
];
CalcbyA("ans",cmdL,[""]); <-- execute command of Risa/Asir
Mxtex("1",ans_1,["Disp=n"]); <-- make tex form
);
if(1==1,
cmdL=[
      "A1:solve",[ans_1+"=0","r"], <-- hand a result of Risa/Asir
      "ans1:ev",["r","A1[1]"],"ans1:ratsimp",["ans1"],
      "ans2:ev",["r","A1[2]"],"ans2:ratsimp",["ans2"],
      "ans3:ev",["r","A1[3]"],"ans3:ratsimp",["ans3"],
```

```
      "ans4:ev",["r","A1[4]"],"ans4:ratsimp",["ans4"],
      "ans1::ans2::ans3::ans4",[]
];
CalcbyM("ans1",cmdL,["","All=y"]); <-- execute command of Maxima
Mxtex("8",ans1_4,["Disp=n"]); <-- make tex form
);
hankei(a):=parse(ans1_4); <-- make function
a=dist(A,B); <-- initial values
r=hankei(a);
Circledata("1",[A,A+[r,0]]);
Circledata("2",[[0,r],[0,r]+[2*r,0]]);
Circledata("3",[[0,(-1)*r],[0,(-1)*r]+[2*r,0]]);
Circledata("4",[[0,2*r],[0,2*r]+[r,0]]);
Circledata("5",[[0,(-2)*r],[0,(-2)*r]+[r,0]]);
Circledata("6",[[3*r,r],[3*r,r]+[r,0]]);
Circledata("7",[[3*r,(-1)*r],[3*r,(-1)*r]+[r,0]]);
Circledata("8",[[(-3)*r,r],[(-3)*r,r]+[r,0]]);
Circledata("9",[[(-3)*r,(-1)*r],[(-3)*r,(-1)*r]+[r,0]]);
Ellipseplot("1",[[sqrt(a^2-9*r^2),0],[(-1)*sqrt(a^2-9*r^2),0],2*a]);
```

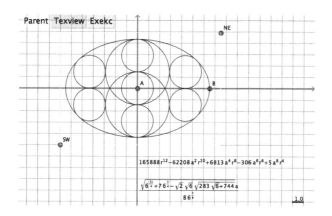

Fig. 6. KeTCindy and CAS(2).

A user can display a figure and calculation result together (Fig. 6). It becomes the interesting teaching materials by combining a figure with a solution using the computer algebra system.

4 Conclusion

We have presented examples of two SANGAKU. We were able to draw a beautiful figure by Cinderella with KetCindy and Computer Algebra System and to demonstrate that they can be used as interesting mathematics teaching materials.

When using KeTCindy, we can relate Computer Algebra System to Cinderella (dynamic geometry software). Such a method can be used in various locations and situations.

References

1. Fukagawa, H., Pedoe, D.: Japanese Temple Geometry Problems Sangaku. Winnipeg, Canada (1989)
2. Makishita, H.: Development of the activated teaching materials in the high school course in mathematics. RIMS Kokyuroku, Japan (2015)

CindyGL: Authoring GPU-Based Interactive Mathematical Content

Aaron Montag$^{(\boxtimes)}$ and Jürgen Richter-Gebert

Technical University of Munich, Munich, Germany
{montag,richter}@ma.tum.de
http://www-m10.ma.tum.de

Abstract. *CindyJS* is a framework for creating interactive (mathematical) content for the web. The plugin *CindyGL* extends this framework and leverages WebGL for parallelized computations.

CindyGL provides access to the GPU fragment shader for *CindyJS*. Among other tasks, the plugin *CindyGL* is used for real-time colorplots.

We introduce the main principles, concepts and application of *CindyGL* and describe the encountered technical challenges. Special focus is put on a novel visualization scheme that uses feedback loops, which were among the motivating forces of developing CindyGL. They can be used for a wide range of applications. Some of them are numerical simulations, cellular automatons and fractal generation, which are described here.

Keywords: Interactive visualization · Web technologies · WebGL · Transpiler · CindyScript · GLSL · OpenGL · Shader based colorplots · Feedback loops on GPU · Fractals · Limit sets · IFS · Kleinian groups

1 Introduction

The *CindyJS* project is a system for authoring dynamic mathematical web content (see [7]). It allows web based prototyping of mathematical experiments and visualizations which can be used for research and demonstration. *CindyScript* is a scripting language for *CindyJS*, that can be directly used in the HTML code. For the design principles of *CindyScript* we refer to [4]. Its language specifications are presented in the Cinderella 2 handbook [5].

In this article the plugin *CindyGL* for *CindyJS* is introduced. *CindyGL* is a plugin for *CindyJS* which provides the high-level mathematically oriented user with access to the shader language of the GPU.

In most other scenarios, knowledge of JavaScript and a shader language is required and many lines of "boilerplate-code" have to be written in order to build even small shader examples in WebGL. On the other hand, they often could be described with only few words. One aim of the WebGL integration into

J. Richter-Gebert—Supported by the DFG Collaborative Research Center TRR 109, "Discretization in Geometry and Dynamics".

© Springer International Publishing Switzerland 2016
G.-M. Greuel et al. (Eds.): ICMS 2016, LNCS 9725, pp. 359–365, 2016.
DOI: 10.1007/978-3-319-42432-3_44

CindyScript through the *CindyGL* plugin is overcoming the technical obstacles that are typically inevitable in usage of OpenGL technologies on the web. While writing *CindyScript* code, the user should not even become aware of using WebGL.

The second aim of *CindyGL* is providing a simple fast-prototyping tool for *feedback loops on the GPU*, that can be used for various novel algorithms. No other web project that overcomes both of this difficulties is known to us.

The technical core of this plugin is a transcompiler, which can translate *CindyScript* to OpenGL Shading Language (GLSL). Aside general-purpose computations on the GPU, the transcompiler is so far used for rendering 2D-colorplots on the GPU. If required, the 2D-colorplots can be animated in real-time as well.

This function is accessible via the `colorplot` command of *CindyScript*. A set of running examples (with their source code) can be viewed on http://cindyjs.org/CindyGL/.

2   The Colorplot Command

The *CindyScript* primitive operations for accessing *CindyGL* were designed such that the boiler plate for creating WebGL-applications is minimized. For instance, an animated plot of the interference of two circular waves can be rendered by evaluating the following *CindyScript* code at every animation step:

```
t = seconds() - t0;
colorplot(
   (sin(|A,#|-t) + sin(|B,#|-t)+2) * (1/2, 1/3, 1/4)
);
```

A static image of the animated result is depicted in Fig. 1(a). The expression inside the `colorplot` command maps pixel coordinates (at position #) to colors (encoded as a 3-component rgb vector). Here `|A,#|` and `|B,#|` are the distances between the current pixel coordinate and the two points on the *CindyJS* canvas, that can be interactively repositioned by drag and drop.

A phase portrait for the complex function $f : \mathbb{C} \to \mathbb{C}$, $z \mapsto z^7 - 1$ can be rendered as follows (the concept of complex phase portraits is explained in [9]):

```
f(z) := z^7 - 1;
colorplot(
   hue(im(log(f(complex(#))))) / (2*pi));
);
```

This program outputs a GPU rendered image as in Fig. 1(b). The argument of `f(complex(#))` determines the color for the pixel with the coordinate #. Note that the computation of complex numbers was inherently carried to the GPU, which has no native support for complex calculus.

Furthermore, sophisticated colorplots are possible as well. As an example, a raycaster for algebraic surfaces can be written as a colorplot in *CindyScript* as depicted in Fig. 1(c). Following the approach of [6], we compute the intersection of each view ray with the algebraic surface as the root of a polynomial that is determined by an interpolation process. In *CindyScript* the interpolation – a linear function that maps lists of evaluated function values to a vector of polynomial coefficients – can be easily described as a matrix multiplication. This matrix computation is coded high-level in *CindyScript* and transpiled to the GPU.

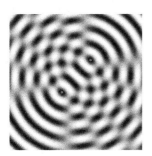

Fig. 1. Screen shots of animations generated by the `colorplot` command: (a) Interference of waves, (b) a complex phase portrait and (c) a raycaster for algebraic surfaces

3   Feedback Loops via *CindyGL*

The `colorplot` command was designed in a way that it is possible to write on textures by a function passed as an argument. The textures then in turn can be read in consecutive calls of `colorplot` with the `imagergb`-command. The possibility to read and write texture data, immediately enables the creation of *feedback loops on the GPU.*

By the term *feedback loop* we mean a system of a set of images that are iteratively re-generated by using themselves. A "physical" example of a single-image feedback loop is the "infinite tunnel" that becomes visible if one points a camera at a screen which directly displays a live video recorded by the camera.

An example for a *feedback loop* in *CindyGL* can be seen here:

```
colorplot("julia", // plots to texture "julia"
  z = complex(#);  // if |z| < 2, take the color from texture "julia"
  if(|z|<2,        // at position z² + c and make it slightly brighter
    imagergb("julia", z^2+c) + (0.01, 0.02, 0.03),
    (0, 0, 0)      // if |z| ≥ 2, (z, f(z), f²(z),....) is not bounded;
  )                // display black.
);
```

If the code is executed several times, a picture of a Julia-fractal for the function $f : z \mapsto z^2 + c$ progressively emerges on the texture `julia`. After roughly 50 iterations a picture as in Fig. 2(a) becomes visible. The Julia-fractal for a specific function depicts the points which remain bounded if the function is iteratively applied to them. The fact that only one iteration per pixel for each rendering step is computed makes the rendering process very fast and enables a real-time escape-time based fractal visualization. Here, a direct and smooth interaction with users changing the parameter c on the fly is possible, even on mobile devices.

Figure 2(b) shows a picture of Conway's Game of Life, where cells of a 2-dimensional grid can either be alive or dead. In each computation step, a cell can die or be reborn according to the number of its living neighbors. Using black (0) and white (1) pixels for dead and living cells respectively, this cellular automaton can be simulated with *CindyGL* as follows:

```
// a function that reads the 80x80 texture "gol",
// assuming a torus like world.
get(x, y) := imagergb("gol", (mod(x, 80), mod(y, 80))).r;
newstate(x, y) := ( number = // number of living neighbors
get(x-1, y+1) + get(x, y+1) + get(x+1, y+1) +
get(x-1, y) +                 get(x+1, y) +
get(x-1, y-1) + get(x, y-1) + get(x+1, y-1);
if(get(x,y)==1, // if the cell lives and it has less than 2
                // or more than 3 neighbors, it will die.
  if((number < 2) % (number > 3), 0, 1),
                // if a cell was dead, then 3 neighbors
                // are required to be born.
  if(number==3, 1, 0)
)
);
```

```
colorplot("gol", newstate(#.x, #.y)); //plots to texture "gol"
```

Here a texture `"gol"`, which encodes the previous state, will be reused as a basis for the computation of all the new states, which will be written to the texture `"gol"` again.

Figure 2(c) shows a simulation of a reaction-diffusion system using feedback loops. It serves as an example how numerical simulations of 2-dimensional partial differential equations can be computed in real time on the GPU. In this example, and also many other numerical simulations, a very fine time discretization is demanded. Since a single iteration step utilizing a feedback loop construction can be computed very fast on the GPU, many iterations of the feedback loop can be done before displaying a single frame. On today's average hardware, decent frame rates are still possible.

Feedback loops also give a natural framework to render limit sets on the GPU. Visualizations of the limit sets of two dimensional *iterated function systems* (IFSs), which are described in [1], can be generated by iteratively applying a slightly modified Hutchinson operator to a texture: A texture is iteratively re-built as a composition of deformed copies of itself. This can be considered as a feedback loop of a single texture and results of *CindyGL* implementations are shown in Fig. 3(a) and (b).

By extending the feedback loop system containing a single texture to a system containing multiple textures that are linked in a sophisticated manner, it is also possible to visualize limit sets of certain Kleinian groups in real time. An example image of such an CindyGL generated limit set of a Kleinian group

 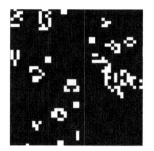 

Fig. 2. Visualizations generated by feedback loops: (a) a progressively built up Julia fractal, (b) Conway's Game of Life and (c) a reaction-diffusion model

is depicted in Fig. 3(c). The required techniques are derivated and described in detail in [2]. Summarizing the generation of Fig. 3(c), two Möbius transformations were chosen by "grandma's recipe" from [3] to generate a free group. Then a deterministic finite automaton was built that accepts the regular language of the geodesic words of the language of the free group, i.e. the shortest words consisting of the two generators and their inverses describing the group elements. By transferring the states of this automaton to textures and the transition between them to corresponding Möbius transformations that are used to generate each of these textures, a complex interlinked system of textures is generated. Now by iterating simultaneously the generation of these textures, one can prove that in the limit an image of the limit set of the Kleinian group is attained.

CindyGL is a tool that can be used to built such interlinked systems of textures with relatively little effort.

4   Technical Aspects

CindyJS is licensed under the Apache 2 license and can be obtained from https:// github.com/cindyjs. The plugin *CindyGL* is integrated into the *CindyJS* project.

One development aim for *CindyGL* was obtaining a performance that is comparable with the one of native WebGL applications. During real time animations,

Fig. 3. Images of different limit sets generated by feedback loops: (a) an IFS generated by two affine transformations, (b) an IFS generated by circle inversions and (c) a Kleinian group

the `colorplot` command is called many times within a second. Typically, the syntactic expression within the argument remains the same – only the values of variables might change.

Performance preservation was mainly achieved by doing all the computations that are demanded by an additional layer between a native WebGL application and a *CindyJS* application only at the first time the `colorplot`-command is called. During successive calls of `colorplot` (with the same arguments), a native shader program is executed.

OpenGL Shading Language (GLSL) is the language which is used to run specific programs on the GPU in WebGL. It is strongly typed. In contrast, *CindyScript* has dynamic typing.

We have developed a transcompiler that is able to translate *CindyScript* primitives into GLSL. During this process types of terms and variables in *CindyJS* (e.g. real numbers, complex numbers, matrices, ...) are – if possible – automatically detected and modeled to corresponding data structures on the GPU (e.g. `float`, `vec2`, `mat4`, ...).

A partial order on the types has been introduced in order to capture subtype relations between types. A type is defined to be a subtype of another type (for instance, real numbers are a subtype of complex numbers), if there is a inclusion function from values of the subtype to the other type such that every function having multiple signatures for different types commutes with all the inclusion functions. Hence, the "weakest possible" type can always be chosen for the calculations in order to save resources and obtain good performance.

When the function `colorplot` is called for the first time, the syntax tree of the color expression and functions that are called within this expression are traversed recursively in order to find out the terms that depend on the varying pixel variable #. Those terms are suitable for a massive parallelization on the GPU and are translated to GLSL via the introduced transcompiler. A fragment shader is built in WebGL, that computes the corresponding expressions for each

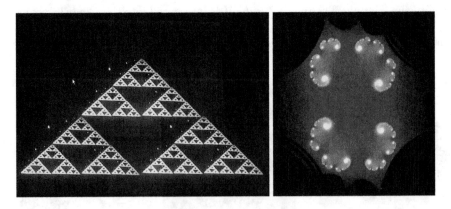

Fig. 4. Two fractals generated by analog feedback loops. (Using an integrated webcam and a mirror)

pixel, while the other terms that are independent from # are calculated just once on the CPU and passed to the GPU as uniforms. Since the segmentation in parallelized code and CPU code is created automatically, this on the on hand eases the work of the programmer. On the other hand, it very often creates the most general and performant split of the code.

A *ping-pong approach* is used for the feedback loops. If `colorplot` tries to read and write on a texture at the same time, the texture will be stored twice: One texture for reading and another target texture for writing. After the function call, the two textures will be swapped. In the next call of `colorplot` then recently written texture can be used as input texture.

5   Conclusion and Outlook

Overall, the *CindyGL* project aims to provide an easy-to-use technical backbone for a wide range of different mathematical visualizations.

CindyGL is not finished yet. Some primitive operations and data structures from *CindyScript* are still missing. Also, an integration of the ideas of [8] is planned. In particular, enabling live access to a camera is possible. Using a live image of a camera-picture as input texture for a `colorplot` opens the door for new educational concepts. The image can be easily deformed using *CindyScript*. An analog design setup where a camera points to the currently displayed image can be used to explain the concept of feedback loops. For a valuable educational experience, the real world, that might consist of persons, patterns or a system of mirrors for example, can be included in the setting. Results of a prototypical setting are shown in Fig. 4.

References

1. Barnsley, M.F.: Fractals Everywhere. Academic Press, Boston (2014)
2. Montag, A.: Interactive image sequences converging to fractals. Bachelors Thesis. http://aaron.montag.info/ba/main.pdf
3. Mumford, D., Series, C., Wright, D.: Indra's Pearls: The Vision of Felix Klein. Cambridge University Press, Cambridge (2002)
4. Richter-Gebert, J., Kortenkamp, U.: The power of scripting: DGS meets programming. Acta didactica Napocensia **3**(2), 67–78 (2010)
5. Richter-Gebert, J., Kortenkamp, U.: The Cinderella.2 Manual: Working with the Interactive Geometry Software. Springer, Heidelberg (2012)
6. Stussak, C.: Echtzeit-Raytracing algebraischer Flächen auf der GPU. Ph.D. thesis, Diploma thesis, Martin Luther University Halle-Wittenberg (2007)
7. von Gagern, M., Kortenkamp, U., Richter-Gebert, J., Strobel, M.: CindyJS - Mathematical visualization on modern devices. Submitted to ICMS 2016 Berlin (unpublished)
8. von Gagern, M., Mercat, C.: A library of OpenGL-based mathematical image filters. In: Fukuda, K., Hoeven, J., Joswig, M., Takayama, N. (eds.) ICMS 2010. LNCS, vol. 6327, pp. 174–185. Springer, Heidelberg (2010)
9. Wegert, E.: Visual Complex Functions: An Introduction with Phase Portraits. Springer Science & Business Media, Berlin (2012)

Theoretical Physics, Applied Mathematics and Visualizations

Haiduke Sarafian$^{(\boxtimes)}$

The Pennsylvania State University, University College, USA
has2@psu.edu

Abstract. The conceptual aspects of the majority of physical phenomena readily are comprehensible, yet their analysis conducive to justifiable output require mathematical justifications. Applied mathematics is the backbone of theoretical physics. No field in physics in particular and science in general is immune. Within the last couple of decades advances in computer science introduced a fresh pathway, *computational physics*, augmenting the field. The offspring of these innovations is the scientific software capable of performing operations that could not be accomplished traditionally. The impact of these spectacular innovative technologies is evidence in scientific literature. The focus of this article is to demonstrate the graphical usefulness of one such scientific software, Mathematica analyzing the electrostatic features of discrete charge distributions. This is an example of a theoretical physics problem focusing on the overlap of physics, graphics and math. Ever since its birth a quarter century ago, Mathematica steadily has been growing in popularity and practicality. This article embodies the codes compatible with the latest version of the software including one, two and three dimensional *sliders*. Practitioner physicists, interested individuals and mathematicians may adjust the code to meet their needs.

Keywords: 2D and 3D electrostatic potential · Theoretical physics · Mathematica

1 Introduction

Electrostatics is a well established branch of physics – however, comprehension of some of its abstract concepts relies on visualization of quantities such as scalar potentials and vector fields. Visualizing the potentials could be challenging; this by itself contributes to the challenges of understanding the related concepts – leave alone the associated vector fields. It is not a common practice to display the potentials, the majority of the standard texts [1,2] conveniently have ignored them. In a few cases, 2D contour plots are displayed [3], seldomly 3D plots of the contours are considered.

Since the aforementioned texts have been published, computer technology has progressed tremendously. Along with technological advances, powerful software programs capable of performing symbolic and graphic scientific computation have been developed. By adapting one such program, *Mathematica* [4], the

© Springer International Publishing Switzerland 2016
G.-M. Greuel et al. (Eds.): ICMS 2016, LNCS 9725, pp. 366–370, 2016.
DOI: 10.1007/978-3-319-42432-3_45

author by way of examples revisited a few basic cases. For a systematic approach
to graphics review the text by the author [5].

The 2D and 3D potential curves and surfaces created by charged particles
possess somewhat artistic features. Scientifically speaking, a charge distorts the
homogeneity of the space. To study the shape of the distortion and the interaction between the charges, the electric potential is introduced. The electric potential at point r from a set of scattered discrete point charges $q_i, (i = 1, 2, 3, \cdots, n)$
positioned at r_i is $V(r) = \sum_{i=1}^{n} V(|r - r_i|)$, where $V(|r - r_i|) = \frac{kq_i}{|r-r_i|}$ with
$|r - r_i|$ being the distance between r and r_i and k is a constant. By applying
this mathematical function to various situations in the following sections a few
fundamental cases are discussed.

2   Case Studies

We skip the electrostatic character of a single point charge and begin with two
charged particles. As stated each charge generates its own potential subject
to expressions given in the Introduction. We place two charges symmetrically
at $(a,0)$ and $(-a, 0)$ about the origin. The associated electrostatic potential is
$V(x, y) = \frac{kq_1}{\sqrt{(x-a)^2+y^2}} + \frac{kq_2}{\sqrt{(x+a)^2+y^2}}$. For demonstration purpose we set $kq_1 = 1$,
and $kq_2 = 0.2$ and conveniently set $a = 0.5$ units. The 2D contour plot of
equipotentials are displayed in Fig. 1. Its generating code is:

To investigate the impact of placing the charges at different positions we
modify the code applying Manipulate. The output of the code allows interactiv-
ity. Its function shown on the right panel of Fig. 1. Its generating code is given
below:

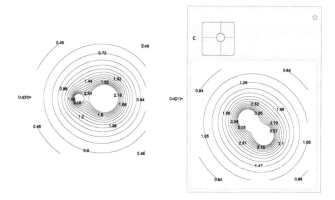

Fig. 1. Contour curves of two charged particles along with their associated numeric
contour values. The right panel is an interactive version of the left generated with a
modified code above.

$$\text{Manipulate}\left[\text{ContourPlot}\left[\frac{1.0}{\sqrt{(x-c[\![1]\!])^2+(y+c[\![1]\!])^2}}+\frac{1.0}{\sqrt{(x+c[\![2]\!])^2+(y-c[\![2]\!])^2}},\right.\right.$$

$\{x, -2, 2\}, \{y, -2, 2\}, \text{Frame} \rightarrow \text{False}(\text{*ColorFunction}\rightarrow\text{Hue*}),$

$\text{ContourShading} \rightarrow \text{None}, \text{Contours} \rightarrow 15,$

$\left.\text{ContourLabels} \rightarrow \text{True}, \text{ImageSize} \rightarrow 250\right],$

$\left.\{c, \{-2, -2\}, \{2, 2\}\}\right]$

The square at the upper left corner of the output is a 2D slider. It is an interactive slider, meaning placing the mouse pointer at the crossing round circle of the intersecting axes allows interactively to move the charges. The action of the slider is instantaneous. This is a useful feature and one may utilize in design mode.

We have upgraded the functionality of the design by adding options to change the value of the charges. This is done by inserting two additional control parameters. The output has three interactive sliders. The top square slider allows two dimensional movement and the vertical sliders control the values of the individual charges.

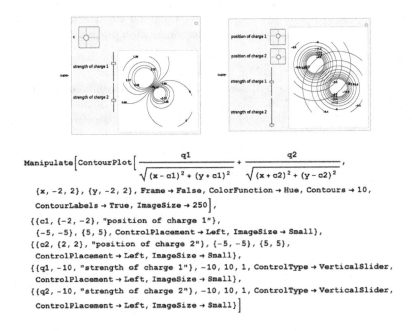

$$\text{Manipulate}\left[\text{ContourPlot}\left[\frac{q1}{\sqrt{(x-c1)^2+(y+c1)^2}}+\frac{q2}{\sqrt{(x+c2)^2+(y-c2)^2}},\right.\right.$$

$\{x, -2, 2\}, \{y, -2, 2\}, \text{Frame} \rightarrow \text{False}, \text{ColorFunction} \rightarrow \text{Hue}, \text{Contours} \rightarrow 10,$

$\left.\text{ContourLabels} \rightarrow \text{True}, \text{ImageSize} \rightarrow 250\right],$

$\{\{c1, \{-2, -2\}, \text{"position of charge 1"}\},$

$\{-5, -5\}, \{5, 5\}, \text{ControlPlacement} \rightarrow \text{Left}, \text{ImageSize} \rightarrow \text{Small}\},$

$\{\{c2, \{2, 2\}, \text{"position of charge 2"}\}, \{-5, -5\}, \{5, 5\},$

$\text{ControlPlacement} \rightarrow \text{Left}, \text{ImageSize} \rightarrow \text{Small}\},$

$\{\{q1, -10, \text{"strength of charge 1"}\}, -10, 10, 1, \text{ControlType} \rightarrow \text{VerticalSlider},$

$\text{ControlPlacement} \rightarrow \text{Left}, \text{ImageSize} \rightarrow \text{Small}\},$

$\{\{q2, -10, \text{"strength of charge 2"}\}, -10, 10, 1, \text{ControlType} \rightarrow \text{VerticalSlider},$

$\left.\text{ControlPlacement} \rightarrow \text{Left}, \text{ImageSize} \rightarrow \text{Small}\}\right]$

Fig. 2. Similar to Fig. 1 and includes a 2D slider.

The right panel of Fig. 2 is the refined version of the left panel; position of each charge is controlled individually.

These examples show the power of the program and its usefulness. Following the same steps one may add additional point-like charges displaying their corresponding 2D associated contours and if needed their density plots.

The natural extension of the shown examples is the corresponding 3D electrostatic potential surfaces. We consider three charges placing them at three interactive locations. Their associated potential surface is shown in Fig. 3.

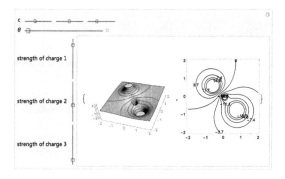

Fig. 3. A 3D potential surface and its 2D contours for a three point-like charges.

As in 2D cases the program is interactive. With the assist of the slider one may study the impact of the values of the individual charges and their corresponding geometric locations.

Figure 3 includes two different plots, a 3D side-by-side with a 2D contour plot. An addition options added to view the surfaces at different desired angles. The observation angle is automated so that the surfaces can be viewed at different angles.

If needed one may place charges not all on the horizontal plane but on any position in a 3D space. This is done by rotating the plane of the charges. For instance we consider four charges, two on the xy-plane and the other two on the yz-plane. Their potential is,

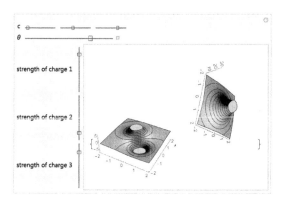

Fig. 4. 3D plots of four different charges; two in a xy-plane (left) and the other two on the yz-plane (right).

For practical applications one may add an automation displaying the potential from any desired observation angle. The graph shown in Fig. 4 includes two 3D potential, the one on the left is static and the one on the right may be either manually and/or automatically rotated.

3 Summary and Conclusion

By way of examples it is shown that it is useful to display the abstract mathematical functions describing distorted charged space. Characters of distorted space are shown plotting 2D and 3D associated electrostatic potentials. Described methods along with accompanied codes can readily be extended to study cases of interest. Reviewing these plots may intrigue a physicist to think about the implicit artistic features of the distorted space or an artist conversely may be fascinated about the way the nature works.

References

1. Halliday, D., Resnick, R., Walker, J.: Fundamental of Physics, 6th edn. John Wiley, New York (2001)
2. Purcell, E.M.: Electricity and Magnetism, Berkeley Physics Course, vol. 2. McGraw-Hill, New York (1965)
3. Reitz, J.R., Milford, F.J.: Foundations of Electromagnetic Theory. Addison-Wesley, Reading (1960)
4. Wolfram, S.: Computational software program to do scientific computation. Wolfram Research, Champaign (2014)
5. Sarafian, H.: Mathematica Graphic Example Book for Beginners. Scientific Research Publishing Inc., USA (2015). http://www.scirp.org

What is and How to Use KETCindy – Linkage Between Dynamic Geometry Software and LATEX Graphics Capabilities –

Setsuo Takato[✉]

Toho University, Funabashi, Japan
takato@phar.toho-u.ac.jp
http://ketpic.com

Abstract. We introduce KeTCindy, which is the right combination of Cinderella and KeTpic we developed to produce high-quality LaTeX graphics using free mathematical software such as Scilab and R. In KeTCindy, Cinderella works as a GUI of KeTpic, so the interactive operation on PC display can be reflected directly on the generated image on LaTeX final output. The generated image can be finely tuned using KeTCindy commands embedded into CindyScript, the scripting language of Cinderella. KeTCindy can be regarded as a prominent scheme to establish an effective linkage between visualization tools and editing tools. Moreover, KeTCindy enables the importation of data calculated or simulated using other mathematical software such as Maxima, Fricas, Risa/Asir and R, and to combine them with the graphical data, so that an extremely wide range of mathematical objects can be presented. We will show some eminent capabilities of KetCindy and also its usage.

Keywords: KeTpic · KeTcindy · Cinderella · LaTeX

1 Introduction

In mathematics classes given at the collegiate level, printed materials are often distributed. Many teachers produce such materials with LATEX because LATEX produces the best looking typeset text and mathematical formulas. They are accustomed to using it. However, few teachers insert figures in their printed materials despite the fact that figures are important and that they are almost indispensable for students to understand course contents. This lack of use of figures appears to be attributable to teachers' lack of abilities for using tools to insert figures into LATEX documents. Alternatively, such tools in themselves might be insufficient for teachers' needs. The normal way is to make figures with other software and insert them using the command includegraphics. However, differences between the text body and these figures will cause students to feel incompatible. Use of TiKZ is another means of producing figures. However, TiKZ has abundant commands that are somehow excessive to produce printed

© Springer International Publishing Switzerland 2016
G.-M. Greuel et al. (Eds.): ICMS 2016, LNCS 9725, pp. 371–379, 2016.
DOI: 10.1007/978-3-319-42432-3_46

materials distributed in mathematics classes. Moreover, the scripts are not only difficult to write but are also difficult to read. As a simple example, it is necessary to try to make graphs of $y = \sin x$ and $y = x$.

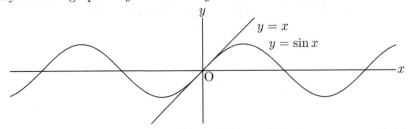

An example of TiKZ scripts is shown below, which shows poor readability.

```
\begin{TiKZpicture}
  \draw[->, ultra thick, opacity=0.7] (0,2) -- (16,2) node[right] {$x$};
  \draw[->, ultra thick, opacity=0.7] (8,0) -- (8,4) node[above] {$y$};
  \draw[domain=-7.5:7.5, xshift=8cm, yshift=2cm, very thick, samples=80]
      plot[id=sin] function{sin(x)} node[above right] {$y=\sin x$};
  \draw[domain=-2:2, xshift=8cm, yshift=2cm, very thick, samples=80]
      plot[id=x] function{x} node[above right] {$y=x$};
  \node [xshift=8cm, yshift=2cm] (0) at (0,0) [label=225:$0$] {};
  \foreach \x in {-6,-4,-2,2,4,6}
    \fill [radius=1.5pt, xshift=8cm, yshift=2cm] (\x, 0)
        circle node[below] {$\x$};
  \foreach \y in {-1,1}
    \fill [radius=1.5pt, xshift=8cm, yshift=2cm] (0, \y)
        circle node[left] {$\y$};
\end{TiKZpicture}
```

For these reasons, we developed KᴇTpic, which is a macro package of mathematical software such as Maple, Mathematica, Scilab, R, and the first Maple version of KᴇTpic was released in 2006. Recently, Scilab has become the most often used of this software.

The flow of generating and inserting graphs with KᴇTpic is the following.

1. One describes KᴇTpic and Scilab commands in Scilab editor and executes them.
2. Scilab generates a LATEX file composed of codes for drawing figures.
3. One inputs the file into a LATEX document with \input command.
4. Compiling the document, one can obtain the pdf file.

Scripts for the example presented above are the following. As might be readily apparent, the readability of the scripts of KᴇTpic is much better than those of TiKZ.

```
Setwindow([-7.5,7.5],[-2,2]);
A=[2.5,1]; B=[2.0,1.5];
Setax(7,"se");
```

```
gr1=Plotdata("sin(x)","x");
gr2=Plotdata("x","x","Num=1");
Openfile("figsin");
  Drwline(gr1,gr2);
  Expr(A,"e","y=\sin x",B,"e","y=x");
Closefile('1');
```

With LATEX and K_ETpic, collegiate teachers were able to produce printed materials with figures easily and on a daily basis. However, they must write all scripts in the editor before confirming on the screen that the figure is desirable. This fact sometimes caused some hesitation of teachers, deterring them from using K_ETpic. As a result, many have desired a graphical user interface (GUI) for use with K_ETpic.

Cinderella [1] is a dynamic geometry software (DGS) package. We had been exploring the possibility of using Cinderella as the GUI of K_ETpic. In 2014, we invited Professor Kortenkamp, who is a main developer of Cinderella, and had a research meeting with him. The first version of K_ETCindy, a combination of Cinderella and K_ETpic, was released on September, 2014. The package can be downloaded freely from a link page of our website. ketpic.com or directly from https://www.dropbox.com/sh/kzt2bgaz07n7dr0/AABZRvOrqqCp5Tn1JZYpnvSQa?dl=0.

2   Flow of K_ETCindy

The next image shows a Cinderella screen. The display screen of figures is shown at left. The editor screen of CindyScript is shown at right. The message display console of CindyScript is portrayed at the lower right.

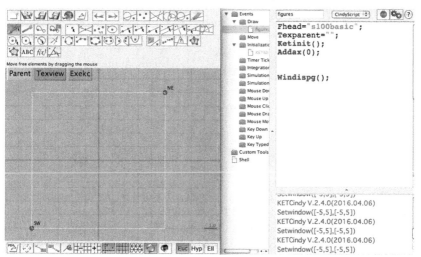

The display screen is similar to that of other DGSs. A user arranges geometric elements such as points, segments, lines, and moves them freely on this screen.

CindyScript is the programming language of Cinderella. It distinguishes Cinderella from other DGSs. Actually, KETCindy is a macro package of CindyScript, uses the display screen only to put points, draw auxiliary segments/lines/circles, to decide the area for LATEX drawing, and to confirm the figure beforehand.

Scripts used to make a figure with KETCindy are simpler than those of KETpic, for example, to produce a graph of $y = \sin x$.

1. Put points A, B on the display screen.
2. Write scripts using the CindyScript editor as shown below.

```
Fhead="sin";
Ketinit();
Setax([7,"se"]);
Plotdata("1","sin(x)","x");
Plotdata("2","x","x",["Num=1"]);
Expr(["A","e","y=\sin x","B","e","y=x"]);
Windispg();
```

3. Press the button Texview and Exekc in order on the display screen.
4. Then one can obtain both a LATEX file "fig.tex" to be input other LATEX sources and a pdf file "figmain.pdf" to be generated for confirmation.

The image below shows the flow of producing a figure with KETCindy.

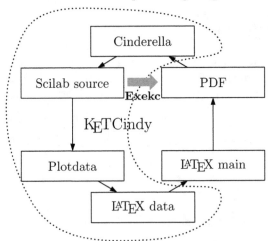

3   Examples

With the latest version of KETCindy, one can produce geometric figures, graphs of functions, various tables, figures of Bézier curves, figures of space curves, polyhedra, surfaces. Moreover, we can call computer algebra systems (CASs) Maxima, Fricas, Risa/Asir from KETCindy. The results are returned to KETCindy, which is useful for producing figures or for additional calculations.

In these sections, we present examples of the above with main part of scripts and figures. For some of them, brief explanations will be added.

3.1 Geometric Figures and Graphs of Functions

Ex 1 : Triangle and Inscribed Circle

In preparation, put points A, B, C, draw triangle ABC and the Inscribe circle on the display screen.

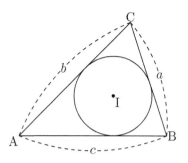

```
Listplot([A,B,C,A]);
Circledata([D,E]);
Bowdata([A,B],[1,0.5,"Expr=c","da"]);
Bowdata([B,C],[1,0.5,"Expr=a","da"]);
Bowdata([C,A],[1,0.5,"Expr=b","da"]);
Pointdata("I",D,["size=4"]);
Letter([A,"sw","A",B,"ne","B",C,"se","C"]);
Letter([D,"s1e1","I"]);
```

Ex 2 : Graph of Solution of Differential Equation

One can apply sliders to decide coefficients of the equation as shown below. Scripts are very simple as described below.

```
Deqplot("2","y''=-L.x*y'-G.x*y","t=[0,XMAX]",0,[C.y,0],["Num=200"]);
Expr([M,"e","\dfrac{d^2 x}{dt^2}+"+L.x+"\dfrac{dx}{dt}+"+G.x+"x=0"]);
```

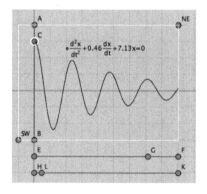
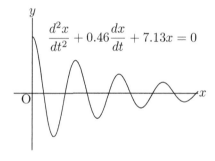

$$\frac{d^2x}{dt^2} + 0.46\frac{dx}{dt} + 7.13x = 0$$

3.2 Making Tables

It is troublesome work to insert tables into the LATEX document. LATEX has poor ability to produce tables, and only tabular or array environments. Using KᴇTCindy, one can produce such tables easily. Moreover, one can move each rule that is, each interval, by moving points on the display screen.

Ex 3 : Simple Table

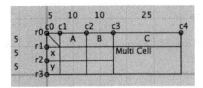

	A	B	C
x			Multi Cell
y			

Scripts are as follows.

```
xL=[5,10,15,20]; // Set intervals between vertical rules
yL=[5,5,5]; // Set intervals between horizontal rules
rL=["r2c3c4"];   // Set rules removed
Tabledata("",xL,yL,rL);
Tlistplot(["c0r0","c1r1"]);
Putrow(1,"c",["","A","B","C"]);
Putcolexpr(1,"c",["","x","y"]);
Putcell("c3r1","c4r3","l1t1","Multi Cell");
```

Ex 4 : Special Form of Matrix/Determinant

Matrix is a kind of table. It can therefore be input into the LaTeX document easily with KₑTCindy.

```
xL=5*[1,1,1,1]; yL=5*[1,1,1,1]; rL=[];
Tabledata("",xL,yL,rL,["notex"]);
Putrowexpr(1,"c",["a_{11}","a_{12}",
   "\cdots","a_{1n}"]);
Putcolexpr(1,"c",["","0","\vdots","0"]);
Putcell("c1r1","c4r4","c","\Large$A_{n-1}$");
Tlistplot(["c0r0","c0r4"]);
Tlistplot(["c4r0","c4r4"]);
Tlistplot(["c0r1","c4r1"],["do"]);
Tlistplot(["c1r0","c1r4"],["do"]);
```

$$\begin{vmatrix} a_{11} & a_{12} \cdots a_{1n} \\ 0 & \\ \vdots & A_{n-1} \\ 0 & \end{vmatrix} = a_{11}|A_{n-1}|$$

3.3   Bézier Curves

KₑTCindy supports several commands related to Bézier curves. Here we give examples of Ospline and Mkbezierptcrv.

Ex 5 : Ospline Command

Oshima [2] devised an algorithm to improve the degree of fitness of a spline curve to the origin curve. We have implemented it to KₑTCindy as a command "Ospline". In the following figures, a dotted curve is the original circle, a dashed curve is a Catmull–Rom spline, and a solid curve is an Oshima spline. Knots, which are control points on the curve, are freely movable.

```
Circledata([A,B],["do"]);
CRspline("1",[B,C,D,B],["da"]);
Ospline("1",[B,C,D,B]);
```

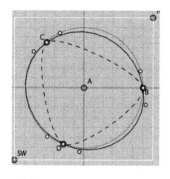 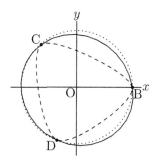

Ex 6 : Mkbezierptcrv Command

The command "Mkbezierptcrv" takes the initial position of control points on trisection points of each segment as the lower-left of the following figures.

```
drawimage([0,0],"bezieroriginal.png",scale->2,alpha->0.4);
Mkbezierptcrv([A,B,C,D,E,A]);
```

Here we use a functionality of Cinderella to display an image with png or jpeg format on the screen. Moving those control points, fitting the curve to the original picture, we can get the lower-right LaTeX figure.

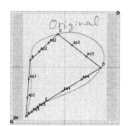 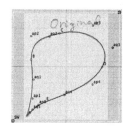 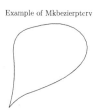

Example of Mkbezierptcrv

3.4   Calling CASs from K$_E$TCindy

Cinderella does not support symbolic computation or calculation of special functions. Especially the former is often necessary for mathematics education. Therefore, we have implemented K$_E$TCindy to call CASs as the following chart.

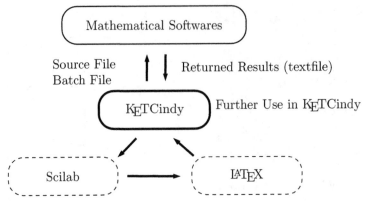

Here we give examples of calling Maxima.

Ex 7 : Calling Single Command of Maxima

```
Mxfun("1","diff",["1/(x^2+1)","x"]);
Plotdata("1","1/(x^2+1)","x",["da"]);
Plotdata("2",mx1,"x",["Num=100"]);
Mxtex("1",mx1); Mxtex("f","1/(x^2+1)");
Expr([A,"e",txf,B,"e",tx1]);
```

Mxfun and Mxtex are commands of calling Maxima.

Ex 8 : Calling Several Commands of Maxima

```
mat="[1,1,1],[1,2,2],[2,1,1]";
cmdL=["batch",[Dq+mxL+Dq],
 "A:matrix",[mat],
 "A0:rowadd",["A",2,1,-1],
 "A1:rowadd",["A0",3,1,-2],
 "A2:rowadd",["A1",3,2,1],
 "A::A0::A1::A2",[] ];
CalcbyM("ans",cmdL);
```

$$\begin{pmatrix} 1 & 1 & 1 \\ 1 & 2 & 2 \\ 2 & 1 & 1 \end{pmatrix} \quad \begin{pmatrix} 1 & 1 & 1 \\ 0 & 1 & 1 \\ 2 & 1 & 1 \end{pmatrix}$$

$$\begin{pmatrix} 1 & 1 & 1 \\ 0 & 1 & 1 \\ 0 & -1 & -1 \end{pmatrix} \quad \begin{pmatrix} 1 & 1 & 1 \\ 0 & 1 & 1 \\ 0 & 0 & 0 \end{pmatrix}$$

4   Future Work

We can also produce space figures using KETCindy.
Improvement of the algorithm to speed up hidden line processing is a pressing issue.

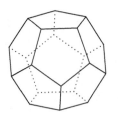

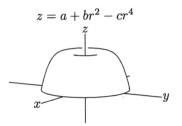

$$z = a + br^2 - cr^4$$

Acknowledgment. This work was supported by JSPS KAKENHI Grant Numbers 25350370, 15K01037, 15K00944.

References

1. Cinderella. http://www.cinderella.de/tiki-index.php
2. Oshima, T.: Drawing curves. In: Symposium MEIS 2015: Mathematical Progress in Expressive Image Synthesis. MI Lecture Notes 2015, vol. 64, pp. 117–20. Kyushu University (2015). ISSN: 2188–200

How to Generate Figures at the Preferred Position of a TeX Document

Hisashi Usui[✉]

National Institute of Technology, Gunma College, Maebashi, Japan
usui@nat.gunma-ct.ac.jp

Abstract. When we use TeX to edit a document, it is sometimes necessary to place the figure of a preferred shape into a suitable position. In this presentation, we propose a method using KeTCindy for this purpose. KeTCindy is a plug-in to Cinderella that converts the procedure to generate geometric shapes into TeX readable code to generate the corresponding image on TeX final output. One merit of using KeTCindy is its interactive character. On the Cinderella screen, a user can control the shape of the figure as desired. When we place the resulting image at the exterior side of text part, simple conversion to TeX graphical image through KeTCindy is sufficient. However, when it is necessary to place it onto the text part, some extra elaboration is necessary to ensure that both the text part and the generated figure are finely balanced. The key idea is making the screen of Cinderella semi-transparent using software named feewhee.

Keywords: KeTCindy · KeTpic · Cindellera · Semi-transparent

1 Introduction

We use TeX to create a document with mathematical expressions. When we create a document, we sometimes want to add a simple figure into the text part. At such times, it is necessary to place the figure of preferred shape into a suitable position. For example, we might want to create a TeX document like this:

$$A\boldsymbol{x} = \lambda\boldsymbol{x} \quad (\boldsymbol{x} \neq \boldsymbol{0}) \Longleftrightarrow \lambda : \text{an eigen value} \ , \ \boldsymbol{x} : \text{an eigen vector}$$
$$(A - \lambda E)\boldsymbol{x} = \boldsymbol{0} \qquad\qquad\qquad\qquad \text{contradict}$$
$$\exists (A - \lambda E)^{-1}(\Longleftrightarrow |A - \lambda E| \neq 0) \Longrightarrow \boldsymbol{x} = \boldsymbol{0}$$
$$|A - \lambda E| = 0 \ : \text{the characteristic equation of A}$$

From the perspective of reuse and modification, it is better to do so in TeX system than using some other method such as a PDF editor or manual drawing. In this paper, we propose a method using KeTCindy [4]. KeTCindy is a plug-in to Cinderella [2] that converts the procedure to generate geometric shapes into TeX readable code to generate the corresponding image on TeX final output. One

© Springer International Publishing Switzerland 2016
G.-M. Greuel et al. (Eds.): ICMS 2016, LNCS 9725, pp. 380–385, 2016.
DOI: 10.1007/978-3-319-42432-3_47

merit of using KETCindy is its interactive character. On the Cinderella screen, a user can control the the figure shape as desired. When placing the resulting image at the exterior side of text part, simple conversion to TEX graphical image through KETCindy is sufficient. However, when it is necessary to place it onto the text part, some extra elaboration is necessary to ensure that both the text part and the generated figure are finely balanced. KETCindy has a ketlayer environment, which enables a user to put figures into a TEX document at the desired position ([1,3]). The idea is explained below. First, we lay the screen of Cinderella on the screen of the PDF file made from the TEX source file. Second, we make the Cinderella screen semi-transparent using software named feewhee [5]. Then we adjust the shape of the figure to the desired shape and produce a figure file by KETCindy. By compiling the TEX source file written to import the figure file in advance, one can obtain a document with figures as desired.

2   Underlying Technique

- KETCindy ([2,4])

  KETCindy is a plug-in to Cinderella which converts the procedure to generate geometric shapes into TEX readable "tpic" code to generate the corresponding image on TEX final output. One of the merit of using KETCindy is its interactive character. On the screen of Cinderella, we can control the shape of the figure as we want. A user can also draw a free type curve such as a Bezier curve. Of course, one can control the curve shape by moving the control points. A user can make a figure file that is useful by the input command in a TEX document. A PDF file of the figure can be made even if the system does not support tpic.

- ketlayer ([1,3])

  KETCindy has a ketlayer environment of TEX. The ketlayer environment enables a user to put a figure or letters at a preferred position of the document.

  When we use ketlayer, we show the scale guide grid lines at first.

```
\begin{layer}{100}{20}
\end{layer}
```

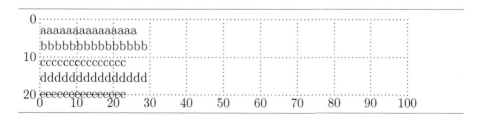

Then we input the position data to fix a figure or letters at the desired position.

```
\begin{layer}{100}{20}
\putnotese{45}{6}{Here}
\putnotenw{98}{17}{There}
\putnotec{70}{10}{$\bigcirc$}
\end{layer}
```

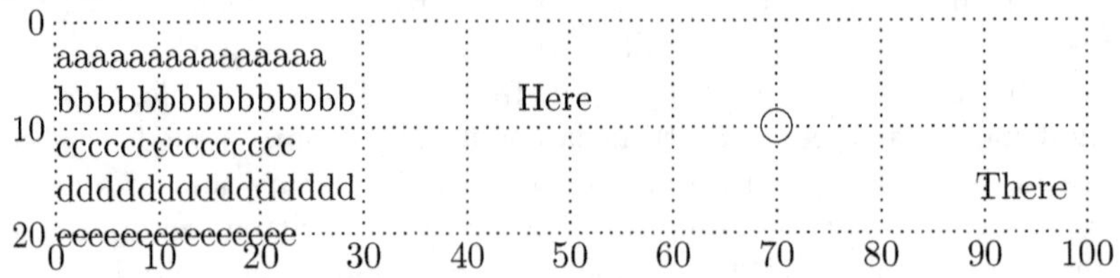
aaaaaaaaaaaaaaa
bbbbbbbbbbbbbbb
Here
ccccccccccccccc
ddddddddddddddd
There
eeeeeeeeeeeeeee
0
10
20
0 10 20 30 40 50 60 70 80 90 100

Finally, we erase the scale guide grid lines.

```
\begin{layer}{100}{0}
\putnotese{45}{6}{Here}
\putnotenw{98}{17}{There}
\putnotec{70}{10}{$\bigcirc$}
\end{layer}
```

Change to 0

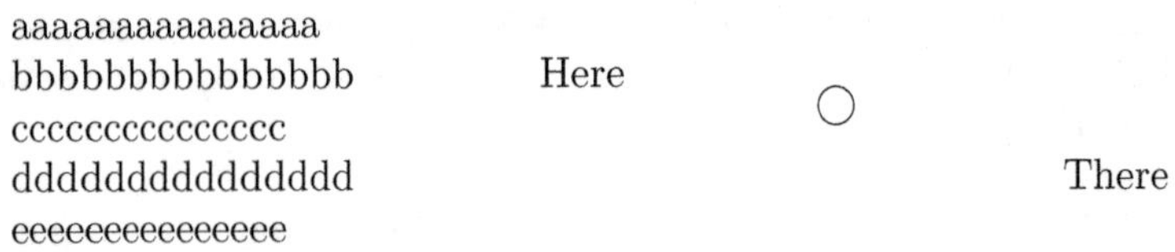
aaaaaaaaaaaaaaa
bbbbbbbbbbbbbbb
Here
ccccccccccccccc
ddddddddddddddd
There
eeeeeeeeeeeeeee

– feewhee.exe ([5])
  We use feewhee.exe to control the transparency of the Cinderella screen. The feewhee.exe software was developed by Nattyware for Windows (Microsoft Corp.). This cannot be done on a Macintosh (Apple Computer Inc.) device. We could use this method on a Macintosh device if there were software for Macintosh to make the Cinderella screen semi-transparent.

3 Procedure

1. Use style files ketpic.sty and ketlayer.sty.
2. To use ketlayer environment, write

```
\begin{layer}{100}{20}
\end{layer}
```

   a little before where we want to add a figure in the TEX source file.

3. Display the PDF file with scale guide grid lines of ketlayer. When we use pdfLaTeX, which does not support tpic, we show the scale guide grid lines made in advance.

```
\begin{layer}{100}{20}
\putnotese{0}{0}{\includegraphics{guide.pdf}}
\end{layer}
```

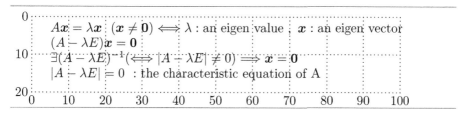

4. Lay the screen of Cinderella on the screen of the PDF file.
5. Using feewhee.exe, make the screen of Cinderella semi-transparent.

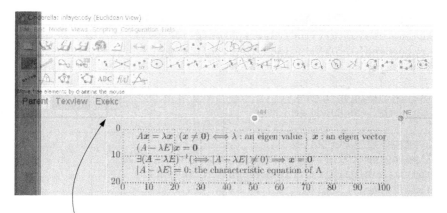

6. Moving the screen of Cinderella or using Translate View of Cinderella, we fit the point (0, 0) of Cinderella to the point (0, 0) of ketlayer. Then move the point HH to the point (100, 0) of ketlayer. The unitlength of the figure is calculated using the x-coordinate of the point HH to fit the text.

7. Draw a figure on the Cinderella screen. Seeing both the figure and the text, adjust the figure shape.

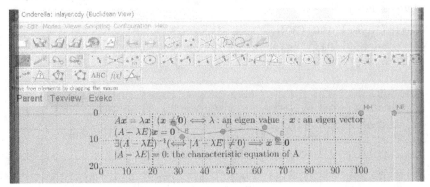

8. Using KETCindy, make a figure file "fig.pdf".
9. Input the figure file into TEX by includegraphics command of TEX.

```
\begin{layer}{100}{20}
\putnotese{0}{0}{\includegraphics{guide.pdf}}
\putnotese{0}{0}{\includegraphics{fig.pdf}}
\end{layer}
```

10. Adjust the figure shape.
11. Erase the scale guide grid lines.

```
\begin{layer}{100}{0}
%\putnotese{0}{0}{\includegraphics{guide.pdf}}
\putnotese{0}{0}{\includegraphics{fig.pdf}}
\end{layer}
```

12. If another figure is needed, repeat same procedure.

4 Another Example and Vision

Using this method, a user can indicate some part of a figure, connect the figure and the text part, and produce a hand-drawn like figure.

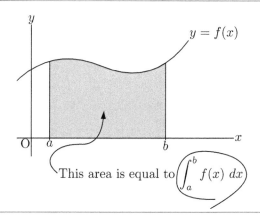

Then we can create a more effective document easily. It is also important from an educational viewpoint. Another important task ahead is informing many people of this method and presenting useful examples of its application.

We can use this method only for Windows now. We could apply this method for Macintosh if there were software for Macintosh to make the Cinderella screen semi-transparent. Another method to do something similar is that we set the PDF file as the background image of Cinderella. It would be a more efficient and intuitive method if we could do that easily. It would also be beneficial if Cinderella had a function to control the screen transparency.

References

1. Kaneko, M., Takato, S.: The extension of $\mathcal{K}_E$Tpic functions-meta commands and their applications. Asian J. TEX 4(2), 111–120 (2010)
2. http://www.cinderella.de/tiki-index.php
3. http://ketpic.com
4. https://sites.google.com/site/ketcindy/home
5. http://www.nattyware.com/feewhee.php

The Programming Style for Drawings from KETpic to KETCindy

Satoshi Yamashita[(⊠)]

National Institute of Technology, Kisarazu College, Kisarazu, Japan
yamasita@kisarazu.ac.jp

Abstract. To produce class materials with figures using TeX, the KeT-pic Development Group (KDG), comprising S. Takato, the author and several Japanese mathematics education researchers, completed KeTpic in 2011 as a plug-in for the Scilab numerical analysis software package. We describe KeTpic programs with the command line user interface Scinotes. KeTpic users produce KeTpic programs based on their original programming styles. This leads other KeTpic users to a shortcoming that renders it difficult to use their KeTpic programs. To resolve this situation, KDG has developed the KeTpic programming style for drawings in 2013. KeTpic programs include three parts: a preamble part that describes setting commands, a part for making plot data and a part for extracting plot data into a figure TeX file. Since 2014, KDG has improved KeTCindy as a plug-in for an interactive geometry software Cinderella. Cinderella has two screens: the interactive geometric screen and a screen that describes Cinderella commands called Script Editors. When a KeTCindy command is run in Script Editors, the corresponding KeTpic commands are extracted to the proper position of the three parts in a Scilab executable file described above. This paper explains the KeTCindy system from the viewpoint of the programming style for drawings and introduces the author's related website, which describes the utilization of KeTCindy.

Keywords: TeX · Scilab · KeTpic · Cinderella · KeTCindy

1 Introduction

The TeX typesetting system enables a user to produce high-quality documents with mathematical formulae, but it is difficult to insert accurate and understandable figures, which are not mere images, into the documents. To mitigate or resolve this shortcoming of TeX, the *KETpic* Development Group (KDG), comprising S. Takato, the author, and several Japanese mathematics education researchers, developed in 2006 a plug-in for mathematical software, i.e., Maple, Mathematica, Maxima, Matlab, Scilab, and R, and KDG has completed it as a plug-in for Scilab in 2010. KETpic produces 2D or 3D figures by accurate line drawings. It is also equipped with a function of the page layout, a function of

© Springer International Publishing Switzerland 2016
G.-M. Greuel et al. (Eds.): ICMS 2016, LNCS 9725, pp. 386–393, 2016.
DOI: 10.1007/978-3-319-42432-3_48

making TEX macro, and a function of the tabulation where each cell has a given width and height. KETpic is suitable to produce our original class materials with figures by TEX because it can make accurate figures and carry out a page layout into a TEX document as desired. KETpic is a useful tool for mathematics teachers, but KETpic users have the following dissatisfaction.

- KETpic is input by Command line User Interface (CUI), not a Grahmmphic User Interface (GUI).
- It is hard for KETpic users to use other users' program because it is based on the user's original programming style.

The author established the KETpic programming style for drawings in 2013 [1]. A KETpic program is divisible into the following three parts. The author has found nine requirements for describing a KETpic program in these three parts. In Sect. 2, the author introduces the KETpic system for Scilab and the KETpic programming style for drawings.

In 2014, KDG has established KETCindy through cooperation of KETpic and an interactive geometry software Cinderella [2]. The interactive geometric screen, called the *main screen*, of Cinderella is used as a GUI of KETpic and a screen describing scripts, that is called *Script Editors*, is used as CUI of KETpic (see Fig. 1). In Sect. 3, the author introduces the KETCindy system and explains the difference of a KETpic program and a KETCindy program. In Sect. 4, the author introduces future works.

2   KETpic System for Scilab

KETpic is a library folder called ketpicsciL5, which consists of many functions defined by Scilab commands. We shall prepare to use KETpic system. Herein, we use KETpic system of Macintosh edition; additionally there is Windows edition. We download Scilab 5.5.2 from the following website.

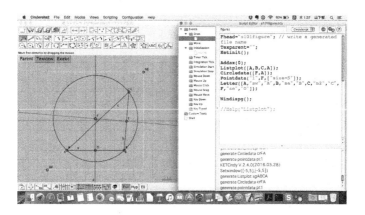

Fig. 1. The main screen (left) and Script Editors (right) of Cinderella.

http://www.scilab.org/download/latest

We produce a folder called KeTpic, where we use KₑTpic system, in Applications folder of Macintosh HD, and make the folder called ketwork, where we produce a figure TₑX file, in Ketpic folder. We go to the following website called KₑTpic.com.

http://ketpic.com/?page_id=18#_13

After opening Scilab folder in this website, we download the zip file called ketpicsciL5_1_6c.zip and defrost it. The ketpicsciL5_1_6c folder holds a folder called ketpicsciL5 and two TₑX style files respectively called ketpic.sty and ketlayer.sty. We move ketpicsciL5 folder into Ketpic folder and move ketpic.sty and kelayer.sty into ketwork folder. Our preparations are finished.

We shall make a KₑTpic program for drawings. We start Scilab 5.5.2 and open SciNotes from Application of the pull-down menu. Using SciNotes, we describe a KₑTpic program in a Scilab executable file called fig.sce as follows (see Fig. 2).

Fig. 2. The main screen (left) and SciNotes (right) of Scilab.

```
1    cd("/Applications/KeTpic/ketwork");
2    Ketlib=lib("/Applications/KeTpic/ketpicsciL5");
3    Ketinit();
4    Fnametex="fig.tex";
5
6    Setax(7,"se");
7    Setwindow([-5,5], [-2,2]);
8    G1=Plotdata("sin(x)","x");
9    D1=Listplot([Xmin(),1],[Xmax(),1]);
10   D2=Listplot([Xmin(),-1],[Xmax(),-1]);
11   Windisp(G1);
12
13   Openfile(Fnametex,"1cm");
```

```
14   Drwline(G1);
15   Dashline(D1,D2);
16   Htickmark(-%pi,"-\pi",%pi,"\pi");
17   Vtickmark(-1,"sw","-1",1,"nw","1");
18   Closefile("1");
```

A K$_E$Tpic program can be divided into the following three parts.

- The *preamble part*, which describes setting commands.
- The *part for making plot data*.
- The *part for extracting plot data* into a figure T$_E$X file.

In the program above, the preamble part extends from line 1 through line 4. Line 1 changes the directory to the ketwork folder to produce a figure T$_E$X file called fig.tex. Line 2 reads the library folder ketpicsciL5. Line 3 initializes K$_E$Tpic. Line 4 defines the name of a figure T$_E$X file as fig.tex.

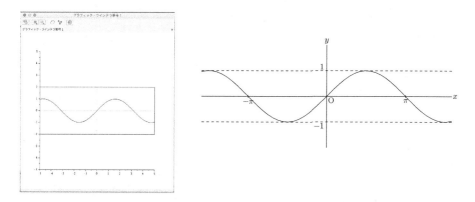

Fig. 3. Graphic window (left) of Scilab and PDF output (right) of fig.tex.

The part for making plot data extends from line 6 through line 12. Line 6 sets the position of the origin name "O" southeast from [0, 0]. Line 7 sets the drawing region in the area that $-5 \leq x \leq 5$ and $-2 \leq y \leq 2$. Line 8 makes plot data of the graph of $y = \sin x$ in the drawing region. Line 9 and line 10 respectively mean making plot data of the straight lines $y = 1$ and $y = -1$. Line 11 previews the graph of $y = \sin x$ using Graphic window (See Fig. 3).

The part for extracting plot data extends from line 13 through line 18. Line 13 describes opening fig.tex and setting the unit length of drawings as 1 cm. Line 14 and line 15 extract plot data into fig.tex, and line 16 and line 17 mean extracting the data of scales of two coordinate axes. Line 18 extracts data of the axes and closing fig.tex.

In 2013, KDG established the *K$_E$Tpic programming style* for drawing. This style has the following nine requirements including five basic requirements and four applied requirements [1]:

- Basic 5 Requirements are following:
  - The producer must arrange a command in a suitable position.
  - The producer must attach the suitable name for a variable or plotting data (see line 8, line 9, and line 10 in the above program).
  - The producer must use the calculation function of CAS.
  - The producer must use KETpic commands appropriately.
  - The producer must divide a program into readable blocks (see line 5 and line 12 in the above program).
- Applied 4 Requirements are the following:
  - The producer must use a reference point to arrange a character and an expression in a suitable position (see line 16 and line 17 in the above program).
  - The producer must use the list structure appropriately.
  - The producer must use syntax appropriately.
  - The producer must define local variables.

The above program is produced using some of these requirements.

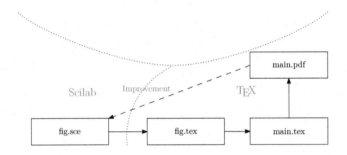

Fig. 4. KETpic system.

When we execute Scilab, we generate fig.tex. We insert fig.tex into the main TEX document using the TEX command called \input. We can obtain the PDF output by compiling TEX (see Fig. 4).

3   KETCindy System for Cinderella

KETCindy is a library folder called ketlib that consists of some folders (including ketpicsciL5) and numerous executable files defined by Cinderella commands. We shall prepare to use KETCindy system. Herein, we use the KETCindy system of Macintosh edition. A Windows edition is available. We proceed to the following website titled "Making of Teaching Materials by KETCindy ".

http://www65.atwiki.jp/ketcindy-eng/pages/1.html

When we click "KeTCindy Install" in "Link" of the left column, we go to the following website: Ketinstall-Dropbox. We download two zip files called Install-forMac.dmg and ketcindycontents.zip. Then we defrost them. In the KeTCindyM folder of the InstallforMac folder, we move the empty folder called KeTCindy into the alias of Applications folder. We defrost texliveF.dmg and generate the texlive folder in the Desktop. We copy the texlive folder into KeTCindy folder. We move all contents of the Ketcindycontents folder into KeTCindy folder. In the ketwork folder of KeTCindy folder, we delete the shell file called kc.sh. In s1figure folder (including KeTCindy/ ketsample/samples), we open the Cinderella file called s101figure.cdy and click the button called Texview. A new shell file kc.sh is generated in the ketwork folder. We start the terminal and change the directory to the ketwork folder. Then we execute the following line.

```
chmod +x kc.sh
```

When we return to s101figure.cdy and click the button called Exekc, we obtain the PDF output. Our preparations are finished.

We produce a KETCindy program for drawing. We save s001basic, cdy as fig.cdy and open fig.cdy, We open Script Editors from Scripting of the pull-down menu. Using Script Editors, we describe a KETCindy program as shown below.

```
1   Fhead="fig";
2   Texparent="fig-clip";
3   Ketinit();
4
5   Setax([7,"se"]);
6   Addax(1);
7
8   Plotdata("1","sin(x)","x");
9   Listplot("1",[[XMIN,1],[XMAX,1]],["da"]);
10  Listplot("2",[[XMIN,-1],[XMAX,-1]],["da"]);
11
12  Htickmark([-pi,"-\pi",pi,"\pi"]);
13  Vtickmark([-1,"sw","-1",1,"nw","1"]);
14
15  Figpdf();
16  Windispg();
```

When executing Script Editors, one can generate three files: kc.sh, fig.sce and figmain.tex. The Scilab executable file fig.sce is described as the following KETpic program:

```
1   cd('/Applications/ketcindy/ketwork');
2   Ketlib=lib('/Applications/ketcindy/ketlib/ketpicsciL5');
3   Ketinit();
4   disp('KETpic '+ThisVersion())
5   Fnametex='fig.tex';
6   Fnamesci='fig.sce';
```

```
 7   Fnamescibody='figbody.sce';
 8   Fnameout='fig.txt';
 9   pi=%pi; i=%i;
10   arccos=acos; arcsin=asin; arctan=atan;
11
12   Setwindow([-5,5], [-2,2]);
13   Assignadd('pi',%pi);
14   Assignadd('XMIN',Xmin());
15   Assignadd('XMAX',Xmax());
16   Assignadd('YMIN',Ymin());
17   Assignadd('YMAX',Ymax());
18   Setax(7,"se");
19   gr1=Plotdata(Assign('sin(x)'),Assign('x'));
20   sg1=Listplot([[-5,1],[5,1]]);
21   sg2=Listplot([[-5,-1],[5,-1]]);
22   PtL=list();
23   GrL=list();
24   //if length(fileinfo(Fnamescibody))>0
25   //   Gbdy=ReadfromCindy(Fnamescibody);
26   //   execstr(Gbdy)
27   //end;
28
29   //Windisp(GrL,'c');
30
31   if 1==1 then
32
33   Openfile(Fnametex,'1cm');
34     Drwline(gr1);
35     Dashline(sg1);
36     Dashline(sg2);
37     Htickmark(-3.14159,"-\pi",3.14159,"\pi");
38     Vtickmark(-1,"sw","-1",1,"nw","1");
39   Closefile('1');
40
41   end;
42
43   quit();
```

In the program above, the preamble part is that section from line 1 to line 10.
The part for producing plot data extends from line 12 through line 29. The part
for extracting plot data extends from line 31 through line 41. Line 43 closes
Scilab. When executing line 8 of the K$_E$TCindy program, one obtains line 19
and line 34 of the K$_E$Tpic program. Exactly one line of the K$_E$TCindy program
generates two lines of the K$_E$Tpic program: one is in the part of making plotdata;
the other is in the part of extracting plot data. It follows that the K$_E$TCindy
program is simpler to use than the K$_E$Tpic program.

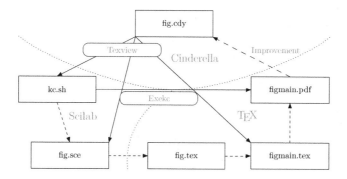

Fig. 5. K_ETCindy system.

When clicking the Exekc button, a user can treat execution fig.sce and compiling figmain.tex with lumping. Then the user can obtain PDF output (see Fig. 5). Therefore, the K_ETCindy system is easy for beginners to use.

4   Future Works

KDG has developed K_ETCindy using K_ETpic system and K_ETpic programming style. The author describes future work as presented below.

- The author intends to investigate how to produde PDF output to which sounds and simulation are added.
- The author shall examine how to use PDF class materials with figures (printed matter, slides, animation, etc.) produced by K_ETCindy , to encourage the students' active learning.

Acknowledgement. This research has been funded by Grant-in-Aid for Japan Scientific Research (No. 25350370). The author extends special appreciation to Setsuo Takato (Toho University) for constructive comments and numerous valuable suggestions.

References

1. Yamashita, S., Maeda, Y., Usui, H., Kitahara, K., Makishita, H., Ahara, K.: Establishment of K_ETpic programming styles for drawing. In: Hong, H., Yap, C. (eds.) ICMS 2014. LNCS, vol. 8592, pp. 641–646. Springer, Heidelberg (2014)
2. Kaneko, M., Yamashita, S., Kitahara, K., Maeda, Y., Nakamura, Y., Kortenkamp, U., Takato, S.: K_ETCindy – Collaboration of cinderella and K_ETpic reports on CADGME2014 conference working group. Int. J. Technol. Math. Educ. **22**(4), 179–186 (2015)

Information Services for Mathematics: Software, Services, Models, and Data

The Software Portal swMATH: A State of the Art Report and Next Steps

Hagen Chrapary[1,2(✉)] and Yue Ren[3]

[1] FIZ Karlsruhe/Zentralblatt MATH, Franklinstr. 11, 10587 Berlin, Germany
hagen@zentralblatt-math.org
[2] Zuse Institute Berlin (ZIB), Takustr. 7, 14195 Berlin, Germany
[3] Fachbereich Mathematik, Technische Universität Kaiserslautern,
Postfach 3049, 67653 Kaiserslautern, Germany

Abstract. swMATH with its web interface www.swmath.org is an open access portal for mathematical software and mathematical research data. After 5 years in operation, it provides information on more than 12,500 items in all mathematical fields and lists nearly 120,000 scientific publications citing the software (May 2016). A unique and novel feature of swMATH besides its scope is the so-called publication-based approach, which uses the information of the scientific database Zentralblatt MATH (zbMATH) for identifying mathematical software and extracting relevant information about them.

Keywords: Database · Software · Publication-based approach · MSC · Web service · zbMATH · swMATH

1 Introduction

The role of software in mathematics is steadily growing. In addition to its importance for applications, its relevance for fundamental research and mathematical education is gradually expanding. Incidentally, mathematical software is under permanent development. The swMATH service [6] is an attempt to develop and establish a powerful information service on mathematical software and mathematical research data. The goal is to improve the visibility of mathematical software and strengthen the subject as a whole. Up to now there exist a lot of portals, repositories, and websites for specific mathematical subjects but a service for a complete overview is missing. Some reasons are the dynamic character, the distributed development and deployment of software, and the missing standards for metadata and description. This means that also manual maintenance of a mathematical software portal is cost-intensive. To avoid this barrier swMATH persues a so-called publication-based approach.

swMATH contains information on software covering all mathematical fields, ranging from large general purpose systems to small specialized packages. It also includes benchmarks, programming and specification languages, data sets, web services, etc. The information model is an extension of the Dublin Core metadata set [5], and includes a short description of the software, an autogenerated

G.-M. Greuel et al. (Eds.): ICMS 2016, LNCS 9725, pp. 397–402, 2016.
DOI: 10.1007/978-3-319-42432-3_49

keyword cloud and a list of publications citing the software, see Fig. 1 for the page on the computer algebra system SINGULAR.

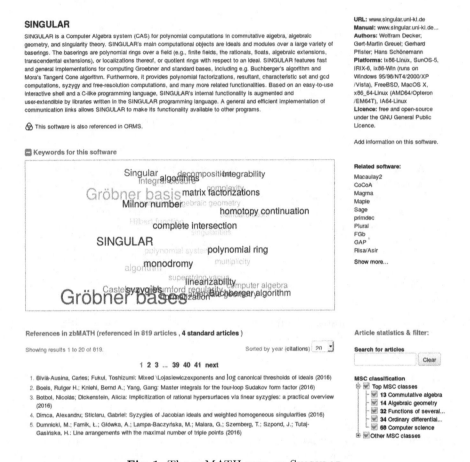

SINGULAR

SINGULAR is a Computer Algebra system (CAS) for polynomial computations in commutative algebra, algebraic geometry, and singularity theory. SINGULAR's main computational objects are ideals and modules over a large variety of baserings. The baserings are polynomial rings over a field (e.g., finite fields, the rationals, floats, algebraic extensions, transcendental extensions), or localizations thereof, or quotient rings with respect to an ideal. SINGULAR features fast and general implementations for computing Groebner and standard bases, including e.g. Buchberger's algorithm and Mora's Tangent Cone algorithm. Furthermore, it provides polynomial factorizations, resultant, characteristic set and gcd computations, syzygy and free-resolution computations, and many more related functionalities. Based on an easy-to-use interactive shell and a C-like programming language, SINGULAR's internal functionality is augmented and user-extendible by libraries written in the SINGULAR programming language. A general and efficient implementation of communication links allows SINGULAR to make its functionality available to other programs.

This software is also referenced in ORMS.

URL: www.singular.uni-kl.de
Manual: www.singular.uni-kl.de...
Authors: Wolfram Decker; Gert-Martin Greuel; Gerhard Pfister; Hans Schönemann
Platforms: Ix86-Linux, SunOS-5, IRIX-6, Ix86-Win (runs on Windows 95/98/NT4/2000/XP /Vista), FreeBSD, MacOS X, x86_64-Linux (AMD64/Opteron /EM64T), IA64-Linux
Licence: free and open-source under the GNU General Public Licence.

Add information on this software.

Keywords for this software

Singular decomposition integrability
integral closures
algorithms complexity
Gröbner basis matrix factorizations
Milnor number algebraic geometry
homotopy continuation
Hilbert function singularities
complete intersection
SINGULAR
polynomial systems polynomial ring
monodromy multiplicity
algorithm
superstring vacua
linearizability
Castelnuovo Mumford regularity computer algebra
Gröbner bases syzygies normalization Buchberger algorithm

Related software:
Macaulay2
CoCoA
Magma
Maple
Sage
primdec
Plural
FGb
GAP
Risa/Asir

Show more...

References in zbMATH (referenced in 819 articles , **4 standard articles**)

Showing results 1 to 20 of 819.   Sorted by year (citations) 20

1 2 3 ... 39 40 41 next

1. Bivià-Ausina, Carles; Fukui, Toshizumi: Mixed \Lojasiewiczexponents and log canonical thresholds of ideals (2016)
2. Boels, Rutger H.; Kniehl, Bernd A.; Yang, Gang: Master integrals for the four-loop Sudakov form factor (2016)
3. Botbol, Nicolás; Dickenstein, Alicia: Implicitization of rational hypersurfaces via linear syzygies: a practical overview (2016)
4. Dimca, Alexandru; Sticlaru, Gabriel: Syzygies of Jacobian ideals and weighted homogeneous singularities (2016)
5. Dumnicki, M.; Farnik, Ł.; Główka, A.; Lampa-Baczyńska, M.; Malara, G.; Szemberg, T.; Szpond, J.; Tutaj-Gasińska, H.: Line arrangements with the maximal number of triple points (2016)

Article statistics & filter:

Search for articles
Clear

MSC classification
Top MSC classes
13 Commutative algebra
14 Algebraic geometry
32 Functions of several...
34 Ordinary differential...
68 Computer science
Other MSC classes

Fig. 1. The swMATH page on SINGULAR

The swMATH service was designed and developed by the Mathematisches Forschungsinstitut Oberwolfach (MFO) and FIZ within a joint project of the German Leibniz Association. Currently, it is a project under the auspices of the MODAL research campus of the Konrad-Zuse-Zentrum (ZIB) and the Freie Universität Berlin.

2   The Publication-Based Approach

swMATH follows a *publication-based approach*, which means that its main source of information are scientific publications.

For this, it closely collaborates with zbMATH [7], the world's most comprehensive and longest-running abstracting and reviewing service in pure and applied mathematics, which currently covers more than 3.5 million bibliographic entries with reviews or abstracts drawn from more than 3 000 journals and serials, and 170 000 books. swMATH clones and extends some of the data, e.g. it builds a index from the reference lines of all articles. This database is very large and continuously increasing, already containing more than 18 million entries. As of now, both services are fully integrated: links from software-relevant articles to swMATH as well as back links from swMATH to zbMATH articles are provided. In addition, there is a special software tab on the zbMATH homepage through which swMATH can be called.

For mathematical software related to education, swMATH also relies on the Mathematics Education Database [8], which is currently the only international reference database offering a world-wide overview of literature on research, theory and practice in mathematics education.

The publication-based approach inherits many advantages from the peer-reviewing process. They are expanded upon in Sects. 4, 5 and 6.

3 Software Identification

The core of the publication-based approach is the ability to reliably identify which publications use software and which do not.

As a first step, the titles of zbMATH publications are analyzed to identify the names of mathematical software. Heuristic methods are developed to search for characteristic patterns in the titles of the articles, e.g., 'toolbox', 'software', 'package', 'solver', 'implementation', 'framework' in combination with an artificial or capitalized first word. Examples are

- SCIP: solving constraint integer programs [1],
- KNITRO: an integrated package for nonlinear optimization [2],
- LANCELOT. A Fortran package for large-scale nonlinear optimization [3],
- Plural – a computer algebra system for noncommutative polynomial algebras [4].

These publications, describing a certain software in detail, are called *standard articles* and appropriately highlighted on swMATH.

In a second step, the reference lines of all articles are used as a starting point for searching for *application articles* of the identified software. Searching in the publication abstracts is done only for very specific names, i.e. 'Microsoft Excel' or 'PolyBoRi'; short acronyms or ambiguous words like 'singular' and 'reduce' are not appropriate.

Besides identifying application articles for known software names, the reference lines are used for identifying new software, too. The lack of a widely accepted citation standard of software is a main disadvantage. The development of an accepted standard combined with technical LaTeX tools would decisively influence the process for an easy and valid software identification. In cooperation with a partner project we develop a proposal which is based on Biber [11], a BibTeX replacement for users of BibLaTeX.

A main problem of a peer reviewed publication process is the time lag between writing and the public presentation of an article. In order to present new software as well, we have started to analyze current ArXiv articles. As these entries are not created through zbMATH articles and not much information can be collected, a special link to a web-interface for expanding the data is provided. Because swMATH stores the standard article specification, a later citation in zbMATH publications will expand the initially empty articles list of the software. This process is done automatically, no manual work is necessary.

Other sources for finding new software packages are well maintained repositories with a rigorous submission policy and a homepage structure suitable for an automatic extraction. The most important example is *The Comprehensive R Archive Network* [15] which is a collection of several thousand R packages.

The websites of software are the definite source for detailed information. These URLs - if existing - are identified by a web search and linked in swMATH. Unfortunately, the websites are heterogeneous in content and structure. We have started to work on automatic methods to analyze the information, e.g. versions, hard- and software requirements, legal rights, etc.

4   Software Quality

An advantage of the publication-based approach is the quality which is guaranteed by a rigorous peer-reviewing control system. The references to a specific software in the database zbMATH can therefore be used as an indirect criterion for its quality: A high amount of references from peer-reviewed articles is a strong indication for the relevance of the software in question.

Moreover, there exist journals with a special focus on mathematical software, in which both the submitted article as well as the underlying software will receive a review. Any submission to these journals must be accompanied by a detailed software description, installation instructions, implementation details and necessary data sets. Any publication in these journals guarantees that independent tests have been carried out to verify and replicate its results. Two notable journals are the *Journal of Statistical Software* [13] and the *Mathematical Programming Computation* [12], the latter covering optimization software. A software package that has undergone such a review is highlighted on swMATH with a link to the respective journal. For an example, see the swMATH entry of SCIP (http://www.swmath.org/software/1091).

swMATH maintains a second direct quality marker: Packages which are part of the *Oberwolfach References on Mathematical Software* (ORMS) [14] are labeled with a link to their respective entry. The ORMS presents high-quality entries of carefully selected mathematical software. For an example, see the swMATH entry of SINGULAR (http://www.swmath.org/software/866).

swMATH contains only software families without versions, but the evaluation of the computer code is based on a fixed version in a fixed environment at a fixed time. Later bugfixes, software updates and 'disimprovements' can falsify the former review, therefore the quality mark must be considered carefully.

5   Use of MSC and Keywords

The piece of information, that any information service on mathematical software has to deliver, are the general scope and possible applications of each software in its database. Ideally, this information should be usable in the search for specific software and in the ranking of different software based on relevance. So while most software comes with a short description from its developers, a more differentiating and independent source is desirable.

A feature that is always present in any mathematical publication is the canonical use of MSC 2010 tags [9]. MSC stands for "Mathematics Subject Classification" and it is widely used to assign publications to their mathematical subjects. Mathematical reviewing services such as zbMATH or MathSciNet [10] rely on them by default. Since swMATH systematically links software packages to relevant mathematical publications, the collected MSC information can be used to build a short MSC profile.

Although the knowledge about MSC is widely accepted in the mathematical research community, its understanding is limited for many application users. Fortunately a list of describing keywords is provided in scientific publications. These important items are presented as an eyecatcher in a clickable rotating keyword cloud in the centre of the software detail page. This distinctive design is a well-known feature of swMATH. In the industrial and economic area it would be called a marketing gag.

6   Operation and Sustainability

The technical swMATH infrastructure consists of the web service, a duplicate test system, and the production system for handling the input data. The web service itself is updated weekly, all other systems are availabe in daily update.

Sustainability and easy maintenance of swMATH have been important features and important design goals from the very beginning. The publication-based approach allows for a wide usage of automatic methods, hence the resources needed for the maintenance of the service are limited. Therefore, feature requests that come with high maintenance costs are currently postponed or even rejected. Tracking version numbers, links to the homepages of each individual author, and developing an explicit typing schema for software are examples of these tasks. The developments in design and functionalities are carried out in close coordination with the corresponding developments in zbMATH.

The support by the Berlin Forschungscampus MODAL and the integration of the database swMATH in the information services of FIZ Karlsruhe/Zentralbatt MATH ensure the longstanding existence of the swMATH service. The last two years of operation with limited manpower but an increasing amount of data have proven the feasibility of this concept.

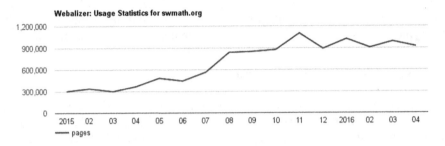

Fig. 2. Usage statistics for swmath.org by WEBALIZER (pages)

The acceptance of swMATH is shown by an increasingly high amount of user traffic, see Fig. 2. We hope that swMATH will become a valuable tool for the mathematical community and gives an appropriate credit to the developer of mathematical software.

References

1. Achterberg, T.: SCIP: solving constraint integer programs. Math. Program. Comput. **1**(1), 1–41 (2009)
2. Byrd, R.H., Nocedal, J., Waltz, R.A.: KNITRO: an integrated package for nonlinear optimization. In: Di Pillo, G., Roma, M. (eds.) Large-Scale Nonlinear Optimization. Papers Based on the Presentation at the Workshop on Large Scale Nonlinear Optimization, pp. 35–59. Springer, Heidelberg (2006)
3. Conn, A.R., Gould, N.I., Toint, P.L.: LANCELOT. A Fortran Package for Large-scale Nonlinear Optimization (Release A). Springer, Heidelberg (1992)
4. Levandovskyy, V., Schöneman, H.: Plural – a computer algebra system for non-commutative polynomial algebras. In: Proceedings of the 2003 International Symposium on Symbolic and Algebraic Computation, ISSAC 2003, Philadelphia, PA, USA, 3–6 August 2003, pp. 176–183. ACM Press (2003)
5. The Dublin Core Metadata Initiative (DCMI). http://dublincore.org/
6. swMATH: an information service for mathematical software. http://www.swmath.org
7. zbMATH: the first resource for mathematics. https://zbmath.org
8. Mathematics Education Database. https://www.zentralblatt-math.org/matheduc/en
9. MSC2010. https://zbmath.org/classification/
10. MathSciNet, Mathematical Reviews. http://www.ams.org/mathscinet/
11. Biber. http://biblatex-biber.sourceforge.net/
12. Submission guidelines of the Mathematical Programming Computation. http://mpc.zib.de/MPC/information/authors.html
13. Submission guidelines of the Journal of Statistical Software. https://www.jstatsoft.org/information/authors
14. Oberwolfach References on Mathematical Software. http://orms.mfo.de/
15. The Comprehensive R Archive Network. https://cran.r-project.org/

The polymake XML File Format

Ewgenij Gawrilow[1]($\boxtimes$), Simon Hampe[2]($\boxtimes$), and Michael Joswig[2]($\boxtimes$)

[1] TomTom International BV, Berlin, Germany
egawrilow@gmail.com
[2] TU Berlin, Berlin, Germany
{hampe,joswig}@math.tu-berlin.de

Abstract. We describe an XML file format for storing data from computations in algebra and geometry. We also present a formal specification based on a RELAX-NG schema.

Keywords: XML · RELAX-NG · Polymake

1 Introduction

polymake is an open source software system for computing with a wide range of objects from polyhedral geometry and related areas [5]. This includes convex polytopes and polyhedral fans as well as matroids, finite permutation groups and ideals in polynomial rings. As a key feature polymake is designed as an extensible system, where each new version comes with new objects and new data types. It is crucial to be able to store these objects in files since they themselves or part of the information on them result from costly computations. The purpose of this note is to explain the general concept for polymake's file format which is powerful enough to be able to grow with extensions to the software.

It is safe to say that the Extensible Markup Language (XML) is the de facto standard for exchanging data across platform and implementation boundaries. XML imposes a tree structure on any kind of text, and it comes with a large array of tools which allow to process an XML file independent from the software which generated that file. The tree structure makes it easy to ignore part of the data on input without losing consistence by pruning of subtrees. Part of the realm of XML tools are transformation style sheets (XSLT) which, e.g., allow for simplified versioning or even translating into non-XML documents. This makes XML especially useful for the long-term storage of data; see also [6, Sect. 1.1].

XML file formats for storing mathematical content are ubiquitous. The most widely used is MathML [1] whose initial purpose was the presentation of mathematics in web pages. However, by now there is also Content MathML and OpenMath [2] which focus on the semantics. Our goal here is to describe a simple XML format which is useful for the serialization of data which occur in computations in algebraic and polyhedral geometry.

© Springer International Publishing Switzerland 2016
G.-M. Greuel et al. (Eds.): ICMS 2016, LNCS 9725, pp. 403–410, 2016.
DOI: 10.1007/978-3-319-42432-3_50

2   The File Format by Example

We start out by looking at one short `polymake` example XML file which stores a
square and some of its properties, including a triangulation. The formal description in terms of RELAX-NG [3] is deferred until Sect. 3 below.

```
 1  <?xml version="1.0" encoding="utf-8"?>
 2  <?pm chk="56e977e8"?>
 3  <object name="square" type="polytope::Polytope&lt;Rational&gt;"
 4      version="3.0"
 5      xmlns="http://www.math.tu-berlin.de/polymake/#3">
 6    <description><![CDATA[cube of dimension 2]]></description>
 7    <property name="VERTICES">
 8      <m>
 9        <v>1 0 0</v>
10        <v>1 1/3 0</v>
11        <v>1 0 1/3</v>
12        <v>1 1/3 1/3</v>
13      </m>
14    </property>
15    <property name="FACETS"
16        type="SparseMatrix&lt;Rational,NonSymmetric&gt;">
17      <m cols="3">
18        <v> <e i="1">1</e> </v>
19        <v> <e i="0">1/3</e> <e i="1">-1</e> </v>
20        <v> <e i="2">1</e> </v>
21        <v> <e i="0">1/3</e> <e i="2">-1</e> </v>
22      </m>
23    </property>
24    <property name="LINEALITY_SPACE"><m /></property>
25    <property name="BOUNDED" value="true" />
26    <property name="N_FACETS" value="4" />
27    <property name="N_VERTICES" value="4" />
28    <property name="VOLUME" value="1/9" />
29    <property name="TRIANGULATION">
30      <object name="unnamed#0">
31        <property name="FACETS">
32          <m>
33            <v>0 1 2</v>
34            <v>1 2 3</v>
35          </m>
36        </property>
37        <property name="F_VECTOR">
38          <v>4 5 2</v>
39        </property>
40      </object>
41    </property>
42  </object>
```

Listing 1.1. A polymake XML file, encoding a square.

Mathematical Background. A *(convex) polytope* is the convex hull of finitely
many points in a Euclidean space or, equivalently, the bounded intersection

of finitely many affine halfspaces. In `polymake` points are encoded in terms of *homogeneous coordinates* to allow for a consistent treatment of both polytopes and polyhedral cones. Therefore, the polytope $\text{conv}(S)$ for $S \subset \mathbb{R}^n$ finite is encoded as the cone spanned by $\{1\} \times S \subseteq \mathbb{R} \times \mathbb{R}^n$. That is, in practical terms, the coordinates of points are always prepended with a 1. In our example, we are considering the unit square scaled by $1/3$. Its *vertices*, which form the unique generating set which is minimal with respect to inclusion, are written as $(1, 0, 0), (1, 1/3, 0), (1, 0, 1/3), (1, 1/3, 1/3)$. Linear inequalities are encoded in a similar fashion. The vector $(a_0, a_1, \ldots, a_n)$ ought to be read as $a_0 + a_1 x_1 + \ldots a_n x_n \geq 0$. In this way, a point p given in homogeneous coordinates fulfills an inequality given by a vector a, if and only if the scalar product $p \cdot a$ is nonnegative. See [7] for an introduction to polytope theory from an algorithmic point of view.

Now we will walk the user through the Listing 1.1 line by line.

The Parent Object (Lines 3–6). The mathematical entities relevant to `polymake` occur as *objects* each of which has a *type*. It will tell the parser what properties to expect and how to interpret them. In this case the type describes a convex polytope with rational coordinates. The *version number* refers to a specific `polymake` version. Via XSLT this allows for automatic updates from one object or file format version to the next. Optional names and descriptions provide additional human-readable information for quick identification.

Properties and Matrices (Lines 7–24). Every object is made up of various *properties*, which are identified by their names; their types are implicit. `polymake` keeps track of the type of each property. The combination of the property's name with the version number (see above) uniquely determines the type.

However, properties may also be encoded in a more involved way. In our example the property named `FACETS` comes with the type `SparseMatrix` explicitly given. This can be useful for saving space. In general, it is legal to specify types which can be converted into the defined type of a property. For sparse data types that conversion is only implicit, i.e., the matrix is never expanded into a dense matrix. Most of the time the user will not notice the difference.

Any `polymake` matrix is stored as a sequence of row vectors. If it is sparse only the nonzero entries are written down. The column of each entry is encoded in the `i` attribute, and the `cols` attribute of the matrix indicates the total number of columns of the matrix. In this case property `FACETS` encodes the matrix with the row vectors $(0, 1, 0)$, $(1/3, -1, 0)$, $(0, 0, 1)$, $(1/3, 0, -1)$, and this yields the non-redundant exterior description

$$x \geq 0, \quad x \leq 1/3, \quad y \geq 0, \quad y \leq 1/3.$$

If `polymake` encounters a property with an unknown name, that property is discarded — but a backup file is created.

Primitive Properties (Lines 25–28). Simple properties containing, e.g., numbers (integer, rational or float) or boolean values are stored in an XML attribute named `value`.

Subobjects (Lines 29–41). An object may have properties which are again objects themselves (and which, in turn, may have further subobjects, etc.). Again the object types are identified via the name of that property of the parent object. Here TRIANGULATION is a SimplicialComplex. This mechanism allows for rather elaborate constructions.

The maximal cells of the triangulation (called FACETS) are specified as subsets of the vertices of the polytope. Each number refers to the corresponding row of the property VERTICES of the parent object.

Notice that a convex polytope can be triangulated in more than one way. Therefore, TRIANGULATION is a property of a Polytope object which may contain several objects (of type SimplicialComplex). The various triangulations are distinguished by their unique names. These can be set by the user or are generated automatically (like here).

3   Format Specification in RELAX NG

The features presented above only provide a partial view of what can be expressed in polymake's XML. The full formal specification is expressed via RELAX NG (or RNG for short) [3]; see the Listings 1.2 and 1.3 below. RNG is a rather simple XML schema language based on the theory of hedge automata [8]. Table 1 contains a short overview of the compact RNG syntax [4]. The full specification file, which complies with the official RNG standard and contains some additional explanatory annotations can be found in any current polymake distribution under [polymake_folder]/xml/datafile.rng.

Table 1. RELAX NG compact syntax

start = ...	Defines the pattern for the root element.
PatternName = ...	Defines a pattern with a chosen name.
element, attribute	Define XML tags / attributes.
{ ... }	Describes the content of an element or attribute.
|	Pattern alternatives.
&	Combine two patterns in arbitrary order.
?, +, *	Quantifiers: At most one, one or more, zero or more.

Listing 1.2 contains pattern definitions for the higher level elements in polymake's XML. Each file either contains exactly one object or one data element as its root. The pattern ObjectContent specifies that any object may contain multiple property and attachment elements. Here we focus on objects, while loose data and attachments are discussed briefly at the end of this section.

```
1   start = TopObject | LooseData
2
3   TopObject = element object { TopAttribs , ObjectContent }
4
5   TopAttribs = attribute type {
6       xsd:string { pattern = "[a-zA-Z][a-zA-Z_0-9]*::.*" } },
7       attribute version { xsd:string { pattern = "[\d.]+" } }?,
8       attribute tm { xsd:hexBinary }?
9
10  ObjectContent =
11      attribute name { text }?,
12      attribute ext { text }?,
13      element description { text }?,
14      element credit { attribute product { text }, text }*,
15      ( Property* & Attachment* )
16
17  Property = element property {
18      SimpleName ,
19      attribute ext { text }?,
20      ( ( attribute undef { "true" }, empty )
21      | ( attribute type { text }?, PropertyData )
22      | Text | SubObject+ ) }
23
24  SubObject = element object
25      { attribute type { text }?, ObjectContent }
26
27  Attachment = element attachment {
28      SimpleName , attribute ext { text }?, AttachmentData }
29
30  LooseData = element data {
31      TopAttribs , attribute ext { text }?,
32      element description { text }?, PropertyData }
33
34  SimpleName = attribute name
35      { xsd:string { pattern = "[a-zA-Z][a-zA-Z_0-9]*" } }
```

Listing 1.2. Format specification, Part 1: top-level elements

Listing 1.3 contains the pattern definitions for elements that encode actual content. A `property` or `data` element can contain either a simple value stored in an attribute, a reference to another property, a container or a list of subobjects. The `polymake` XML format knows three container patterns: `Vector`, `Matrix` and `Tuple`. The precise syntactical differences are somewhat subtle. A `Matrix` is an array of containers of the same type. In Listing 1.1 the `VERTICES` (Lines 7–14) and the `FACETS` (Lines 15–23) of the square as well as the `FACETS` of the triangulation subobject (Lines 31–36) are matrices. A `Vector` encodes an array of homogeneous content. In Listing 1.1 the rows of the matrices mentioned above occur as vectors; additionally we have the `F_VECTOR` of the triangulation (which counts the cells of the triangulation by dimension). Sparse vectors employ the `e` element to specify the non-zero entries. The final container pattern `Tuple` establishes records of heterogeneous content. For maximum flexibility the three container types can be nested recursively.

```
39   PropertyData = ( attribute value { text }, empty )
40     | IdReference | Complex | element m { SubObject+ }
41
42   AttachmentData =
43     ( attribute type { text }?, attribute value { text }, empty )
44     | ( attribute type { text }, attribute construct { text }?,
45       Complex ) | Text
46
47   Text = attribute type { "text" }, text
48
49   Complex = Vector | Matrix | Tuple
50
51   VectorContents = text
52     | ( attribute dim { xsd:nonNegativeInteger }?,
53         ( element e { ElementIndex, text }*
54         | element t { ElementIndex?, TupleContents }+ ) )
55
56   ElementIndex = attribute i { xsd:nonNegativeInteger }
57
58   IdReference = element r {
59     attribute id { xsd:nonNegativeInteger }?, empty }
60
61   Vector = element v { VectorContents }
62
63   MatrixContents =
64     ( attribute cols { xsd:nonNegativeInteger }?, Vector* )
65     | ( attribute dim { xsd:nonNegativeInteger },
66         element v { ElementIndex, VectorContents }*)
67     | Matrix+ | Tuple+
68
69   Matrix = element m { MatrixContents }
70
71   TupleContents = attribute id { xsd:nonNegativeInteger }?,
72     ( text | ( Vector | Matrix | Tuple
73               | IdReference | element e { text } )+ )
74
75   Tuple = element t { TupleContents }
```

Listing 1.3. Format specification, Part 2: Content elements

Attachments. Attachments provide a mechanism for storing essentially arbitrary data with an object — regardless of its type and the current version of poly- make. They can be primitive data types as well as more complex types such as matrices and sets. Object types such as Polytope are not allowed. This can, for example, be used to store unrecognized data from pre-XML polymake files or to keep track of relevant context data in an involved computation without having to create multiple files. Every attachment is identified by a unique name.

Loose Data. The polymake XML file format can also be used to store data which are not a full object. The object node is then replaced by a data node and no properties or attachments may appear. Otherwise, the format is essentially the same. Listing 1.4 encodes an array whose single entry is the polynomial $\sqrt{5}/5x^2 - y^3$. Since the coefficients lie in the quadratic field extension $\mathbb{Q}(\sqrt{5})$, each of them is encoded as a `Tuple` (a, b, c) which is to be read as $a + b \cdot \sqrt{c}$. The polynomial again is a `Tuple` where the first entry encodes the terms (which form a `Matrix` of `Tuple` elements) and the second one the names of the variables.

```
1   <data type="Array&lt;Polynomial&lt;QuadraticExtension&gt;&gt;"
2        version="3.0"
3        xmlns="http://www.math.tu-berlin.de/polymake/#3">
4     <v>
5       <t>
6         <m>
7           <t>
8             <v dim="2"> <e i="0">2</e> </v>
9             <t>0 1/5 5</t>
10          </t>
11          <t>
12            <v dim="2"> <e i="1">3</e> </v>
13            <t>-1 0 0</t>
14          </t>
15        </m>
16        <t id="1">
17          <v>x y</v>
18        </t>
19      </t>
20    </v>
21  </data>
```

Listing 1.4. A file representing an array which contains one polynomial.

Element References. To avoid writing the same data multiple times, an element can be replaced by a reference tag `<r>`, which points to another element using an identification number. This is useful, for example, when storing multiple polynomials which all share the same variable names. An example of an element using the `id` attribute can be seen in Listing 1.4, line 16. By referencing to this id, e.g., one can express that another polynomial is contained in the same ring.

4   Concluding Remarks

A key design decision is that the `polymake` RNG schema does not restrict the types of objects and their properties in any way. It provides a simple syntax to recursively structure mathematical data in terms of vectors, matrices and tuples as it occurs in computations. The precise type information relies on the implementation of the `polymake` version specified. In this way `polymake` can be extended easily by adding new objects, new properties and new types. The long-term sustainability of the data relies on the extra flexibility which comes from XSLT transformation style sheets.

It should be emphasized that this file format is by no means a replacement of existing standards such as OpenMath or (Content) MathML. While MathML focuses on the *presentation* of mathematical content, OpenMath and Content MathML are comprehensive frameworks for defining the *semantics* of arbitrary mathematical information. The `polymake` XML format aims at something more modest: It provides a simple mechanism for storing *concrete* mathematical data in a well-structured manner which still allows for extensions and modifications without breaking the overall concept.

`polymake`'s release documentation at

http://polymake.org/release_docs/3.0/

is automatically generated. This contains the complete list of objects, properties and their types. We intend to enhance the mechanism for the documentation generation to export this information again as `RNG` schema files. This will allow third party developers to access `polymake` data without relying on our software.

References

1. Mathematical Markup Language (MathML) version 3.0 2nd edition. https://www.w3.org/TR/MathML3/
2. OpenMath. http://www.openmath.org/
3. RELAX NG Specification, Technical report, The Organization for the Advancement of Structured Information Standards (OASIS), December 2001. http://relaxng.org/spec-20011203.html
4. RELAX NG Compact syntax specification, Technical report, The Organization for the Advancement of Structured Information Standards (OASIS), November 2002. http://relaxng.org/compact-20021121.html
5. Gawrilow, E., Joswig, M.: polymake: a framework for analyzing convex polytopes. In: Kalai, G., Ziegler, G.M. (eds.) Polytopes—Combinatorics and Computation (Oberwolfach, 1997), vol. 29, pp. 43–73. Birkhäuser, Basel (2000)
6. Joswig, M., Mehner, M., Sechelmann, S., Techter, J., Bobenko, A.I.: DGD Gallery: Storage, sharing, and publication of digital research data. In: Bobenko, A.I. (ed.) Advances in Discrete Differential Geometry. Springer, Heidelberg (2016)
7. Joswig, M., Theobald, T.: Polyhedral and Algebraic Methods in Computational Geometry. Universitext. Springer, London (2013). Revised and updated translation of the 2008 German original
8. Murata, M.: Hedge automata: a formal model for XML schemata. Technical report, Fuji Xerox Information Systems (1999). http://www.horobi.com/Projects/RELAX/Archive/hedge_nice.html

Semantic-Aware Fingerprints of Symbolic Research Data

Hans-Gert Gräbe[(✉)]

Leipzig University, Leipzig, Germany
graebe@informatik.uni-leipzig.de
http://www.zv.uni-leipzig.de/en/university/profile-and-management.html

Abstract. One of the goals of the SYMBOLICDATA Project is to set up a navigational structure on the research data associated with the project. In 2009 we started to refactor the data and metadata along standard semantic web concepts based on the Resource Description Framework (RDF) thus opening the door to the Linked Open Data world.

One of the main metadata concepts used for navigational purposes is that of *semantic-aware fingerprints* as semantically sound invariants of the given data. We applied this principle, first used to navigate within polynomial systems data, to the data sets on polytopes and on transitive groups newly integrated with SYMBOLICDATA version 3, and also within the recompiled version of test sets from integer programming.

The RDF based representation of fingerprints allows for a unified navigation and even cross navigation within such data using the SPARQL query mechanism as a generic web service, a clear advantage compared to metadata management traditionally in use within the domain of computer algebra.

In this paper we discuss merely the conceptual background of our *fingerprinting approach* and refer to the SYMBOLICDATA wiki for more details and examples how to use that service.

Keywords: Semantic technology · RDF · Computer algebra · Metadata management · SPARQL query mechanism

1 Introduction

The section "Information Services for Mathematics" addresses a more complex target compared to the title "Mathematical Software" of this conference at large since mathematical software can be considered as part of a whole infrastructure for mathematical research. Nowadays such an infrastructure goes much beyond the classically hawked "paper and pencil" or "chalk and blackboard" claimed to be sufficient – together with access to the work of colleagues within an, nowadays also not self-evident, information and communication infrastructure – to pursue advanced mathematical research. The themes "software, services, models, and data" point to at least four dimensions to enhance the mathematical research

© Springer International Publishing Switzerland 2016
G.-M. Greuel et al. (Eds.): ICMS 2016, LNCS 9725, pp. 411–418, 2016.
DOI: 10.1007/978-3-319-42432-3_51

infrastructure in the era of ubiquitous computing and increasingly important digital interconnectedness.

This paper addresses the dimension of *research data* in more detail, in particular aspects of public availability of reliable and well curated *research input data* that is important for the coherence of research questions addressed by communities and thus for the formation of specific research communities themselves.

We discuss relevant questions in the specific context of intra- and intercommunity communication within the specific research domain of *symbolic and algebraic computations* (CA) coarsely defined by the MSC 2010 classification code 68W13. We analyze the situation of public availability of research data in that area on the background of almost 20 years of experience with research data management in that domain within the SYMBOLICDATA Project [18]. We address the special challenges to small scientific communities as the CA community compared to larger ones as the whole mathematical community, that nevertheless splits into a number of CA subcommunities. These CA subcommunities are organized around special research topics and in many cases already managed to organize and consolidate their own intracommunity research infrastructures. We discuss lessons to be learned from these activities and hurdles and obstructions to generalize such experience to an intercommunity level within the CA domain.

In Sect. 2 we develop a more detailed view on the interplay between (digital) research data and research infrastructures and discuss the situation of the mathematical digital research infrastructure compared to other sciences.

In Sect. 3 we give a short report about SYMBOLICDATA activities in the CA domain during the years. In particular we emphasize the importance of a redesign of the SYMBOLICDATA basics during the last years towards standard semantic web concepts and the implementation of an RDF based infrastructure to manage descriptions ("fingerprints") of research data collections of different CA subcommunities and thus to open them for the Linked Open Data world.

Sections 4 and 5 are devoted to a more detailed explanation of the notion of *fingerprints* of research data and our conceptual background of data and metadata management. Further we discuss the advantages of an RDF based approach to metadata management compared to approaches traditionally in use within the CA domain.

2   Research Data and Digital Research Infrastructures

Digital change and the accelerated development of a (seemingly) universally interconnected digital universe lead to an essential reshaping of many areas of life. Also the world of scientific research is affected by these mainly technologically triggered social changes. The public availability and easy accessibility of very detailed descriptions and information about research processes leads to a strong increase of transparency and provides a basis for completely new cooperation forms whose importance for the future hardly can underestimated.

Such a development started to change research methods already in the *computer age* since the 1960th complementing established forms of intermediation of scientific results by journal papers and preprints with computer simulations and scientific software[1] as an essentially new form of scientific knowledge production. Within the upcoming *networking age* a number of questions of scientific knowledge production have to be addressed anew. Three themes related to simulation are of particular importance: (1) the input data, (2) the simulation procedures (scientific software) and (3) the output data.

Not only with the digital universe the free availability of data for public research from each of these three thematic areas plays an important but scientific-sociologically different role:

(1) The public availability of reliable and well curated *input data* is important for the coherence of research questions addressed by the community and thus for the formation of a specific research community itself around its central research problems.

(2) The public availability of newly developed *simulation methods, procedures and techniques* is relevant for the traceability of the proposed scientific approaches and increasingly accompanies classical forms of description of scientific advancement by academic papers.

(3) The public availability of *output data* is important for the independent reproduction of results and thus of essential importance for the process of academic quality assurance.

It is in the nature of the scientific process that output data is the starting point for new research questions and thus output data mutates to input data. In most of the cases such a mutation happens not immediately but is mediated by a community-internal interpersonal transformation process that transforms the often large output data (or a whole bundle of such data) into (one or several) more compact input data adapted to the new research question(s).

Within the digital change the different scientific communities are faced with the challenge to adapt their research and communication infrastructure to these new socio-technical opportunities. Of central importance – beside a culture of public access – is the allocation of resources for such a mainly non-academic business to restructure this highly technical research infrastructure of the community and keep it running. After many years of community-driven grassroot activities of academic self-organization (e.g., ArXiv) this topic begins to move into the focus of research and political administrations at different levels and is reflected in different calls and rules at German wide (e.g., [3]) or EU level (e.g., [4,14]).

Other scientific communities (e.g., with programs as TextGrid, DARIAH, CLARIN-PLUS) act very successful to acquire EU funding to upgrade their

---

[1] *Scientific software* is written to run *computer simulations* – we use this notion in an appropriate broad meaning – and if software is not used in such a way it is of less academic interest. Moreover, computer simulations often require the interplay of several *scientific packages* bundled within an *application,* hence computer simulation is the broader notion and we use it throughout this paper instead of scientific software.

research data infrastructure mainly at the theme (1) level – in particular to set up a sustainable environment for text corpora (e.g., "Deutsches Textarchiv") as the central research data form within Digital Humanities. The mathematical community is much less successful within the EU Research Infrastructures Program [14] (but see the OpenDreamKit Project [11]) and concentrates with projects as swMath [19], sagemath [15] and also this conference on the theme (2) level of sustainably available scientific software. Note that the application of the OpenDreamKit Project was successful also due to the fact that it does *not* address mathematical software as such but *successful cooperate practices* using mathematical software.

Efforts to secure a research infrastructure for mathematical data at the theme (1) or even theme (3) levels are lost in the brushwood of everlasting (for at least a decade) debates about reliable formal but semantically expressive formats as MathML or OpenMath for data resulting from calculi, that are already highly formalized – at least at an informal level – by the internal nature of the research topics themselves. The situation reminds the Tower of Babel Project, since subcommunities are digitally already well established, developed their own formalizations for their own research data at theme (1) level and apply such formalizations very successful within their intracommunity communication processes.

3   The SYMBOLICDATA Project

The SYMBOLICDATA Project is a small project initiated at the end of the 1990th as an intracommunity project in the area of *Polynomial Systems Solving* to secure a research data infrastructure at the theme (1) level built up within the EU funded PoSSo [13] and FRISCO [5] projects. It grew up from the Special Session on Benchmarking at the 1998 ISSAC conference in a situation where the research infrastructure built up within these projects – the Polynomial Systems Database – was going to break down. After the end of the projects' fundings there was neither a commonly accepted process nor dedicated resources to keep the data in a reliable, concise, sustainably and digitally accessible way. Even within the ISSAC Special Session on Benchmarking the community could not agree upon a further roadmap to advance that matter.

The SYMBOLICDATA Project was set up by a small number of volunteers not involved within the EU funded projects, but strongly interested in the public availability of this research data as reference that can be used as input data (1) for certified benchmark activities on specialized mathematical software that was written to run simulations (2) in a special domain of Algebraic Geometry. At those times almost 20 years ago most of the nowadays well established concepts and standards for storage and representation of research data did not yet exist – even the first version of XML as a generic markup standard had to be accepted by the W3C. It was Olaf Bachmann and me who developed during 1999–2002 with strong support by the Singular group concepts, tools and data structures for a structured representation and storage of this data and prepared about 500 instances from *Polynomial Systems Solving* and *Geometry Theorem Proving* to be available within this research infrastructure, see [1].

The main conceptional goal was a nontechnical one – to develop a research infrastructure that is independent of (permanent) project funding but operates based on overheads of its users. This approach was inspired by the rich experience of the Open Culture movement "business models" to run infrastructures. It was an early attempt to emphasize the advantage of an explicitly elaborated concept of a community-based solution to the "tragedy of the commons" [8] within the CA community and to apply such a concept to run a part of its research infrastructure.

Even 15 years later it remains difficult to keep the SYMBOLICDATA Project running on such a base, and for many years we concentrate our efforts to secure the sustainable public digital availability of the research input data within our collections and to develop appropriate concepts and tools to manage, search and filter this data. In 2009 we started to refactor the data along standard semantic web concepts based on the Resource Description Framework (RDF). With SYMBOLICDATA version 3 released in September 2013 we completed a redesign of the data along RDF based semantic technologies, set up a Virtuoso based RDF triple store and an SPARQL endpoint as Open Data services along Linked Data standards, and started both conceptual and practical work towards a semantic-aware Computer Algebra Social Network [7].

Since then we continued that development. On March 1, 2016, version 3.1 of the SYMBOLICDATA tools and data was released. The new release contains

- new resource descriptions ("fingerprints") of remotely available data on transitive groups (*Database for Number Fields* of Gunter Malle and Jürgen Klüners [10]) and polytopes (databases of Andreas Paffenholz [12] within the *polymake* project [6]),
- a recompiled and extended version of test sets from integer programming – work by Tim Römer (*normaliz* group [2]) –,
- an extended version of the *SDEval benchmarking environment* – work by Albert Heinle [9] – and
- a partial integration (SYMBOLICDATA People database, databases of upcoming and past conferences) of data from the Computer Algebra Social Network subproject.

Moreover, the github account https://github.com/symbolicdata was transformed into an organizational account and the git repo structure was redesigned better to reflect the special life-cycle requirements of the different parts and activities within SYMBOLICDATA. We provide the following repos

- *data* – the data repo with a single master branch mainly to backup recent versions of the data,
- *code* – the code directory with master and develop branches,
- *maintenance* – code chunks from different tasks and demos as best practice examples how to work with RDF based data,
- *publications* – a backup store of the LaTeX sources of SYMBOLICDATA publications,

– *web* – an extended backup store of the SYMBOLICDATA web site that provides useful code to learn how RDF based data can be presented.

The main development is coordinated within the SYMBOLICDATA *Core Team* (Hans-Gert Gräbe, Ralf Hemmecke, Albert Heinle) with direct access to the organizational account. We refer to the SYMBOLICDATA Wiki [18] for more details about the project's organization and the new release.

4   Research Data and Metadata

From the internal perspective of a research community a special aspect of every research data collection is the design of management, search and filter functionality. For this purpose data is usually enriched with metadata that collect important relevant information of the individual data records in a compact manner. We denote such metadata for an individual data record as its *fingerprint*.

Similar to a hash function a fingerprint function computes a compact metadata record (*resource description* in the RDF terminology) to each individual data record (*resource* in the RDF terminology). As with a hash function one can use the fingerprints to (almost) distinguish different data records within the given collection and to match new records with given ones. But there is an essential difference between (classical) hash functions and well designed fingerprints: fingerprint functions exploit not only the textual representation of the data record as meaningless syntactical character string but convey semantically important information or even compute such information from the string representation. Fingerprints are in this sense *semantic-aware* and can even be designed in such a way that they map ambiguities in the textual representation of records (e.g., polynomial systems given in different polynomial orders and even in different variable sets) to *semantic invariants*.

The design of appropriate fingerprint signatures is an important *intracommunity* activity to structure its own research data collections. Such fingerprint signatures are also very useful for the *intercommunity* usage of research data collections, since they allow to navigate within the (foreign) research data collection without presupposing the full knowledge of the "general nonsense" of the target research domain, i.e., the informal background knowledge required freely to navigate as scientist in that domain. Hence well designed fingerprint signatures are to be considered also as a first class service of a special research community to a wider audience to inspect their research data collections without using the community-internal tools to access the resources themselves.

5   Working with Semantic-Aware Fingerprints

Usually the research data collections (resources in the RDF terminology) of a certain community are stored in a specially designed community-internal format, often as plain text (e.g., the Normaliz Collection [2]), in a special XML notation (e.g., the Polymake Collection [6]) or as SQL database (e.g., the Database for Number Fields [10]). Such formats usually employ special formal semantics agreed within the community as an effective way to store domain specific input and output data and used by commonly developed tools with appropriate parsing functionality.

Usually such formats are extended to store research metadata, i.e., fingerprints or *resource descriptions* in the RDF terminology, together with the research data. This has one benefit and two drawbacks:

- *Benefit:* A fingerprint can be computed immediately by the commonly used tools or with their slight extension, and can be stored with the resource itself.
- *First Drawback:* Metadata unfold its full expressiveness only if one can search and navigate within it. A storage together with the resource itself implies high extraction costs for metadata navigation and access to the research data collection.
- *Second Drawback:* The very different formats prevent an easy combination of metadata from different communities and even from different sources.

The first drawback can be addressed if the metadata are extracted into a database – either a central one or delivered with the tools for local use – and the commonly used intracommunity tools provide search and navigational functionality within that metadata representation. Such an approach based on a web interface was realized for the *Database for Number Fields* [10] and a tool integration based on a Mongo-DB for the *Polymake Database* [6]. But such a solution has two further drawbacks:

- *Drawback 1a:* The search and navigational functionality is not or only in a restricted way adapted for machine-readable interaction and thus cannot be integrated into more comprehensive search and navigational processes.
- *Drawback 1b:* The search and navigational functionality can't be adapted by the user for its own needs.

A general solution that avoids these drawbacks proposes to extract the metadata information from the resource data and to transform it into RDF. RDF – the Resource Description Framework – is the conceptual basis of Linked Open Data as a worldwide distributed database that can be globally queried and navigated using the SPARQL query language in a similar unified way as SQL allows to navigate in local relational databases.

We applied this approach, first used within the SYMBOLICDATA Project to navigate within polynomial systems data, to the data sets on polytopes and on transitive groups newly integrated with SYMBOLICDATA version 3, and also within the recompiled version of test sets from integer programming. We store these fingerprints in our RDF data store [16] thus allowing for a unified navigation and even cross navigation within such data using the SPARQL query mechanism as a generic Web service provided by our SPARQL endpoint [17].

We refer to the SYMBOLICDATA wiki [18] for detailed information and examples how to use that service.

References

1. Bachmann, O., Gräbe, H.-G.: The SymbolicData Project - Towards an Electronic Repository of Tools and Data for Benchmarks of Computer Algebra Software. Reports on Computer Algebra 27, Centre for Computer Algebra, University of Kaiserslautern (2000)
2. Bruns, W., Ichim, B., Römer, T., Sieg, R., Söger, C.: Normaliz: Algorithms for Rational Cones and Affine Monoids, 08 March 2016. https://www.normaliz. uni-osnabrueck.de
3. DFG verabschiedet Leitlinien zum Umgang mit Forschungsdaten. DFG-Magazin "Information für die Wissenschaft" Nr. 66 (2015). http://www.dfg.de/foerderung/ info_wissenschaft/2015/info_wissenschaft_15_66/index.html. 07 May 2016
4. Strategy Report on Research Infrastructures. Roadmap 2016. Published by the European Strategy Forum for Research Infrastructures (ESFRI), Brüssel (2016). http://www.esfri.eu/roadmap-2016 . 16 March 2016
5. FRISCO - A Framework for Integrated Symbolic/Numeric Computation (1996–1999). http://www.nag.co.uk/projects/FRISCO.html. 19 February 2016
6. Gawrilow, E., Joswig, M.: Polymake: a framework for analyzing convex polytopes. In: Kalai, G., Ziegler, G.M. (eds.) Polytopes–Combinatorics and Computation. DMV Seminar, vol. 29, pp. 43–73. Birkhäuser, Basel (2000)
7. Gräbe, H.-G., Johanning, S., Nareike, A.: The SYMBOLICDATA project– towards a computer algebra social network. In: Workshop and Work in Progress Papers at CICM 2014, CEUR-WS.org, vol. 1186 (2014)
8. Hardin, G.: The tragedy of the commons. Science **162**(3859), 1243–1248 (1968). doi:10.1126/science.162.3859.1243
9. Heinle, A., Levandovskyy, V.: The SDEval benchmarking toolkit. ACM Commun. Comput. Algebra **49**(1), 1–10 (2015)
10. Klüners, J., Malle, G.: A Database for Number Fields, 08 March 2016. http:// galoisdb.math.uni-paderborn.de/
11. OpenDreamKit: Open Digital Research Environment Toolkit for the Advancement of Mathematics, 16 March 2016. http://opendreamkit.org/, http://cordis.europa. eu/project/rcn/198334_en.html
12. Paffenholz, A.: Polytope Database, 08 March 2016. http://www.mathematik. tu-darmstadt.de/~paffenholz/data/
13. The PoSSo Project. Polynomial Systems Solving - ESPRIT III BRA 6846 (1992–1995)
14. Research Infrastructures, including e-Infrastructures, 16 March 2016. http://ec. europa.eu/programmes/horizon2020/en/h2020-section/research-infrastructures-including-e-infrastructures
15. The SageMath Project, 16 March 2016. http://www.sagemath.org/
16. The SYMBOLICDATA RDF Data Store, 15 March 2016. http://symbolicdata.org/ Data
17. The SYMBOLICDATA SPARQL Endpoint, 19 February 2016. http://symbolicdata. org:8890/sparql
18. The SYMBOLICDATA Project Wiki, 13 March 2016. http://wiki.symbolicdata.org
19. swMATH–a new Information Service for Mathematical Software, 07 March 2016. http://www.swmath.org/

Linking Mathematical Software in Web Archives

Helge Holzmann[1(✉)], Mila Runnwerth[2], and Wolfram Sperber[3]

[1] L3S Research Center, Appelstr. 9a, 30167 Hannover, Germany
holzmann@L3S.de
[2] German National Library of Science and Technology (TIB),
Welfengarten 1b, 30167 Hannover, Germany
Mila.Runnwerth@tib.eu
[3] zbMATH, FIZ Karlsruhe - Leibniz Institute for Information Infrastructure,
Franklinstr. 11, 10587 Berlin, Germany
wolfram@zentralblatt-math.org

Abstract. The Web is our primary source of all kinds of information today. This includes information about software as well as associated materials, like source code, documentation, related publications and change logs. Such data is of particular importance in research in order to conduct, comprehend and reconstruct scientific experiments that involve software. swMATH, a mathematical software directory, attempts to identify software mentions in scientific articles and provides additional information as well as links to the Web. However, just like software itself, the Web is dynamic and most likely the information on the Web has changed since it was referenced in a scientific publication. Therefore, it is crucial to preserve the resources of a software on the Web to capture its states over time.

We found that around 40 % of the websites in swMATH are already included in an existing Web archive. Out of these, 60 % of contain some kind of documentation and around 45 % even provide downloads of software artifacts. Hence, already today links can be established based on the publication dates of corresponding articles. The contained data enable enriching existing information with a temporal dimension. In the future, specialized infrastructure will improve the coverage of software resources and allow explicit references in scientific publications.

Keywords: Scientific software management · Web archives

1 Introduction

Providing specialized information services for software (SW) is challenging for various reasons: SW is highly dynamic, references in literature are often not declared as SW, and structured metadata is sparse. Repositories as well as directories typically represent SW in an abstract manner with provided information

---

This work is partly funded by the German Research Council under FID Math and the European Research Council under ALEXANDRIA (ERC 339233).

G.-M. Greuel et al. (Eds.): ICMS 2016, LNCS 9725, pp. 419–422, 2016.
DOI: 10.1007/978-3-319-42432-3_52

corresponding not to a specific state or only to the current version. In contrast to these representations, mentions of SW in scientific articles often refer to the SW's state at the time of use or publication. However, SW is dynamic and often a version has been updated since it was referenced. With changes of the SW, its website is likely to be updated as well. Thus, corresponding information as well as associated materials of the referenced version might not be available anymore. This makes it difficult to trace the development process and to obtain information about a previous version of the SW, although this can be necessary to reproducing published scientific results. While open source SW facilitates access to different states of the source code, this does not always include associated materials, such as a documentation.

To fix this temporal gap between publications and representations of SW on the Web in the long run, a sophisticated Web archiving infrastructure is required to preserve available information about SW at the time when it is referenced. In the short term, however, establishing links to existing Web archives at the publication time of an article will be a first step in this direction. SW directories can serve as a bridge in this endeavor. swMATH[1], a directory for mathematical SW (MSW), follows a publication-based approach to link SW and corresponding scientific publications (s. Sect. 2). From all MSW listed on swMATH, we found the websites of around 40 % to be already existent in the *Internet Archive*'s Web archive[2], with many of them providing associated materials (s. Sect. 2).

2   Publication-Based Approach of swMATH

swMATH is one of the most comprehensive information services for MSW [1]. Based on simple heuristics, swMATH identifies MSW in scientific articles from its underlying bibliographic database zbMATH[3], consisting of nearly 120,000 publications referring to MSW. Currently, swMATH contains more than 12,500 SW records linked to corresponding articles, as shown in Fig. 1.

One of the major challenges of swMATH is the identification of MSW in scientific publications. In many articles, only names are mentioned, while versions or explicit labels as MSW are missing. swMATH addresses this by scanning titles, abstracts, as well as references for typical terms, such as *solver*, *program*, or simply *software*, in combination with a name. After that, a manual intervention step verifies the recognized SW to ensure a high quality of the service. As part of this, additional metadata, such as a website, the authors, technical requirements, dependencies, licenses, documentations and more, is looked-up through a regular Web search and added if available.

Similar to other SW directories and repositories (s. Sect. 1), the focus of swMATH is to provide time-agnostic information about SW rather than specific versions. Therefore, included websites are periodically checked and outdated links as well as related information are removed or replaced. While this ensures an up-to-date representation of the SW, it introduces inconsistencies with included

---

[1] http://www.swmath.org.

[2] http://web.archive.org.

[3] http://www.zbmath.org.

Fig. 1. Record of the mathematical software Singular on swMath

publications, which are annotated with the year of publication and constitute temporal witnesses of different SW states. Therefore, it has been considered to integrate temporal information in order to represent the different versions of a SW over time and match publications. Web archives can serve as source for this information in the future.

3   Linking Web Archives

Web archives have recently been of growing interest in research, however, they have either been used as scholarly sources or have been a subject of research themselves, with questions focusing on coverage and evolution [2,3]. To the best of our knowledge, Web archives have never been used to recover information and associated materials of former states of SW. We tackle this by linking articles on SW with available resources in Web archives. As proxy serve the website URLs as well as the publication dates of corresponding articles as listed on swMATH. An initial analysis of this data unveiled what to expect from integrating Web archives as source for temporal information about MSW.

As shown in Fig. 2a, we found that around 60 % of the analyzed websites contain some kind of documentation, almost 50 % link to publications, more than 40 % even provide downloads of SW artifacts with 30 % being open source, and 10 % could be identified to publish updates or news, such as a changelog. Although not all of this is currently being preserved by Web archives, already today around 40 % of the URLs under investigation are included in the considered Web archive with at least one capture, as shown by the *archived* bars in Fig. 2b dissected by the year of the highest cited publication of a SW. However, looking at the fraction with captures in the corresponding year, the *past archived* bars unveil a clear growth over time with relatively low numbers in the early times of Web archiving. Hence, we are able to successfully link around 25 % of the analyzed MSW with their top publication in 2013, but only very few before 2000. The reason for this is two-fold, while the coverage of Web archives has drastically improved over time, as pointed out in Sect. 2, outdated links in swMATH have been replaced with new URLs, which might not have existed at the time of the top publication yet.

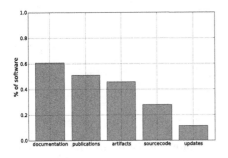 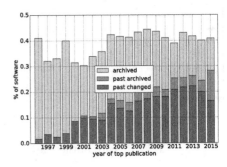

(a) Information on mathematical software pages.

(b) Pages changed since top publication.

Fig. 2. Mathematical software in Web archives.

4   Conclusion and Outlook

Linking MSW in Web archives will help to recover information as well as associated materials of previous versions referenced in scientific publications. As shown by the third category in Fig. 2b, *past changed*, almost all of the websites that were archived in the year of the top publication mentioning a SW have changed, which indicates the need of our approach. Moreover, the tools used for the analysis of Web archives are a first step towards a machine-based content analysis of the websites of a MSW. This opens up new possibilities to enrich the information in swMATH.

In order to overcome the challenge of identifying former URLs of SW resources in a Web archive (s. Sect. 3), temporal tags may be incorporated in the future, as demonstrated by Holzmann and Anand [4].

References

1. Greuel, G.-M., Sperber, W.: swMATH – an information service for mathematical software. In: Hong, H., Yap, C. (eds.) ICMS 2014. LNCS, vol. 8592, pp. 691–701. Springer, Heidelberg (2014)
2. Ainsworth, S.G., Alsum, A., SalahEldeen, H., Weigle, M.C., Nelson, M.L.: How much of the web is archived? In: JCDL 2011 (2011)
3. Holzmann, H., Nejdl, W., Anand, A.: The dawn of today's popular domains - a study of the archived german web over 18 years. In: JCDL 2016 (2016)
4. Holzmann, H., Anand, A.: Tempas: temporal archive search based on tags. In: WWW 2016 Companion (2016)

Mathematical Models: A Research Data Category?

Thomas Koprucki and Karsten Tabelow$^{(\boxtimes)}$

Weierstrass Institute (WIAS), Mohrenstr. 39, 10117 Berlin, Germany
karsten.tabelow@wias-berlin.de
http://www.wias-berlin.de

Abstract. Mathematical modeling and simulation (MMS) has now been established as an essential part of the scientific work in many disciplines and application areas. It is common to categorize the involved numerical data and to some extend the corresponding scientific software as research data. Both have their origin in mathematical models. In this contribution we propose a holistic approach to research data in MMS by including the mathematical models and discuss the initial requirements for a conceptual data model for this field.

Keywords: Research data · Mathematical modeling and simulation · Mathematical knowledge management

1 Introduction

In recent years the handling of research data as part of the scientific practice has created vivid discussions within the scientific community, at research institutions as well as in funding agencies. Specifically, the importance of research data and its storage in view of new digital technologies is emphasized by the recent adoption of the "DFG Guidelines on the Handling of Research Data" by the Deutsche Forschungsgemeinschaft [1], the Open Research Data Pilot within the EU Horizon 2020 program [2], or the development of principles for research data handling within the german scientific organizations, the Leibniz Association, the Max-Planck-, Helmholtz-, and Fraunhofer Society by the Priority Initiative "Digital Information"[1].

The importance of appropriate handling of research data is increasingly recognized in view of its rising amount. It is central part of the discussion on Open Data and a prerequisite of the scientific method. In the face of the emerging digital science agenda research data proves to be an essential foundation for scientific work. Driven by such considerations universities and scientific institutions started creating policies for the handling of research data. These include rules for the full data life-cycle including generation, storage, preparation for subsequent re-use, publication and curation of data. However, the nature of research data is as diverse as the scientific disciplines requiring specific discussions and concepts.

---

[1] http://www.allianzinitiative.de/en/core-activities/research-data.html.

© Springer International Publishing Switzerland 2016
G.-M. Greuel et al. (Eds.): ICMS 2016, LNCS 9725, pp. 423–428, 2016.
DOI: 10.1007/978-3-319-42432-3_53

2   Research Data in Mathematical Modeling and Simulation

Mathematics is one of the foundations of today's key technologies and science. Mathematical methodology is required for interdisciplinary modeling of a research problem, for its mathematical treatment and solution, and for the transfer of the results into practice. In the last decade *mathematical modeling and simulation* (MMS) has been established alongside experiment and theory and is now essential part of the scientific work in many disciplines and application areas.

Research in the area of MMS is characterized by *mathematical models, scientific software* for their treatment, and *numerical data* related to computations (input, output, parameters), see Fig. 1. Here, we propose to categorize these three parts as the research data in MMS as they are jointly required to understand and verify research results, or to build upon them.

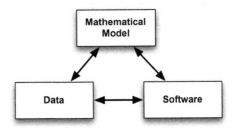

Fig. 1. This figure illustrates the three different components constituting research data in MMS: mathematical models, simulation software and numerical data (e.g., input parameters and data, output of simulation software).

Specifically, numerical data is generally regarded as research data in common sense and data repositories and information services such as DataCite [3] or RADAR [4,5] exist or are emerging. Increasingly, software is categorized as reasearch data [1] and a world-leading information service on mathematical software, swMath [6], has already been developed.

Yet, communication in MMS suffers from the absence of a unified concept including mathematical models: instead of considering mathematical models as entities of their own class, which can be uniquely identified, cited and categorized, they are rather found as plain text with a mixture of mathematical notation and common language. This potentially leads to ambiguity, cites to different original work, incompleteness, and "re-invention of the wheel".

A comprehensive approach to research data in MMS should cover all three aspects in a similar manner. While for software and numerical data the above mentioned services are reasonable starting points for the implementation of such a concept, a corresponding definition and service for mathematical models is missing.

A similar system for computational models of biological processes has been introduced at the BioModels Database[2].

In this contribution we discuss the initial requirements for a conceptual data model for mathematical models, starting from a simple and widely-used example. This is a first step towards the creation of a semantic corpus for mathematical models in MMS, which can serve as a standardized access to mathematical models with cross-links to software, data repositories, and publications.

3   Mathematical Models: The Heat Transport Problem

As an example for mathematical modeling we consider the heat transport problem. Modeling and simulation of heat transport is a common task in many technical applications ranging from large heat exchangers to heating effects in small semiconductor devices. For our discussions on a formalization concept for mathematical models we outline the corresponding description of the heat transport model and its ingredients as plain text, which might be similarly found in typical publications on this topic or in software documentations.

Heat transport model. We describe heat conduction by Fourier's law

$$q = -\lambda \nabla T, \tag{1}$$

where q denotes the *heat flux*, λ represents the *heat conductivity* and ∇T is the *temperature* gradient. In a bounded spatial domain Ω the time evolution of the *temperature distribution* $T(x,t)$ is then governed by the *heat flow equation*:

$$\frac{\partial}{\partial t}(C(x)T(x,t)) - \nabla \cdot (\lambda(x)\nabla T(x,t)) = f(x) \text{ in } \Omega, \tag{2}$$

with a *heat source* $f(x)$, the *heat capacity* $C(x)$ of the material, and boundary conditions

$$-\nu \cdot q = \nu \cdot (\lambda(x)\nabla T(x,t)) = \kappa(x)(T(x,t) - T_a(x)) \text{ on } \partial\Omega, \tag{3}$$

where ν denotes the outer normal vector, κ the *heat transfer coefficient* to the environment and T_a is the *ambient temperature*. In studies of time-dependent heating phenomena the time evolution of the temperature and thus the heat flow is given by the solution of the boundary value problem (2)–(3) with the initial value $T_0(x) = T(x, t = 0)$. In contrast, one is often only interested in the stationary heat paths, e.g., in studying the heat flow from a device. Then it suffices to solve the *stationary heat equation*

$$-\nabla \cdot (\lambda(x)\nabla T(x,t)) = f(x) \text{ in } \Omega \tag{4}$$

subject to the boundary conditions (3).

---

[2] https://www.ebi.ac.uk/biomodels-main/.

4   Towards a conceptual data model for mathematical models

The recognition of mathematical models as part of research data in MMS can be established by the creation of a semantic digital corpus of mathematical models. An information service for the registration and retrieval of mathematical models is then necessary for the adoption of the approach by the MMS community and for navigation, indexing and searching the model corpus.

The creation of such a corpus cannot just rely on a plain text description as above, instead one has to develop a normal or canonical form. A similar normalization is common in general mathematical texts where definitions, lemmas, theorems, proofs, corollaries, propositions help to structure the content. Similar to the approach of the semantical annotation of mathematical texts a normal form for mathematical models needs to be represented in a modeling-oriented markup-language, which can be based on LaTeX or MathML. In contrast to a pure plain text description such a mark-up can be used to generate relations between the entities of the formal description.

The encoded entities should contain the main characteristics of the model, such as the equation, the domain, boundary conditions, material laws and constitute a signature for the mathematical model. However, the complexity of the task is far above simple one-to-one mappings as it is possible, e.g., for special functions. A mathematical model is an abstract notion relying on a mathematical equation combined with semantic binding. Despite the fact that typically multiple notations for the same equation exist, the task is further complicated by the non-trivial question which entities are to be considered as atoms of the description. For example, the definition of the heat flux in the heat transport model can be itself considered as a model. The same applies to the material laws such as the heat conductivity or the heat capacity where a constant or linear dependence can be described by a single parameter. Finally, the replacement of the boundary conditions of the heat transfer (3) on parts of the boundary by a model for heat radiation (T^4-law) leads to further variants of the original model with specific properties.

A data model for mathematical models must reflect a sufficient level of complexity of the formal description to cover a large number of models while avoiding unnecessary duplications in their encoding. It is a-priori not clear whether such a description exists. The problem can be mitigated by appropriate relations between different entries of the model corpus. In its final form an information service for mathematical model should not only include models characterized by partial differential equations, but also statistical or discrete models, as well as systems of ordinary differential equations and many more.

Beside the plethora stemming from different specializations of a certain model as introduced above two further dimensions of a data model are essential which we introduce as *math bindings* and *application bindings*.

Math Bindings. The mathematical notation of a specific model is everything but unique. Even for the non-dimensionalized heat equation with constant coefficients ($\lambda(x) = const.$, $C(x) = const.$) there exists a whole diversity of possible mathematical notations such as a notation with Nabla calculus as above, a representation in Cartesian coordinates, simplification to a Laplacian, a notation with *div-* and *grad-* operators, weak formulations, or formulations as a gradient flow.

It is common to classify linear, second order partial differential equations as elliptic, parabolic or hyperbolic. For instance the transient heat equation (1) is mathematically classified as a *parabolic partial differential equation*, whereas the stationary heat flow problem (4) constitutes a partial differential equation of *elliptic* type. The classification provides useful hints for their mathematical treatment and for the characterization of their solutions. The mathematically precise formulation of the model equations relates the assumptions on the data of the problem, e.g., regularity of coefficient functions or smoothness of the domain and its boundaries, to mathematical theory.

Furthermore, for the numerical solution of the model equations different computational methods can be used. This introduces another aspect related to the model description and the utilized software, which might also be regarded as a math binding.

Application Bindings. The universality of mathematical models allows for transferring models from one application area to a different context. For example the heat flow model above can be re-interpreted as model of diffusion processes of particles. In this case the quantities get a new semantic meaning together with a new notation: the temperature T is the particle density u, the heat conductivity λ becomes the diffusion coefficient D.

A second aspect is the usage of models as building blocks to describe coupled phenomena like in thermistor models, which couple thermal and electric transport, or heat treatment of steel which couples heat transport with phase transitions and elasticity. In these cases coefficient functions are defined by solutions of supplemental differential equations, e.g., for thermistors the Joule heat generated by a current flow enters the heat equation (2) as a source term $f(x)$.

Both aspects are key features of mathematical modeling which are related to the abstraction given by the mathematical language. They are the basis for the strength of mathematical modeling and for the success of MMS as a third discipline between theory and experiment.

Connection to Software and Data. The application of a mathematical model, such as the heat transport model, to a specific technical problem requires a mapping of mathematical objects such as coefficient functions to *properties*, or more precisely *material parameters*, of the involved *materials* and boundaries. In our example these are the heat conductivity $\lambda(x)$, the heat transfer coefficient $\kappa(x)$ and the heat capacity $C(x)$. The numerical solution of the heat transport problem requires the approximation of the continuous problem (2) by discretization methods. This involves a *geometric description* of the simulation domain by suitable meshes. Typically, the simulation results in numerical values of the temperature distribution T on the numerical mesh constituting the output of the simulation software. Correspondingly, *initial values* and the *material data* are the input for the software. Both, simulation results and input data, constitute the numerical data part of the research data in MMS, see Fig. 1. Certain mathematical objects occurring in mathematical model have a semantical binding, namely of T being the temperature, λ being the heat conductivity etc., but they also link input and output of the utilized software and its interpretation.

5   Conclusions

We proposed to categorize mathematical models, scientific software and numerical data as the research data in MMS requiring suitable information services for their management and handling. For numerical data and scientific software the awareness of this fact in the MMS community is growing and suitable concepts and information services are emerging. However, for the category of mathematical models, a corresponding definition and service is missing. We highlighted the initial conceptual requirements for the definition of a suitable data model and its difficulties on the basis of the heat transport model. A unified approach to research data management that not only includes numerical data and scientific software but also mathematical models can help to enhance future MMS publications by making them more concise. A digital corpus of mathematical models together with a suitable information service is necessary to reduce the additional effort for the authors. On success its creation will be an important contribution of applied math to the digital science agenda. Furthermore, it has the potential to reduce today's language barriers between disciplines and requires an interdisciplinary effort.

Acknowledgments. The authors are grateful for many fruitful discussions with W. Sperber who helped to shape their knowledge on the topic.

References

1. Deutsche Forschungsgemeinschaft : DFG Guidelines on the Handling of Research Data. Adopted by the Senate of the DFG at 30 September 2015
2. European Commission: Guidelines on Open Access to Scientific Publications and Research Data in Horizon 2020, Version 2.1, 15 February 2016
3. Brase, J.: DataCite - a global registration agency for research data. In: Fourth International Conference on Cooperation and Promotion of Information Resources in Science and Technology, COINFO 2009, pp. 257–261, Beijing (2009)
4. Razum, M., Neumann, J., Hahn, M.: RADAR - Ein Forschungsdaten-Repositorium als Dienstleistung für die Wissenschaft. Zeitschrift für Bibliothekswesen und Bibliographie **61**, 18–27 (2014)
5. Kraft, A.: RADAR - a repository for long tail data. In: Proceedings of the IATUL Conferences. Paper 1 (2015)
6. Greuel, G.-M., Sperber, W.: swMATH – an information service for mathematical software. In: Hong, H., Yap, C. (eds.) ICMS 2014. LNCS, vol. 8592, pp. 691–701. Springer, Heidelberg (2014)

Mathematical Research Data and Information Services

Wolfram Sperber$^{(\boxtimes)}$

zbMATH, FIZ Karlsruhe - Leibniz Institute for Information Infrastructure,
Franklinstr. 11, 10587 Berlin, Germany
`wolfram@zentralblatt-math.org`

Abstract. In the last centuries mathematical research results were published on paper, as articles, reports, monographs, etc. The permanently increasing number of mathematical publications required to develop powerful information management services and tools for search and access to mathematical knowledge. 1868, the 'Jahrbuch über die Fortschritte der Mathematik' was founded to inform the mathematical community and interested scientists on the progress of mathematical knowledge. Also today, its successors, the bibliographic databases zbMATH and MathSciNet, provide an overview about recent developments in mathematics. But the digital era has changed the situation dramatically. All kinds of information are stored in digital form. Mathematical research results are not longer limited to mathematical publications. New types and formats of mathematical data and knowledge tackle new challenges also to information services. The talk addresses the subject of Mathematical SoftWare (MathSW) in the context of Mathematical Research Data (MathRD) and describes some challenges to create information services for this data from a personal view.

Keywords: Mathematical research data · Mathematical software · Mathematical services · Mathematical models

1 Introduction: What is Mathematical Research Data?

Today, a lot of research results will be achieved with the help of computers and SoftWare (SW). For these research results, reproduction, verification and reuse involves also SW and data which are used. Up to now there is no satisfying definition of Research Data (RD). But RD is gaining attention in the contemporary discussion about the scientific infrastructure. The library of the University Boston gives the following more or less tautological definition: 'Research data is data that is collected, observed, or created, for purposes of analysis to produce original research results' [1]. Of course, this definition could be easily adapted to mathematics by adding the attribute 'mathematical' to it. But such a definition is very general. RD is immediately connected with the research process and the objects and concepts declared as RD are specific for each science. Therefore we discuss some criteria for MathRD and then address some important classes of MathRD. MathRD

© Springer International Publishing Switzerland 2016
G.-M. Greuel et al. (Eds.): ICMS 2016, LNCS 9725, pp. 429–433, 2016.
DOI: 10.1007/978-3-319-42432-3_54

- covers distinct objects of mathematical research
- has an unambiguously defined mathematical content
- is necessary for the reproduction, verification and reuse of mathematical research results

Some remarks:

- These criteria address especially also computational tools as MathSW.
- These properties are applicable to many heterogeneous kinds of resources which are inside (e.g. mathematical models) or outside (e.g. MathSW) of mathematical publications.
- Also data collections with a strong relatedness to real-world problems, e.g. measurement data, which are used for mathematical research are MathRD.

Roughly, MathRD is all data which is used in mathematical research plus tools for handling this data, MathRD in the narrower sense is data of mathematical research provided in databases and services.

2   MathSW, Services, Data, and Models

2.1   Mathematical SoftWare (MathSW)

MathSW plays an exceptional role for the following reasons and is a first candidate for a special MathRD class. MathSW is an own object of mathematical research which is strongly connected with computer science. MathSW is written in a formal language, not in a natural language. Another difference to other forms of mathematical knowledge, especially publications, is its dynamic character resulting from the dependencies from hard- and software, user interfaces, etc. The dynamic character of MathSW manifests itself in versions, bug-fixings, etc.

MathSW requires new concepts for maintaining and content analysis. The source code is inappropriate to give human audience information about the content of MathSW. Typically, documentations or manuals independent from source code and written in natural language accompany the source code. Documentations and manuals describe the functionalities of MathSW, the algorithms behind, the implementation, specify technical requirements, the user interface etc. Documentations and manuals of MathSW must be updated when a new version is available.

'Regular' publications on MathSW as journals articles or books refer to a special – possibly outdated – version of MathSW.

Some mathematical journals, e.g., 'ACM Transaction on Mathematical Software' (ACM TOMS) [2], the 'Journal of Statistical Software' (JSS) [3], or the Journal 'Mathematical Programming Computation' (MPC) [4] which have their focus on MathSW, have defined policies for evaluation which involve also the software code (of the analyzed version). Typically, the evaluation requires an independent implementation of MathSW and the data used as well as the reproduction of the achieved research results.

2.2   Further Classes of MathRD

MathSW and data are closely linked. Data is the input for MathSW and its output. The verification of research results achieved with MathSW requires data. Data can be plain tables of numbers, e.g. measurements results, which are analyzed by statistical methods, but also complex representations of mathematical objects as functions, sequences, differential equations, matroids, groups or ideals, etc.

Data used must be available in a unique and standardized way. This involves especially a formalization and semantic encoding of data. The development of languages and formats for presenting data has led to specific data formats for MathSW, see e.g. the Polymake format [5].

Benchmarks, test sets, are used for analyzing the performance of MathSW.

Collections of MathRD are often accessible via services. Services providing MathSW can be regarded as a special class of MathSW. There exists a row of popular mathematical services in the Web, e.g. the On-Line Encyclopedia of Integer Sequences (OEIS) [6] or the NIST Digital Library of Mathematical Functions (DLMF)[7]. Mathematical services can be databases of mathematical information or MathSW, e.g. cloud computing, or a combination of both.

Mathematical Models (MathMs) are MathRD of their own interest and an unique feature of mathematics. MathMs play a significant role in all phases of mathematical research. They are the result of mathematical modeling processes or mathematical transformations or the starting point for the use of MathSW. Typically, MathMs are part of mathematical publications but a separate presentation would be useful, especially for applications of mathematics. The presentation of research models is challenging which show first prototypes of model databases in other science because research models are often complex and up to now there exist no standards for a semantic presentation. MathMs will be discussed in a separate talk in more detail.

Remarks:

- It is not the intention to describe the complete spectrum of MathRD. There are other important classes of MathRD, e.g. algorithms, simulations, encyclopeadias and glossaries of mathematical terms and concepts, etc. The talk should give some impulses for the future discussion of the topic.
- The borderline between different MathRD classes is fuzzy. Interactive mathematical services, e.g. OEIS which cover also calculations of data, contain both data and tools. Nevertheless, a classification of MathRD and a semantic linking between these data, e.g. between models and software would be helpful for the retrieval. Our impetus for typing of MathRD was caused by swMATH [8] activities. swMATH is an information service for MathSW, with the focus on MathSW. But the publication-based approach for the identification of MathSW doesn't allow a secure identification of the type of resources. So swMATH also contains entries describing benchmarks, services, models, and other resources. A classification of the type would improve the retrieval precision.

– Different kinds of research data are in close connection. Context information is very helpful for humans as well as automatic processing of information. As an example we refer to the references to MathSW in publications which are used by the swMATH service for identification and maintaining the service.

3   Information Services and Open Platforms for MathRD

Information services on MathRD are those which provide search for, navigation in, and access to MathRD. Therefore qualified information, especially semantic metadata, about the MathRD are necessary. MathRD services are under development. Up to now, the existing services are often focused to special subjects and are widely distributed.

We start with brief overview on information services on MathSW because the majority of existing mathematical information services on MathRD are focused on MathSW and the associated data as benchmarks etc.

The information in the Web on MathSW consists of sources which address MathSW directly, especially Web sites, repositories, directories and also indirect sources, publications which refer to MathSW, mathematical models which use MathSW, or associated data to a software (benchmarks, etc.).

With the advent of MathSW some activities started to make MathSW searchable and usable. The Netlib [9] is an early and well-known example of a MathSW repository. Repositories are archives. They provide the source code and metadata of MathSW. Often MathSW repositories are focused to individual mathematical subjects developed in a special programming language. An excellent example for this type of repositories is CRAN [10], the Comprehensive R Archive Network, collecting statistical computing packages written in R. It contains currently more than 8,300 packages. Classification schemes as the GAMS index [11] help the users to find relevant MathSW. The providers define the policy of the repository, that means specifying the mathematical subjects, criteria and standards for the input (formats, metadata), check the consistency and completeness of the input, administrating and handling the MathSW data, and provide the information on the resources in a uniform manner. In general, repositories are well accepted entry points with a standardized access and improve the visibility of MathSW listed.

On the other hand repositories capture the existing MathSW only partially and the standards for the description of resources are proprietary. Therefore, a lot of MathSW developers run local Web sites for their MSW with special structure and design. The intention of MSW directories, e.g. swMATH, is to provide a fast and comprehensive search and navigation to MathSW resources in some or all mathematical subjects in the Web. Directories are collections on MathSW providing metainformation and links to further information on and the source code of MathSW. Metadata of MathSW results from analyzing direct or indirect resources on MathSW. swMATH is an example for a directory on MathSW which analyzes indirectly MathSW resources by heuristic means (especially information about MathSW in mathematical publications).

The SymbolicData project [12] has the goal to provide an enhanced access to MathRD for the Computer Algebra (CA) community. Therefore methods for the content analysis of CA data have been developed basing on Semantic Web concepts, especially Resource Description Framework (RDF) [13] and Ontology Web Language (OWL) [14].

An future trend for enhanced informnation services in mathematics is the development of open platforms for computational mathematics. Initiatives as SAGEMATHCLOUD [15] or the recent EU project OpenDreamKit [16] aim for to support collaborative work in computational mathematics. Virtual research environments which integrate various MathSW and other MathRD allow a seamless use of different MathSW and data. This requires new concepts for the integration of different resources and formats.

But for many other kinds of MathRD, e.g. MathMs, specialized information services are not existing.

4 Summary

The availability of MathRD plays a central role for the development of computational mathematical research and application of mathematics. MathRD is a problem-oriented view on models, tools, associated data, etc. A differentiated presentation of MathRD classes opens new perspectives for presentation and dealing with MathRD. But up to now a holistic concept for a systematic and comprehensive availability and access of MathRD is missing because MathRD is very heterogeneous and multifacted. The mathematical community is challenged to develop and realize concepts for a distributed system of information services for MathRD which extend, enhance, integrate, and link the existing ones.

References

1. http://www.bu.edu/datamanagement/background/whatisdata/
2. ACM Transaction on Mathematical Software (ACM TOMS). http://toms.acm.org
3. Journal of Statistical Software (JSS). http://www.jstatsoft.org
4. Mathematical Programming Computation (MPC). http://mpc.zib.de
5. Polymake website. https://polymake.org
6. On-Line Encyclopedia of Integer Sequences (OEIS). https://oeis.org/
7. NIST Digital Library of Mathematical Functions (DLMF). http://dlmf.nist.gov/
8. swMATH database. http://www.swmath.org
9. Netlib Repository. http://www.netlib.org/
10. CRAN the Comprehensive R Archive Network. https://cran.r-project.org/
11. Guide to Available Mathematical Software (GAMS). https://www.gams.com/
12. SymbolicData project. http://wiki.symbolicdata.org/Main_Page
13. Resource Description Framework (RDF). https://www.w3.org/RDF/
14. Ontology Web Language. https://www.w3.org/TR/owl2-overview/
15. SAGEMATHCLOUD. https://cloud.sagemath.com/
16. OpenDreamKit. http://opendreamkit.org/about/

SemDML: Towards a Semantic Layer of a World Digital Mathematical Library

Stam's Identities Collection: A Case Study for Math Knowledge Bases

Bruno Buchberger[✉]

Research Institute for Symbolic Computation (RISC),
Johannes Kepler University, Linz, Austria
bruno.buchberger@risc.jku.at
http://www.risc.jku.at/people/buchberg/

Abstract. In the frame of the work of the Working Group "Global Digital Mathematical Library", Jim Pitman proposed Aart Stam's collection of combinatorial identities as a benchmark for "digitizing" mathematical knowledge. This collection seems to be a challenge for "digitization" because of its size (1300 pages in a .pdf file) and because of the fact that, for the most part, it is hand-written. However, after an in-depth analysis, it turns out that the real challenges are of mathematical and logical nature. In this talk we discuss what digitization of such a piece of mathematics means and report on various tools that may help in this endeavor. The tools range from technical tools for typing formulae all the way to sophisticated algebraic and reasoning algorithms. The experiments for applying these tools to Stam's collection are currently carried out by two of the working groups at RISC.

1   The Problem

Aart Stam's collection of combinatorial identities (Stam 2012) consists of hundreds of identities that show how formal sums involving bionomial coefficients can be simplified. The collection also explains how these identies can be proved using various proof techniques.

In the context of the "Global Digital Math Library" project, Jim Pitman (Pitman 2015) proposed to consider this collection as a benchmark for the "digitization" of mathematical knowledge. We are faced with the challenge how the extremely valuable knowledge contained in such a collection can be transformed to a form in which the individual identities can be stored, accessed, and processed by algorithmic tools over the web. One might think that the task should and could be decomposed into a first step of (automated) translation of the hand-written formulae into LATEX or any other mathematical expression format and a second step of processing the LATEX formulae by sophisticated algorithms and tools from computer algebra and automated mathematical reasoning. However, given current technologies, we will show that this may not be the most reasonable approach. In fact, we will see that it already can be questioned whether the individual identities need to be stored or, alternatively, may be generated and / or proved on demand!

© Springer International Publishing Switzerland 2016
G.-M. Greuel et al. (Eds.): ICMS 2016, LNCS 9725, pp. 437–442, 2016.
DOI: 10.1007/978-3-319-42432-3_55

2   The RISC Approach

For looking into the feasibility of formalizing Stam's paper, we installed a seminar at our RISC institute (my Theorema group and Peter Paule's Symbolic Combinatorics group) for working together on the formalization of Stam's paper. Our team consists of:

- Theorema Group: Bruno Buchberger, Alexander Maletzky, Wolfgang Windsteiger.
- From the Symbolic Combinatorics Group: Peter Paule, Christoph Koutschan, Clemens Raab, Silviu Radu, Carsten Schneider.

After in-depth discussion, we came up with the following decomposition of work and distinction between the various aspects of the problem:

a. *Translating the formulae into predicate logic form (or any variation of this form), but still in the usual nice two-dimensional appearance used in math papers and typing the formula in this format:*
This can be easily done in Theorema. I did some timing experiments and my estimate is that, for the approximately 1200 formulae in the section "Tables" (the kernel of Stam's paper) I would need approximately 60 h. After this, all formulae would be available in correct logical form that could be translated to any other logic form automatically. Also, after the formalization, hyperlinks to all formulae would be available. Here is a view to the first few formulae in Theorema notation (which could be changed according to the taste of users):

$\forall n \in \mathbb{N}$

From D(19):

$$\left(\sum_{k=0,\ldots,n} \binom{n}{k} \right) == 2^n \tag{1}$$

From D(19):

$$n \geq 1 \Rightarrow \left(\left(\sum_{k=0,\ldots,n} (-1)^k \binom{n}{k} \right) == 0 \right) \tag{2}$$

$$\left(\left(\sum_{k=0,\ldots,n} (-1)^k \binom{x}{k} \right) == (-1)^n \binom{x-1}{n} \right) \tag{3}$$

The important thing is that, internally, the complete parse tree of the formulae as a *Mathematica* nested expression is available. Thus, automated translation to any other formalization, to the input format of arbitrary reasoners, and of course also to pretty-print LaTeX, would be possible. This is in clear distinction to formulae presented, first, in LaTeX, which does not display the logical structure of the formulae and from where automated translation to formulae in logic is *not* possible.

For example, the *Mathematica* formula:

$$\sum_{i=1}^{n} k^2$$

has the internal representation

Sum[Power[k,2], List[i,1,n]]

which reveals the structure completely. Theorema formulae, internally are also *Mathematica* nested expressions but with a structure that is closer to some common forms of predicate logic.

b. *Automated proofs of all formulae using the "old-fashioned" proof methods:*
   Some of these methods (simplification, induction, summation quantifier inference rules as described in (Buchberger 1980)), are already implemented in Theorema. Some adjustments are necessary though. Stam lists approximately 15 "old-fashioned" proof methods. I estimate that we can implement all of them in Theorema with an effort of about one person year. Most of them, however, are superseded by the "modern" proof methodology, see remark c., or already have some flavor of the "new-fashioned" proof methods, e.g. Ergoychev's method. See however also remark g.
   - reduction to known formulae
   - rearranging factorials
   - Fibonacci
   - Lucas
   - poly of convolution type
   - specialization in general summation formulae
   - the complex argument
   - induction over naturals
   - recurrence
   - finite differences
   - Newton interpolation
   - inverse relations (to do with convolution)
   - inclusion - exclusion
   - multisection of sums
   - expansion of factor in the summand
   - the beta integral
   - generating functions
   - partial fractions
   - Egorychev' s method.

c. *Automated proofs of the formulae using "new-fashioned" proof methods, I call them "Algebraic Simplification Methods":*
   These methods proceed by translating the formulae into objects in suitable algebraic (polynomial) domains and sophisticated new simplification techniques based on new math results for these poly domains (e.g. the noncommutative Gröbner bases methodology). This is prominent research activity of about 15 people in the world over the past 20 years. In this talk and

extended abstract, I cannot give a fair account of the individual contributions of the key players in this research field. Ground was laid by Doron Zeilberger (the "holonomic systems approach to special function identities") together with Herb Wilf, George Andrews and Marko Petkovsek. Today the methods (with software implementation) by Peter Paule and his former PhD students Manuel Kauers, Christoph Kouschan, Carsten Schneider et al., and by Frederic Chyzak seem to be the most advanced. The literature on the field is contained in the recent monograph (Kauers 2011).

After detailed inspection of all the formulae in Stam's paper, we are pretty sure that 95% or more of these formulae can be proved completely automatically with the methods available in Paule's group. After having typed all formulae, the actual proof (verification) of all these identities, in typical cases, is a matter of a few seconds per formula.

d. *Formal, and maybe automated, proof of the correctness of the algebraic theory (like Gröbner bases etc.) which is behind the methods in c.:*
This is a major task, which goes far beyond Stam's paper but would of course be an essential and interesting part of a future comprehensive paper on combinatorial identities. For the commutative case of Gröbner bases theory, I am working on this with one of my PhD students, see (Maletzky 2016) and, in fact, this theory is now completely formalized *and* formally proved. This includes formalization and formal proof of my algorithm for computing Gröbner bases within the same logic in which the formalization of the rest of the theory is done. In fact, the execution of the algorithm on concrete input is also done within the same logic. Many more theses etc. would be necessary for formalizing and formally proving all current theory behind symbolic combinatorics. I consider research of this type as the essential goal of future formal math. Thinking further ahead, the question arises if, with today's mathematical knowledge and methodology of type c., it would at all make sense to type all the formulae in Stam's paper. My answer comes in the next items:

e. *Automated generation of combinatorial identities:*
One could write a "conjecture generator" that automatically generates all (and more of) the formulae listed in Stam's paper as conjectures and then submits the conjectures to the methods in c., keeping those that are true. I have ideas for this and did this already on a smaller scale for a different area.

f. *Proof of identities on demand:*
Alternatively, one just would not any more generate tables of identities but would "wait for the user" who has a particular instance of any of the formulae in the table and wants to get an automated verification by the methods in c. Even more attractively, by the methods in c., one can obtain automatically a simplified right-hand side if one provides a complicated left-hand side. This is similar to the situation in symbolic integration: We do not any more store integral tables in math systems like Mathematica, Maple etc. but, rather, one uses Risch' algorithm (Risch 1969) or extensions thereof for generating the integrals on demand.

g. *Providing "old-fashioned" proof methods in the presence of "modern" proof methods:*

My personal view on the question of "old-fashioned" proof methods versus "modern" proof methods is as follows: The ("manual" or automated) proof of formulae by some "older" proof techniques needs extra "handcrafting" for each individual formula. Often (95% or more), proofs of these formulae can be obtained, without extra hand-crafting, completely automatically by a "newer" proof technique. However, still, there may be reasons why a mathematician working in a particular area, as for example statistics or algorithm complexity, may want to see proofs generated by newer *and* older methods. Also, he may want to see "complete tables of identities" (like Stam's collection) even if they would not be necessary any more in the presence of newer methods. The reason for this may be:

- Proofs generated by various different methods may give various different insights about the formulae proved.
- The use of older or newer proof techniques and the desire of seeing "tables" of formulae may depend on the particular application of the identities in other fields of mathematics (Pitman 2015).
- The relation between "older" and "newer" proof techniques is, in fact, as old as mathematics. However, so far, in the lifetime of a mathematician, the proof techniques in his field did not really change. Logically, proceeding from "older" to "newer" proof techniques is an important ingredient of mathematics. We pointed this out, for example, in (Buchberger 2012). In essence, the transition from "old" to "new" is the transition from the "object" to the "meta" level of mathematics. Therefore, we advocate that modern math proving systems have to provide means for proceeding from the object to the meta level (e.g. level b. to d. and then to the application of d. to c.). In Theorema, this is an important design principle and we showed its feasibility in the work on formalizing Gröbner bases theory.

In the talk, I will report on the current state of the joint work of our team on the various aspects above.

Acknowledgements. The work described in this talk is carried out in a team at RISC consisting of B. Buchberger, C. Koutschan, A. Maletzky, P. Paule, C. Raab, S. Radu, C. Schneider, and W. Windsteiger.

References

Buchberger, B., Lichtenberger, F.: Mathematics for Computer Scientists. Springer, Heidelberg (1980). (in German)

Buchberger, B., Mathe is meta. In: Invited Talk at the Summer School "Summation, Integration and Special Functions in Quantum Field Theory", 9–13 July. RISC. Johannes Kepler University, Castle of Hagenberg, Austria (2012)

Kauers, M., Paule, P.: The Concrete Tetrahedron. Texts and Monographs in Symbolic Computation. Springer, Vienna (2011)

Maletzky, A.: Formalization of Gröbner Bases Theory in Theorema (working title). Ph.D. thesis, July 2016, to appear

Pitman, J.: Personal Communcation to the Working Group "Global Digital Math Library", December 2015

Risch, R.H.: The problem of integration in finite terms. Trans. Am. Math. Soc. **139**, 167–189 (1969)

Stam, A.: Binomial Identities with Old-fashioned Proofs, Manuscript, University of Groningen (2012)

The GDML and EuKIM Projects: Short Report on the Initiative

Bruno Buchberger[✉]

Research Institute for Symbolic Computation (RISC),
Johannes Kepler University, Linz, Austria
bruno.buchberger@risc.jku.at
http://www.risc.jku.at/people/buchberg/

Abstract. We give a report on the EuKIM project, which was recently submitted to the EU Horizon 2020 program, INFRAIA-02-2017 (Integrating Activities for Starting Communities) topic, by a consortium of twelve European research groups. The project aims at building up a "Global Digital Math Library" (knowledge base) integrating and extending current efforts worldwide. A central part of the project is the design and implementation of a software system that organizes open and one-stop access to mathematical knowledge and to various tools for processing mathematical knowledge. Recent progress in automated reasoning is an important issue for achieving more sophisticated levels in this endeavor.

In 2014 the International Mathematical Union (IMU) called for creating a "Global Digital Mathematics Library" (GDML) (GDML 2016) – a comprehensive, extensible, machine-processable knowledge base of mathematics providing one-stop open access for researchers, teachers, students and all users of mathematics in science, technology, and industry.

Existing digital libraries provide web access and search of entire papers, but the fine-grain units of knowledge in mathematics are statements expressing concepts, theorems, conjectures, problems, algorithms, proofs and more. A mathematical knowledge base must, therefore, allow users to store, retrieve, verify, and even invent knowledge at the level of individual statements and theories (collections of statements) from which new statements can be derived.

In the past twenty years, major advances in automating mathematical reasoning by using algorithmic intelligence in interaction with artificial intelligence have been made with important contributions by European research groups. This was a strong motivation for the recent decision of twelve European academic institutes to form a consortium for establishing the proposed digital knowledge base. The name of this initiative is "European Knowledge Infrastructure for Mathematics" (EuKIM). Roughly, the EuKIM infrastructure should do for precise, human AND machine processable mathematical knowledge what Wikipedia has done for imprecise human-only processable general knowledge.

B. Buchberger—Speaker of the EuKIM Consortium and representative of the RISC Institute.

G.-M. Greuel et al. (Eds.): ICMS 2016, LNCS 9725, pp. 443–446, 2016.
DOI: 10.1007/978-3-319-42432-3_56

The first joint action of the EuKIM Consortium was the formulation and submission (March 2016) of a long-term EU-Proposal "EuKIM" in the frame of the EU-Program Horizon-2020-INFRAIA-02-2017. The EuKIM project has three aspects:

– joined research on the topics relevant for creating GDML
– building up and extending a (European and global) network of research institutes that want and are able to contribute to the goal of building up a GDML
– designing and developing a software system that can serve as the one-stop open-access interface to all existing mathematical knowledge bases and to all existing processing tools for mathematical knowledge and, at the same time, will provide new mathematical knowledge bases (with an emphasis on formal knowledge bases) and new processing tools for mathematics (with an emphasis on automated reasoning tools).

In this talk we focus on the software aspect of the EuKIM project. The following, very rough, diagram (Fig. 1) gives an overview on the software to be designed and implemented.

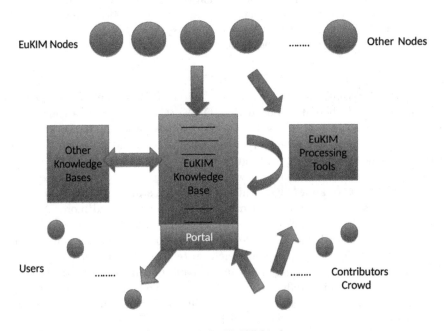

Fig. 1. Concept of the EuKIM infrastructure

We start from the side of the typical users (math researchers, teachers, students, people who apply math in science, engineering, etc.). The users should be able to input queries on mathematics in various levels of sophistication: On the first level, simple (natural language) keywords and combinations of keywords can represent the input, like for ordinary search engines. On the second level, natural language sentences describing problems, questions, or conjectures in context should be presentable in a style similar to what is possible in very recent natural language question answering systems for everyday situations (e.g.: the new version of Siri from Apple presented at Disrupt NY, May 9–11, 2016). The third level (which is specific to mathematics) will include mathematical formulae in unambiguous formalization in some logic language (some version of predicate logic) as part of the input.

The EuKIM software should analyze the request and decide which existing web-accessible mathematical libraries and knowledge bases contain relevant information on the questions or problem and should give this information back in a good structure with a navigator through the items found. (This information may include links to papers whose download may request paying a fee.)

However, in addition, the system should direct the user also to all (open-access and commercial) tools for processing the knowledge found in relation to the request submitted. This will include all the recent automated reasoning tools that may help in finding logically equivalent, similar, or relevant mathematical knowledge, in deciding or semi-deciding the validity of formulae, in finding the logical dependence of formulae from each other, in using algorithms for processing mathematical data etc. Important examples of such tools are the existing mathematical algorithm libraries like Mathematica, Maple, etc., and the existing automated reasoning systems like HOL (Nipkow et al. 2015), Coq (Affeldt and Kobayashi 2008), Isabelle (Nipkow et al. 2015), Theorema (Buchberger et al. 2016), etc. The system will also deliver complete mathematical theories in formal representation as a starting point for original research, as far as such theories are available on the web.

In addition to existing tools for processing mathematical knowledge and existing mathematical theories, the EuKIM software will also add more and more mathematical theories and processing tools and will motivate research groups (and also users) to contribute to theories and tools under the structural guidance and tools-support of the system.

Over the years, the EuKIM software system should grow into an intelligent dialogue machine for mathematics: Not by re-invention of all the existing beautiful and powerful web-accessible mathematical paper collections, formulae collections, algorithm collections, but by integration of all these tools into an open-access (as much as possible) and one-stop portal and by the addition of new tools of high mathematical, i.e. automated reasoning, intelligence.

References

Affeldt, R., Kobayashi, N.: A Coq library for verification of concurrent programs. Electron. Notes Theor. Comput. Sci. **199**, 17–32 (2008). Proceedings of the Fourth International Workshop on Logical Frameworks and Meta-Languages (LFM 2004)

Buchberger, B., Jebelean, T., Kutsia, T., Maletzky, A., Windsteiger, W.: Theorema 2.0: computer-assisted natural-style mathematics. J. Formalized Reasoning **9**(1), 149–185 (2016). Special Issue: Twenty Years of the QED Manifesto

The Global Digital Mathematics Library Working Group. International Mathematical Knowledge Trust Charter (2016). http://mathontheweb.org/gdml/IMKT-Charter-final.pdf

Nipkow, T., Paulson, L.C., Wenzel, M.: Isabelle/HOL - A Proof Assistant for Higher-Order Logic. Springer, Heidelberg (2015)

Math-Net.Ru Video Library: Creating a Collection of Scientific Talks

Dmitry Chebukov$^{(\boxtimes)}$, Alexander Izaak, Olga Misyurina, and Yuri Pupyrev

Steklov Mathematical Institute of Russian Academy of Sciences, Moscow, Russia
tche@mi.ras.ru
http://www.mi.ras.ru

Abstract. In this article we give a brief description of the method for constructing a collection of conference talks applied in the Math-Net.Ru project.

Keywords: Math-Net.Ru · Scientific conference talk · Video record

1 Introduction

The Math-Net.Ru project collects information about leading mathematical events occurring in Russia, along with publications in Russian mathematics journals [1,2]. The events database [3] includes talks at scientific conferences, seminars and public and educational lectures. For researchers (particularly young researchers and students) it is hardly possible to attend all conferences and seminars they may be interested in. Sometimes unpublished results given in a specific talk may be required for current research. This is why a collection of talks powered by a search engine and supplied with extended metadata is of great importance to the scientific community. The most important thing is to supply a video record of the talk if possible. The collection of scientific events which occurred in the period from 1991 to 2016 and were indexed in the Math-Net.Ru project counts about 12500 talks and 4500 of them are supplied with video records.

2 Metadata Structure

The way of arranging the collection of conference talks is similar to the method for constructing the publication database. The main metadata fields of a journal article and a conference talk are shown in Table 1. Of course, both journal paper and conference talk metadata include the same common fields, namely: authors (or speakers) names, affiliations, title, abstract, keywords, references. Other fields differ but have a similar meaning: the conference is specified by its title, date and place, which are the analogues of the journal title and ISSN. The talk is specified by its location, date and time, which correspond to the year, the volume, the issue and the pages of a journal paper. The video record of the talk corresponds

© Springer International Publishing Switzerland 2016
G.-M. Greuel et al. (Eds.): ICMS 2016, LNCS 9725, pp. 447–450, 2016.
DOI: 10.1007/978-3-319-42432-3_57

Table 1. Main metadata fields of a journal paper and a conference talk

Journal paper	Conference talk
Journal title	Conference title
ISSN	Conference date and place
Founder, publisher	Organizers
Editorial board	Program Committee
Editorial staff	Organizing Committee
Paper title	Talk title
Author ("Persons" database)	Speaker ("Persons" database)
Affiliation ("Organizations" database)	Affiliation ("Organizations" database)
Paper abstract	Talk abstract
Keywords	Keywords
References	References
Year, volume, issue, page numbers	Date, time
Full-text (PDF file)	Video record (MP4 file or YouTube link)

to a full-text PDF file. Thus we can apply the way developed for a journal database to create a database of scientific events.

The basic element of the journal database is the database record for a paper. A journal volume or issue is formed as a set of records for papers. The basic element of the event database is a database record for a talk. The conference program is formed as a set of talks. The authors of journal papers and the authors of talks (speakers) are collected into a single database of persons and have their own unique IDs. The talk web page and the paper web page have links to their authors' personal pages. A personal web page shows the list of papers indexed in the journal database and the list of talks indexed in the events database. There are cross-links between web pages of talks, publications and persons. Of course, there are many systems like YouTube which offer video hosting. All of them are designed to show a video supplied with a limited number of description fields. Several videos can be collected into a playlist. The Math-Net.Ru VideoLibrary provides an expanded metadata structure aimed at being used with scientific events and fully integrated with other Math-Net.Ru data. In Math-Net.Ru a video record is just an option, a very important one.

Journal paper metadata is just a short description of the full text, as well as conference talk metadata is a description of the talk. For end users metadata may be helpful in finding the full data (PDF file or video record) but cannot replace them. A conference website which has a list of conference talks without video records looks like a journal website without full texts. This is why an important part of our work consists in filming scientific talks, processing the videos and post-production.

3   Processing Video

At the Steklov Institute of the Russian Academy of Sciences we use professional video equipment including Full HD Panasonic AG-AC160A video camera, a Shure PGX14/PG30 wireless lapel microphone and a TH-650 DV tripod stand. Filming without a lapel microphone results in a very low sound quality. The camera has 2 audio input channels and we use both: one for the lapel microphone to record the speaker's voice and the second for external microphones installed in the conference hall. The camera is equipped with a SDI output connected with the video server to arrange live online streaming. The camera outputs Full HD 1080p video stream. The server is powered by Intel (R) Xeon (R) CPU-2450, 32 Gb RAM and equipped with a Blackmagic DeckLink SDI card. Adobe Flash Media Live Encoder 3.2 is video capture software which enables the broadcast of live video to a media server. Various software to set up a media server is available on the market, but we use YouTube Live, which provides a stable, reliable and free service to arrange live streaming.

Video is processed in Adobe Premier Pro CS6. We prepare MP4 video files in three resolutions: 1080p (5500 Mbps), 720p (3500 Mpps) and 480p (900 Mbps). These resolutions allow viewing videos smoothly on mobile devices and desktops. End users can switch between resolutions. Video files are stored on the local server. Video is embedded to the conference web pages with the help of JWPlayer [4]. Pseudo-streaming [5] is arranged by means of the H264 streaming module of LightHTTPD server [6]. Pseudo-streaming enables fast rewind/forward options. Video files can be also loaded to YouTube and then embedded in conference web pages.

Statistics is counted for every talk indexed in the database. We follow the COUNTER project standards [7], so we do not count repeat views and bots activity. Lists of videos that have been most viewed during a week, a month or all the time are available. Table 2 shows the number of video records and the number of views for some events which occurred in 2015.

Table 2. Number of video records and its views in 2015

Event	Videos number	Views number	Average views
Scientific conferences	378	11561	30
Summer math. school for students	152	8814	60
Courses and lectures for students	123	10969	90
Steklov Institute scientific seminars	31	6420	210
Popular lectures	6	2630	440
Memorial events	6	525	88

The most viewed video records are lectures given at educational events, summer schools. Video records of Steklov Institute scientific seminars are very popular. In general, for all events the number of views significantly exceeds the number of visitors attending the event.

Thus in the Steklov Mathematical Institute we apply a full-cycle system to film, broadcast online, edit and distribute scientific and educational talks and lectures.

Acknowledgments. This work is supported by the Russian Foundation for Basic Research (grant no. 16-07-01281-a).

References

1. Zhizhchenko, A.B., Izaak, A.D.: The information system Math-Net.Ru. application of contemporary technologies in the scientific work of mathematicians. Russ. Math. Surv. **62**(5), 943–966 (2007)
2. Chebukov, D.E., Izaak, A.D., Misyurina, O.G., Pupyrev, Y.A., Zhizhchenko, A.B.: Math-Net.Ru as a digital archive of the russian mathematical knowledge from the XIX century to today. In: Carette, J., Aspinall, D., Lange, C., Sojka, P., Windsteiger, W. (eds.) CICM 2013. LNCS, vol. 7961, pp. 344–348. Springer, Heidelberg (2013)
3. http://www.mathnet.ru/eng/video
4. http://www.jwplayer.com
5. https://support.jwplayer.com/customer/portal/articles/1430518-pseudo-streaming-in-flash
6. http://www.lighttpd.net
7. http://www.projectcounter.org

The SMGloM Project and System: Towards a Terminology and Ontology for Mathematics

Deyan Ginev[1], Mihnea Iancu[1], Constantin Jucovshi[1], Andrea Kohlhase[1],
Michael Kohlhase[1], Akbar Oripov[1], Jürgen Schefter[2], Wolfram Sperber[2(✉)],
Olaf Teschke[2], and Tom Wiesing[2]

[1] Computer Science, Jacobs University Bremen, Bremen, Germany
[2] Zentralblatt Math, Berlin, Germany
wolfram@zentralblatt-math.org
http://kwarc.info
http://zbmath.org

Abstract. Mathematical vernacular – the everyday language we use to communicate about mathematics is characterized by a special vocabulary. If we want to support humans with mathematical documents, we need to extract their semantics and for that we need a resource that captures the terminological, linguistic, and ontological aspects of the mathematical vocabulary. In the SMGloM project and system, we aim to do just this. We present the glossary system prototype, the content organization, and the envisioned community aspects.

1 Introduction

One of the challenging aspects of mathematical language is its special terminology of technical terms that are defined in various mathematical documents. To alleviate this, mathematicians use special glossaries, traditionally lists of terms in a particular domain of knowledge with the definitions for those terms. Originally, glossaries appeared as alphabetical lists of new/introduced terms with short definitions in the back of books to help readers understand the contents. Another kind of resource that deals with terminology of mathematics are "dictionaries", which align mathematical terms in different languages by their meaning – originally without giving a definition.

In the last decades the term "glossary" has also been applied to digital vocabularies (online encyclopedias, thesauri, dictionaries, etc.), which have become important resources in knowledge-based systems. This is especially true for vocabularies that have a (*i*) semantic aspect – i.e. some of the relations between the terms and the concepts, objects, and models they denote are made explicit and machine-actionable, they are also called "ontologies" – or (*ii*) that are multilingual. Digital vocabularies can be hand-curated, or machine-generated/collected; an example of the former is the WordNet lexical database for English, [WN] an example of the latter is DBPedia, [DBP13] but they can also be hybrid, e.g. the UWN/Menta project [YAGO] generates a multilingual WordNet by automatically adding other languages by crawling Wikipedia.

© Springer International Publishing Switzerland 2016
G.-M. Greuel et al. (Eds.): ICMS 2016, LNCS 9725, pp. 451–457, 2016.
DOI: 10.1007/978-3-319-42432-3_58

We present the SMGloM project, which aims to create a semantic, multilingual glossary for mathematics. This resource combines the characteristics of dictionaries and glossaries, with those of ontologies, but restricts the content to definitions and the relations to the lexical ones to keep the task manageable. Here we give a high-level overview over the data model, the SMGloM system, organizational and legal issues, possible applications, and the state of the effort of seeding the glossary.

2   The SMGloM System

Data Model and Encoding. We build the data model of SMGloM on top of the one of OMDoc/MMT [Koh06, RK13], which provides views, statements, and theories. In a nutshell – see [Koh14] for details, a **glossary entry** consists of one **symbol**, its **definition**, and a set of **verbalizations** and **notations**. A symbol is a formal identifier of a mathematical object/concept (i.e. a formal object). The verbalizations relate it to lexical entries (identified by the stem of the head), which we call **glossary terms**. The definitions could be written down in a formal logic, but in the SMGloM, we write them down in mathematical vernacular (common mathematical language) and encode them in sTEX, a variant of LATEX with semantic annotations [sTeX]). Figure 1 shows the concrete form of a glossary entry. It consists of a **module signature**, which specifies the formal part: here the module name graphconnected in **mhmodsig** environment and the symbol connected via the **\symi** macro. The module signature also specifies the conceptual dependency on paths in graphs by importing the grpahpath module via **\gimport**.

```
\begin{mhmodsig}[creators=hwk]{graphconnected}
\gimport{graphpath}
\symi{connected}
\end{mhmodsig}

\begin{mhmodnl}[creators=hwk]{graphconnected}{en}
\begin{definition}
   A non−empty \trefi[graph]{graph} $G$ is said to be \defi{connected} if any two
   of its \trefis[graph]{node} are linked by a \trefi[graphpath]{path} in $G$.
\end{definition}
\end{mhmodnl}

\begin{mhmodnl}[creators=hwk]{graphconnected}{de}
\begin{definition}
   Ein \mtrefi[graph?graph]{Graph} $G$ hei"st \defi[connected]{zusammenh"angend}
   wenn je zwei seiner \mtrefi[graph?node]{Knoten} durch einen
   \mtrefi[graphpath?path]{Weg} verbunden sind.
\end{definition}
\end{mhmodnl}
```

Fig. 1. A glossary entry for "connected" graphs encoded in sTEX

The informal – and language-specific – part is given in the **language binding** given in the **mhmodnl** environment. In Fig. 1 this consists of a single definition, where the definiendum is marked up via the **defi** macro and the concepts of a "graph", a "node", and a "path" are via the **trefi**, **trefi** (for plurals), and **mtrefi** (for multilinguality) macros. Their optional argument specifies the glossary module they are imported from and – after the ? the symbol name. In the German language binding we can appreciate the difference between the symbol name connected and its **verbalization** – the word "zusammenängend" used to denote it in German (this is the content of the **defi** macro). We will use the fact that symbols coordinate verbalizations in our multilingual glossary and mathematical dictionary (see Sect. 3 below).

In general Glossary entries are grouped into a **glossary module**, which is represented as $n + 1$ OMDoc/Mmt theories: the module signature and n language bindings. Figure 2 shows graphconnected glossary module in the center and its 1-neighborhood wrt. the imports relation.

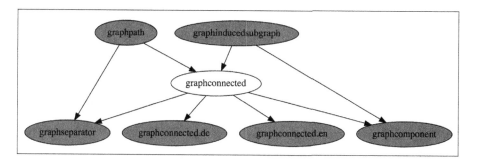

Fig. 2. The module graph around "Connected Graph"

A Terminology of Mathematics. In fact, we have all the information we need for a mathematical terminology:

(i) we can identify **semantic fields** – here mathematical concepts, objects, and models. Oher terminologies like WordNet use "synsets": sets of synonyms for this,
(ii) we can link semantic fields to technical terms,
(iii) and we can relate concepts to each other via **terminological relations** like synonymy, hyper/hyponymy, and meronymy.

The last point needs some elaboration. In SMGloM we identify certain symbols as **primary** in their module, e.g. graph is primary in the grpahs module while node and edge that are defined in the same module are not. The terminological relations can be read off from the imports relation (see [Koh14] for details): For instance, if t verbalizes a primary symbol s in module m, which imports a module m' with primary symbol s', which has verbalization t', then t' is a hyponym

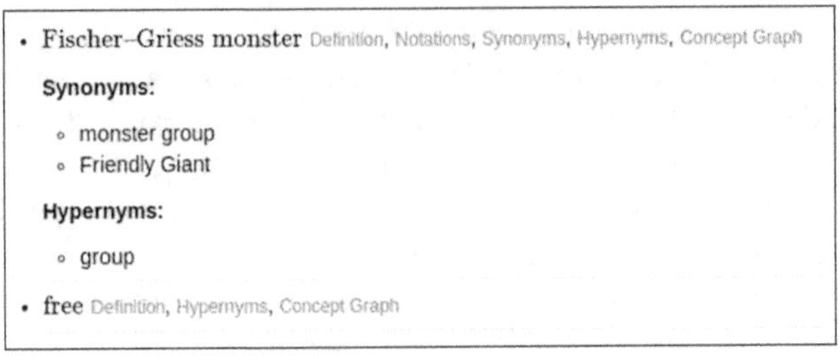

Fig. 3. Terminological relations in the glossary

of t. For instance, "tree" is a hyponym of "graph". Figure 3 shows computed terminological relations in SMGloM.

An Ontology of Mathematics. As Fig. 2 already suggests, the SMGloM data induces not only a mathematical terminology, but an ontology of mathematics as well. But note that again, the concepts of the ontology are not the glossary modules themselves, but the mathematical concepts, objects, and models, i.e. the symbols. But again, the taxonomic information can be gleaned from the module graph structure: primary symbols become ontological concepts, whereas non-primary ones become roles – detalis are mainly due to their linguistic forms. For instance the definition of a "graph" as a pair $\langle V, E \rangle$ of "vertices" V and "edges" $E \subseteq V^2$ leads to the concept graph with two functional roles nodes: graph→set and vertices: graph→set. Together with the definition of the adjective "connected" on graphs in Fig. 1, we get the sub-concept connected–graph which inherits these two roles from graph.

Induced Terminological and Ontological Relations via Views. The imports relation in the module graph gives rise to direct terminological and onto-logical relations. But experienced mathematicians recognize more relations, for instance, that the set E of "edges" of a "graph" $G = \langle V, E$ form a "relation on" V. We call this ability **mathematical literacy** in ??; they can be modeled by a new form of edge in the module graph in OMDOC/MMT: views. Epistemically, these behave like the imports relation, but their truth-preserving nature has to be proven by proof obligations. If we generalize the computation of terminolog-ical and ontological relations to allow views, then we obtain **induced** relations that are recognized – and utilized by mathematically literate users.

Organizing a Communal Resource. The ultimate cause of the SMGloM project and system is to facilitate the establishment of a knowledge resource for mathematics. We need to take appropriate organizational measures to support this. We are currently establishing a wiki-like archive submission system for glossary modules on MATHHUB [MH] and thinking of a quality assurance system that is based on a community/karma-driven approval system. Openness and semantic stability are ensured by a special licensing and publication regime: The SMGloM license [SPL] protects symbols against non-conservative changes while allowing derived works.

3   Applications of the **SMGloM**

The main advantage of SMGloM over existing terminological resources for mathematics is that it makes important linguistic and ontological relations explicit that these do not. This extension makes a large variety of applications feasible without requiring full formalization, the cost of which would be prohibitive. We will sketch some of the applications here.

Glossary of Mathematical Terms. An interface that presents SMGloM like a traditional glossary, i.e. as a (sorted) list of glossary entries. In addition, the semantic information in SMGloM can be used to adequately mark up references to as well as relations with (e.g. "synonym of", or "translation of") other entries. See Fig. 4 for the current interface. There can be sub-glossaries, for certain areas of mathematics, for certain languages, etc.

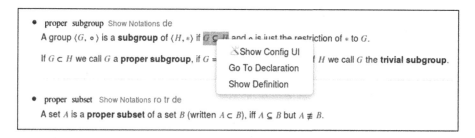

- proper subgroup Show Notations de

  A group $\langle G, \circ \rangle$ is a **subgroup** of $\langle H, * \rangle$ if $G \subseteq H$ and $\circ$ is just the restriction of $*$ to G.

  If $G \subset H$ we call G a **proper subgroup**, if $G =$ H we call G the **trivial subgroup**.

     ⚙ Show Config UI
     Go To Declaration
     Show Definition

- proper subset Show Notations ro tr de

  A set A is a **proper subset** of a set B (written $A \subset B$), iff $A \subseteq B$ but $A \not\equiv B$.

Fig. 4. The glossary interface at https://mathhub.info/mh/glossary

Mathematical Dictionaries. The mathematical terminology is synchronized by content symbols in SMGloM, therefore a mathematical dictionary is simply an interface problem; see Fig. 5. Again, all terms are hyperlinked to their definitions.

Math Dictionary

The math dictionary on this page is a service based on the SMGloM terminology. To translate mathematical terms, select the source and target languages and enter them in the text window on the left (autocompletion). The translations are hyperlinked to their respective definitions for convenience.

| From: | en | ▾ | To: | de | ▾ | Translate |

| clopen | | | | abgeschloffen, |

Fig. 5. The dictionary interface at https://mathhub.info/mh/dictionary

Flexible Styling/Presentation. If we have formulae in content markup (i.e. in content MathML e.g. in OMDoc or sTeX), then we can adapt the rendering of formulae with symbols that having multiple notations in SMGloM to the user's preferences. Then, each user can state their notational preferences (in terms of SMGloM notation definitions), and the formulae in SMGloM will be rendered using these, adapting to the preferences of the reader.

Notation-Based-Parsing. The notation definitions from SMGloM can be seen as user-contributed grammar rules. Therefore, they can be used for parsing formulae from presentation to content markup [Tol16]. This will lead to a context-sensitive formula parser, where "context" is defined by the SMGloM glossary modules currently in focus – here the data model in term of OMDoc/MMT theories directly contributes to the applications of the SMGloM.

More Semantic Search. As SMGloM declares symbols together with notations, definitions and verbalizations it provides an unique opportunity for applying semantic search services based on it in a variety of settings:

1. notation-based parsing in the input phase could make formula entry into an interactive disambiguation process. For instance, a user enters e^?x, and the system ask her: "with e, do you mean Euler's number?", and also: "Is $e^{?x}$ a power operation?". The answers will then help refine the search.
2. Alternatively, search could use disambiguation as a facet in the search to refine the results or for clustering the results.
3. Furthermore, the SMGloM information could be used for query expansion (both visible or automatic): if the user searches for e, then the query could be expanded e.g. by (*i*) the string Euler's Number (there is an interesting question about what to do with the language dependency here) and even (*ii*) the formula $\lim_{?n\to\infty}(1+\frac{1}{?n})^{?n}$ (?n is a query variable).

Verbalization-Based Translation. One of the most tedious parts of translating mathematical documents is the correct use of technical terms. A semantically preloaded text (i.e. one that has all formulae in content markup and many semantic objects explicitly marked up) can be term-translated automatically using the translation relation induced by SMGloM. Of course, synonyms must be resolved consistently (there has to be an interface for this). This (and related semantic tasks) are for domain specialists. The intervening text can be done by lesser trained individuals (or even a variant of google translate). This will make translations much cheaper and will make math available in more languages.

Wikifiers like NNexus. Wikifiers are systems that given a glossary of terms create definitional links in documents. A math-specific example is the NNexus system [GC14], it can already use the SMGloM glossary.

4   Conclusion and State

We have described a project to establish a public, semantic, and multilingual termbase for mathematics. We have a first prototype that supports authoring of glossary entries and glossary management at https://mathhub.info/smglom. The

SMGloM system partially automates editing, management, refactoring, quality control, etc.; for more information see https://mathhub.info/help/main.html.

To make public contributions to SMGloM feasible, it must already contain a nucleus of (basic) entries that can be referenced in other glossary components. The SMGloM project is currently working towards a basic inventory of glossary entries, and has almost arrived at the first milestone of 700 entries – most with two language bindings (English and German), some with 6 (+ Romanian, Chinese, Turkish, and Bulgarian). The current glossary contains

(i) ca. 300 glossary entries from elementary mathematics, to provide a basis for further development
(ii) ca. 400 are special concepts from number theory to explore the suitability of the SMGloM for more advanced areas of mathematics.
(iii) ca. 30 views that generate induced terminological and ontological relations.

Acknowledgements. Work on the SMGloM system has been partially supported by the Leibniz association under grant SAW-2012-FIZ_KA-2 and the German Research Foundation (DFG) under grant KO 2428/13-1.

References

[DBP13] DBpedia. 17 September 2013. http://dbpedia.org. Acccessed 21 Feb 2014

[GC14] Ginev, D., Corneli, J.: NNexus reloaded. In: Watt, S.M., Davenport, J.H., Sexton, A.P., Sojka, P., Urban, J. (eds.) CICM 2014. LNCS, vol. 8543, pp. 423–426. Springer, Heidelberg (2014). http://arXiv.org/abs/1404.6548

[Koh06] Kohlhase, M.: Communication with and between mathematical software systems. In: Kohlhase, M. (ed.) OMDoc – An Open Markup Format for Mathematical Documents [version 1.2]. LNCS (LNAI), vol. 4180, pp. 75–79. Springer, Heidelberg (2006). http://omdoc.org/pubs/omdoc1.2.pdf

[Koh14] Kohlhase, M.: A data model and encoding for a semantic, multilingual terminology of mathematics. In: Watt, S.M., Davenport, J.H., Sexton, A.P., Sojka, P., Urban, J. (eds.) CICM 2014. LNCS, vol. 8543, pp. 169–183. Springer, Heidelberg (2014). http://kwarc.info/kohlhase/papers/cicm14-smglom.pdf

[MH] MathHub.info: Active Mathematics. http://mathhub.info. Accessed 28 Jan 2014

[RK13] Rabe, F., Kohlhase, M.: A scalable module system. Inf. Comput. **230**, 1–54 (2013). http://kwarc.info/frabe/Research/mmt.pdf

[SPL] Kohlhase, M.: The SMGloM Public License (SPL) version 0.1. https://mathhub.info/help/spl0.1.html

[sTeX] KWARC, sTeX. https://github.com/KWARC/sTeX. Accessed 15 May 2015

[Tol16] Toloaca, I.: MathSemantier - a notation-based semantication study. B.Sc. Thesis, Jacobs University Bremen (2016)

[Wat+14] Watt, S.M., Davenport, J.H., Sexton, A.P., Sojka, P., Urban, J. (eds.): CICM 2014. LNCS, vol. 8543. Springer, Heidelberg (2014). ISBN: 978-3-319-08433-6

[WN] WordNet: a lexical database for English. https://wordnet.princeton.edu/. Accessed 26 May 2013

[YAGO] Towards a Universal Multilingual Wordnet. http://www.mpi-inf.mpg.de/yago-naga/uwn/. Accessed 26 May 2013

The Effort to Realize a Global Digital Mathematics Library

Patrick Ion[✉]

AMS, University of Michigan, Ann Arbor, MI, USA
ion@ams.org pion@umich.edu
http://www-personal.umich.edu/~pion/

Abstract. A decade after a resolution in 2006 by the International Mathematical Union endorsing the notion of a global digital mathematics library, and following a thorough report on possibilities written under the auspices of the US National Research Council in 2012, an 8-person Working Group, set up in 2014, is still working toward implementations of some of the ideas. There are difficulties with mobilizing the mathematical community toward building worthwhile infrastructure in times that are both perilous and well off, depending on where you stand. But progress continues.

1 Introduction and History

Vignettes from history help to emphasize the way that members of the mathematics community have long wished access to more of the world's mathematics to have a better understanding of it. The name of Giuseppe Peano (1858–1932) is universally well known among students of mathematics as the author of an axiom system for the natural numbers. That this remarkably inventive and productive Italian mathematician, who has recently been the subject of renewed historical interest [Dolecki&Greco:2016, Dolecki&Greco:2010a, Dolecki&Greco:2010b] also was a leading light of the push for the international auxiliary language Interlingua is less familiar [Peano:1903]. This can be seen in the mathematical context as feeding into the efforts at pasigraphy, a writing system where each symbol represents a concept. Indeed it was in connection with such efforts at representing mathematics by well-designed formulas that Peano developed his natural number axioms [Peano:1894].

At the ICM in 1897 there was a session under the chairmanship of Peano on how to encode mathematical knowledge. It remains striking to me that Ernst Schröder, the algebraist and logician from Karlsruhe, delivered a plenary address "On pasigraphy, its current status and the pasigraphic movement in Italy" [Schröder:1897]. By the last phrase he meant the work of Peano and his followers. Schröder disagreed with the distinguished chairman and suggested that the new system he had developed with a small number of basic symbols, somewhat in the vein that had been worked on by the American C. S. Peirce, was what was wanted. He began his remarks with the ringing statement that if there

© Springer International Publishing Switzerland 2016
G.-M. Greuel et al. (Eds.): ICMS 2016, LNCS 9725, pp. 458–466, 2016.
DOI: 10.1007/978-3-319-42432-3_59

were any topic that really belonged at an International Conference of Mathematicians, then it was pasigraphy. He was sure that pasigraphy would take its rightful place on the agenda of all succeeding such conferences.[1]

At the 1928 ICM in Bologna there was active discussion of how to provide comprehensive bibliographic resources for mathematics to everyone, especially in regard to publication of the comprehensive catalogue of mathematical work prepared by the German Georg Valentin (which was later bombed out of existence at Unter den Linden in Berlin in February 1942) [Valentin:1900, GöbelSperber:2010]. From 1931 on there were publications such as Zentralblatt [ZM], Mathematical Reviews [MR] and Referativny Zhurnal [RZ] indexing and abstracting the mathematical literature, but full text could only be had in the traditional way. In 2006 the IMU saw the possibility of realizing the imagined Global Digital Mathematical Library, or World Digital Mathematical Library, which had been discussed again [BallBorwein:2005,IMU/CEIC:2006, Jackson:2003], and put out a resolution to this effect [IMU:2006]. But that was all.

The adjective 'digital' is important here as it is the new digital technologies that allow better access to the resources of mathematical knowledge than ever before. Surely in 16th century Europe, or even earlier in 14th century Korea, when printing from metal type was a brand-new technology, the possibilities of new forms of books for the recording and dissemination of knowledge were welcomed by some. That they were right we all now know. That sort of opportunity is open to us again now. Of course, printing revolutionized in other ways: in Europe printing of indulgences was a labor-saving device for the church that may have changed religion, and printed money changed economics in China.

The adjective 'global' is important too. We all think the truths of our subject to be global, independent of location in this world.[2,3] Think of the IMU's membership from over 120 countries. A GDML can have a global reach as a result of digital technology, particularly the internet. It should be a shared global good. We are looking for global support for the idea and expect that there can be contributions to a GDML from all over the world. The benefits will be felt all over the world.

---

[1] Zur Diskussion auf einem internationalen Mathematiker-Kongresse dürfte sich kaum ein Thema in höherem Maße empfehlen als das der Pasigraphie, und ich bin berzeugt, daß der Gegenstand von der Tagesordnung künftiger Kongresse nicht mehr verschwinden wird.

Ist doch das Ziel dieser neuen Disziplin kein geringeres als: die endgültige Festlegung einer wissenschaftlichen Universal-Sprache, die völlig frei von nationalen Eigentümlichkeiten und bestimmt ist, durch ihre Konstruktion die Grundlage zur wahren, nämlich exakten Philosophie zu liefern!

[2] Naturally there are ethnomathematical considerations to be discovered, and different developments in mathematics to be related to sociology. Also Marcus and Watt point out that there are linguistic differences even in what is an equation [MarcusWatt:2012].

[3] See also a discussion by Gray of natural languages used for the expression of mathematical ideas [Gray:2002].

In 2010, IMU President Ingrid Daubechies and Peter Olver, chair of the IMU's Committee on Electronic Information and Communication (CEIC) took the initiative to work toward a WDML or GDML through consultations with a broad-based expert group [WDML Blog]. This culminated in a comprehensive report from a Workshop at the US National Academy of Sciences [NAS:2012]. At the 2014 ICM a small working group of 8 persons was given the task of working toward realizing some of the dreams [Seoul CEIC Panels, Seoul DML Panel]. The resources need to be found to begin building the GDML. It will not be an easy task [Bouche:2014, PitmanLynch:2014].

2 Challenges

One can distinguish four facets of the GDML initiative:

– The mathematical community
– The mathematical literature
– Mathematical knowledge management
– Management of the enterprise

They are all discussed in the NRC report [NRC/USA:2014]. Some parts of the GDML require work that is understood, or already done in part, but that just takes much time and effort to complete. Other parts require serious investigation and prototyping which will take time, although the general ideas and development paths may seem clear. But it turns out that each facet, though many would think their meanings pretty clear, leads a number of people to hold differences of opinion as to how each should be understood. It's in those details that the sticking points lie in realizing the public good we crave. Nonetheless, the WG and its successors will keep moving forward, probably with progress showing stick-slip characteristics that are hard to describe.

3 Goals

The goals of the GDML effort are still of considerable generality. To achieve them is going to be a longer process. Setting up any new organization for the public good which is to be sustained over a long period is a slow business. But we need to do it since the openness of mathematical knowledge to all who need it is very important.

To this end the GDML WG is founding an International Mathematical Knowledge Trust [IMKT], which it is hoped will function eventually as an organizing center for the knowledge base that will be the GDML. Setting out a charter for this proposed trust, and getting it endorsed by the IMU has been one of the tasks of the WG. It has proposed the present goals:

> The purpose of the International Mathematical Knowledge Trust, IMKT, is to establish a mathematical knowledge commons — a public resource consisting of mathematical knowledge represented in non-proprietary, machine-readable formats and an international network of knowledge providers, information systems, and semantic services based on it, that is, a global digital mathematical library.

The mission of the IMKT is to construct a mathematical knowledge commons as a global public good, an effective knowledge base of open mathematical knowledge data, encompassing the world's mathematics through collaborations deploying both present and future technology, and to foster a supporting community. In particular, the IMKT should work to

- enhance accessibility of all mathematical knowledge world-wide, present, past and future,
- serve people in research mathematics, education, and applications of mathematics,
- promote the creation of open standards and best practices for management of mathematical data, and encourage the use of such standards,
- facilitate the development of open source tools and open mathematical data repositories,
- facilitate creation, dissemination and open archiving of semantically rich forms of mathematical data,
- encourage the collaborative development of open services based on representations of mathematical knowledge.

The mathematical knowledge commons resulting from these efforts of the IMKT and affiliated organizations should be a truly global resource, which matches the highest possible standards of independence, of reliability, and of data protection.

4 Achievements

Aside from working toward the problems of global existence and governance of an IMKT and associated international efforts, the GDML WG has been paying attention to building wider support for a GDML in the mathematical community and to planning and beginning projects that start constructing tools and structures to underpin a GDML. An activity of the support-building type was a Special Session "Mathematical Information in the Digital Age of Science" at the Joint Mathematics Meetings in Seattle Jan. 6–9, 2016 with 18 speakers over 11 h [JMM:2016]. This is a large gathering with this year about 6,300 registrants. These meetings are organized by the Mathematical Association of America (MAA) and the American Mathematical Society (AMS), and host additional sessions for the Association for Symbolic Logic (ASL), the Association for Women in Mathematics (AWM), the National Association for Mathematicians (NAM), and the Society for Industrial and Applied Mathematics (SIAM).

It was an excellent venue for speaking to the mathematical community on matters related to GDML efforts.

The materials collection efforts for a digital library, such as a world-wide extension of what EuDML did [EuDML], and building on their experience is being considered, but as yet have not been proposed to any funding agency. A European consortium has, however, begun the process of applying for significant funds under a European Union program.

The GDML WG helped organize a workshop to start on on of the infrastructure developments with long-range promise for a GDML. The Semantic Representation of Mathematical Knowledge Workshop was held 3–5 February 2016 at the Fields Institute, located at the University of Toronto, Toronto, ON CA [Fields:2016]. The support of a grant from the Alfred P. Sloan Foundation, staff and resources provided by the Wolfram Foundation, and staff and resources for local arrangements provided by the Fields Institute made possible a very successful workshop. The workshop was organized by the Wolfram Foundation, represented by Michael Trott and Eric Weisstein, by the Fields Institute and its Director Ian Hambleton, and by the GDML WG.

To paraphrase the application for support to the Alfred P. Sloan Foundation, the goals of the workshop were to pool the knowledge and experience of a group of experts to agree on design principles leading the way toward implementation of a semantic capture language applicable to all mathematics. Such a semantic encoding is expected to help realize one of the goals of a Global Digital Mathematical Library.

Through a program of talks and discussions the workshop was to work toward consensuses enabling the creation of a semantic language both for mathematics as a whole and for its sub-disciplines. The workshop was intended to produce (1) a white paper outlining the structure of the proposed semantic language, (2) a concrete plan for creating an explicit semantic language that will be used to mark up results in a specific area of mathematics, and (3) internet publications for all to see.

A great number of opinions were offered and a good deal of intense discussion ensued amongst the 40 participants of the workshop. Part 3 of the intentions is well covered by materials on the Fields Institute and Wolfram Foundation websites. However, production of a white paper and of definite plans proved difficult. Though discussions were started it turned out there was more to learn than expected. The first difficulty was the fact that there were several meanings of the word semantics in use, which could be seen as contradictory and were often confusing to one participant or another. The resulting short white paper is online, and linked to a wiki for further discussion [SRMwiki].

After term distinctions there are, for instance, possibilities for different views of the literature of mathematical interest, different levels of formalization, different kinds of semantically explicit mathematics, different audiences for mathematical communications, and even different levels of readability required.

When it comes to semantic capture language design there are language design issues, use cases to list and satisfy, issues of organizing design when a large vocabulary is presumably involved, and the whole matter of the relationship of semantic levels to formal proofs which can be machine checked.

There were some ideas for possible projects exploring the semantic capture aspect of recording the mathematical literature, whether old or newly created. One can try and accord various mechanized reasoning styles with each other, and tie them to more conventional computer algebra systems. There is lots of scope for exploring particular subjects and trying to capture, to various degrees, the special peculiarities of, say, algebra versus analysis, or geometry and probability.

One particularly attractive opportunity is the area of orthogonal polynomials and special functions (OPSF). This is a classical subject which is still being explored and has fascinating connections with other parts of mathematics. However, the basics of the subject are generally thought to be well understood. The fact that current computer algebra systems do not always agree on definitions, and so can produce strongly contrasting results to a fairly simple calculation, shows this is not so. The need for a concordance of special functions seems clear, and will be a good test case for proposed techniques of capturing mathematical semantics.

Another useful beginning will be to take a subject area and to try and capture all the significant theorems in it with a full set of definitions. A good prototype area might be geometry, since geometrical thinking can be contrasted with algebraic although much of classical geometry can be done by algebraic calculation; and there are sometimes insights to be gained into algebra from a geometrical view.

This leads on naturally to the matter of ontology creation. In the medical and life sciences use of mechanizable ontologies has proved useful. A lot of resources and attention have gone into the Gene Ontology, for instance. However, the use of the term ontology is also one fraught with possibilities for incomprehension, it seems. Indeed opposition to the idea that an ontology could be useful can provoke the sort of heat that political opinions are more known for. Nonetheless mathematics may be able to benefit from the practical experience and developments in other fields, where researchers have developed software that, *mutatis mutandis* could be useful for mathematics.

5   Prognosis

I expect to see the IMKT founded, and that it will work toward spreading the ideas of cooperation to achieve a GDML. We should see other regional mathematical Knowledge Trusts formed. There will come a wider awareness of what well-organized mathematical knowledge resources can bring both rich and poor communities. With better communication about how our subject's knowledge can be managed there will be a chance that it will not be lost to most people, as could happen for plausible commercial reasons. If the mathematical community can organize itself a little better, it can hope to avoid duplication of effort, and to achieve more. To a perhaps surprising extent many of the problems in implementing a GDML are social ones, though there are intellectual problems enough in trying to clarify what mathematical knowledge is and to make machinery to help us with it.

References

[BallBorwein:2005] Ball, J., Borwein, J., ACCESS: Who gets what access, when and how? In: MSRI Digitizing Mathematics Workshop (2005). http://www.mathunion.org/fileadmin/CEIC/Publications/ MSRI.pdf

[Bouche:2014] Bouche, T.: The digital mathematics library as of 2014. Not. Am. Math. Soc. **61**(2014), 1085–1088 (2014)

[Dolecki&Greco:2010a] Dolecki, S., Greco, G.H.: Tangency vis-à-vis differentiability by Peano, Severi and Guareschi. J. Convex Anal. (2010). http:// arXiv.org/abs/1003.1332

[Dolecki&Greco:2010b] Dolecki, S., Greco, G.H.: Towards historical roots of necessary conditions of optimality. Regula of Peano. http://arxiv.org/ abs/1002.45811002.4581

[Dolecki&Greco:2016] Dolecki, S., Greco, G.H.: The astonishing oblivion of Peano's mathematical legacy (I). Youthful achievements, foundations, arithmetic, vector spaces, (13 pages). http://dolecki.perso. math.cnrs.fr/dolecki_greco_pro_Peano_I.pdf   (Submitted). Dolecki, S., Greco, G.H.: The astonishing oblivion of Peano's mathematical legacy (II). Analysis and geometry, (18 pages). http://dolecki.perso.math.cnrs.fr/dolecki_greco_pro_Peano_II. pdf (Submitted). Dolecki, S., Greco, G.H.: The astonishing oblivion of Peano's mathematical legacy (III). Measure theory and topology, (15 pages). http://dolecki.perso.math.cnrs.fr/ dolecki_greco_pro_Peano_III.pdf (Submitted)

[EuDML] EuDML - The European Digital Mathematics Library. https:// eudml.org/

[Fields:2016] Semantic Representation of Mathematical Knowledge Workshop Videos. http://www.fields.utoronto.ca/video-archive/ event/2053,   http://www.fields.utoronto.ca/activities/ workshops/semantic-representation-mathematical-knowledge-workshop

[GöbelSperber:2010] Göbel, S., Sperber, W.: Bibliographische Klassifikationen in der Mathematik : Werkzeuge der inhaltlichen Erschlieung und für das Retrieval. Forum der Berliner Mathematischen Gesellschaft. **12**, 77–99 (2010)

[Gray:2002] Gray, J.J.: Languages for mathematics and the language of mathematics in a world of nations. In: Parshall, K.H., Rice, A.C. (eds.) Mathematics Unbound: The Evolution of an International Mathematical Research, pp. 201–228. Providence, Rhode Island, American Mathematical Society and London Mathematical Society, London (2002)

[IMKT] International Mathematical Knowledge Trust. http://imkt. org/

[IMU:2006] Digital Mathematics Library: A Vision for the Future, International Mathematical Union (2006). http://www.mathunion. org/fileadmin/CEIC/Publications/dml_vision.pdf

[IMU/CEIC:2006] Some Best Practices for Retrodigization, International Mathematical Union (2006). http://www.mathunion.org/fileadmin/ CEIC/Publications/retro_bestpractices.pdf

[Jackson:2003] Jackson, A.: The digital mathematics library. Not. Am. Math. Soc. **50**, 918–923 (2003)

[JMM:2016] AMS Special Session on Mathematical Information in the Digital Age of Science, 6-7 January 2016. http://jointmathematicsmeetings.org/meetings/national/jmm2016/2181_program_ss65.html

[MarcusWatt:2012] Marcus, S., Watt, S.M.: What is an equation? In: 2012 Proceedings of the 14th International Symposium on Symbolic and Numeric Algorithms for Scientific Computing, SYNASC 2012, pp. 23–29. IEEE Computer Society, Washington, DC, USA (2012). ISBN: 978-0-7695-4934-7, doi:10.1109/SYNASC.2012.79, http://dl.acm.org/citation.cfm?id=2477708

[MR] Mathematical Reviews: MathSciNet. http://www.ams.org/mathscinet/

[NAS:2012] Symposium Wiki on The Future World Heritage Digital Mathematics Library. National Academy of Sciences (2012). http://ada00.math.uni-bielefeld.de/mediawiki-1.18.1/index.php/Main_Page

[NRC/USA:2014] National Research Council. Developing a 21st Century Global Library for Mathematics Research. The National Academies Press (2014)

[Peano:1894] Peano, G.: Formulaire de mathématiques. t. I-V. Turin, Bocca frères, Ch. Clausen (1894–1908)

[Peano:1903] Peano, G.: De Latino Sine Flexione. Lingua Auxiliare Internationale [1], Revista de Mathematica (Revue de Mathématiques), Tomo VIII, pp. 74–83. Fratres Bocca Editores: Torino (1903). https://en.wikipedia.org/wiki/Latino_sine_flexione

[PitmanLynch:2014] Pitman, J., Lynch, C.: Planning a 21st century global library for mathematics research. Not. Am. Math. Soc. **61**, 776–777 (2014)

[RZ] Referativny Zhurnal. http://www.viniti.ru/pro_referat.html

[Schröder:1897] Schröder, E.: Über Pasigraphie, ihren gegenwärtigen Stand und die pasigraphische Bewegung in Italien. In: Verhandlungen des ersten Internationalen Mathematiker-Kongresses in Zürich, pp. 147–162, vom 9. bis 11 August 1897. English translation in The Monist, vol. **9**, 44–62 (1899). (Corrigenda p. 320). http://www.mathunion.org/ICM/ICM1897/Main/icm1897.0147.0162.ocr.pdf

[SRMwiki] Semantic Representation of Mathematics White Paper and Wiki. http://www.fields.utoronto.ca//sites/default/files/whitepaper.pdf, http://www.wolframfoundation.org/programs/whitepaper.pdf, http://imkt.org/Activities/SemanticMathematics/Workshops/2016-02-03-Fields/whitepaper.pdf, http://imkt.org/Activities/SemanticMathematics/wiki/index.php?title=Main_Page

[Seoul CEIC Panels] http://www.mathunion.org/ceic/resources/icm-2014-panels/

[Seoul DML Panel] https://www.youtube.com/watch?v=OERXmv2oIyU

[Valentin:1900] Valentin, G.H.: Die Vorarbeiten für die allgemeine mathematische Bibliographie. Bibl. Math. **1**(3), S. 237–245 (1900) (JFM 31.0003.04). http://hdl.handle.net/2027/hvd.32044102938867

[WDML Blog] http://blog.wias-berlin.de/imu-icm-panel-wdml/

[ZM] Zentralblatt für Mathematik und ihre Grenzgebiete; zbMATH. http://zbmath.org

Formula Semantification and Automated Relation Finding in the On-Line Encyclopedia for Integer Sequences

Enxhell Luzhnica and Michael Kohlhase[⊠]

Computer Science, Jacobs University Bremen, Bremen, Germany
{e.luzhnica,m.kohlhase}@jacobs-university.de

Abstract. The On-line Encyclopedia of Integer Sequences (OEIS) is an important resource for mathematicians. The database is well-structured and rich in mathematical content but is informal in nature, so knowledge management services are not directly applicable. In this paper we provide a partial parser for the OEIS that leverages the fact that, in practice, the syntax used in its formulas is fairly regular. Then, we import the result into OMDoc to make the OEIS accessible to OMDoc-based knowledge management applications. We exemplify this with a formula search application based on the MATHWEBSEARCH system and a program that finds relations between the OEIS sequences.

1 Introduction

Integer sequences are important mathematical objects that appear in many areas of mathematics and science and are studied in their own right. The On-line Encyclopedia of Integer Sequences (OEIS) [Inc15] is a publicly accessible, searchable database documenting such sequences and collecting knowledge about them. Sequences can be looked up using a text-based search functionality that OEIS provides, most notably by giving the name (e.g. "Fibonacci") or starting values (e.g. "$1, 2, 3, 5, 8, 13, 21$"). However, given that the source documents describing the sequences are mostly informal text, more semantic methods of knowledge management and information retrieval are limited.

In this paper we tackle this problem by building a formula parser for the source documents and exporting them in content MathML, the pertinent XML-based standard. This opens up the OEIS library to knowledge management applications, which we exemplify by a semantic search application based on the MATH-WEBSEARCH [HKP14] system that permits searching for text and formulas and by a relation finder that induces new relations from the parsed formulae. This paper is based on [Luz16] to which we refer for details we had to elide.

2 The OEIS

The OEIS is a web information system about integer sequences. Started in 1964 by Neil Sloane, an active community now curates over $250\,000$ sequences,

© Springer International Publishing Switzerland 2016
G.-M. Greuel et al. (Eds.): ICMS 2016, LNCS 9725, pp. 467–475, 2016.
DOI: 10.1007/978-3-319-42432-3_60

collecting their starting values, literature references, implementations, and formulae that encode representations and relations between sequences. This data is stored in a line-keyed ASCII documents internally. We introduce this by way of the snippet in Fig. 1 – we will use the Fibonacci numbers as the running example. There we see a document fragment with identification (%I), values (%S), name (%N) and reference (%D) lines, followed by three formula lines (%F) and the author line (%A). The formula lines are the main object of interest in this paper.

```
%I A000045 M0692 N0256
%S A000045 0,1,1,2,3,5,8,13,21,34,55,89,144,233,377,610,987
%N A000045 Fibonacci numbers: F(n) = F(n−1) + F(n−2) with F(0) = 0 and F(1) = 1.
%D A000045 V. E. Hoggatt, Jr., Fibonacci and Lucas Numbers. Houghton, Boston, MA, 1969.
%F A000045 F(n) = ((1+sqrt(5))^n−(1−sqrt(5))^n)/(2^n*sqrt(5))
%F A000045 G.f.: Sum_{n>=0} x^n * Product_{k=1..n} (k + x)/(1 + k*x). − _Paul D. Hanna_,
        Oct 26 2013
%F A000045 This is a  divisibility  sequence; that is,  if n divides m, then a(n)  divides  a(m)
%A A000045 _N. J. A. Sloane_, Apr 30 1991
```

Fig. 1. The OEIS sources for sequence A000045 (Fibonacci Numbers)

The bulk of formulae in the OEIS consist of **generating functions** used to represent sequences efficiently by coding the terms of a sequence as coefficients of powers of a variable x in a formal power series. Unlike an ordinary series, this formal series is allowed to diverge, meaning that the generating function is not always a true function and the "variable" is actually an indeterminate.

There are different types of generating functions, including **ordinary generating functions**, **exponential generating functions**, **Lambert series**, **Bell series**, and **Dirichlet series**. The **ordinary generating function** (or just **generating function**) of a sequence $a_0, a_1, a_2 \ldots, a_{n-1}, a_n, \ldots$ is the infinite series:

$$GF(x) = a_0 + a_1 x + a_2 x^2 + \ldots + a_{n-1} x^{n-1} + a_n x^n + \ldots = \sum_{i=0}^{\infty} a_i x^i \quad (1)$$

For example, the sequence $A000012 = (1, 1, 1, 1, 1, \ldots)$ can be represented as:

$$1 + x + x^2 + \ldots = \frac{1}{1 - x} \quad (2)$$

We know that the equation above only holds for $|x| < 1$ but we ignore the issues of convergence, as already mentioned. Thus, the ordinary generating function of this sequence can be written as $\frac{1}{1-x}$.

3   Parsing OEIS Formulae

We built a partial parser for OEIS formulas by identifying and analyzing well-behaved formulas to produce a workable grammar. We leverage the fact that, although there is no standardized format for OEIS formulas, many of them use a sufficiently regular syntax. At the core, the parser uses a rather standard grammar for infix, suffix, and prefix operators and binding operators with precedences. Instead of presenting it in detail we discuss some of the challenges we encountered:

Open Set of Primitives. Since the formulas are not standardized, not only is the syntax flexible, but so is the set of primitive operators that are used. For instance, the formulas in Fig. 1 (on lines 5–6) use square root, power, as well as the sum (Σ) and product (Π) binders. The challenges arise because of the many different notations used for such primitives. For instance, in line 6 of Fig. 1 the range for sum and product is given in two different ways. Similar problems appear with limits and integrals as well as numerous atypical infix and suffix operators. In order to parse these correctly, we investigate the documents and the grammar failures manually and incrementally extend the grammar.

Ambiguity. As it is often the case with informal, presentation-oriented formulas, there can be ambiguity in the parsing process when there exist several reasonable interpretations. Since the OEIS syntax is not fixed, this is quite common, so we do additional disambiguation during parsing to resolve most of the ambiguities. Some common ambiguities are:

- Implicit multiplications: `a*(x+y)` is usually written as `a(x+y)` which is ambiguous since it can also be parsed as a function application.
- Natural way of using the power operator: `T^2(y)` is used for $(T(y))^2$, however `T^y(x^2+2)` is ambiguous.
- Unbracketed function applications: `sin x` is a common way of writing $sin(x)$. However, this form of function application can also be parsed as multiplication between variable `sin` and `x`, as in `Pi x`.

We employ heuristics based on a type system that we use to assign types to each of the parsed terms to resolve these ambiguities.

Delineating Formulas. OEIS formula lines freely mix text and formulas so it is required to correctly distinguish between text and formula parts within the lines in order to accurately parse each line. For instance, line 6 in Fig. 1 starts with the text `G.f.:` (meaning "Generating function:") and continues with the formula. The line then has the author and date, separated from the formula by a dash (`-`) which could also be interpreted as a minus and, therefore, a continuation of the formula. In the extraction of the formulas we use the help of a dictionary. The text in the OEIS documents has words that are not found in the dictionaries since it contains many technical terms so we first run a pre-parsing procedure which enriches the dictionary. The final grammar tries to parse words until it fails and then tries to parse formulas; this process repeats.

Formula Parsing. The formula parser is implemented using the Packrat Parser for which Scala provides a standard implementation. Packrat parsers allow us to write left recursive grammars while guaranteeing a linear time worst case which is important for scaling to the OEIS.

There are 223866 formula lines in OEIS and the formula parser succeeds on 201384 (or 90 %) of them. Out of that, 196515 (or 97.6 %) contain mathematical expressions. Based on a manual inspection of selected formulas we determined that most parser fails occur because of logical connectives since those are not yet supported. Other failures include wrong formula delineation because of unusual mix of formulas and text. We did a manual evaluation of the parsing result for 40 randomly selected OEIS documents and evaluated 85 % of successfully parsed formulas as semantically correct.

The importer is implemented in Scala as an extension for the MMT system and consists of about 2000 lines of code. It is available at https://svn.kwarc.info/repos/MMT/src/mmt-oeis/.

There are 257654 documents in OEIS totaling over 280 MB of data. The OMDOC/MMT import expands it to around 9 GB, partly due to the verbosity of XML and partly due to producing the semantic representation of formulas. The total running time is around 1 h 40 min using an Intel Core i5, 16 GB of RAM and a SATA hard drive.

Search. MATHWEBSEARCH (MWS) is an open-source, open-format, content-oriented search engine for mathematical expressions. We refer to [HKP14] for details.

To realize the search instance in MWS we need to provide two things:

1. A *harvest* of MATHML-enriched HTML files that the search system can resolve queries against. The content-MATHML from the files will be used to resolve the formula part of the query while the rest of the HTML will be used for the text part. The harvest additionally requires a configuration file that defines the location in the HTML files of MWS-relevant metadata such as the title, author or URL of the original article. This, together with the HTML itself is used when presenting the query results.
2. A *formula converter* that converts a text-based formula format into MATHML. This will be used so that we can input formulas for searching in a text format (in our case OEIS-inspired ASCII math syntax) rather than writing MATHML directly.

To produce the harvest of the OEIS library for MWS we export the HTML from the content imported into MMT. We reuse the MMT presentation framework and only enhance it with OEIS-specific technicalities such as sequence name or OEIS link. For the formula converter we use the same parser used for OEIS formulas and described above, except extended with one grammar rule for MWS *query variables*. Figure 2 shows (a part of) the current interface answering a query about Fibonacci numbers. The search system is available at http://oeis.search.mathweb.org.

4   Relation Finding

Part of the mathematical interest in the OEIS is that it gives interpretations of sequences and provides a basis for establishing relations between them. Consequently extending the latter has been an important concern. As the initial values of the sequences were the only machine-actionable part of the OEIS, relation-finding has concentrated on them. However it is important to note that

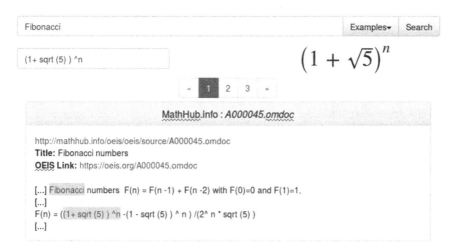

Fig. 2. Text and formula search for OEIS

even an exact match of initial subsequences can never verify a relation, thus any numeric match can only be a relation conjecture. An extreme example of two sequences that match for 777451915729367 terms but are not equal, is $\lfloor \frac{2n}{\log(2)} \rfloor$ and $\lceil \frac{2}{2^{1/n}-1} \rceil$ [NJ12]. Ralf Stephan found 117 conjectures from which 17 of them are still open [Ste04].

Our database of parsed formulae allows us to do better: we can directly look for relations between the formula representations, most prominently between the generating formulae. The approach we follow is mathematically simple. We will show two methods, the second building on top of the first. Refer to [Luz16] for a more elaborate discussion and another method.

Method 1. In this case, we *normalize* the ordinary generating functions of the sequences and check for equality between the normalized expressions. The normalization rules are defined as follows:

$$\frac{cG \quad c\text{ - constant} \quad G\text{ - generating function}}{G} \quad (\text{CONST})$$

$$\frac{x^n G \quad x\text{ - the indeterminate of } G \quad G\text{ - generating function}}{G} \quad (\text{UNSHIFT})_n$$

$$\frac{P/Q \quad P, Q\text{ - polynomials}}{(\sum_{i=0}^{n} p_i x^i)/(\sum_{i=1}^{m} q_i x^i) \quad p_n > 0 \quad q_n > 0} \quad (\text{SORT})$$

Intuitively, in this case we are checking if sequences are scaled and/or shifted versions of each other. These relations are not meant to be interesting or new.

Method 2. In this case we check if a sequence can be expressed as a sum of other sequences existing in the OEIS, possibly *transformed* and/or *normalized.*

A simplified algorithm roughly follows this pseudocode:

```
foreach sequence
  foreach ogf in ordinaryGeneratingFunction(sequence)
    add normalize(ogf) to hashSet
foreach sequence
  foreach ogf in ordinaryGeneratingFunction(sequence)
    pdf = partialFractionDecomposition(ogf)
    partialFractions = decompose(partialFractions(pfd))
    relationsExists =
    forall pgf in partialFractions
      transformedPartialFractions =
          normalize(applyTransformations(pgf))
      transformedPartialFractions.intersection(hashSet).length > 0
    if (relationsExists)
      print relations
```

We will now explain the functions that we are using above.

Let GF_n be one of the ordinary generating functions of sequence A_n. The partial fraction decomposition (*partialFractionDecomposition*) would leave us with $GF_n = \sum_{j=1}^{n} G_j$ where G_j is also an ordinary generating function. The function *partialFractions* extracts the summands, in this case, the partial fractions (ordinary generating functions) themselves. The function *decompose* does a further step of decomposition. If $G_j = \frac{P}{Q}$ where P, Q are polynomials ($P = \sum_{i=0}^{n} a_i x^i$) then it rewrites $G_j = \sum_{i=0}^{n} \frac{a_i x^i}{Q}$. These summands are then considered partial fractions too.

The transformations are *integration, differentiation* and *unit.* The transformations are selected such that expressions that match under these transformations can be easily related both mathematically and semantically.

The relation finder is implemented in Scala and is available at https://github. com/eluzhnica/OEIS. The page will be kept up to date with results. The implementation of the *normalization* rules makes use of the parsing tree of the expression. The *transformations* are done using SageMath [Dev16] as a math engine. Our Scala code communicates with a local SageMath server using a REST API.

We show below some examples of the relations found from each method.

Method 1. This is more of a sanity check of the data. Due to the nature of these relations, these are self-evident relations. Additionally, these relations can be effectively searched utilizing a numerical method. An instance that our algorithm finds is that sequence $A001478(n) = -A000027(n)$. Sequence $A001478$ is the sequence of the negative integers, while $A000027$ is the sequence of positive integers.

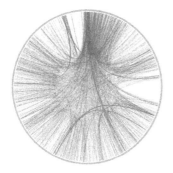 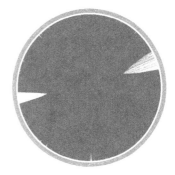

(a) OEIS Relation Graph of Current Relations   (b) OEIS Relation Graph after Method 2

The points around the circle represent the theories and the blue lines views between them. The theories presented here are only the ones for which we have parsed the generating functions.

Fig. 3. OEIS relation graphs (Color figure online)

Method 2. An example of this method, which we have submitted and it is accepted in the OEIS (https://oeis.org/A037532), is as follows.

$$A037532(n) = \frac{5}{57}A049347(n-1) + \frac{3}{57}A049347(n) + \frac{29}{171}A000420(n) - \frac{2}{9} \quad (3)$$

There is one subtlety that needs to be explained. The sequence with ordinary generating function $\frac{1}{1-x}$ is the sequence $(1,1,1,\ldots)$. However, for simplicity we write down $\frac{2}{9}$ instead of $\frac{2}{9}A000012(n)$.

Since our parser runs over all the formulas of OEIS, we have extracted the existing explicit relations in OEIS and made a graph (Fig. 3a) showing the existing connections between sequences. The second method enriches the theory graph as shown in Fig. 3b.

We converted the parsed generating functions to the SageMath syntax and checked if SageMath can compute with the expressions. From manual inspection, we found out that most of the unaccepted cases were referencing functions defined somewhere within the document. For instance, $1 + Q(0)$ where the function $Q(n)$ is defined later on in the sequence document. We currently do not resolve these references (Table 1).

Method 1 reports 4859 relations of that kind. However, in total only 853 sequences can be normalized to other existing sequences.

It is noticeable that there are a lot of relations generated from the second method. This is due to the number of relations found using the normalization rules (Method 1). Take for instance, $G = A + B + C$, and say that each of A, B, C is an OEIS sequence and is related with 3 other sequences under the normalization rules. Then the number of relations that we can form is actually 4^3. For this reason, we also report the number of relations when we remove the relations that come due to the normalization. So, in the example above the relation would count only once, instead of 4^3.

Table 1. Evaluation of the relation finder

Parsed generating functions	43 005
SageMath verified generating functions	16 065
Parsed ordinary generating functions	35 953
SageMath verified ordinary generating functions	13 400
Method 1 relations	4 859
Sequences in method 1 relations	853
Method 2 relations	297 284 646
Method 2 relations without normalization	66 427

Out of three submissions, two relations are already accepted in the OEIS. One of them has already been presented in Eq. 3 and the other relation is $A001787(n) = A007283(n)\frac{n}{6}$ which can be found at https://oeis.org/A001787. The unaccepted submission was not perceived to add new information since a similar relation was already present. The submitted relations were selected randomly.

5   Conclusions and Future Work

We improved the digitalization of the OEIS by parsing the formulae. Even though our parser can definitely be improved, it already supports two important added-value services. First, the MATHWEBSEARCH instance on OEIS which allows the users to search the OEIS by text formula queries. Second, a way of generating knowledge from OEIS, specifically, relations between sequences. The relation finding experiment presented above only uses very simple mechanisms for finding relations between generating functions. We make the parsed and induced formulae in content MathML form at https://github.com/eluzhnica/OEIS to allow other parties to extend our methods and find even more relations.

Acknowledgements. We acknowledge financial support from the OpenDreamKit Horizon 2020 European Research Infrastructures project (#676541), and thank the OEIS community for support Neil Sloane for giving us access to a full OEIS dump and Jörg Arndt for fruitful discussions. All the work reported in this paper has only been possible, since the OEIS foundation had the foresight to license the contents under a CreativeCommons license that allows derivative works.

Sustainability. To make our work sustainable, we would need to (*i*) periodically re-run our system on future versions of the OEIS and (*ii*) feed the results back into the knowledge base. The first needs a setup which facilitates change management, minimally a way to query the OEIS for changes like the OAI-PMH, but even better, maintaining the OEIS sources in a revision control system like GIT. For the second we note that with the huge volume of induced formulae, manual submission to the OEIS cannot be the answer. Automated submission – while simple to implement – would overwhelm the OEIS editors.

References

[HKP14] Hambasan, R., Kohlhase, M., Prodescu, C.: MathWebSearch at NTCIR-11. In: Kando, N., Joho, H., Kishida, K. (eds.) NTCIR Workshop 11 Meeting, pp. 114–119. NII, Tokyo (2014)

[Inc15] The OEIS Foundation Inc. The On-Line Encyclopedia of Integer Sequences (2015). http://oeis.org/

[Luz16] Luzhnica, E.: Formula Semantification and Automated Relation Finding in the OEIS. B.Sc. Thesis. Jacobs University Bremen (2016). https://github.com/eluzhnica/OEIS/doc/Enxhell_Luzhnica_BSC.pdf

[NJ12] Sloane, N.J.A.: The On-Line Encyclopedia of Integer Sequences (2012). http://neilsloane.com/doc/eger.pdf

[Dev16] The Sage Developers. SageMath, the Sage Mathematics Software System (Version 7.1) (2016). http://www.sagemath.org

[Ste04] Stephan, R.: State of 100 Conjectures From The OEIS (2004). http://www.ark.inberlin.de/conj.txt

Mathematical Videos and Affiliated Supplementaries in TIB's AV Portal

Mila Runnwerth[✉]

German National Library of Science and Technology (TIB),
Welfengarten 1B, 30167 Hannover, Germany
Mila.Runnwerth@tib.eu

Abstract. Scientific videos often are supplementary material to otherwise published research material or contain additional material on their own. Especially in the mathematical context there are four main categories of accompanying media: (1) visual demonstrations of numerical simulations, (2) video recordings of conference talks, (3) video abstracts for journal submissions, and (4) lecture videos. TIB's AV portal links its videos to all kinds of supplementary research information if it is freely available. Furthermore, a user's query to the portal is automatically expanded to a query in TIB's discovery system in order to retrieve supplementary material for further reading indexed therein. This article discusses the four categories of media combinations listed above and how to interlink them in order to guarantee easy access.

Keywords: Video platform · Scientific video · Information retrieval

1 Introduction

Publishing the latest research is no longer restricted to text-based journal articles. Media packages consisting of published articles (including preprints), research data, conference contributions, and software are submitted to scientific libraries. This publication behaviour reflects both the technical potential to provide almost all documentation of the research process and the implementation of the guidelines for good scientific practice. The challenge for libraries is to make the data bundle accessible both in its entirety as well as individually via interlinking information. This contribution illustrates the challenge of combining the formal requirements of library indexing and referencing information in new and less restricted ways to provide access to library systems and search engines alike.

We present four categories of scientific videos with supplementary material and how we provide access to them: (1) visualisations of numerical simulations, (2) conference or workshop videos, (3) video abstracts to visually promote journal articles, and (4) video lectures with visual information on slides or in handwritten form on blackboard or whiteboard.

ⓒ Springer International Publishing Switzerland 2016
G.-M. Greuel et al. (Eds.): ICMS 2016, LNCS 9725, pp. 476–481, 2016.
DOI: 10.1007/978-3-319-42432-3_61

2 TIB's AV Portal

TIB's AV Portal[1] is a free, web-based platform for audiovisual media. It provides access to about 6.700 scientific videos[2] with emphasis to science and engineering. The mathematics collection alone comprises altogether more than 1.000 videos from notable institutions such as IHÈS, ICM, or Stanford University. The portal enables extensive search functionalities, allows reliable citation and sharing via Digital Object Idenitifier (DOI) and functions as an archive to preserve audio-visual contributions to science as cultural heritage (Neumann and Plank 2014). The portal is developed by the Competence Centre for Non-Textual Materials, a subdivision at the German National Library of Science and Technology (TIB), in cooperation with the Hasso Plattner Institute for Software Systems Engineering (HPI). It combines multimedia retrieval techniques and semantic analysis. Enhanced video analysis features are automatic scene, speech, text and object recognition. Enhanced search features are concept mapping and semantic search (Arraiza and Strobel 2015). Figure 1 illustrates the workflow each video passes in order to be provided in the portal.

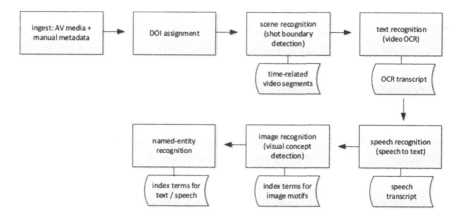

Fig. 1. Process chain of the automated video analysis (Strobel and Arraiza 2015).

Services and functionalities include:

Time-Stamped Citation Link: Every video is associated with a DOI. Additionally, by using the open standard Media Fragment Identifier (MFID), every video time segment can be cited individually.

Legally Secure Handling: When publishing a scientific video in TIB's AV Portal the customer specifies the legal terms of publication by concluding a license agreement. Thus, the customer determines the terms under which the portal may process, supply and further distribute the video.

---

[1] https://av.tib.eu/.
[2] Current status as of May 2016.

Metadata: A video's metadata set consists of intellectually maintained metadata like title, file size, abstract, keywords provided by the customer. Additionally, automatically generated metadata from the video analyses mentioned above enrich the formal metadata (Lichtenstein et al. 2014).

3   Mathematical Media Packages

Especially research output that involves mathematical modelling or numerical simulations is submitted to libraries more and more as data bundles or media packages. They typically contain scientific software, experimental (measurement) data, processed data, visualisations as images or film, documentation and the resulting publication(s). Most audiovisual information, however, is not obviously interlinked to additional materials because the customer is not aware of uploading the data in a bundled package or because there are still more media to appear in the course of a project.

A basic approach for simple maintenance is to assign a DOI to a superior project archive and store the DOIs assigned to its associated components in the metadata of the package. Without further processing that does not help for the retrieval, though.

3.1   Visual Simulation Data and Their Related Information

Data Description: Videos of visual simulation data can be characterised by being short (less than two minutes per experiment) and by consisting of exclusively visual information. Current supplementaries are documentation papers, abstracts explaining the experiment, a project web page of the affiliated research institution or a published paper that references the video explicitly if it is freely available. Visualised simulations themselves are generally supplementary appendices.

State of the Art at TIB's AV Portal: In general simulation data are uploaded with additional materials to provide the scientific context. In the manual editing process of the metadata the semantic connections can be considered intellectually. The scientific librarian who approved of the video in the first place also decides how and where to provide the accompanying materials. The simplest form to supply a reference to the video is to add a link to the accompanying materials in the metadata set to a web page or the library system. A more sophisticated approach is a reference to add the data to the library system including interconnections. In TIB's AV Portal the metadata section shows a paragraph called *Accompanying Material* with all linked – although not interlinked – supplementaries.

Outlook: The primary goal is a more comprehensive retrieval for both the video and its associated information from various distributed sources. Cross-linking DOIs is a step to connect project data formally. Matching the visual analyses data against analysis data of the supplementary materials – ideally fulltexts – may give hints to semantically link them to the film. Feining a superior data set for the abstract project and linking all associated data to it may help to inherit classical subject indexing of related information for the child instances.

3.2 Conference Recordings and Their Corresponding Proceedings

Data Description: An increasing number of conference hosts produce videos of contributions like lectures and talks. The added value in comparison with the associated article in the conference proceedings can be the documented discussion about the topic with critical input from third-parties, bringing together slide or blackboard information as well as the spoken text which generally contains subtle subtext information.

State of the Art at TIB's AV Portal: Conference videos are especially popular in computer science and mathematics. TIB's AV Portal contains a large collection of computer science conferences that do not publish in printed form at all. In mathematics, most conference videos can be associated to a published article in textual conference proceedings. These, however, cannot be linked together automatically. The current workflow for subject librarians is to check a conference video collection against the textual conference stock. If the affiliated article within the proceedings volume can be confirmed, the video (or even the collection) can be linked in a comment to the volume in the library system. Accordingly, a reference can be included in the video's metadata set to point out the associated library stock, for example by referencing the DOIs of article and film. From a librarian's point of view the relation of an article to the talk on it is not precisely defined. They are strongly related but neither the same format nor of identical content.

Outlook: The intuitive way to link video and textual information would be to link the whole video collection to the proceedings volume (where each video has a match to an article, and vice versa). On a finer scale we need to link a video talk to an individual article. The prominent challenges to address here are to find the matching pairs (manually and eventually automatically) and connecting them in a sound way. In mathematics, a significant number of conference contributions are published as preprint on arXiv.org. Thus, we are able to link the open access preprint to the video as accompanying material with a link to the repository page.

3.3 Video Abstracts and the Associated Article

Data Description: There is an upcoming trend of submitting video abstracts in addition to or instead of textual abstracts for scientific journals[3]. Together

---

[3] A prominent example is given by Elsevier's Journal of Number Theory (http://www.journals.elsevier.com/journal-of-number-theory/video-abstracts/).

with an open access preprint as well as a final, published article a video abstract may present the entity of the publication. Video abstracts are short summaries with various image information: mostly with slides or handwritten content on a blackboard, but also environments that express the author's personality best, for example summarising the article during a stroll on the beach or while sitting on a lawn.

State of the Art at TIB's AV Portal: Authors themselves do usually not directly supply a video abstract to TIB's AV Portal. The library acquisition team researches them on the web and gets in touch with the publishing house. The subject librarians check for further publications or preprints and recommend to link them as accompanying material for direct download (if freely accessible) or link them via a comment in the library system. In both cases, once the additional material is found, all associated materials are retrievable irrespective of the access portal.

Outlook: The main challenge is to keep track of the publication process: where there is a video abstract there is at least a paper in waiting. We would like to pick users up at the video abstract and naturally forward them to its associated preprint or article. If these are open access we can easily provide them via link. Otherwise a direct link to the library system is needed with the article's disposability information attached.

3.4   Lecture Videos and Their Scripts

Data Description: Lecture recordings are the most prominent form of Open Educational Resources. They are often accompanied by an informal script by the lecturer herself or the students. Most lectures in TIB's AV Portal are mathematics or physics lectures with the specific characteristic that they contain handwritten formulae. The largest collection of tutorial videos in TIB's AV Portal are Professor Jörn Loviscach's[4] short presentations on undergraduate topics in mathematics. Additional materials are often handouts or scripted fragments of the lecture. The duration of this video category ranges from short tutorials with only a few minutes up to lectures of 90 min. The two main subcategories are khan style videos[5] (handwritten tutorials on electronic blackboards) and classroom sceneries with a lecturer in front of a blackboard.

State of the Art at TIB's AV Portal: For most lecture videos there exists an inofficial script. Publishing a script as accompanying material is only feasible with the lecturer's consent. The Loviscach Collection not only provides scripts but also further teaching materials like homework or question sheets to several videos. This material is not relevant for the library stock. Hence, it is not entered to the library system. To retrieve these documents specifically, the video must be retrieved beforehand or the lecture web page is found via a web search engine.

---

[4] http://www.j3l7h.de/videos.html.

[5] The choice of expression makes allowance to the pioneering mathematical learning platform https://www.khanacademy.org/.

Outlook: A trivial challenge is to make the additional documents retrievable as well. If fulltext search is possible it should be provided. On a more challenging level associated materials to lectures provide an opportunity to enhance retrieval techniques and formula search. It is the most promising category to exploit semantical information via different documents. Especially mathematical lectures of 90 min cover many aspects which might be relevant for a user to search within: a certain theorem or proof, an example or illustration for a mathematical phenomenon. This calls for a sophisticated table of contents based on information of the video analyses techniques that are already in use.

4   Conclusion and Outlook

In TIB's AV Portal occur mainly four categories of scientific videos with a focus on mathematics with additional material: visual simulations, conference talks, video abstracts, and lectures. Each brings diverse types of material that must be referenced in different ways to optimise retrieval and access. Although we already automised most of the analyses features for the videos themselves, finding associated information and the deciding on how to interlink it must be mainly organised intellectualy.

Our future task is to discover and link associated materials automatically by using semantic search or analysis of formal metadata. Also, to optimise and develop video analyses techniques, additional materials are a data basis for further information with respect to machine learning.

References

Neumann, J., Plank, M.: TIB's portal for audiovisual media new ways of indexing and retrieval. IFLA J. **40**(1), 17–23 (2014)

Arraiza, P.M., Strobel, S.: The TIB|AV portal as a future linked media ecosystem. In: WWW (Companion Volume), pp. 733–734 (2015)

Strobel, S., Arraiza, P.M.: Metadata for scientific audiovisual media: current practices and perspectives of the TIB|AV-portal. In: Garoufallou, E., Hartley, R.J., Gaitanou, P. (eds.) MTSR 2015. CCIS, vol. 544, pp. 159–170. Springer, Heidelberg (2015). http://dx.doi.org/10.1007/978-3-319-24129-6_14

Lichtenstein, A., Plank, M., Neumann, J.: TIB's portal for audiovisual media: combining manual and automatic indexing. Cataloging Classif. Q. **52**(5), 562–577 (2014)

Miscellanea

Complexity of Integration, Special Values, and Recent Developments

James H. Davenport[(✉)]

University of Bath, Bath, UK
J.H.Davenport@bath.ac.uk
http://staff.bath.ac.uk/masjhd

Abstract. Two questions often come up when the author discusses integration: what is the complexity of the integration process, and for what special values of parameters is an unintegrable function actually integrable. These questions have not been much considered in the formal literature, and where they have been, there is one recent development indicating that the question is more delicate than had been supposed.

Keywords: Integration · Complexity · Parameters

1 Introduction

The author is often asked two questions about integration.

1. "What is the complexity of integration?".
2. "My integrand $f(x, a)$ is unintegrable. For what special a is it integrable?"

These questions have rather different answers for purely transcendental integrands and for algebraic function (or mixed) integrands. In fact, they are essentially unexplored for mixed integrands, given the difficulties of the two special cases.

Integration of $f(x)$, in the sense of determining a formula $F(x)$ such that $F'(x) = f(x)$, is a process of differential algebra. There is then a question of whether this formula actually corresponds to a continuous function $\mathbf{R} \to \mathbf{R}$. This is an important question in terms of usability of the results, but a rather different one than we wish to consider here: see [7].

2 Transcendental Integration

In order to use differential algebra, the integrand f is written (itself a non-trivial procedure: see [9], generally known as the **Risch Structure Theorem**) in a suitable field $K(x, \theta_1, \ldots, \theta_n)$ where each θ_i is transcendental over $K(x, \theta_1, \ldots, \theta_{i-1})$ with $K(x, \theta_1, \ldots, \theta_i)$ having the same field of constants as $K(x, \theta_1, \ldots, \theta_{i-1})$ and each θ_i being either:

G.-M. Greuel et al. (Eds.): ICMS 2016, LNCS 9725, pp. 485–491, 2016.
DOI: 10.1007/978-3-319-42432-3_62

(l) a *logarithm* over $K(x, \theta_1, \ldots, \theta_{i-1})$, i.e. $\theta_i' = \eta_i'/\eta_i$ for $\eta_i \in K(x, \theta_1, \ldots, \theta_{i-1})$;
(e) an *exponential* over $K(x, \theta_1, \ldots, \theta_{i-1})$, i.e. $\theta_i' = \eta_i'\theta_i$ for $\eta_i \in K(x, \theta_1, \ldots, \theta_{i-1})$.

This process may generate special cases: for example $\exp(a \log x)$ lives in such a $K(x, \theta_1, \theta_2)$ with $\theta_1' = \frac{1}{x}$ (θ_1 corresponds to $\log x$) and $\theta_2' = \frac{a}{x}\theta_2$ (θ_2 corresponds to $\exp(a \log x)$), *except* when a is rational, when in fact we have x^a. However, this is generally not what is meant by the "special values" question, and in general we assume that parameters are not in exponents.

2.1   Elementary Transcendental Functions

Here we have a decision procedure, as outlined in [8]. The proof of the procedure proceeds by induction on n, the ingenuity lying in the induction hypothesis: we suppose that we can:

(a) "integrate in $K(x, \theta_1, \ldots, \theta_{n-1})$", i.e. given $g \in K(x, \theta_1, \ldots, \theta_{n-1})$, either write $\int g dx$ as an elementary function over $K(x, \theta_1, \ldots, \theta_{n-1})$, or prove that no such elementary function exists;
(b) "solve Risch differential equations in $K(x, \theta_1, \ldots, \theta_{n-1})$", i.e. given elements $F, g \in K(x, \theta_1, \ldots, \theta_{n-1})$ such that $\exp(F)$ is transcendental over $K(x, \theta_1, \ldots, \theta_{n-1})$ (with the same field of constants), solve $y' + F'y = g$ for $y \in K(x, \theta_1, \ldots, \theta_{n-1})$, or prove that no such y exists.

We then prove that (a) and (b) hold for $K(x, \theta_1, \ldots, \theta_n)$.

2.2   Logarithmic θ_n

If θ_n is logarithmic, the proof of part (a) is a straightforward exercise building on part (a) for $K(x, \theta_1, \ldots, \theta_{n-1})$: see, e.g. [3, Sect. 5.1]. Unintegrability manifests itself as the insolubility of certain equations, and any special values of the parameters will be found as special values rendering these equations soluble.

It is also straightforward (though as far as the author knows, not done) to prove that, if all θ_i are logarithmic, then the degree in each θ_i of the integral is no more than it is in the integrand, and that the denominator of the integral is a divisor of the denominator of the integrand. Hence, in the dense model, the integral is, apart from coefficient growth, not much larger than the integrand, and the compute cost is certainly polynomial.

In a sparse model, the situation is very different.

$$\int \log^n x dx = x \log^n x - nx \log^{n-1} x + \cdots \pm n! x,$$

so an integrand requiring $\Theta(\log n)$ bits can require $\Omega(n)$ bits for the integral. The same is true for $\int x^n \log^n x dx$, but $\int x^n \log^n (x+1) dx$ shows that $\Omega(n^2)$ bits can be required. As far as the author knows, it is an open question whether the problem is even in EXPSPACE, though it probably is.

2.3   Exponential θ_n

Here the problem is different. Suppose $\theta_n = \exp(F)$. $\int g \exp(F) \mathrm{d}x = y \exp(F)$ where $y' + F'y = g$ (and can be nothing else if it is to be an elementary function). Hence solving (a) in $K(x, \theta_1, \ldots, \theta_n)$ reduces (among other things) to solving (b) in $K(x, \theta_1, \ldots, \theta_{n-1})$. In general, the solution to (b) proceeds essentially by undetermined coefficients, which is feasible as $y' + F'y$ is linear in the unknown coefficients. Before we can start this, we need to answer two questions: what is the denominator of y, and what is the degree (number of unknown coefficients)? In general, the answers are obvious: if the denominator of g has an irreducible factor p of multiplicity k, y' will have the same, so the denominator of y will have a factor of (at most) p^{k-1}, and F' can only reduce this. Similarly, if g has degree d, y' will have degree at most d, so y will have degree $d+1$, and again F' can only reduce this. The complication is when there is cancellation in $y' + F'y$, so that this has lower degree, or smaller denominator, than its summands. [8] shows how to resolve this problem, and does not pay it much attention, not being interested in the complexity question.

In [2] it is noted that these complications come from what one might loosely call "eccentric" integrands. For example

$$y' + \left(1 + \frac{5}{x}\right) y = 1 \tag{1}$$

has solution
$$y = \frac{x^5 - 5x^4 + 20x^3 - 60x^2 + 120x - 120}{x^5}, \tag{2}$$

(and in general $y' + \left(1 + \frac{n}{x}\right) y = 1$ will have a solution with denominator x^n) but this comes from

$$\int \exp(x + 5 \log x) \mathrm{d}x, \tag{3}$$

which might be more clearly expressed as

$$\int x^5 \exp(x) \mathrm{d}x. \tag{4}$$

However, the integrand in (3) has total degree 1, whereas that in (4) has total degree 6, consistent with the degrees in (2). Ultimately, the point is that the dense model is not applicable when we can move things into/out of the exponents at will.

We do have a result [2, Theorem 4] which says that, provided $K(x, \theta_1, \ldots, \theta_n)$ is *exponentially reduced* (loosely speaking, doesn't allow "eccentric" integrands) then we have natural degree bounds on the solutions of (b) equations. As stated there, "this is far from being a complete bounds on integrals, but it does indicate that the worst anomalies cannot take place" here.

Again, the complexity is still an open question, but the author is inclined to conjecture that it is no worse than EXPSPACE.

What of special values of parameters? These come in two kinds.

1. As in the logarithmic case, we can get proofs of unintegrability because certain equations are insoluble. For example $(x + a)\exp(-bx^2 + cx)$ is integrable if, and only if, $c = -2ab$, and this equation arises during the undetermined coefficients process.
2. More complicated are those that change the "exponentially reduced" nature of the integrand. For example, $\int \exp(x + a \log x)\mathrm{d}x$ does not have an elementary expression *except* when a is a non-negative integer, when we are in a similar position to (3). These values are similar to those that change the Risch Structure Theorem expression of the integrand.

3   Algebraic Functions

The integration of algebraic functions [1,11] is a more complex process. If $f \in K(x,y)$ where y is algebraic over $K(x)$, the integral, if it is elementary, has to have the form $v_0 + \sum c_i \log(v_i)$, where $v_0 \in K(x,y)$, the c_i are algebraic over K, and the $v_i \in L(x,y)$ where L is the extension of K by the c_i (and possibly more algebraic numbers added by the algorithm, though these should be irrelevant). So far, this is the same as the integration of rational functions, and the challenge is to determine the c_i and v_i.

3.1   The Logarithmic Part

Looked at from the point of view of analysis, the $\sum c_i \log(v_i)$ term is to represent the logarithmic singularities in $\int f\mathrm{d}x$, which come from the simple poles of f: in a power series world c_i would be the residue at the pole corresponding to v_i. Hence an obvious algorithm would be

1. Compute all the residues r_j at all the corresponding poles p_j (which might include infinity, and which might be ramified: the technical term would be "place"). Assume $1 \le j \le m$.
2. Let c_i be a **Z**-basis for the r_j, so that $r_j = \sum \alpha_{i,j} c_i$.
3. For each c_i, let v_i be a function $\in L(x,y)$ with residue $\alpha_{i,j}$ at p_j for $1 \le j \le m$ (and nowhere else). The technical term for this residue/place combination is "divisor", and a divisor with a corresponding function v_i is termed a "principal divisor".
   * Returning "unintegrable" if we can't find such v_i.
4. Having determined the logarithms this way, find v_0 by undetermined coefficients.

The problem with the correctness of this algorithm is a major feature of algebraic geometry. It is possible that D_i is not a principal divisor, but that $2D_i$, or $3D_i$ or ... is principal. In this case, we say that D_i is a torsion divisor, and the corresponding order is referred to as the torsion of the divisor. If, say, $3D_i$ is principal with corresponding function v_i, then, although not in $L(x,y)$, $\sqrt[3]{v_i}$ corresponds to the divisor D_i, and we can use $c_i \log \sqrt[3]{v_i}$, or, more conveniently and fitting in with general theory, $\frac{c_i}{3} \log v_i$ as a contribution to the logarithmic part.

3.2 Complexity

There are three main challenges with complexity theory for algebraic function integration.

1. The first is that it is far from clear what the "simplest" form of an integral of this form is. The choice of c_i is far from unique, and a "bad" choice of c_i may lead to large $\alpha_{i,j}$ and complicated v_i.
2. The second is that the r_j are algebraic numbers, and there are no known non-trivial bounds for the r_j, or the $\alpha_{i,j}$.
3. The third is that there is very little known about the torsion. This might seem surprising to those who know some algebraic geometry, and have heard of, say, Mazur's bound [6]. This does indeed show that, if the algebraic curve defined by y is elliptic (has genus 1) *and* the divisor is defined over $\mathbf{Q}$, then the torsion is at most 12. The trouble is that this requires the divisor to be defined over $\mathbf{Q}$, and not just f. For elliptic curves, a recent survey of the known bounds is given in [10].

Hence it appears unrealistic to think of complexity bounds in the current state of knowledge.

4 Two Meis Culpis About Algebraic Integration and Parameters

In the author's thesis (see the expanded version in [1]) we considered the question of whether $f(x, u)\mathrm{d}x$, an algebraic function of x, could have an elementary integral for specific values of u, even if the uninstantiated integral were not elementary.

4.1 The Claim

We began [1, pp. 89–90] with a rehearsal of the ways in which substituting a value for u could change the working of the integration algorithm, and how these could be detected, i.e. given such an unintegrable $f(x, u)$ how one might determine the specific u values for which the integrand *might* have an elementary integral.

1. The curve can change genus: look at the canonical divisor.
2. The [geometry of the] places at which residues occur can change: look at values of u for which numerator/denominator cancel, or roots coincide.
3. The dimension of the space of residues can collapse.
4. A divisor may be a torsion divisor for a particular value of u, even though it is not a torsion divisor in general. These cases can be detected by looking at the roots of SUM in FIND_ORDER_MANIN.
5. the algebraic part may be integrable for a particular u, though not in general. Hence the contradicting equation in FIND_ALGEBRAIC_PART collapses.

As a potential example of case 3, consider

$$\frac{1}{x\sqrt{x^2+1}} + \frac{1}{x\sqrt{x^2+u^2}}$$

whose residues are $\pm 1, \pm u$ and therefore every rational u is a special case.

Lemma 1 ([1, Lemma 6, page 90]). *Let the $\mathbf{Z}$-module of residues r_i of $f(x,u)$ have dimension k, and suppose there are values $(u_1, \ldots, u_k)$ such that $f(x, u_i)$ has an elementary logarithmic part (not in cases 1,2,4,5) and such that the set of vectors $\{(r_i(u_a) : 1 \le i \le k)1 \le a \le k\}$ is of dimension k. Then $f(x,u)\mathrm{d}x$ has an elementary logarithmic part.*

Proof. Some $(n, 0 \ldots, 0)$ can be expressed as a linear combination with integer coefficients of the $(r_i(u_a))$. Hence the divisor d_1 must be a torsion divisor, as nd_1 is a sum of torsion divisors. Similarly the other d_i.

We suppose $f(x,u)$ depends algebraically on u (else it's a new transcendental).

Theorem 1 ([1, Theorem 7, page 90]). *If $f(x,u)\mathrm{d}x$ is not elementarily integrable, then there are only finitely many values u_i of u for which $f(x, u_i)\mathrm{d}x$ has an elementary integral.*

Proof. "Case 3 is the hard one. Lemma 6 disposes of the case where k values generate a full-dimensional space, so there is a linear relationship between the $r_i(u_a)$ which is not true in general, but which is true infinitely often. But the $r_i(u_a)$ are algebraic in u (Proposition 5) and this means we have an algebraic expression which is not identically zero, but which has infinitely many roots, and this establishes the required contradiction" (from [1]).

4.2   The First Problem

[4] observes that $\dfrac{x\mathrm{d}x}{(x^2 - u^2)\sqrt{x^3 - x}}$ is not elementarily integrable, but is integrable whenever the point $(u, ?)$ is of order at least three on the curve $y^2 = x^3 - x$, and this can be achieved infinitely often, at the cost of extending the number field. The simplest example is $u = i$, when $(i, 1 - i)$ is of order 4 and we have

$$\int \frac{x\mathrm{d}x}{(x^2 + 1)\sqrt{x^3 - x}} = \frac{1+i}{16} \ln\left(\frac{x^2 + (2 + 2i)\sqrt{x^3 - x} + 2ix - 1}{x^2 - (2 + 2i)\sqrt{x^3 - x} + 2ix - 1}\right)$$
$$+ \frac{1-i}{16} \ln\left(\frac{x^2 + (2 - 2i)\sqrt{x^3 - x} - 2ix - 1}{x^2 - (2 - 2i)\sqrt{x^3 - x} - 2ix - 1}\right)$$

Unfortunately neither Maple (2016) nor Mathematica (10.0) nor Reduce (build 3562) can actually integrate this elementarily.

The full problem is treated in [5]. It seems that the arguments in [1] are implicitly assuming a fixed number field, but a full analysis awaits the publication of [5].

4.3   The Second Problem

The assertion that the case of transcendental u is trivial, if true at all, is certainly not trivial, and probably false, if we also allow transcendental constants in f, for they and u can then "collide" [4].

Acknowledgements. I am immensely grateful to Professors Masser and Zannier for devoting such time to an obscure corner of an old and obscure thesis. I am also grateful to Barry Trager for his comments, to ICMS for allowing me to state the problems, and to Christoph Koutschan for his careful editing.

References

1. Davenport, J.H.: On the Integration of Algebraic Functions. Springer, Heidelberg (1981). vol. 102 of Springer Lecture Notes in Computer Science. Heidelberg New York (Russian ed. MIR Moscow 1985)
2. Davenport, J.H.: On the risch differential equation problem. SIAM J. Comp. **15**, 903–918 (1986)
3. Davenport, J.H., Siret, Y., Tournier, E.: Computer Algebra, 2nd edn. Academic Press, London (1993)
4. Masser, D.W.: Integration Update. Private Communications to JHD, February– March 2016
5. Masser, D.W., Zannier, U.: Torsion points on families of abelian varieties. Pell's equation and integration in elementary terms (2016). In Preparation
6. Mazur, B.: Rational points on modular curves. In: Serre, J.-P., Zagier, D.B. (eds.) Modular Functions of One Variable V. Lecture Notes in Mathematics, vol. 601, pp. 107–148. Springer, Heidelberg (1977)
7. Mulders, T.: A note on subresultants and the Lazard/Rioboo/Trager formula in rational function integration. J. Symbolic Comp. **24**, 45–50 (1997)
8. Risch, R.H.: The problem of integration in finite terms. Trans. A.M.S. **139**, 167–189 (1969)
9. Risch, R.H.: Algebraic properties of the elementary functions of analysis. Amer. J. Math. **101**, 743–759 (1979)
10. Sutherland, A.V.: Torsion subgroups of elliptic curves over number fields (2012). https://math.mit.edu/~drew/MazursTheoremSubsequentResults.pdf
11. Trager, B.M.: Integration of Algebraic Functions. Ph.D. thesis, M.I.T. Dept. of Electrical Engineering and Computer Science (1984)

An Algorithm to Find the Link Constrained Steiner Tree in Undirected Graphs

Luigi Di Puglia Pugliese[1], Manlio Gaudioso[2], Francesca Guerriero[1([⊠])],
and Giovanna Miglionico[2]

[1] Department of Mehcanical, Energy and Management Engineering,
University of Calabria, Rende, Italy
{luigi.dipugliapugliese,francesca.guerriero}@unical.it
[2] Dipartimento di Ingegneria Informatica, Modellistica,
Elettronica e Sistemistica, University of Calabria, Rende, Italy
{gaudioso,gmiglionico}@dimes.unical.it

Abstract. We address a variant of the classical Steiner tree problem defined over undirected graphs. The objective is to determine the Steiner tree rooted at a source node with minimum cost and such that the number of edges is less than or equal to a given threshold. The link constrained Steiner tree problem ($\mathcal{LCSTP}$) belongs to the NP-hard class. We formulate a Lagrangian relaxation for the $\mathcal{LCSTP}$ in order to determine valid bounds on the optimal solution. To solve the Lagrangian dual, we develop a dual ascent heuristic based on updating one multiplier at time. The proposed heuristic relies on the execution of some sub-gradient iterations whenever the multiplier update procedure is unable to generate a significant increase of the Lagrangian dual objective. We calculate an upper bound on the $\mathcal{LCSTP}$ by adjusting the infeasibility of the solution obtained at each iteration. The solution strategy is tested on instances inspired by the scientific literature.

Keywords: Constrained Steiner tree · Lagrangean relaxation

1 Introduction

Let $G(V, E)$ be a graph where V is the set of nodes and E is the set of edges. Let T be a subset of V. The Steiner Tree for T in G is a set of edges $\bar{E} \subseteq E$ such that the graph $(V(\bar{E}), \bar{E})$ contains a path between each pair of nodes in T, where $V(\bar{E})$ is the set of nodes incident to the edges in $\bar{E}$. A cost (weight) c_e is associated to each edge $e \in E$. The Steiner tree problem ($\mathcal{STP}$) aims at finding a minimum cost (weight) Steiner tree. The $\mathcal{STP}$ belongs to the class of $\mathcal{NP}$-hard problem (see, e.g., [3]). Due to its theoretical complexity and its practical importance, several ideas and solution strategies have been developed to address the $\mathcal{STP}$. The scientific literature provides both heuristic and exact solution methods. For a comprehensive survey on the $\mathcal{STP}$, the reader is referred to [5,10].

© Springer International Publishing Switzerland 2016
G.-M. Greuel et al. (Eds.): ICMS 2016, LNCS 9725, pp. 492–497, 2016.
DOI: 10.1007/978-3-319-42432-3_63

Real-life applications require additional restrictions on the structure of the Steiner tree. The scientific literature focuses on three main additional constraints: (1) hop constraints [6,7,11]; (2) diameter constraints [2,4]; and (3) delay constraints [8,9].

In this paper we consider a variant of the $\mathcal{STP}$ in which the number of edges involved in the Steiner tree is constrained to be less than or equal to a given upper limit K, that is, $\bar{E} \leq K$. We call this variant Link Constrained $\mathcal{STP}$ ($\mathcal{LCSTP}$).

In [1], the authors consider the $\mathcal{LCSTP}$ defined over directed graph. A heuristic procedure is tested on instances originated from 3D placement of unmanned aerial vehicles used for multi-target surveillance.

We propose a solution approach for determining a valid upper bound that relies on the solution of a Lagrangian problem. An ad-hoc procedure is defined in order to solve the dual Lagrangian problem. A feasible solution is determined by solving a restricted $\mathcal{LCSTP}$ at each iteration.

In the next section, we introduce a formulation for the $\mathcal{LCSTP}$, together with a possible Lagrangian relaxation. In Sect. 3, a multiplier update procedure is described. In Sect. 4 we report the computational results, while Sect. 5 concludes the paper.

2   A Lagrangean Relaxation for the $\mathcal{LCSTP}$

The $\mathcal{LCSTP}$ can be formulated on a bi-directed graph $B(V, A)$, where an edge e, incident to nodes i and j, is replaced by two arcs, that is, (i, j) and (j, i), with $c_{ij} = c_{ji} = c_e$. Finding a Steiner tree on the graph B, means to obtain a Steiner arborescence rooted at node 1 and containing a directed path from node 1 to every other terminal node in $T/\{1\}$. A commodity k for each node $k \in T/\{1\}$ is considered. Given the undirected graph $G(V, E)$ and the corresponding bi-directed graph $B(V, A)$, for each arc $a = (i, j)$, the variable f_{ij}^k represents the flow of commodity k from node i to node j. The variable x_e is defined for each edge $e \in E$ and takes value equal to 1 is edge $e \in \bar{E}$, 0 otherwise.

The formulation for the $\mathcal{LCSTP}$ is as follows:

$$\min \sum_{e \in E} c_e x_e \tag{1}$$

$$\sum_{\{j \in V\}} f_{ij}^k - \sum_{\{j \in V\}} f_{ji}^k = \begin{cases} 1 & \text{if } i = 1 \\ -1 & \text{if } i = k \\ 0 & \text{otherwise} \end{cases} \quad \forall k \in T/\{1\} \quad i \in V \tag{2}$$

$$f_{ij}^k \leq x_e \quad \forall e = (i, j) \in E, k \in T/\{1\} \tag{3}$$

$$\sum_{e \in E} x_e \leq K \tag{4}$$

$$f_{ij}^k \geq 0 \tag{5}$$

$$x_e \ \textit{binary} \tag{6}$$

A Lagrangian relaxation of problem (1)–(6) can be obtained by associating to constraints (3) the multipliers $\mu_{ek} \geq 0$ thus obtaining

$$Z_{LR}(\mu) = \min \sum_{e \in E} c_e x_e - \sum_{e \in E} \sum_{k \in T} \mu_{ek}(x_e - f_{ij}^k)$$

$$= \sum_{e \in E} \sum_{k \in T} \mu_{ek} f_{ij}^{(k)} + \sum_{e \in E}\left(c_e - \sum_{k \in T} \mu_{ek}\right) x_e$$

s.t.

$$(2), (4), (5), (6)$$

We observe that $Z_{LR}(\mu)$ can be decomposed into two problems, that is, $Z_{LR}^{(1)}(\mu)$ and $Z_{LR}^{(2)}(\mu)$.

$$\left\langle \begin{array}{c} Z_{LR}^{(1)}(\mu) = \min \sum_{k \in T} \sum_{e \in E} \mu_{ek} f_{ij}^{(k)} \\ s.t. \\ (2), (5) \end{array} \right\rangle \qquad \left\langle \begin{array}{c} Z_{LR}^{(2)}(\mu) = \min \sum_{e \in E}\left(c_e - \sum_{k \in T} \mu_{ek}\right) x_e \\ s.t. \\ (4), (6) \end{array} \right\rangle$$

Given a vector of Lagrangian multipliers μ, solving $Z_{LR}^{(1)}$ requires the solution of $|T|$ minimum path problems, whereas $Z_{LR}^{(2)}$ can be solved by inspection.

3   Solving the Lagrangian Dual Problem

In order to find the best values for the Lagrangian multipliers, we need to solve the Lagrangian Dual Problem, that is $z_{LD} = \max_{\mu \geq 0} z_{LR}(\mu)$.

To this aim, we developed an ad-hoc multiplier updating procedure which provides, at each iteration, either increase or bounded deterioration of $z_{LR}(\mu)$.

In particular, let $\{x_e = x_e(\mu), \ f_{ij}^k = f_{ij}^k(\mu)\}$ be any optimal solution to $Z_{LR}(\mu)$, we update the multipliers by considering the following cases:

(Case 1) $\exists \bar{e} : f_{ij}^k = 0 \ \forall k$ and $x_{\bar{e}} = 1$;
(Case 2) $\exists \bar{e} : f_{ij}^k > 0$ for some k and $x_{\bar{e}} = 0$;

(Case 1). It holds $c_{\bar{e}}(\mu) = c_{\bar{e}} - \sum_{k \in T} \mu_{\bar{e}k} < 0$. Let $e_{min} = \arg \min_{\{e | x_e(\mu)=0\}} c_e(\mu)$ the index of the smallest coefficient $c_e(\mu)$ having associated an $\bar{x}_e = 0$. Set $\Delta_\mu = c_{e_{min}}(\mu) - c_{\bar{e}}(\mu)$ with $c_{e_{min}}(\mu) = (c_{e_{min}} - \sum_{k \in T} \mu_{e_{min}k})$ and $c_{\bar{e}}(\mu) = (c_{\bar{e}} - \sum_{k \in T} \mu_{\bar{e}k})$. Select a new set of non negative multipliers $\mu'_{\bar{e}k}$ such that $\sum_{k \in T} \mu'_{\bar{e}k} = \sum_{k \in T} \mu_{\bar{e}k} - \Delta_\mu - \epsilon$. The objective function $Z_{LR}^{(1)}(\mu)$ decreases at most of $\Delta_\mu + \epsilon$, whereas $Z_{LR}^{(2)}(\mu)$ increases of Δ_μ.

(Case 2). Calculate $\Delta_\mu = c_{\bar{e}}(\mu) - \max_{\{e | x_e(\mu)=1\}} c_e(\mu) \geq 0$ and select a new set of non negative multipliers $\mu'_{\bar{e}k}$ such that $\sum_{k \in T} \mu'_{\bar{e}k} = \sum_{k \in T} \mu_{\bar{e}k} + \Delta_\mu + \epsilon$. The objective function $Z_{LR}^{(1)}(\mu)$ does not decrease and $Z_{LR}^{(2)}(\mu)$ decreases of ϵ.

To find a feasible solution for the original problem, starting from the solution obtained by using the multipliers updating procedure, we devise a repairing procedure shortly described in the following.

Given $\{x_e = x_e(\mu), f_{ij}^k = f_{ij}^k(\mu)\}$, we set $x_e = 1$ if $f_{ij}^k > 0$ for some k, $\forall e = (i,j)$. If $\sum_{e \in E} x_e \geq K$, then we define a sub-graph G_x induced by x where the $\mathcal{LCSTP}$ is solved.

In particular, let $V_x = \{i,j | x_e = 1, e = (i,j)\}$ be the set of nodes associated with the activated edges in solution x. G_x is defined as $G_x(V_x, E_x)$, where $E_x = \{e = (i,j) \in E | i,j \in V_x\}$. Set E_x contains all edges incident to nodes belonging to V_x.

4   Computational Results

In this section, we evaluate the performance of the proposed approach. It is implemented in Java language and it is tested on an Intel(R) Core(TM) i7-4710HQ 2.50 GHz 16 GB RAM.

We have generated instances of $\mathcal{LCSTP}$ starting from the SteinLib benchmarks (http://steinlib.zib.de). In particular, we have considered sparse graphs with random weights (B), sparse graphs with incidence weights ($I080$), complete graphs with random weights ($P4Z$), and grid graphs with holes, L1 weights (MSM).

We define $K = \alpha * \bar{K}$, where $\bar{K}$ is the number of edges in the optimal solution of the minimum hops path problem. The value of α is chosen in the set $\{0, 0.2\}$.

Summarizing, we have 36 instances of type B, 200 instances of type $I080$, 16 instances of type $P4Z$, and 36 instances of type MSM.

The multiplier update procedure is interleaved by sub-gradient optimization. In particular, the algorihms performs, 4 times, 30 multiplier updates and 60 sub-gradient iterations and ends by executing the multipliers updating procedure (30 iterations). Thus, the total number of performed iterations is equal to 390 (i.e., (30+60)*4+30). The value of ϵ is set equal to 0.1 and 10.

Table 1 shows the average numerical results obtained for the undirected instances, that is B, $I080$, $P4Z$, and MSM. In particular, we report the average upper bound under column UB, the optimality gap under column gap, the execution time in seconds to perform 390 iterations under column time, the iteration in which the best upper bound is found under column iterUB, and the execution time in seconds to find the best upper bound under column timeUB.

The subscript associated to the value of UB indicates the number of instances for which the best upper bound is the optimal solution. A lower bound is derived by solving to optimality the linear relaxation of problem (1)–(6).

The numerical results underline that the proposed approach is able to obtain the optimal solution for the 97.2 % of the instances generated from B, the 75.5 % of those generated from $I080$, the 100.0 % of those generated from $P4Z$, and the 65.3 % of those generated from MSM. It is worth observing that the percentages are under estimated, since the optimality is certified only by the gap. On average, the gap is around 4 % with a peack of 16 % for the instances generated

Table 1. Average results for undirected networks.

ϵ	Test	UB	Gap	Time	iterUB	timeUB
0.1	B	140.36^{35}	0.0003	57.79	7.58	1.34
	I080	3109.69^{151}	0.0018	619.47	33.73	21.42
	P4Z	1115.19^{16}	0.0000	2999.74	6.31	70.58
	MSM	834.67^{25}	0.1600	2016.91	188.92	982.17
AVG		1299.98	0.0405	1423.48	59.13	268.88
10	B	140.36^{35}	0.0003	61.31	8.39	1.59
	I080	3109.28^{152}	0.0017	594.57	26.22	21.14
	P4Z	1115.19^{16}	0.0000	2855.40	4.94	70.52
	MSM	834.44^{22}	0.1598	2138.27	163.53	902.06
AVG		1299.82	0.0404	1412.39	50.77	248.83

from MSM. On average, the best upper bound is obtained at about the 55-th iteration.

Referring to ϵ, Table 1 shows a better behaviour of the algorithm for $\epsilon = 10$ than that observed for $\epsilon = 0.1$. Indeed, it is required a less computational effort and a smaller number of iterations to obtain the best upper bound with $\epsilon = 10$ than in case $\epsilon = 0.1$.

5 Conclusions

In this paper, we have studied a constrained version of the $\mathcal{STP}$. In particular, an upper bound on the number of links is imposed.

The problem is solved via a Lagrangean relaxation approach. A dual ascent update multipliers procedure is defined for solving the dual Lagrangean problem. Infeasible solutions are repaired by solving a restricted instance of the $\mathcal{LCSTP}$ at each iteration.

The proposed approach has been tested on instances derived from $\mathcal{STP}$ benchmarks defined over undirected graphs.

The computational results highlight the goodness of the relaxed approach that provides the optimal solution for more than a half of the considered instances

References

1. Burdakov, O., Kvarnstrom, J., Doherty, P.: Local search heuristics for hop-constrained directed Steiner tree problemy. In: Examining Robustness and Vulnerability of Networked Systems, IOS Press (2014)
2. Ding, W., Lin, G., Xue, G.: Diameter-constrained Steiner tree. In: Wu, W., Daescu, O. (eds.) COCOA 2010, Part II. LNCS, vol. 6509, pp. 243–253. Springer, Heidelberg (2010)

3. Garey, M.R., Graham, R.L., Johnson, D.S.: The complexity of computing Steiner minimal trees. SIAM J. Appl. Math. **32**(4), 835–859 (1977)
4. Gouveia, L., Magnanti, T.L.: Network flow models for designing diameter-constrained minimum-spanning and Steiner trees. Networks **41**(3), 159–173 (2003)
5. Hwang, F.K., Richards, D.S.: Steiner tree problems. Networks **22**(1), 55–89 (1992)
6. Kang, J., Kang, D., Park, S.: A new mathematical formulation for generating a multicast routing tree. Int. J. Manag. Sci. **12**(8), 63–69 (2006)
7. Kang, J., Park, K., Park, S.: Optimal multicast route packing. Eur. J. Oper. Res. **196**, 351–359 (2009)
8. Leggieri, V., Haouari, M., Layeb, S., Triki, C.: The Steiner tree problem with delays: a tight compact formulation and reduction procedures. Technical report, University of Salento, Lecce (2007)
9. Leggieri, V., Haouari, M., Triki, C.: An exact algorithm for the Steiner tree problem with delays. Electr. Notes Discrete Math. **36**, 223–230 (2010)
10. Oliveira, C.A.S., Pardalos, P.M.: A survey of combinatorial optimization problems in multicast routing. Comput. Oper. Res. **32**, 1953–1981 (2005)
11. Voß, S.: The Steiner tree problem with hop constraints. Ann. Oper. Res. **86**, 321–345 (1999)

The Pycao Software for 3D-Modelling

Laurent Evain[✉]

The Pycao Software, University of Angers, Angers, France
laurent.evain@math.univ-angers.fr

Abstract. Describing a three dimensional object requires a computer code whose maintenance is difficult. Part of the problem is the gap between the 3D-software languages based on coordinates and the natural geometric description of the object which is primarily coordinate free.

Pycao is a software built to reduce the gap between the natural language and the software language in 3D-modelisation. The Pycao language is designed to avoid coordinates as much as possible. The available concepts include CSG geometry, the framework of affine geometry in a massic space, an intuitive "box model", and several systems of measurements which mimic the operations in the workshop. It is developed as a python module to get a code compact and easy to read.

Keywords: Modeller · 3D-software

1 Introduction : Overview of the Problematic

A scene can be illustrated using a "natural language", or using coordinates. To illustrate the difference between the two approaches, consider the situation of the simple image below: a room with a floor, two walls, a table, and a cylindrical flower pot at the center of the table.

The scene may be described with usual words without ambiguity using "the carpenter paradigm", that is with a dynamic description telling how simple objects could be assembled to reproduce the geometry: the carpenter cuts parallelepipeds of appropriate size for the legs and the tray of the table. The legs are positioned and glued on the tray corners. The table is oriented with the legs down and moved to its final position. Finally, the flower pot is placed on the center of the table. This description uses few numbers besides the dimensions of the objects and the location of the table.

In contrast, the code for the existing software often includes a lot of numbers. The wavefront .obj [OBJ] format is a quite standard specification which illustrates this remark. The code is basically a long list of coordinates corresponding to the positions of the material points which characterize the objects. Other specifications [IGES, STL] are also focused on the manipulations of coordinates.

The carpenter description and the .obj description are in some sense opposite. The carpenter description is a short geometrical high-level description.

© Springer International Publishing Switzerland 2016
G.-M. Greuel et al. (Eds.): ICMS 2016, LNCS 9725, pp. 498–504, 2016.
DOI: 10.1007/978-3-319-42432-3_64

The vision is dynamic in the sense that objects are moved and assembled. The .obj is a lengthy coordinate based low-level description of motionless objects.

Obviously, it is difficult for a human to read or produce a .obj file without a modeller. The existing software includes tools to simplify the coordinates manipulations, introducing various concepts such as geometrical transformations, grouping, CSG operations, or genealogy systems to improve readability. But, as far as I know, no systematic effort has been done to avoid coordinates as much as possible and to reduce the code to its minimal possible size. The language is usually a mix between a coordinate description of the objects and some paradigms to simplify the description. Cad software is usually oriented towards versatility and power through elaborated interfaces rather than readability [SALOME].

The programs do not come to a consensus on the relevant paradigms. Blender and Povray [BL, POV] are for instance two software applications which are quite different in their approach and philosophy . An effort of the scientific community is still required to reach consensus. In some sense, one looks for an analog of the SQL language in the realm of 3D-modelisation which could be universal and shared among different software.

The architecture could ideally be closed to the following form:

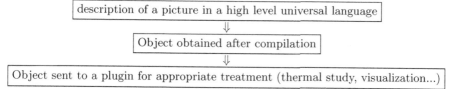

The Pycao software is developed as a step in this direction. The focus is on a coordinate free description of 3D-objects and on the minimization of the size of the descriptive code. There is at present a unique plugin, which allows the visualization of the scene using Povray.

A final word on graphical interfaces. Some software, like Blender and Sketchup [BL, SK] and many others, have graphical modellers. This graphical approach makes it difficult to connect with other software components in a scientific environment. Even on a standalone project, the unavoidable difficulty is the localization of a point in a 3D-space with a 2D-screen. Selecting a point requires an orientation of the camera, zooming, removing objects which block the view of the target, switching to an other view to define the required point as an intersection of two lines. Pycao does not address these problems. We look for an efficient text description of a 3D-scene in the form of a formalized language. We do not consider the graphical interface problematics, although graphical interfaces could be compatible with the Pycao language.

I warmly thank the developers of Blender and Povray. I learned many things from their software.

2   Example

To see Pycao concretely in action, consider the following code with the comments embedded in the code. This is the code to describe the above table with the flower pot.

```
# Some skipped preamble
###############

# The dimensions/constants used are in the next 6 lines
###############
tableTrayDimensions=Vector(1,.5,.05)
tableLegDimensions=Vector(.2,.03,.8)
placementVector=X+Y
flowerPotRadius=0.1
flowerPotHeight=0.3
flowerPotThickness=0.01

# Describing the scene starts here
###############
ground=plane(-Z,origin) # a plane with normal vector Z=(0,0,1) through the origin
wall1=plane(X,origin)
wall2=plane(Y,origin)
tableTray=Cube(tableTrayDimensions)
tableLeg1=Cube(tableLegDimensions)
tableLeg2=tableLeg1.copy()
tableLeg3=tableLeg1.copy()
tableLeg4=tableLeg1.copy()

# The next 2 lines correspond to the movement of tableLeg1 to the
# corner of the tray  and to the bonding of the leg on the tray
###############
tableLeg1.move_against(tableTray,Z,Z,X,X,adjustEdges=-X-Y)
tableLeg1.transplant_on(tableTray)
tableLeg2.move_against(tableTray,Z,Z,X,X,adjustEdges=X+Y)
tableLeg2.transplant_on(tableTray)
tableLeg3.move_against(tableTray,Z,Z,X,X,adjustEdges=-X+Y)
tableLeg3.transplant_on(tableTray)
tableLeg4.move_against(tableTray,Z,Z,X,X,adjustEdges=X-Y)
tableLeg4.transplant_on(tableTray)

# The tray is moved and the legs follow because of the bonding
###############
bottomOfTheLeg1=tableLeg1.point(0,0,0,"aap")
tableTray.translate(origin-bottomOfTheLeg1) # vertical move: legs on the floor
tableTray.translate(placementVector) # horizontal move

# The flower pot is described as a difference between 2 cylinders then
# placed on the table
###############
flowerPot=Cylinder(origin,origin+flowerPotHeight*Z,radius=flowerPotRadius)
toCut=Cylinder(origin+flowerPotThickness*Z,origin+2*flowerPotHeight*Z,radius=flowerPotRadius-flowerPotThickness)
flowerPot.amputed_by(toCut)
topCenterOfTable=tableTray.point(0.5,0.5,1,"ppp")
flowerPot.translate(topCenterOfTable-origin)

# Some color and cameras parameters skipped
###############

# Calling the plugins for rendering
###############
camera.shoot # takes the photo, i.e. creates the Povray file called camera.file
camera.show # shows the photo, i.e. calls Povray to render camera.file
```

3   Comparison with Existing Software and Objectives for Pycao

Comments below are relative to Blender and Povray [BL,POV] which are the leading 3D-software projects in the open source community. Other software applications seem interesting from their documentation. The absence of comments just indicates that I have not used them actively enough.

Blender comes both with a python API and a graphical interface. This complementarity is useful as it makes it easy to check graphically the objects produced by the code. The API is not designed to be a universal language for 3D description. It is a Python access to the low level layers of the Blender code. In particular, there are limitations in the API, which arise as "context errors". The context errors correspond to gates to preserve the compatibility between the graphical interface and the python API, since both of them may simultaneously access the low level objects of Blender. It is usually very difficult to circumvent these errors.

The Povray language is a language with a very complete and clear documentation. On the other hand, the Povray language is a very low level language. The syntax is rigid, including many curly braces. It is a much more low level language than FORTRAN for instance. It is painful to produce and maintain the required code. No complaint about Povray which achieves its mission: Povray is a ray Tracer rather than a modeller. The language is built to feed the ray Tracer, not to simplify the work of the developer.

The Pycao projects started in a applied research project whose goal is to test the performance of innovative bikes with a geometry very different from the usual bikes. After building prototypes, a more precise 3D-modelisation of the bikes was needed before building them. The Pycao project is an attempt to construct the tools that I missed for these steps: a language with a rich vocabulary, focused on the simplification of the description, documented, free from any graphical interface and from any integrated environment. Formalizing these needs, the goal of Pycao is to offer a language with the following characteristics by order of decreasing importance.

- a readable code, easy to check and maintain in a high level language
- a documentation with precise mathematical descriptions of the tools
- a language allowing the shortest possible description of the scene in terms of number of lines
- a language of moderate size, avoiding redundancies or rarely used features, usable by people who are not full time cad developers
- a strict separation between the description of the scene, and the plugins using the scene (in particular visualization is an autonomous plugin)
- a possibility to build and share library of objects among users.

Pycao is written in Python, which is a language allowing rapid development, and thus integrates well with the philosophy of Pycao. An other advantage of Python is the access to a large number of mathematical software (numpy, scipy, matplotlib, sage).

4   Technical Details

4.1   Massic Space and Affine Geometry

The framework of affine geometry is implemented in Pycao. Points and vectors are objects with different types: the point $(1, 1, 1)$ and the vector $(1, 1, 1)$ are distinct. This allows compilation checking: A translation requiring a vector as an input will raise an exception if the input is a point.

The classical arithmetic rules involving points and vectors are possible : the addition $p + v$ of a point p and a vector v is a point, and the difference $p_1 - p_2$ between two points p_1 and p_2 is a vector. This arithmetic provides flexibility for the definition of points and vectors. But it also speeds up the input process. Here are some examples:

- If f is a transformation of the affine space with underlying linear map ϕ_f, and if v is a vector, then $f(v)$ returns $\phi_f(v)$ which is the only sensible operation in this context
- Consider a curve defined using a list of points as an input. The input points can be described in absolute terms or relative terms (with respect to the preceding point using a vector). As an example, if p_1, p_4 are points and v_1, v_2 are vectors, then the lists $[p_1, p_2, p_3, p_4]$ and $[p_1, v_1, v_2, p_4]$ define the same curve if $p_2 = p_1 + v_1$ and $p_3 = p_2 + v_2$.
- Barycenters are constructed using the natural arithmetic without calling specific functions. If p_1 and p_2 are points, then $p_3 = 0.3p_1 + 0.7p_2$ is a well defined point whereas $v = 0.3p_1 - 0.3p_2 = 0.3(p_1 - p_2)$ is a well defined vector

To keep a moderate size of the Pycao code, we have unified the constructions on the points and the vectors by considering the massic space, which is a 4-dimensional vector space containing both the 3-dimensional affine space and the 3-dimensional vector space. This massic space is sometimes considered in some textbooks [V] This unification makes the cost of implementation of this affine framework very low. For instance, both a change of frame in the affine space and a base change in the vector space can be realized with appropriate identifications as base changes in the 4-dimensional vector space.

To speak concretely, a point $p = (x, y, z, t)$ of the massic space has weight t by definition. If $t = 0$, then p is identified with the 3D-vector with coordinates (x, y, z). If $t = 1$, p is identified with the point with coordinates (x, y, z). These identifications are only useful in the dialog between Pycao and the end-user. Behind the scene, triples of coordinates are never used. Pycao manipulates only massic points with four coordinates and all the computations are performed in the 4-dimensional massic space.

4.2   Boxes and Frames Everywhere

Some objects in Pycao are called frameBoxes. They are used both as a box to carry objects and as frames to describe position of points. FrameBoxes are attached to objects. They are automatically created along with primitive objects.

They follow the objects in their moves and vice-versa. We may think a Framebox as an imaginary box around the object.

To move objects to their position, we carry the objects in their Frameboxes. The instruction

$$tableLeg1.move\_against(tableTray, Z, Z, X, X, adjustEdges = -X - Y)$$

corresponds to a move of the box containing the tableLeg1 against the box containing the table. A similar instruction makes sense for non cubic objects since it applies to the frameBoxes, not to the object itself.

The geometry of pair of boxes is intuitive, however it includes many possibilities. If the first box is kept fixed, and if the center of a second box is kept fixed, then there are 24 ways to move the second box parallel to the first. When the dimension of the faces are different, there are several ways to move a box against an other: aligning the centers of the faces of contact or aligning the borders, with possible offsets. When 2 faces have a contact in their center, there are still two possibilities, since the vectors normal to the faces may be positively or negatively proportional. All these possibilities are formalized as options in the *move_against* method.

Frameboxes are used to make measurements. There are several possibilities to proceed corresponding to the natural measurements done in a workshop. A point may be 5 cm from the left face ("absolute measurement") or 5 cm from the right face ("negative measurement") or at equal distance of the two faces ("proportional measurement"). In other words, Frameboxes carry various systems of coordinates corresponding to different strategies of measurement. As an example, in the instruction

$$topCenterOfTable = tableTray.point(0.5, 0.5, 1, "ppp")$$

the option "ppp" means "p(roportional)-p(roportional)-p(roportional)", i.e. all three axes use proportional measurement.

4.3   Markers

Markers are objects automatically created with objects to simplify the positioning of objects. For instance, once we have built a cylinder, it is sometimes handful to consider the central segment to perform intermediate computations. We could let the developer define the segment himself with a few lines of code. But the philosophy of Pycao is to shorten the code length. On the other hand, if we define the segment automatically with each cylinder, and more generally if we define many markers, this creates a long list of unnecessary variables partially recomputed at each movement of an object.

The strategy followed in Pycao to solve this dilemma is to define the markers as functions. With many cylinders, a unique segment function is necessary for all the cylinders, which does not change when the cylinders move. A simple call to the function computes the segment on demand when it is needed in the computations.

In Pycao, we introduce a marker for each interesting characteristic element which cannot be defined with direct readable code.

4.4 Unions and Intersections

Pycao supports two kinds of union, both classical. The first union (the genealogy system) is asymmetric. When the child moves, the parent stays fixed, but when the parent moves, the child moves and follows (recursively, the child of the child follows the move, and so on). The colors/material of each part are defined independently. This type of union is important for animations. For instance, suppose that we modelize a bike movement using a differential equation and that we want to visualize the results. We need an asymmetric link between the objects to be able to move the wheel (the child) without moving the entire bike (the parent).

The other type of union (the Compound class) is symmetric. Objects are aggregated in a compound. Any change on the compound affects each of its object. As compounds may be nested, this is a useful concept to build libraries.

4.5 Visibility

Objects (including the camera) carry a visibility index. An object with no parents and no children is seen by the camera if its visibility is at least the visibility of the camera. This concept helps to concentrate on some parts of the scene.

There are some rules elaborated to make the behavior natural when a genealogy system, or unions, intersection, differences are involved. For instance, to minimize the changes required in the code to display a small part of the scene, it is natural to hide the children if the parents are not visible.

References

[OBJ]   https://en.wikipedia.org/wiki/Wavefront_.obj_file
[IGES]   https://en.wikipedia.org/wiki/IGES
[STL]   https://en.wikipedia.org/wiki/STL_(file_format)
[SALOME] http://www.salome-platform.org/user-section/about/geometry
[BL]   https://www.blender.org/manual/modeling/index.html
[POV]   http://www.povray.org/documentation/3.7.0/
[V]   Vienne, L.: Prsentation algbrique de la gomtrie classique (1996). ISBN:978-2711788385
[SK]   https://www.sketchup.com/

Normal Forms for Operators via Gröbner Bases in Tensor Algebras

Jamal Hossein Poor, Clemens G. Raab, and Georg Regensburger$^{(\boxtimes)}$

Johann Radon Institute for Computational and Applied Mathematics (RICAM),
Austrian Academy of Sciences, Linz, Austria
{jamal.hossein.poor,clemens.raab,georg.regensburger}@ricam.oeaw.ac.at

Abstract. We propose a general algorithmic approach to noncommutative operator algebras generated by linear operators using quotients of tensor algebras. In order to work with reduction systems in tensor algebras, Bergman's setting provides a tensor analog of Gröbner bases. We discuss a modification of Bergman's setting that allows for smaller reduction systems and tends to make computations more efficient. Verification of the confluence criterion based on S-polynomials has been implemented as a Mathematica package. Our implementation can also be used for computer-assisted construction of Gröbner bases starting from basic identities of operators. We illustrate our approach and the software using differential and integro-differential operators as examples.

Keywords: Operator algebra · Tensor algebra · Noncommutative Gröbner basis · Reduction systems

1 Introduction

For an algorithmic treatment of many common operator algebras, like differential and difference operators, skew polynomials are a well-established algebraic construction; see e.g. the survey [4] describing also implementations in computer algebra systems. However, not all common operator algebras are covered by this setting. For example, integral operators cannot be constructed that way.

The principle that can always be applied is construction by generators and relations. In practice, normal forms are needed for effective computation. Finding and proving the structure of normal forms is a difficult task, the general problem is even undecidable. For skew polynomials, normal forms are given by the standard polynomial basis. For tensor algebras, Bergman's paper [1] also provides a framework in which reduction systems and corresponding normal forms can be analyzed, analogous to Gröbner bases. Tensor algebras can be seen as a generalization of free noncommutative polynomial algebras and inherit all their algorithmic obstructions. At the same time, parts of the tensor setting can be automated, in particular, verification of the confluence criterion and subsequent

All authors were supported by the Austrian Science Fund (FWF): P27229.

G.-M. Greuel et al. (Eds.): ICMS 2016, LNCS 9725, pp. 505–513, 2016.
DOI: 10.1007/978-3-319-42432-3_65

computations with normal forms. We provide an implementation in the Mathematica package `TenReS` including our generalization of Bergman's setting [6]: http://gregensburger.com/software/TenReS.zip.

When representing linear operators as tensors, composition of operators is modeled by the tensor product. We briefly describe the main building blocks of the algebraic construction. Over some K-module of basic operators M, we consider the tensor algebra

$$K\langle M \rangle = \bigoplus_{n=0}^{\infty} M^{\otimes n},$$

where K is commutative ring with unit element. Operator identities are encoded as a reduction system Σ for $K\langle M \rangle$. Then the operator algebra is constructed as

$$K\langle M \rangle / I_\Sigma,$$

where I_Σ is a two-sided ideal induced by the reduction system Σ. This setting allows for finite reduction systems even in cases where the module M does not have a finite basis. As our examples illustrate, this approach does not make use of a basis of M at all.

2   Reduction Systems for Tensor Algebras

Using the well-known example of differential operators, we explain the main theoretical notions as well as the most important commands of our package. Usually one defines the differential operators directly via skew polynomials with normal forms $\sum f_i \partial^i$ and noncommutative multiplication defined by

$$\partial f = f \partial + f'.$$

Suppose we do not already know the normal forms of differential operators and we just have the definition of the derivation ∂ as a K-linear operator on some differential K-algebra (R, ∂) obeying the Leibniz rule $\partial f g = (\partial f)g + f \partial g$. (Note that we use operator notation in this paper.) Recall that differential operators with polynomial coefficients (Weyl algebra) over a field $K \supseteq \mathbb{Q}$ can be defined as the quotient algebra $K\langle X, D \rangle / (DX - XD - 1)$ of the free noncommutative polynomial algebra $K\langle X, D \rangle$ modulo the two-sided ideal $(DX - XD - 1)$. Now, we want to do an analogous construction for the differential operators with coefficients in an arbitrary differential K-algebra (R, ∂). To this end, we work in the tensor algebra over the K-module $M = R \oplus K\partial$. First, we explain some main points informally before explaining necessary technical details later.

We interpret elements $f \in R$ as multiplication operators, ∂ as the derivative operator on R. Then we have three basic identities between these operators: Multiplication by $1 \in R$ acts like applying no operator at all. So we can replace it by the empty tensor ϵ, which represents the unit element in the tensor algebra. For $f, g, h \in R$ with $fg = h$ the multiplication operators

$$f \otimes g \quad \text{and} \quad h$$

act in the same way on R. Finally, the Leibniz rule implies that the operators

$$\partial \otimes f \quad \text{and} \quad f \otimes \partial + \partial f$$

act in the same way as well. Deciding to move the differential operators to the right results in the following reduction rules to simplify parts of tensors representing a composition of multiplication operators and the derivative operator:

$$1 \mapsto \epsilon, \quad f \otimes g \mapsto fg, \quad \text{and} \quad \partial \otimes f \mapsto f \otimes \partial + \partial f.$$

A priori it is not clear that we will always end up with the same result if we apply the reduction rules in different ways. Suppose we have a tensor of the form

$$f_1 \otimes f_2 \otimes f_3$$

with $f_1, f_2, f_3 \in R$ arbitrary but fixed. Then we can apply the reduction rule for composition of multiplication operators in different ways obtaining $(f_1 f_2) \otimes f_3$ and $f_1 \otimes (f_2 f_3)$. Trivially, another application of the same rule yields $f_1 f_2 f_3$ in both cases. In general, a minimal case where two (not necessarily distinct) rules can be applied differently to a tensor is called an *ambiguity* and the difference

$$(f_1 f_2) \otimes f_3 - f_1 \otimes (f_2 f_3)$$

is called the corresponding *S-polynomial*. If all S-polynomials of an ambiguity can be reduced to zero by the reduction rules, like above for all f_1, f_2, f_3, then we call the ambiguity *resolvable*.

Analogous to Buchberger's criterion for Gröbner bases [3] and the Composition-Diamond Lemma for Gröbner-Shirshov bases [2], we have a confluence criterion for tensor reduction systems due to Bergman [1]. A reduction system defines unique normal forms if and only if all ambiguities are resolvable. Termination of the reduction process, depends on a compatible Noetherian ordering on words; see [1,6] for details. Throughout this paper and in the package, we tacitly assume that such an ordering exists.

One can distinguish four different types of ambiguities: overlaps and inclusions, each with or without specialization. The ambiguity above is an overlap ambiguity since the factors $f_1 \otimes f_2$ and $f_2 \otimes f_3$ of the tensor $f_1 \otimes f_2 \otimes f_3$ on which the rules act overlap. There is no specialization involved since all cases on which the rules may act actually arise in this way. On the other hand, the tensor

$$\partial \otimes c$$

with $c \in R$ and $\partial c = 0$ may be reduced by the rules $1 \mapsto \epsilon$ or $\partial \otimes f \mapsto f \otimes \partial + \partial f$ to either $c\partial$ or $c \otimes \partial$. Obviously, another application of $1 \mapsto \epsilon$ in the second case gives $c\partial$ as well, so this ambiguity is resolvable again. In this case one factor c of $\partial \otimes c$ on which one rule acts is contained in the other factor $\partial \otimes c$ acted on by the other rule. So we call it an inclusion ambiguity. Moreover, not all cases of the rule $\partial \otimes f \mapsto f \otimes \partial + \partial f$ are needed here, just $f \in K \subsetneq R$, so we say that this ambiguity involves specialization.

2.1   Tensor Algebras

For computation in the tensor algebra $K\langle M \rangle$ over the K-module M, we need to be able to check when objects are contained in K. The user has to implement the function `CoeffQ` returning `True` whenever its argument is contained in K. Based on that, tensors $m_1 \otimes \cdots \otimes m_n \in K\langle M \rangle$ are represented as `Prod[m₁,...,mₙ]` respecting K-multilinearity of the tensor product.

In order to fix domains for the reduction rules, we need corresponding direct sum decompositions of the module M indexed by some finite set (alphabet). In our example above, we define $M_F = R$ and $M_D = K\partial$ so that

$$M = M_F \oplus M_D.$$

Hence, in terms of the word monoid $\langle Y \rangle$ over the alphabet $Y = \{F, D\}$, we have the following decomposition of the tensor algebra into modules $M_W = M_{y_1} \otimes \cdots \otimes M_{y_n}$ for $W = y_1 \ldots y_n \in \langle Y \rangle$:

$$K\langle M \rangle = \bigoplus_{W \in \langle Y \rangle} M_W.$$

For each submodule M_F and M_D of M, the user has to implement a membership test `MemberQ_F` and `MemberQ_D` indexed by the corresponding letter of the alphabet. Moreover, the user has to implement all computations with elements of each module. In particular, this applies to additional operations on a module, like in our example multiplication and derivation on M_F respecting the Leibniz rule in R. For computation with S-polynomials, we need to compute with general elements of non-cyclic modules. For example, in the module M_F, general elements will always be called `F[1]`, `F[2]`, ..., which together with their derivatives also have to be recognized by the membership test.

In order to formalize the rule $1 \mapsto \epsilon$, we need a refinement of the above decomoposition of M by choosing a complement $M_{\tilde{F}}$ of $M_K = K$ such that

$$M_F = M_K \oplus M_{\tilde{F}}.$$

Of course, also for the new submodules, membership tests have to be implemented. All such refinements of submodules have to be stored in the variable

$$\texttt{Specialization=\{F}{\rightarrow}\texttt{\{K,}\tilde{\texttt{F}}\texttt{\}\}}.$$

Altogether, with $X = \{K, \tilde{F}, D\}$ we have two decompositions of M:

$$M = \bigoplus_{x \in X} M_x = \bigoplus_{y \in Y} M_y.$$

For all cyclic submodules the user should store the letter and the generator as a pair in the list `CyclicModules`. In our example we denote ∂ by `Diff`, so that

$$\texttt{CyclicModules=\{\{K,1\},\{D,Diff\}\}}.$$

2.2 Reduction Systems

With the submodules introduced above, the reduction rule $f \otimes g \mapsto fg$ can be formally defined as the pair (FF, h_{FF}) consisting of the word $FF \in \langle Y \rangle$ and the module homomorphism $h_{FF} : M_{FF} \to M_F \subset K\langle M \rangle$. The homomorphism h_{FF} is defined by $f \otimes g \mapsto fg$ and could be implemented by the user as $h_{FF}[f_-, g_-] := Prod[mul[f, g]]$, for example. Similarly, we have the reduction rule (DF, h_{DF}), where $h_{DF} : M_{DF} \to M_{FD} \oplus M_F$ may be implemented as $h_{DF}[Diff, f_-] := Prod[f, Diff] + Prod[Diff[f]]$. Thanks to the refinement, we also can define the reduction rule $1 \mapsto \epsilon$ as the pair (K, h_K) with the homomorphism $h_K : M_K \to M_\epsilon$ ($h_K[1] := Prod[]$), where ϵ is the empty word. These homomorphisms have to be implemented by the user.

Our reduction system over the *combined alphabet* $Z = X \cup Y$ is given by

$$\Sigma = \left\{ (FF, f \otimes g \mapsto fg), \quad (DF, \partial \otimes f \mapsto f \otimes \partial + \partial f), \quad (K, 1 \mapsto \epsilon) \right\}$$

The corresponding definition for our package is

$$\texttt{RedSys} = \{\{\{F,F\}, h_{FF}\}, \{\{D,F\}, h_{DF}\}, \{\{K\}, h_K\}\}.$$

In general, any reduction rule $r = (W, h)$, where h is a K-module homomorphism $h : M_W \to K\langle M \rangle$ reduces tensors $a \otimes w \otimes b$ with $a \in M_A$, $w \in M_W$, and $b \in M_B$ for some $A, B \in \langle Z \rangle$ by $a \otimes w \otimes b \to_r a \otimes h(w) \otimes b$. In other words, similarly to polynomial reduction, we "replace" the "monomial" w by the "tail" $h(w)$ given by the homomorphism h. The *reduction ideal* induced by a reduction system Σ is defined as the two-sided ideal

$$I_\Sigma := (t - h(t) \mid (W, h) \in \Sigma \text{ and } t \in M_W) \subseteq K\langle M \rangle.$$

If one tensor can be reduced to another, then their difference is contained in I_Σ. This is implemented via the command $\texttt{ApplyRules}$. For example, we can reduce $f_1 \otimes \partial \otimes f_2$ to $(f_1 f_2) \otimes \partial + f_1 \partial f_2$ by the reflexive-transitive closure $\xrightarrow{*}_\Sigma$:

```
ApplyRules[Prod[F[1], Diff, F[2]], RedSys]
Prod[mul[F[1], Diff[F[2]]]] + Prod[mul[F[1], F[2]], Diff]
```

2.3 Normal Forms and Confluence

Determination of normal forms and ambiguities can be reduced to problems in the word monoid $\langle Z \rangle$ over the combined alphabet. We introduce the notion of specializing a word $W \in \langle Z \rangle$ to a word in $\langle X \rangle$ by replacing all letters of W from $Y \setminus X$ by corresponding letters from X. For example, the specializations of the word $FFD \in \langle Z \rangle$ are given by $\{KKD, K\tilde{F}D, \tilde{F}KD, \tilde{F}\tilde{F}D\} \subset \langle X \rangle$. In terms of modules, we then have that M_W is the direct sum of all M_V such that V is a specialization of W.

The module $K\langle M \rangle_{irr}$ of *irreducible tensors*, i.e. tensors that cannot be reduced by any reduction rule in Σ, is determined by the set of *irreducible words* $\langle X \rangle_{irr} \subseteq \langle X \rangle$ via

$$K\langle M \rangle_{irr} = \bigoplus_{W \in \langle X \rangle_{irr}} M_W.$$

Irreducible words are those words over the refined alphabet X that do not contain a subword that is a specialization of the word W of any rule $(W, h) \in \Sigma$. In our example one easily sees that the irreducible words are given by D^j and $\tilde{F}D^j$ with $j \in \mathbb{N}_0$. Consequently, irreducible tensors are of the form $\partial^{\otimes j}$ and $f \otimes \partial^{\otimes j}$ with $j \in \mathbb{N}_0$ and $f \in M_{\tilde{F}}$ and K-linear combinations thereof.

In order to show that irreducible tensors already define a normal form of tensors in $K\langle M \rangle$ modulo the reduction ideal I_Σ, we need to verify that the S-polynomials of all ambiguities can be reduced to zero. In our example, there are 5 ambiguities, which we may informally denote by $F\underline{F}F$, $D\underline{F}F$, $\underline{K}F$, $F\underline{K}$, and $D\underline{K}$. Two of them ($F\underline{F}F$ and $D\underline{K}$) already have been dealt with above. The S-polynomials associated to the remaining ambiguities are given by

$$h_{DF}(\partial \otimes f) \otimes g - \partial \otimes h_{FF}(f \otimes g) = f \otimes \partial \otimes g + (\partial f) \otimes g - \partial \otimes (fg)$$
$$h_{FF}(c \otimes f) - h_K(c) \otimes f = 0$$
$$h_{FF}(f \otimes c) - f \otimes h_K(c) = 0$$

for all $c \in M_K$ and $f, g \in M_F$. The first "family" can be reduced to zero by the rules (FF, h_{FF}) and (DF, h_{DF}) and the others are zero anyway.

Using the commands `ExtractReducibleWords`, `GenerateAmbiguities`, `SPoly`, and `ApplyRules` of the package we can verify that all S-polynomials reduce to zero in the following way.

```
ambiguities = GenerateAmbiguities[ExtractReducibleWords[RedSys]]
```
```
{Overlap[{F, F, F}, {F}, {F}],
 Overlap[{D, F, F}, {F}, {D}], SpecialInclusion[{K, F}, {}, {F}],
 SpecialInclusion[{F, K}, {F}, {}], SpecialInclusion[{D, K}, {D}, {}]}
```

```
spolys = Map[SPoly[#, RedSys] &, ambiguities]
```
```
{-Prod[F[1], mul[F[2], F[3]]] + Prod[mul[F[1], F[2]], F[3]],
 -Prod[Diff, mul[F[1], F[2]]] + Prod[Diff[F[1]], F[2]] + Prod[F[1], Diff, F[2]],
 0, 0, -Prod[Diff] + Prod[1, Diff]}
```

```
ApplyRules[spolys, RedSys]
```
```
{0, 0, 0, 0, 0}
```

This process is also available through the command `CheckResolvability`, which returns a list of all ambiguities that are not resolvable together with the reduced from of their S-polynomials, see also its application in the following section. If `CheckResolvability[RedSys]` returns the empty list, then the irreducible tensors w.r.t. the reduction system given by `RedSys` really are normal forms.

3   Applications

In this section, we illustrate how to use the package for constructing a confluent reduction system starting from a given one. For tensor reduction systems, the computer-assisted process is heuristic and users can proceed in different ways. In each step, we add new rules to the reduction system based on S-polynomials, similar to Buchberger's algorithm for computing Gröbner bases [3] and Knuth-Bendix completion [7].

As an example, we consider the algebra of integro-differential operators. Based on the free noncommutative polynomial algebra using a basis of the "function" algebra, it was introduced in [8,9] to study boundary problems; see also [10] for an automated confluence proof relying on free integro-differential algebras. First, we recall the definition of an integro-differential algebra [5,9].

Definition 1. *Let K be a commutative ring and let (R, ∂) be a differential K-algebra such that $1 \in R$ and $\partial R = R$. Moreover, let $\int: R \to R$ be an K-linear operation on R such that*

$$\partial \int f = f$$

for all $f \in R$. Then we call $(R, \partial, \int)$ an integro-differential algebra *over K if the* evaluation E: $R \to K$ *defined by* $E = \mathrm{id} - \int \partial$ *is multiplicative, i.e. for all $f, g \in R$ we have $Efg = (Ef)Eg$.*

We fix an integro-differential K-algebra $(R, \partial, \int)$ with ring of constants K and evaluation $E = \mathrm{id} - \int \partial$. Recall from [5] that in any integro-differential algebra, we have the direct sum decomposition

$$R = K \oplus \int R$$

into constant and non-constant "functions". We consider the corresponding K-modules $M_K = K$ and $M_{\tilde{F}} = \int R$. Note that the elements of M_K and $M_{\tilde{F}}$ are not interpreted as functions but as multiplication operators induced by those functions. For the K-linear operators ∂, $\int$, and E we consider the free modules $M_D = K\partial$, $M_I = K\int$, and $M_E = K E$ generated by them. Now, let

$$M = M_F \oplus M_D \oplus M_I \oplus M_E$$

with $M_F = M_K \oplus M_{\tilde{F}}$ and alphabets $Y = \{F, D, I, E\}$ and $X = \{K, \tilde{F}, D, I, E\}$. In order to compute with these operators we need to collect the identities they satisfy in form of a reduction system. The above definition contains the following basic identities for all $f, g \in R$:

$$\partial fg = f\partial g + (\partial f)g, \quad \partial \int g = g, \quad \int \partial g = g - Eg, \quad Efg = (Ef)Eg.$$

Based on these, we start with the following reduction system:

$$\Sigma = \{(FF, f \otimes g \mapsto fg), (DF, \partial \otimes f \mapsto f \otimes \partial + \partial f), (DI, \partial \otimes \textstyle\int \mapsto \epsilon),$$
$$(ID, \textstyle\int \otimes \partial \mapsto \epsilon - E), (EF, E \otimes f \mapsto (Ef) \otimes E), (K, 1 \mapsto \epsilon)\}$$

Using our package, we determine that out of the 10 ambiguities 6 are resolvable and 4 remain. The reduced forms of the corresponding S-polynomials give rise to additional identities, among tensors as well as among elements of M_F.

```
CheckResolvability[RedSys]

10 ambiguities in total

2 ambiguities have all S-polynomials equal to zero

6 ambiguities are resolvable

{{Overlap[{D, I, D}, {D}, {D}], Prod[Diff, Eval]}, {Overlap[{I, D, F}, {F}, {I}],
    Prod[F[1]] - Prod[Int, Diff[F[1]]] - Prod[Eval[F[1]], Eval] - Prod[Int, F[1], Diff]},
  {Overlap[{I, D, I}, {I}, {I}], -Prod[Eval, Int]},
  {SpecialInclusion[{E, K}, {E}, {}], -Prod[Eval] + Prod[Eval[1], Eval]}}
```

To proceed, we introduce 3 new rules. We also update the implementation of computations in $M_F = R$ correspondingly, including the relation $E1 = 1$.

$$\Sigma_1 := \Sigma \cup \{(DE, \partial \otimes \mathrm{E} \mapsto 0), (EI, \mathrm{E} \otimes \smallint \mapsto 0),$$
$$(IFD, \smallint \otimes f \otimes \partial \mapsto f - \smallint \otimes \partial f - (\mathrm{E}f)\mathrm{E})\}$$

Repeating the steps above, 15 of 20 ambiguities can be resolved. Only 3 of the 5 S-polynomials not reduced to zero are essentially different, and we add corresponding rules to the reduction system.

$$\Sigma_2 := \Sigma_1 \cup \{(EE, \mathrm{E} \otimes \mathrm{E} \mapsto \mathrm{E}), (IFE, \smallint \otimes f \otimes \mathrm{E} \mapsto \smallint f \otimes \mathrm{E}),$$
$$(IFI, \smallint \otimes f \otimes \smallint \mapsto \smallint f \otimes \smallint - \smallint \otimes \smallint f)\}$$

Now, among 37 ambiguities 35 are resolvable, resulting in two new rules.

$$\Sigma_3 := \Sigma_2 \cup \{(IE, \smallint \otimes \mathrm{E} \mapsto \smallint 1 \otimes \mathrm{E}), (II, \smallint \otimes \smallint \mapsto \smallint 1 \otimes \smallint - \smallint \otimes \smallint 1)\}$$

Finally, we get 52 ambiguities which are all resolvable. This means that the derived reduction system Σ_3 is confluent. The function `IrreducibleWords` of the package generates all irreducible words up to a given length. One can prove that the irreducible words are given by $\tilde{F}ED^j$ and $\tilde{F}I\tilde{F}$ with $j \in \mathbb{N}_0$. Consequently, irreducible tensors are K-linear combinations of tensors of the form $f \otimes \mathrm{E} \otimes \partial^{\otimes j}$ and $f \otimes \smallint \otimes g$ with $j \in \mathbb{N}_0$, $f, g \in M_{\tilde{F}}$ where f, g, and E can also be absent.

References

1. Bergman, G.M.: The diamond lemma for ring theory. Adv. Math. **29**, 178–218 (1978)
2. Bokut, L.A., Chen, Y.: Gröbner-Shirshov bases and their calculation. Bull. Math. Sci. **4**, 325–395 (2014)
3. Buchberger, B.: An algorithm for finding the bases elements of the residue class ring modulo a zero dimensional polynomial ideal (German). Ph.D. thesis, University of Innsbruck (1965)
4. Gómez-Torrecillas, J.: Basic module theory over non-commutative rings with computational aspects of operator algebras. In: Barkatou, M., Cluzeau, T., Regensburger, G., Rosenkranz, M. (eds.) AADIOS 2012. LNCS, vol. 8372, pp. 23–82. Springer, Heidelberg (2014)
5. Guo, L., Regensburger, G., Rosenkranz, M.: On integro-differential algebras. J. Pure Appl. Algebra **218**, 456–473 (2014)
6. Hossein Poor, J., Raab, C.G., Regensburger, G.: Algorithmic operator algebras via normal forms for tensors. In: Proceedings of ISSAC 2016, p.8. ACM, New York (2016, to appear)
7. Knuth, D.E., Bendix, P.B.: Simple word problems in universal algebras. Computational Problems in Abstract Algebra, pp. 263–297. Pergamon, Oxford (1970)
8. Rosenkranz, M.: A new symbolic method for solving linear two-point boundary value problems on the level of operators. J. Symbolic Comput. **39**, 171–199 (2005)

9. Rosenkranz, M., Regensburger, G.: Solving and factoring boundary problems for linear ordinary differential equations in differential algebras. J. Symbolic Comput. **43**, 515–544 (2008)

10. Regensburger, G., Tec, L., Buchberger, B.: Symbolic analysis for boundary problems: from rewriting to parametrized Gröbner bases. Numerical and Symbolic Scientific Computing, pp. 273–331. Springer, Vienna (2012)

Robust Construction of the Additively-Weighted Voronoi Diagram via Topology-Oriented Incremental Algorithm

Mokwon Lee[1], Kokichi Sugihara[2], and Deok-Soo Kim[1(✉)]

[1] School of Mechanical Engineering, Hanyang University, Seoul, Korea
mwlee@voronoi.hanyang.ac.kr, dskim@hanyang.ac.kr
[2] Meiji University, Tokyo, Japan
kokichis@meiji.ac.jp
http://voronoi.hanyang.ac.kr,
http://home.mims.meiji.ac.jp/~sugihara/Welcomee.html

Abstract. Voronoi diagrams tessellate the space where each cell corresponds to an associated generator under an *a priori* defined distance and have been extensively used to solve geometric problems of various disciplines. Additively-weighted Voronoi diagrams, also called the Voronoi diagram of disks and spheres, have many critical applications and a few algorithms are known. However, algorithmic robustness remains a major hurdle to use these Voronoi diagrams in practice. There are two important yet different approaches to design robust algorithms: the exact-computation and topology-oriented approaches. The former uses high-precision arithmetic and guarantees the correctness mathematically with the cost of a significant use of computational resources. The latter focuses on topological properties to keep consistency using logical computation rather than numerical computation. In this paper, we present a robust and efficient algorithm for computing the Voronoi diagram of disks using a topology-oriented incremental method. The algorithm is rather simple as it primarily checks topological changes only during each disk is incrementally inserted into a previously constructed Voronoi diagram of some other disks.

Keywords: Topology-oriented · Additively-weighted · Voronoi diagram · Disk · Circle · Robustness · Algorithm

1 Introduction

The Voronoi diagram is a tessellation of space where each element of the tessellation is associated with a generating particle. The ordinary Voronoi diagram VD of a set of points $P = \{p_1, p_2, \ldots, p_n\}$, $p_i \in \mathbb{R}^d$, is the tessellation so that the Voronoi cell of $p_i \in P$ is the set of locations closer to p_i than to $p_j \in P$, $i \neq j$ for a given distance definition [1,2]. Suppose that $p_i \in P$ is assigned a weight $w_i \geq 0$. In other words, $P^w = \{p_1^w, p_2^w, \cdots, p_n^w\}$ where $p_i^w = (p_i, w_i)$.

© Springer International Publishing Switzerland 2016
G.-M. Greuel et al. (Eds.): ICMS 2016, LNCS 9725, pp. 514–521, 2016.
DOI: 10.1007/978-3-319-42432-3_66

Then, the power diagram can be similarly defined with the power distance which is defined as $dist(x, p_i)^2 - w_i$, where $dist()$ denotes the Euclidean distance function. Let $D = \{d_1, d_2, \ldots, d_n\}$ be a set of disks where $d_i = (p_i, r_i)$ is a disk with the center p_i and radius r_i. Let $\mathcal{VC}(d_i)$ denote the V-cell of d_i defined as $\mathcal{VC}(d_i) = \{x \in \mathbb{R}^d \mid dist(x, p_i) - r_i \le dist(x, p_j) - r_j, \ i \ne j\}$. Then, the Voronoi diagram of disks D is defined as $\mathcal{VD}(D) = \{\mathcal{VC}(d_1), \mathcal{VC}(d_2), \cdots, \mathcal{VC}(d_n)\}$ which is also frequently called the *additively-weighted Voronoi diagram*. It turns out that $p_i^w = (p_i, w_i)$ corresponds to a disk $d_i = (p_i, r_i)$ with the center p_i and the radius $r_i = \sqrt{w_i}$. Then, $dist(x, p_i)^2 - r_i^2$ is the square of the tangential distance from x to ∂d_i. The connectivity among the topological entities in the tessellations above are usually represented in a winged-edge data structure.

The Voronoi diagram of disks in $\mathbb{R}^2$ [3,4] is important for both on its own right for diverse applications [5–8] and for its extension to the Voronoi diagram of spheres in $\mathbb{R}^3$ which has many important applications. Despite of many previous studies [3,4,9–14], its robust and efficient computation remains a challenge. This is particularly the case with the Voronoi diagram of disks which has topological properties that do not exist in the ordinary Voronoi diagram. For details, see [3,4].

There are two general approaches to cope with degeneracy and instability in the construction of geometric structures. The exact computation approach employs high-precision arithmetic to guarantee the correctness of geometric decisions and thus avoids inconsistency due to numerical errors [15–17]. It can guarantee the correctness of the behavior of the resulting software, but in return of this it is necessary to pay high cost in computation. The exact computation is widely used in prevailing software such as CORE [18], CGAL [19,20], etc. The topology-oriented approach is based on the observation that numerical computations are imprecise and hence we cannot rely on them [21,22]. We concentrate on the purely topological properties of geometric objects and try to keep their consistency by using logical computation instead of numerical computation. Hence, the implementation of a topology-oriented algorithm is usually rather easy.

In this paper, we present a new topology-oriented incremental algorithm for the Voronoi diagram of disks. It turns out that the proposed algorithm is accurate, efficient, and robust and we believe that the idea can be easily generalized for its counterpart for the additively-weighted Voronoi diagram in $\mathbb{R}^3$.

2   Anomalies of the Voronoi Diagram of Disks

The topological properties of the Voronoi diagram of disks is significantly different from its counterpart for points due to the intriguing cases called *anomaly*. Figure 1 shows the cases that can occur in the Voronoi diagram of disks. Figure 1(a) shows an *ordinary* case where each pair of disks defines at most one V-edge ("V-" denotes "Voronoi-"). Let us call a V-cell of a disk d which is bounded by two V-edges an *anomaly* and the d an *anomaly disk*. Such a case is referred to an anomaly because three disks define two V-vertices, not one as its counterpart of for points. Figure 1(b) shows a case that the V-cell of d_{new} defined by the new disk d_{new} is bounded by only two V-edges. Hence, d_{new} becomes an anomaly disk by itself and this case is referred to a *self-anomaly*. In Fig. 1(c) shows a case that the V-cell of d_{small} becomes bounded by only two V-edges by the new disk d_{new}. Hence, the V-cell of d_{small} becomes an anomaly V-cell. As d_{new} causes a neighbor V-cell to be anomaly, the V-cell of d_{new} is referred to as a *vicinity-anomaly*.

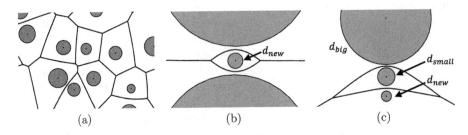

Fig. 1. Three cases in the Voronoi diagram of disks. (a) ordinary case, (b) self-anomaly case, (c) vicinity anomaly case.

3   Topology-Oriented Incremental Construction of the Voronoi Diagram of Disks

The proposed algorithm for the Voronoi diagram of circular disks in $\mathbb{R}^2$ inherits the original topology-oriented incremental algorithm (hereafter, abbreviated as TOI) for the ordinary Voronoi diagram $VD(P)$ of a point set P in $\mathbb{R}^2$ proposed in 1992 [21,22]. Its basic idea was to insert each point $p_i \in P$ into pre-constructed Voronoi diagram VD_{old} of $P' = \{p_1, p_2, \ldots, p_{i-1}\} \subseteq P$ with a priority of topological consistency. Let $VC'(p_j)$ be the V-cell corresponding to the point generator $p_j, j = 1, 2, \ldots, i$, in VD_{new}. In the ordinary Voronoi diagram of points, the new V-cell $VC'(p_i)$ contains at least one old V-vertex of VD_{old} and there exists a V-cell $VC(p_j), j = 1, 2, \ldots, i - 1$, which contains p_i.

Suppose that $VC(p_j)$ contains p_i. We find the V-vertex of $VC(p_j)$ which is contained in $VC'(p_i)$ and insert it to $V_{removed}$, which is the set of V-vertices to be trimmed off by the new V-cell $VC'(p_i)$. Then, for all the V-edges incident to each of the V-vertices in $V_{removed}$, we check if the V-edge should be split into two segment or not: one inside of and the other outside of $VC'(p_i)$. The test can be done by examining the two V-vertices of a V-edge and an inside V-vertex is also inserted into $V_{removed}$ if and only if it passes `Segmentation-Test`. This propagation process continues until when the V-vertices in $V_{removed}$ are all blocked by those cannot be contained in $V_{removed}$.

`Segmentation-Test` is as follows. When $VC'(p_j)$ and $VC'(p_i)$ share one and only one V-edge, $\partial VC(p_j)$ splits into two segments: one to be trimmed by $VC'(p_i)$ and the other not trimmed. If only one segment exists, $\partial VC(p_j)$ is entire removed and $VC'(p_j)$ has an empty region. If more than two segments exist, $VC'(p_j)$ and $VC'(p_i)$ share more than one V-edge and it is abnormal.

The original TOI-algorithm above should be revised so that the anomalies can be taken into consideration in the Voronoi diagram of disks. Let a V-edge of $\mathcal{VD}_{old}$ which twice intersect $\partial VC'(d_i)$ be *anomalizing V-edge* (e^A for short). The intermediate part of an e^A is trimmed in a self-anomaly case or remains in a vicinity anomaly case after a new disk is inserted. While $\mathcal{VC}'(d_i)$ contains at least one old V-vertex in the point set case, Fig. 1(b) shows that the V-cell $\mathcal{VC}'(d_{new})$ of d_{new} may not contain any old V-vertex in the Voronoi diagram of disks. The V-edge trimmed by $\mathcal{VC}'(d_{new})$ is an anomalizing V-edge and is split into two V-edges by inserting a fictitious V-vertex which is contained in $\mathcal{VC}'(d_{new})$. Then, the algorithm can proceed just like the original TOI-algorithm.

Figure 1(c) shows a vicinity anomaly case when d_{new} is most recently inserted. Then, the V-edge between d_{big} and d_{small} remains while its two V-vertices are removed. The V-edge also becomes an anomalizing one and we also split it into two V-edges by inserting a fictitious V-vertex. Note that the two V-vertices are contained in $\mathcal{VC}'(d_{new})$ but the fictitious V-vertex is not. Hence, the two V-edges after the split remain in the new Voronoi diagram. In Fig. 1(c), $\mathcal{VC}'(d_{big})$ and $\mathcal{VC}'(d_tiny)$ share two V-edges. To reflect the observation above, Segmentation-Test should be revised to Revised-Segmentation-Test to check the number of anomalies and the number of segments of each V-cell.

4   Experiments

The proposed TOI-algorithm was implemented in Visual C++ and thoroughly tested. The code is embedded in the BetaConcept program [8] that can be freely available at Voronoi Diagram Research Center at Hanyang University. The experimental was done in the following environment: Intel Core2 Duo 2.93 GHz, 2.93 GHz; 4.00 GB RAM; Windows 7 Ultimate K (32bit); Visual C++ on Microsoft Visual Studio 2010 (32bit).

The test data set TestSet consisted of $5*4=20$ sets of 100 random disk models (i.e. 2,000 disk models) where each model has from 100 to 10,000 disks. Each model consists of disks where the radius of each disk is assigned at random from a uniform distribution following five different ranges: $[r_{min} = 1.0, r_{max} = 3.0]$, $[1.0, 5.0]$, $[1.0, 10.0]$, $[1.0, 50.0]$, and $[1.0, 100.0]$. For convenience, we denote each group by referring to only r_{max} (referring to as the radius range R, for notational convenience) as r_{min} is identically 1.0 for all disks in TestSet. The disks are contained within a sufficiently big enclosing circle. Let ρ be the *packing density* which is defined as the ratio of the union of the area of all disks in a model to that of an enclosing circle. We chose four values of ρ: 0.1, 0.2, 0.3, and 0.4 as a data set for $\rho > 0.5$ seems time consuming or infeasible to produce. We reference each model by MODEL(ρ, r_{max}, n) where ρ and n represent the packing density and the number of disks, respectively. Figure 1(a), (b), and (c) show the constructed Voronoi diagrams of a random disk model, a high anomaly case, and a degenerate case computed by the BetaConcept program using the proposed TOI-algorithm, respectively. Figure 3(a) shows the computation time spent to compute the Voronoi diagrams of the TestSet data. CGAL using filtered exact computation is much slower than TOI-algorithm but produces correct results for all degenerate data sets we tested. However, CGAL "without" the exact computation produces in correct results for many degenerate data sets, as shown in Fig. 3(b), even if it is slightly faster than the proposed TOI-algorithm which has produced always correct results regardless degeneracy level (Fig. 2).

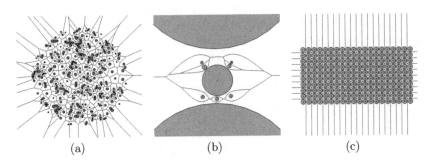

(a)   (b)   (c)

Fig. 2. The result Voronoi diagrams via TOI-algorithm. (a) a random model, (b) an anomaly model, (c) a degeneracy model.

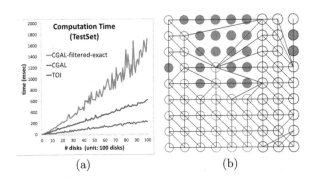

(a)   (b)

Fig. 3. Experimental result. (a) Computation time of the three codes: TOI, CGAL (without exact computation), and CGAL-filtered-exact (with exact computation), (b) Incorrect result produced from CGAL (without exact computation). (Color figure online)

How frequently do anomalies occur? Figure 4 shows the anomaly occurrence per 100 disks for TestSet (In this experiment, we counted the number of V-cells bounded by two, and only two, V-edges). In Fig. 4(a), the horizontal and vertical axes show the density of models and the average number of anomalies per 100 disks occurred in the 100 models for each radius range. Figure 4(b) shows that the number of anomalies per 100 disks remain almost constant and it seems that this measure is independent of the number of disks while the density shows somewhat distinguishable difference. The more dense a model is, the more anomalies a model tend to have even if the influence is not very strong. Let $\mu(\rho, R)$ be the anomaly occurrences per 100 disks, $\mu(., R)$ the average of $\mu(\rho, R)$ over different ρ values, $\mu(\rho, .)$ the average of $\mu(\rho, R)$ over different R values, and $\mu(., .)$ the grand average of $\mu(\rho, R)$ over different ρ and R values. $\mu(0.1, .) = 0.153$; $\mu(0.2, .) = 0.238$; $\mu(0.3, .) = 0.290$; $\mu(0.4, .) = 0.354$. Figure 4(c) shows a similar behavior for the radius range. The larger the radius range is, the more anomalies a model tend to have. We have the following statistics: $\mu(., 3.0) = 0.002$; $\mu(., 5.0) = 0.029$; $\mu(., 10.0) = 0.162$; $\mu(., 50.0) = 0.521$; $\mu(., 100.0) = 0.580$; $\mu(., .) = 0.259$;

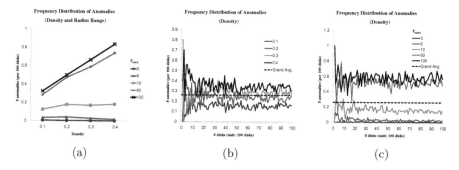

(a)   (b)   (c)

Fig. 4. The number of anomalies per 100 disks on average. (Each point in the graphs denotes the number of anomalies per 100 disks.) The grand average is 0.26 (i.e. the horizontal broken line). (a) With respect to the density (each curve denotes a different radius range), (b) with respect to the number of disks (each curve denotes a different density), and (c) with respect to the number of disks (each curve denotes a different radius range). (Color figure online)

Another interesting issue is whether there is any model without any anomaly at all. Figure 5 shows the frequency of models in TestSet with respect to the number of anomalies. The horizontal axis shows the number of anomalies and the vertical axis the number of models. Figure 5(a) and (b) correspond to the radius range. For example, in Fig. 5(a), the blue curve corresponds to R = 3 (the radius range 1–3). Among 400 models for R = 3, 358 models have no anomaly at all; 33 models have one anomaly; 9 models have two anomaly; there is no model with three or more anomalies. Figure 5(c) and (d) correspond to the density. Among 500 models for $\rho = 0.1$, 113 models have no anomaly and 52 have one anomaly. Among the total 2000 models, 529 models have no anomaly, 194 models have one anomaly, etc. MODEL($\rho = 0.4, R = 100, n = 8900$) has 91 anomalies, the maximum in TestSet.

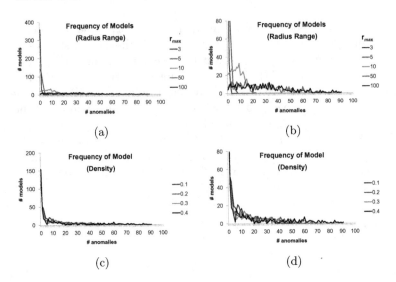

Fig. 5. Distribution of the model corresponding to the range of radius of circle. (a) and (b) For the radius range, and (c) and (d) for the density (Color figure online)

5 Conclusion

The Voronoi diagram of circular disks, also called the additively-weighted Voronoi diagram, is useful for solving important applications and its robust computation has been a challenge. In this paper, we presented a topology-oriented incremental algorithm for the construction of the Voronoi diagram of circular disks, also called the additively-weighted Voronoi diagram. The accuracy, efficiency, and robustness were verified through an extensive test with both random and degenerate data sets. We believe the proposed idea can be easily generalized for its counterpart of three-dimensional spheres.

References

1. Okabe, A., Boots, B., Sugihara, K., Chiu, S.N.: Spatial Tessellations: Conceptsand Applications of Voronoi Diagrams, 2nd edn. Wiley, Chichester (1999)
2. Aurenhammer, F.: Voronoi diagrams - a survey of a fundamental geometric data structure. ACM Comput. Surv. **23**(3), 345–405 (1991)
3. Kim, D.-S., Kim, D., Sugihara, K.: Voronoi diagram of a circle set from Voronoi diagram of a point set: I. topology. Comput. Aided Geom. Des. **18**, 541–562 (2001)
4. Kim, D.-S., Kim, D., Sugihara, K.: Voronoi diagram of a circle set from Voronoi diagram of a point set: II. geometry. Comput. Aided Geom. Des. **18**, 563–585 (2001)
5. Held, M. (ed.): On the Computational Geometry of Pocket Machining. LNCS, vol. 500. Springer, Heidelberg (1991)
6. Kim, D.-S., Hwang, I.-K., Park, B.-J.: Representing the Voronoi diagram of a simple polygon using rational quadratic Bézier curves. Comput. Aided Des. **27**(8), 605–614 (1995)

7. Kim, D.-S., Ryu, J., Shin, H., Cho, Y.: Beta-decomposition for the volume and area of the union of three-dimensional balls and their offsets. J. Comput. Chem. **33**(13), 1252–1273 (2012)
8. Kim, J.-K., Cho, Y., Kim, D., Kim, D.-S.: Voronoi diagrams, quasi-triangulations, and beta-complexes for disks in $\mathbb{R}^2$: The theory and implementation in BetaConcept. J. Comput. Des. Eng. **1**(2), 79–87 (2014)
9. Lee, D., Drysdale, R.: Generalization of Voronoi diagrams in the plane. SIAM J. Comput. **10**(1), 73–87 (1981)
10. Sharir, M.: Intersection and closest-pair problems for a set of planar discs. SIAM J. Comput. **14**(2), 448–468 (1985)
11. Yap, C.-K.: An $O(n \log n)$ algorithm for the Voronoi diagram of a set of simple curve segments. Discrete Comput. Geom. **2**, 365–393 (1987)
12. Fortune, S.: A sweepline algorithm for Voronoi diagrams. Algorithmica **2**, 153–174 (1987)
13. Sugihara, K.: Approximation of generalized Voronoi diagrams by ordinary Voronoi diagrams. Graphical Models Image Process. **55**(6), 522–531 (1993)
14. Karavelas, M., Emiris, I.Z.: Predicates for the planar additively weighted Voronoi diagram, Technical Report ECG-TR-122201-01. INRIA Sophia-Antipolis, Sophia-Antipolis (2002)
15. Sugihara, K., Iri, M.: A solid modelling system free from topological Inconsistency. J. Inf. Process. **12**(4), 380–393 (1989)
16. Sugihara, K.: A simple method for avoiding numerical errors and degeneracy in Voronoi diagram construction. IEICE Trans. Fundam. **E75–A**, 468–477 (1992)
17. Yap, C.-K.: Towards exact geometric computation. Comput. Geom. Theory Appl. **7**(1–2), 3–23 (1997)
18. Yu, J., Zhou, Y., Tanaka, I., Yao, M.: Roll: a new algorithm for the detection of protein pockets and cavities with a rolling probe sphere. Struct. Bioinf. **26**(1), 46–52 (2010)
19. Fabri, A., Giezeman, G.J., Kettner, L., Schirra, S., Schönherr, S.: On the design of CGAL a computational geometry algorithms library. Softw. Pract. Experience **30**(11), 1167–1202 (2000)
20. Goodman, J.E., ORourke, J.: Handbook of Discrete and Computational Geometry. CRC Press, Boca Raton (1997)
21. Sugihara, K., Iri, M.: Construction of the Voronoi diagram for "one million" generators in single-precision arithmetic. Proc. IEEE **80**(9), 1471–1484 (1992)
22. Sugihara, K., Iri, M.: A robust topology-oriented incremental algorithm for Voronoi diagrams. Int. J. Comput. Geom. Appl. **4**(2), 179–228 (1994)

Mathematical Font Art

Joris van der Hoeven[(⊠)]

Laboratoire d'informatique, UMR 7161 CNRS, Campus de L'École polytechnique,
1, rue Honoré d'Estienne d'Orves, Bâtiment Alan Turing,
CS35003, 91120 Palaiseau, France
vdhoeven@lix.polytechnique.fr

Abstract. Currently, only a limited number of fonts are available for high quality mathematical typesetting, such as Knuth's computer modern font, the STIX font, and several fonts from the TEX GYRE family. An interesting challenge is to develop tools which allow users to pick any existing favorite font and to use it for writing mathematical texts. We will present progress on this problem as part of recent developments in the GNU TEX$_{MACS}$ scientific text editor.

1 Introduction

For a long period, most documents with mathematical formulas were typeset using Knuth's COMPUTER MODERN font [5]. Recently, a few alternative fonts were designed, such as the STIX font [7] and the TEX GYRE fonts [3]. These hand-crafted fonts all admit a high quality, but they required an important development effort. Now there exists thousands of fonts for non mathematical purposes. To what extent is it possible to use such fonts for mathematical texts or presentations, or on the web?

In this paper we describe recent developments inside the GNU TEX$_{MACS}$ scientific text editor [1] which aim at a better support of general purpose fonts, thereby making life a bit more colorful. The focus is on fully automatic techniques for using existing fonts inside structured documents with mathematical formulas. Further fine tuning for specific characters in particular fonts is another interesting topic which will not be discussed here.

There are obvious limitations of what we can do with a font if bold and italic declinations or glyphs for various important characters are missing. Nevertheless we will see that quite a lot is often possible even though the resulting quality may be inferior to what can be achieved *via* manual design. Since various special characters or font effects are often only used at a reduced number of places inside actual documents, the occasional loss of quality may remain within acceptable bounds, even for professional purposes.

Our general strategy for turning existing fonts into full fledged mathematical font families is to remedy each of the font's insufficiencies. The most common problems are the following:

– Lack of the most important font declinations as needed in scientific documents:
  Bold, *Italic*, SMALL CAPITALS, Sans Serif, Typewriter.

© Springer International Publishing Switzerland 2016
G.-M. Greuel et al. (Eds.): ICMS 2016, LNCS 9725, pp. 522–529, 2016.
DOI: 10.1007/978-3-319-42432-3_67

- Lack of specific glyphs: non English languages, mathematical symbols, and in particular big operators, extensible brackets and wide accents.
- Inconsistencies: sloppy design of some glyphs that are important for mathematics (such as $-$, $<$, etc.), leading to inconsistencies.

The main countermeasures are *font substitution* and *font emulation*. The first technique (see Sect. 2) consists of borrowing missing glyphs from other fonts. This can either be done on the level of an entire font (e.g. for obtaining bold or italic declinations) or for individual characters (e.g. a missing $\propto$ symbol, or lacking Greek characters). Font emulation consists of combining and altering the glyphs of symbols in a font in order to generate new ones. This can again be done for entire fonts (Sect. 3) or individual glyphs (Sects. 4 and 5).

All techniques described in this paper have been implemented in TEX$_{MACS}$, version 1.99.5 and beyond. The software can freely be downloaded from our website www.texmacs.org. The virtual character definitions described in Sect. 4 below can be found in the `TeXmacs/fonts/virtual` directory; interested users may play with these definitions. Longer examples of what can be obtained using the techniques described in this paper are available here:

http://www.texmacs.org/joris/fontart/fontart-abs.html

In the TEX/LATEX universe, there have also been several efforts towards better support for modern OPENTYPE fonts, most notably XETEX [4] and LUATEX [2]. The first system also contains features that are similar to those described in Sect. 3. However, these systems do not support full mathematical font emulation as presented in this paper. XETEX and LUATEX also tend to diverge from standard LATEX through the introduction of incompatibilities.

2   Font Analysis and Font Substitution

In order to borrow missing characters from other fonts, it is important to be able to determine fonts with a similar design, so that the alien glyphs fit nicely into the main text:

$$\text{The symbols } \alpha, \beta, \gamma \text{ are acceptable inside } x + \alpha + y + \beta + z + \gamma. \tag{1}$$

$$\text{The symbols } \boldsymbol{\alpha}, \boldsymbol{\beta}, \boldsymbol{\gamma} \text{ do not look very well inside } x + \boldsymbol{\alpha} + y + \boldsymbol{\beta} + z + \boldsymbol{\gamma}. \tag{2}$$

Usually, rules for font substitution are specified manually for each individual font. Although this often yields the most precise and predictable results, it can be tedious to write such rules. For this reason, we also implemented a more automatic mechanism in order to determine good substitutes.

A prerequisite for our algorithm for automatic font substitutions is a detailed analysis of the main characteristics of all supported fonts. The results of this analysis are stored in a database. Using this database, we may then compute the distance between two fonts. In the case when a symbol σ is missing in a font F_1, it then suffices to find the closest font F_2 that supports this symbol σ. Notice that the best substitution font may depend on the fonts which are installed on your system.

In our database we both use discrete font characteristics (e.g. sans serif, small capitals, handwritten, ancient, gothic, etc.) and continuous ones (e.g. italic slant, height of an "x" symbol, etc.). Most characteristics are determined automatically by analyzing the name of the font (for some of the discrete characteristics) or individual glyphs (for the continuous ones). Some "font categories" (such as handwritten, gothic, etc.) can be specified manually.

One of the most important font characteristics is the height of the "x" symbol (with respect to the design size). When the font F_1 borrows a symbol from the font F_2 we first scale it by the quotient of these x-heights inside F_1 and F_2. In the example (2) this was done correctly, contrary to (2).

Other common font characteristics are also taken into account into our database, such as the italic slant, the width of the "M" symbol, the ascent and descent (above and above the "x" symbol), etc. In addition, we carefully analyze the glyphs themselves in order to determine the horizontal and vertical stroke widths for the "o" and "O" symbols, the average aspect-ratios of uppercase and lowercase letters, and the average area of glyphs that is filled (how much ink will be used).

Our current implementation manages to find reasonably good font substitutions. Notice that this may even be a problem on certain occasions. For instance, in the example (2) below, the sans serif font is such a good match that it can barely be distinguished from the serif font, thereby defeating its purpose:

This **sample** text is a bit too good.   (3)

3   Poor Man's Font Emulation

Various font alterations such as **Bold**, *Italic* and SMALL CAPITALS can be emulated in rather obvious ways, although with significant loss of quality:

- Emboldening can be achieved through the replacement of pixels by small lines. In addition, it may be worth it to horizontally stretch certain characters such as "m". The appropriate stretching factors are highly font and character dependent, but using the factors corresponding to the computer modern font usually leads to reasonable results.
- Italic fonts can be approximated by slanted fonts, which may be further narrowed for a better result. The most important drawback of this method is that it often falls short of producing the correct italic versions of certain characters ($a/a/a$, $f/f/f$, $g/g/g$, etc.).
- Small capitals can be emulated by rescaling capitals using a factor that roughly turns an "X" into an "x". Instead of conserving the aspect-ratio, we found it more pleasing to slightly widen characters as well. The transformed version of "X" may also be taken slightly higher than "x".

Regular	Bold	Italic	Small Caps	Blackboard Bold	Mathematics
Optima	**Bold**[*]	*Italic*	SMALL CAPS	$\mathbb{C}, \mathbb{N}, \mathbb{Q}, \mathbb{R}, \mathbb{Z}$	$x^2 + f(x, \frac{a}{b+c})$
Cochin	**Bold**[*]	*Italic*[*]	SMALL CAPS	$\mathbb{C}, \mathbb{N}, \mathbb{Q}, \mathbb{R}, \mathbb{Z}$	$x^2 + f(x, \frac{a}{b+c})$
Chartrand	**Bold**	*Italic*	SMALL CAPS	$\mathbb{C}, \mathbb{N}, \mathbb{Q}, \mathbb{R}, \mathbb{Z}$	$x^2 + f(x, \frac{a}{b+c})$
Essays1743	**Bold**[*]	*Italic*[*]	SMALL CAPS	$\mathbb{C}, \mathbb{N}, \mathbb{Q}, \mathbb{R}, \mathbb{Z}^*$	$x^2 \cdot f(x, \frac{a}{b+c})$
Brynr Trxtur	𝕭𝖔𝖑𝖉	𝔍𝔱𝔞𝔩𝔦	𝔖𝔐𝔄𝔏𝔏 𝔠𝔄𝔓𝔖	$\mathbb{C}, \mathbb{N}, \mathbb{Q}, \mathbb{R}, \mathbb{S}$	$x^2 + f(x, \frac{a}{b+f})$
Chalkduster	Bold	*Italic*	SMALL CAPS	$\mathbb{C}, \mathbb{N}, \mathbb{Q}, \mathbb{R}, \mathbb{Z}$	$x^2 + f(x, \frac{a}{b+c})$
Comic Sans	**Bold**	*Italic*	SMALL CAPS	$\mathbb{C}, \mathbb{N}, \mathbb{Q}, \mathbb{R}, \mathbb{Z}$	$x^2 + f(x, \frac{a}{b+c})$
Papyrus	**Bold**	*Italic*	SMALL CAPS	$\mathbb{C}, \mathbb{N}, \mathbb{Q}, \mathbb{R}, \mathbb{Z}$	$x^2 + f(x, \frac{a}{b+c})$

Fig. 1. Emulation of bold, italic, small capitals and blackboard bold. * These declinations are already supported by the original font.

With more work, the above "poor man's" strategies might be further enhanced. For instance, the italic a might be better approximated using a shortened version of d instead of a. In order to improve bold font emulation, we might also replace pixels by small lines of cleverly adjusted lengths.

More elaborate emulation strategies might greatly benefit from a toolkit for "retro-engineering" the design of existing fonts. For instance, given an outline, we might want to determine the curve(s) followed by a "pen" and the size (or shape) of the pen at each point of the curve. This would then make it easy to produce high quality narrowed and widened versions of a font, as well as better emboldened fonts, or variants in which the pen's size is uniform (as needed for sans serif and typewriter fonts). Another interesting question is whether it is possible to automatically detect serifs and to add or remove them.

We have started to experiment with more elaborate emulation algorithms for the generation of "blackboard bold" variants of glyphs. The easiest strategy is to produce an outlined version of the possibly emboldened input glyph. The standard AMS blackboard bold font uses this method ($\mathbb{C}, \mathbb{N}, \mathbb{Q}, \mathbb{R}, \mathbb{Z}$), but we consider the result suboptimal with respect to adding a single stroke ($\mathbb{C}, \mathbb{N}, \mathbb{Q}, \mathbb{R}, \mathbb{Z}$). We implemented an algorithm for the detection of the part of contour to be "double stroked". We next embolden this part and hollow it out (Fig. 1).

4 Virtual Characters

Missing glyphs can be generated automatically from existing ones using a combination of the following main techniques, listed by increasing complexity:

– Superposition of several glyphs: + and − can be combined into ±, and ≪ be obtained by juxtaposing two < symbols.
– Clipping rectangular areas: cutting ↦ and ↠ in their midsts and combining them yields ↦ .

- Linear transformations: combining a crushed O and an I, we may produce the Greek capital Φ. Turning around →, we obtain ↑.
- Simple graphical constructs such as circles and lines. This can for instance be used for producing the missing half circle of ⊂.
- Special *ad hoc* transformations that directly operate on the pixels of a glyph (or on their outlines if possible). For instance, we designed a special "curlyfi-cation" method that turns < into ≺ and < into ≺ . Similarly, we implemented a "flood fill" algorithm for transforming ◁ into ◀.

In a similar vein, we need various querying mechanisms: all glyphs come with logical and physical bounding boxes, but we sometimes may want to compute the exact width of some stroke or obtain other kinds of information.

We developed a small language that can be used for defining new "virtual" characters in terms of existing ones. The design of every new virtual glyph can be regarded as a puzzle: finding a clever way to combine existing glyphs into the desired one using the primitives from the language. Of course, we are looking for robust solutions in the sense that they should work for *any* reasonable font in which the required basic glyphs are available.

Let us consider a few examples. For the construction of arrows, it turns out that the single *guillemets* ‹ and › are often well suited for the heads (the rescaled symbols < and > are acceptable fallbacks). The arrow bars are obtained from the minus sign −, but the determination of an appropriate minus is non trivial. For instance, the width of the dash - is usually too large, so we should avoid using this symbol. The underscore is a better candidate; one may also cut the plus sign into several pieces (avoiding the vertical stroke) and recombine them.

Assuming that we have an appropriate arrow bar and head, we may use the following code for producing an actual arrow:

```
(rightarrow (right-fit arrowbar (align righthead arrowbar * 0.5)))
```

The `align` primitive is used to vertically align the arrow head at the center of the arrow bar. The `right-fit` primitive is less basic and corresponds to sliding the arrowhead from the right to the left until the arrow bar goes past the head on its right. More direct ways to produce arrows turn out to be less robust. Left and left-right arrows can be defined using

```
(leftarrow (left-flip rightarrow))
(leftrightarrow (join (part leftarrow * 0.5) (part rightarrow 0.5 *)))
```

These definitions potentially take advantage of an existing `rightarrow` in the base font. The `part` primitive performs two horizontal clippings between the middle and the extremities, whereas `join` is used for superposition.

An interesting challenge is the emulation of Greek characters. This seems intractable for the lowercase symbols, but is less hopeless for the capitals. For instance, Γ can be obtained by flipping the Roman L upside down and we already mentioned how to obtain a reasonable Φ. More interesting is the case of Π, which

	Γ	Δ	Θ	Λ	Ξ	Π	Σ	Υ	Φ	Ψ	Ω	⊤̸	:=	≈	≍	⊮	→	⇋	≤	⊊	▶
Optima	Γ	Δ	Θ	Λ	Ξ	Π	Σ	Υ	Φ	Ψ	Ω	⊤̸	:=	≈	≍	⊮	→	⇋	≤	⊊	▶
Cochin	Γ	Δ	Θ	Λ	Ξ	Π	Σ	Υ	Φ	Ψ	Ω	⊤̸	:=	≈	≍	⊮	→	⇋	≤	⊊	▶
Didot	Γ	Λ	Θ	Λ	Ξ	Π	Σ	Υ	Φ	Ψ	Ω	⊤̸	:=	≈	≍	⊮	→	⇋	≤	⊊	▶
Cuprum	Γ	Δ	Θ	Λ	Ξ	Π	Σ	Υ	φ	Ψ	Ω	⊤̸	:=	≈		⊮	→	⇋	≤	⊊	▶
Essays 1743	Γ	Δ	Θ	Λ	Ξ	Π	Σ	Υ	Φ	Ψ	Ω	⊤̸	::	≈	≍	⊮	→	⇋	≤	⊊	▶
Am. Typewr.	Γ	Δ	Θ	Λ	Ξ	Π	Σ	Υ	Φ	Ψ	Ω	⊤̸	:=	≈	≍	⊮	→	⇋	≤	⊊	▶
Chalkboard	Γ	Δ	Θ	Λ	Ξ	Π	Σ	Υ	Φ	Ψ	Ω	⊤̸	:=	≈	≍	⊮	→	⇋	≤	⊊	▶
Chalkduster	Γ	Δ	Θ	Λ	Ξ	Π	Σ	Υ	Φ	Ψ	Ω	⊤̸	:=	≈	≍	⊮	→	⇋	≤	⊊	▶
Papyrus	Γ	Δ	Θ	Λ	Ξ	Π	Σ	Υ	Φ	Ψ	Ω	⊤̸	:=	≈	≍	⊮	→	⇋	≤	⊊	▶
Paper Cuts	Γ	Δ	Θ	Λ	Ξ	Π	Σ	Υ	Φ	Ψ	Ω	⊤̸	:=	≈		⊮	→	⇋	≤	⊊	▶

Fig. 2. Emulation of various mathematical symbols in various fonts.

can be obtained from H by moving the horizontal bar to the top. However, extracting this bar is not so easy in some fonts: consider **H**. For a robust method, we therefore cut the H into pieces: we first extract ′′ ·· ·· and recombine them into II. We next take a tiny piece of the central bar, extend it to the desired length, and move it to the top (Fig. 2).

5   Rubber Characters

One specific problem with mathematical fonts is the need for rubber characters. There are essentially four types of them: big operators ($\sum$, $\prod$, $\otimes$), large delimiters $((()))$ $\{\{\}\}\}$, wide accents ($\overset{\frown}{\ }$, $\frown$, $\frown$), and long arrows (—————→ , ⇐======⇒).

We produce these rubber characters using essentially the same techniques as in the previous section. Especially horizontal and vertical scaling are very useful, as well as cutting symbols into several parts and reassembling them appropriately.

For instance, moderately large versions of the bracket (are obtained through magnification, typically with a higher stretch factor in the vertical direction. For larger sizes, this method produces results that are unacceptably thick. In that case, we rather cut the bracket into a top, a bottom, and a tiny middle part. We next repeat the middle part as many times as necessary in order to obtain a bracket of the desired size.

The use of scaling is a very delicate matter. For instance, in the case of square brackets [(and their potential derivatives ⌈ and ⌊), the point where horizontally magnified versions get too fat is usually reached much earlier than for ordinary or curly brackets. In the case of wide accents, we typically need very large horizontal stretch factors, which yield unacceptable results. Magnified versions of $\sum$ and $\prod$ typically look allright, but this is much less so for $\otimes$.

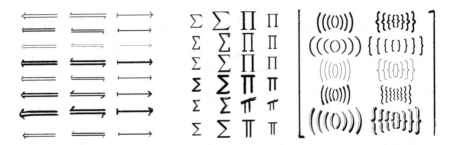

Fig. 3. Assorted rubber symbols from various fonts.

We are still in the process of fine tuning our implementation. For better results, one major challenge is to develop magnification methods with a finer control over the stroke widths. In particular, we need a reliable magnification method that preserves all relevant widths (Fig. 3).

6 Conclusion and Perspectives

After a moderate development investment, we are now able to use a lot of existing fonts for mathematical typesetting. The quality of the obtained results ranges from "better than nothing" to "professional typesetting quality". Our virtual font implementation can be regarded as a genuine "metafont". Paradoxically, and in comparison, Knuth's METAFONT initiative [6] has essentially resulted in the creation of a single mathematical font of extremely high quality.

One interesting question that occurred during our development of a virtual mathematical metafont concerns the "essence of a font": which font features essentially contain all necessary information to reproduce the entire font, and how? For instance, most mathematical symbols can be reconstructed from a few basic glyphs: $-$, $=$, $\sim$, $<$, $\prec$, $\subset$, $>$ (or $\rightarrow$), $.$, $\circ$, $($, $[$ and $\{$. Similarly, the Greek capitals can essentially be reconstructed from E, H, O, X and Z. So what is the real "fingerprint" of a font?

The development of more and better glyph emulation tools might be valuable for font designers. On the one hand, such tools may be used to automatically generate lots of glyphs. On the other hand, they allow designers to compare their own handcrafted glyphs with automatically generated alternatives. This may help to spot errors or increase consciousness about the distinctive features of a personal design.

For the moment, we developed all our font substitution and emulation tools inside TeX_MACS. It might be worthwhile to conceive a separate library with even more systematic tools for font analysis, retro-engineering and glyph emulation. Such a library might come with command line tools for generating mathematically enriched fonts, emboldened or narrowed versions, etc. For the moment, several of our algorithms are also limitated to operating on bitmaps. In the future, it would be nice to systematically work with vector graphics only.

One final issue concerns the purpose of alternative fonts. For instance, certain fonts such as Chalkboard, Chalkduster, Essays1743, Yiggivoo 3D are mainly used in order to produce specific graphical effects: emulate text on a chalkboard or on a blackboard, imitating a degraded retro-font, or producing a 3D sensation. It can be questioned whether these purposes are always best served through the use of a special font. For instance, handwriting might be imitated better by dynamically generating many different versions of the same letter. Better retro and 3D effects might be obtained by applying a suitable graphical filter to an entire portion of text. This might even more be true in the presence of fractions, square roots or geometric pictures.

Disclaimer. Unfortunately, Springer does not yet support GNU TeX_{MACS} for submissions to its journals. For this reason, the original version of this paper had to be converted to LaTeX. Any resulting loss of typesetting quality should be imputed to Springer.

References

1. Gubinelli, M., van der Hoeven, J., Poulain, F., Raux, D.: GNU TeXmacs: towards a scientific office suite. In: Hong, H., Chee, Y. (eds.) ICMS 2014. LNCS, vol. 8529, pp. 562–569. Springer, Heidelberg (2014)
2. Hoekwater, T., Henkel, H., Hagen, H.: Luatex (2007). http://www.luatex.org/
3. Jackowski, B., Nowacki, J., Ludwichowski, J.: The TeX Gyre collection of fonts. http://www.gust.org.pl/projects/e-foundry/tex-gyre/
4. Kew, J.: Xetex (2005). http://tug.org/xetex/
5. Knuth, D.E.: Computer Modern Typefaces. Computers and Typesetting. Addison-Wesley, Reading (1986)
6. Knuth, D.E.: The METAFONTbook. Computers and Typesetting. Addison-Wesley, Reading (1986)
7. STI Pub companies: STIX fonts project (2010). http://www.stixfonts.org/

Author Index

Printed in the United States
By Bookmasters